Media Business and Innovation

Series editor

M. Friedrichsen, Stuttgart Media University, Germany and Humboldt School Stuttgart/Berlin, Germany

More information about this series at http://www.springer.com/series/11520

Artur Lugmayr • Cinzia Dal Zotto
Editors

Media Convergence Handbook - Vol. 2

Firms and User Perspectives

Springer

Editors
Artur Lugmayr
Visualisation and Interactive Media
School of Media, Culture,
and Creative Arts (MCCA)
Curtin University
Bentley/Perth, Western Australia
Australia

Cinzia Dal Zotto
Academy of Journalism and Media
Faculty of Economics and Business
University of Neuchâtel
Neuchâtel
Switzerland

Media Business and Innovation
ISBN 978-3-642-54486-6 ISBN 978-3-642-54487-3 (eBook)
DOI 10.1007/978-3-642-54487-3

Library of Congress Control Number: 2015956388

Springer Heidelberg New York Dordrecht London

Printed on acid-free paper

Springer-Verlag GmbH Berlin Heidelberg is part of Springer Science+Business Media (www.springer.com)

Preface

We are very happy to present our edited book covering the thematic area of convergence from a cross-disciplinary viewpoint. The edited book constitutes an attempt to investigate several aspects of convergence such as the strategic, technological, consumer-driven, and application-oriented convergence. As these aspects of convergence are interrelated, the book shall shed light into the complexity of the topic and act as a reference book and educational resource in the field. Convergence is very often discussed at a rhetorical and political level, without practical convergence process issues or concerns about potential divergence effects being part of the discourse. Today media managers are faced with the need to satisfy users' expectations and at the same time undertake convergence as an industrial strategy to achieve economies of scope and scale. Media technology developers are faced with rapidly emerging new technologies and information systems to quickly develop additional media services. Media scholars are bound to understand technological developments and industry convergence processes in order for their research to be relevant. With this book, we thus attempt to give answers to media managers, media scholars, students, researchers, as well as technology developers and help them get a clearer understanding of this thematic area.

When we launched our edited book initiative, our primary goal was to investigate potential interest in this thematic area through a call for abstracts. We were overwhelmed when we received 87 abstracts, which have been evaluated by us as editors. 73 abstracts were invited to contribute a full book chapter, while 15 abstracts were rejected and not invited for a final manuscript version. To ensure high scientific standards, we implemented a strict blind peer review process for each book chapter. After this first review, we accepted the chapters for which review statements supported the publication. The final version of all chapters has been reviewed one more time in order to check that all review comments were adequately integrated. We ended up with 39 accepted book chapters, including two chapters contributed by us as editors which provide a red line through the book. Thus, the acceptance rate was 45 % in respect to the submitted abstracts and 54 % in respect to the invited abstracts. The chapters were divided between two book volumes: the first volume includes 19 chapters on journalism, broadcasting, and social media aspects of convergence, while the second volume presents 20 chapters covering firm and user perspectives on convergence. Most of the thematic areas that we wanted to include in the book (see Fig. 1) as well as many different media genres are represented within the contributions (see Fig. 2).

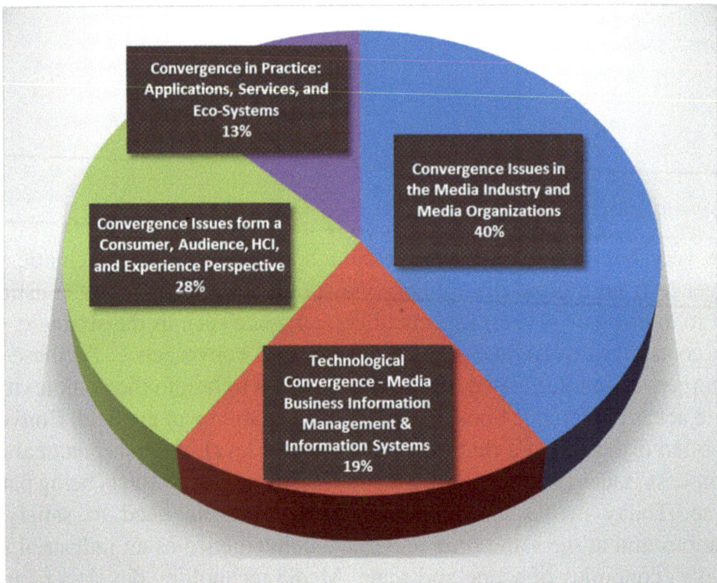

Fig. 1 Contributions to the thematic areas

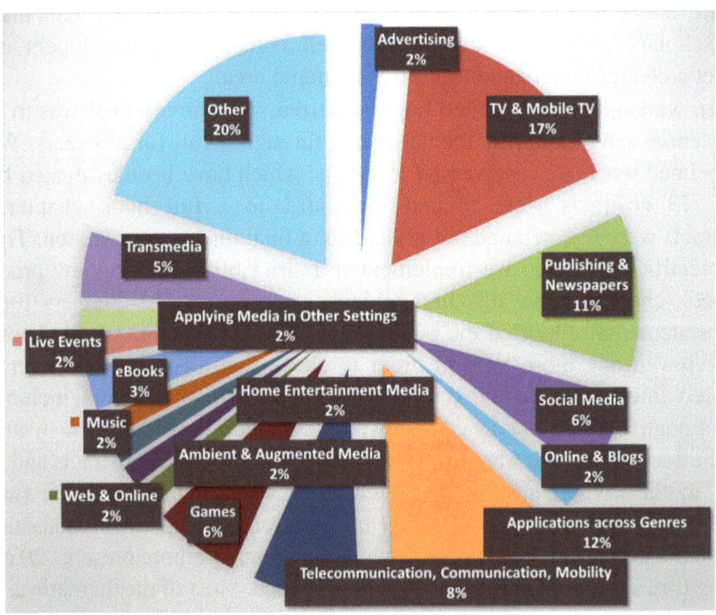

Fig. 2 Overview of the media genres that have been covered by contributions to this edited book

Fig. 3 Tag cloud of contributions

We started to work on the media convergence idea back in 2011, with a common contribution to the European Media Management Conference (EMMA) (Dal Zotto, Galichet, & Lugmayr, 2011). In 2013 we compiled another conference paper about this thematic area (Lugmayr & Dal Zotto, 2013). On both occasions, the audience was rather fascinated by our cross-disciplinary viewpoint on convergence. After many discussions, and in particular in acknowledgment to Prof. Greg Lowe, we started to work on this book project. By the end of 2012, we collected the extensive set of book abstracts which was rather surprising for us as editors. To cope with the high amount of submissions across a wide and thematic area (see Fig. 3 for a tag cloud of the chapters' keywords), Springer-Verlag gave us permission to divide the book into two volumes.

In particular we would like to give our greatest appreciation to and warmly thank Prof. Dr. Greg Lowe, who supported our idea and initially helped us in setting up the book project. He gave us clarifying insights on convergence and very useful hints in which thematic area to develop the book project. We also would like to thank Springer-Verlag and in particular Barbara Fess who was extremely helpful, patient, and helped us through the book editing process with many questions. Last but not least, we definitely would like to thank all numerous reviewers of the book chapters. Without their hard work, we would have not been able to complete this extensive book. The first editor, Prof. Dr. Artur Lugmayr, would especially like to thank the Akerlund Foundation for funding his scientific activities and his professorship.

We would like to make you also aware about our website, which shall act as a collaborative platform for people interested in the thematic area. You can find our group email address, Facebook group, as well as other resources online on: http://www.ambientmediaassociation.org/2012convergence/. Further, we would like to pinpoint to other activities that we are currently conducting within the Association for Information Systems (AIS) in the Special Interest Group (SIG) eMedia: http://aisnet.org/group/SIG-eMedia, or within the International Ambient Media Association (iAMEA): http://www.ambientmediaassociation.org. And finally, we wish you as reader of the book that you will have some useful and clarifying insights into such a complex thematic area as media convergence and gain a deeper understanding of this exciting topic. We also would like to refer to our book website, which contains additional information, email lists, errata, and most importantly course and lecture slides: www.artur-lugmayr.com.

Bentley, WA, Australia — Artur Lugmayr
Neuchâtel, Switzerland — Cinzia Dal Zotto

References

Dal Zotto, C., Galichet, G., & Lugmayr, A. (2011). *Managing media convergence: An economic and editorial project*. European Media Management Conference 2011, Moscow, Russia.

Lugmayr, A., & Dal Zotto, C. (2013). *Convergent divergence – cross-disciplinary viewpoint on media convergence*. European Media Management Education Association Conference 2013 – Digital Transformations and Transactions in Media Industries, Bournemouth, UK.

Contents

Part I

Impact of Convergence on Media Organizations and Markets

Media Convergence as Evolutionary Process

Cinzia Dal Zotto and Artur Lugmayr

1 What do We Understand Under Media Convergence?

One of the challenges of studying media convergence is that the concept is so broad that it has multiple meanings. As a result the academic literature in this area is diverse and underdeveloped from both a theoretical and an empirical perspective (Wirth, 2006). In this introductory chapter we will try to take the different interpretations of media convergence—which will be evident in the various book chapters—into account, but attempt to make them converge into some common ground, which we already investigated in Dal Zotto, Galichet, and Lugmayr, (2011), Lugmayr and Dal Zotto, (2013), AIS SIG eMedia (SIG-eMedia n.d.) http://aisnet.org/group/SIG-eMedia, and the International Ambient Media Association (iAMEA) (Anon n.d.) http://www.ambientmediaassociation.org.

According to the Merriam-Webster's Dictionary the general concept of convergence refers to "the act of converging and especially moving toward union or uniformity" (Mish, 1993). Within the media field, which is the research area that we are addressing in this book, convergence can be identified with the "ability to deliver different media channels via one digital platform" (McPhillips & Merlo, 2008). Previously broadcast media such as radio, television (e.g., Lugmayr, Niiranen, & Kalli, 2004) and the Internet, as well as print media have been distributed via different and well-distinguished platforms. Today content is becoming more and more digitalized: No matter the type of signal, any content can be

C. Dal Zotto (✉)
Academy of Journalism and Media, Faculty of Economics and Business, University of Neuchâtel, Neuchâtel, Switzerland
e-mail: cinzia.dalzotto@unine.ch

A. Lugmayr
Visualisation and Interactive Media, School of Media, Culture, and Creative Arts (MCCA), Curtin University, Bentley/Perth, Western Australia, Australia
e-mail: lartur@acm.org; artur.lugmayr@curtin.edu.au

A. Lugmayr, C. Dal Zotto (eds.), *Media Convergence Handbook - Vol. 2*, Media Business and Innovation, DOI 10.1007/978-3-642-54487-3_1

transformed into undifferentiated bits of data that converge onto the same platform (McPhillips & Merlo, 2008). Media convergence is therefore rather a process more than an outcome. However, digitalized content can be distributed on different digital platforms with little or no impact on costs. Content convergence can lead to distribution and thus consumption divergence. As such, the media convergence concept does not only refer to a technological shift but it includes changes within the industrial, cultural and social paradigms of our environment reflecting both media convergence and divergence processes.

Indeed media convergence alters relationships between technologies, industries, audiences, genres and markets. According to Rheingold (2000) advances in technology enable technological convergence, which in turn has "social-side effects" in that "the virtual, social and physical world are colliding, merging and coordinating". Jenkins (2006) states that media convergence is "the flow of content across multiple media platforms, the cooperation between multiple media industries, and the migratory behavior of media audiences" suggesting that media audiences nowadays play a crucial role in creating and distributing content, and convergence therefore has to be examined in terms of social, as well as technological changes within the society. Thanks to new technologies consumers are encouraged to actively seek out new information and make personalized connections among dispersed media content. Thus, media convergence requires media companies to rethink existing assumptions about media from a consumer's point of view, as these affect not only marketing but also production decisions (Krumsvik, 2013; Urban & Bodoky, 2013). With the proliferation of different media channels and an increasing portability of new telecommunications and computing technologies, we have entered into an era where media are constantly surrounding us. Furthermore, technology has empowered consumers who not only consume content from many different sources but also participate and contribute to content production activities. The subsequent oversupply and commoditization of media content have led to an efficiency decline of the traditional media business model (Picard, 2000; Picard & Dal Zotto, 2006; McPhillips, & Merlo, 2008; Dal, Dichamp, & Sommer, 2012). As direct consequence the initial revolutionary digital paradigm shift presumed old media to be increasingly replaced by new media. However, the currently emerging convergence paradigm suggests that new and old media would interact in more complex ways than previously predicted (Wirtz, Schilke, & Ullrich, 2010).

Already in 2006 Jenkins saw media convergence as an ongoing process that should not be considered as a displacement of the old media, but rather as interaction between different media forms and platforms (Jenkins, 2006). Supporting this argument, Deuze (2011) suggests that media convergence should be viewed as 'cooperation and collaboration' between previously unconnected media forms and platforms. This holds even more true when we consider that, due to the multitude of available content on and off-line, the newly empowered consumers search for trustable media products and services. This has pushed media firms to develop into reputable brands that can offer content in a number of different forms and platforms. Newspapers for instance produce and sell books, or market a radio or web TV under their brand umbrella; the movie industry exploits the name of successful movies to produce and sell books, video games, or cartoons (Tameling & Broersma, 2013).

Branding encourages the expansion of a successful media concept into multiple platforms and therefore stimulates media industry convergence (Matteo, Spolaor, & Dal Zotto, 2013). However, this process does not enhance the creation of new ideas and further leads to technology divergence in the hardware industry: hardware needs to be specific to each function, so it diversifies in order to accommodate media convergence requirements. It thus appears that, in order to regain efficiency and success, media business models should not only react and adapt to the convergence trend, but also—and first of all—proactively factor in causes and effects of the media convergence process: this means to understand the multifaceted meaning of media convergence, including divergence and coexistence effects, develop a more customer oriented vision as well as to innovate by merging traditional with new elements, too.

2 Evolutionary Paths of Media Convergence

The development of media convergence processes is mainly due to the following three factors: (a) digitalization, (b) deregulation of media and communication markets, and (c) changes in user preferences. Digitalization offers new opportunities for the creation, presentation, storage and distribution of media products (Rayport, & Jaworski, 2001). The standardization of storage media that were previously separated enhances the exploitation of economies of scope in preceding and subsequent stages of the value-creation chain (Wirtz, 2011). Further, with the conversion from analog to digital data communication, differing communication networks can be used for data transmission and become substitutable among each other. Internet access is for instance now possible via telephone, cable, satellite, mobile networks as well as wireless LANs.

As to the deregulation, since the middle of the nineties measures have been introduced for liberalizing the telecommunication sectors both in the US and Europe. This has led to the emergence of competitive structures in the information, media and communication industries (Dal & van Kranenburg, 2008; Dal Zotto & Dowling, 2003). As a result the supply of media services has increased while audiences and media consumption were fragmentating. During the last decade consumers have been using an increasing number of different services to satisfy their information and entertainment needs (Rayport & Jaworski, 2001; Aitamurto, Sirkkunen, & Lehtonen, 2011). A clear change in user preferences towards personal information and communication tools can indeed be identified (Wirtz, 2011).

Media convergence has empowered users to produce and distribute their own content, viewing and listening schedules can be personalized via on-demand or search content and mobile services, users can engage with one another via peer to peer technology. Even if consumers can be distinguished between early adopters—the media literate and technology savvy ones—and late adopters—those who are happy with their current media experience—there is no doubt that such changes in user preferences together with the digitalization and deregulation of media and communication markets represent factors with a considerable impact on the competitive environment that media firms are facing (Picard, 2010). Depending on this impact we can distinguish different types and evolutionary paths of media convergence, as we illustrate below.

Technological convergence, that is the tendency of different technological systems to evolve towards performing similar tasks due to digitalization, leads on one hand to the convergence of different functionalities into a new product—*product convergence*—and on the other hand to an aggregation of markets that were previously distinguished from one another (Yoffie, 1997). In the market for communication services cable network and telecommunication companies, as well as mobile and satellite providers compete with each other. In order to differentiate themselves from each other they often pursue integration strategies in the area of content production: by enriching telecommunications products—such as Internet access or voice communication—with content, these operators provide consumers with accessory uses (Wirtz, 2011) and become a competitive threat to media companies. Once users consider products either substitutes or complements and as a consequence such products end up converging into a new product, we talk about *competitive or complementary convergence* (Dowling, Lechner, & Thilmann, 1998; Greenstein & Kanna, 1997; Picard, 2000; 2009).

Product level convergence can lead to *business level convergence* when a company realizes that it would be more advantageous to integrate convergent areas of business either within the company or through outside cooperation. This can stimulate economies of scale and innovations can emerge from a recombination of individual business units' performances (Wirtz, 2011). If the convergence of previously separated services concerns not only a business unit but the whole enterprise, this can lead to cooperations or mergers between previously separated companies. In this case we talk about *strategic or industrial structure convergence* (Chan-Olmsted & Kang, 2003; Thilmann & Dowling, 1999; Wirtz, 1999). Convergence is transforming the media and telecommunication industries from vertical businesses—telephone, television, computers—into horizontal segments—content, packaging, transmission network, manipulation infrastructure, and terminals. Established media firms have for long dominated the entire value chain, from creative inception to production, packaging, marketing and finally distribution. Thanks to technological progress production costs have substantially diminished and the Internet is providing an open-source route to market. Thus content producers are now able to easily bypass mainstream media and distribute their content independently (McPhillips & Merlo, 2008). In this case strategic convergence may lead to industrial structure divergence.

As a consequence of this evolution we have already been witnessing a consolidation phase among technology, media and telecommunication firms. Current structural changes point to a few major media players dominating the mass market and a few niche operators serving specialist areas. Middle size operators seem to be struggling to find scale enough to compete with larger firms. Media convergence seem thus to lead to a polarization of the industry structure with global conglomerates and small independent media owners co-existing (Picard, 2010). Media ownership concentration, that is the control of media concentrated in the hands of a few private owners, is one of the major concerns when examining positive and negative consequences of media convergence (Jenkins, 2006). On the one hand, media ownership concentration—and thus industrial structure convergence—may cause the decline in the diversity of media products and service offerings and result in a tendency not to take into account voices of those lacking economic power (Horwitz, 2005; Valcke, Picard, Süksöd, & Sanders, 2009). On the other hand, it is argued that market driven media owned and controlled by big media corporations, thanks to their disposal of financial resources, can actually improve the value of the service, the plurality of topics and the competence of the contributors as well as enable technological developments, change the elitism of media professionals and create new general awareness (Barwise & Picard, 2012; Dal Zotto & Picard, 2015; Picard & van Weezel, 2008; Valcke, 2011).

3 The Convergence Process Between Advantages and Disadvantages

As mentioned above media convergence seems to represent an evolutionary process that accompanies the development of information technologies, telecommunications, media products and services, media industry and firm structures as well as media professionals. According to Stöber (2004) there are three stages of media evolution: invention, innovation and institutionalization. In the media evolution case of convergence, the **invention** stage consisted in the ability to convert differentiated bits of analogue data (such as voice, text, audio or visual) to undifferentiated bits of digital data. The **innovation** stage refers to the possibility to distribute and consume different media types, whether radio, TV or internet through the same platform and this phase is clearly completed. Indeed, the media industry is now finally moving towards the **institutionalisation** phase of convergence, which is defined as the adaptation to the environment to create a new « system entity » (McPhillips & Merlo, 2008). Such phase acquires more a revolutionary than an evolutionary character, as divergence rather than convergence processes can emerge as a result.

Stöber (2004) argues that four factors of change need to simultaneously occur to shape a new system entity and a viable business model: technological, cultural, political and economic factors. The first three are already in motion. *Advancements in technology* usually correspond to reductions in cost and significant increases in functionality. In the media industry this pattern can be identified with improvements in speed of data transfer for broadband, in quality of mobile devices as well as in costs of data storage. The high rate of adoption of broadband and other digital services confirm improvements of the perceived value of technological advancements. Furthermore, the time spent on consuming media services based on converged technologies such as mobile phone calls and Internet surfing, text messages, traditional and web radio listening as well as TV has rapidly increased and most of these consumption activities take place simultaneously. This is a clear sign of a *cultural shift* away from single source consumption patterns (McPhillips & Merlo, 2008) and denotes a device diverging trend. From a *political* point of view media technology convergence is enhanced by governments' encouragement if not enforcement of digital broadcast adoption.

A fourth critical factor of change towards the institutionalization phase of media convergence is the *economic factor*, which is still evolving. The emergence of an appropriate business model will determine the future of the media industry (McPhillips & Merlo, 2008). For the moment we have just seen media convergence leading to an over-supply and commoditization of media content, the decline of advertising effectiveness due to media fragmentation as well as the over-exposure of audiences to an infinite number of advertising touch points leading to the decline of the current media business model (Dal et al., 2012; Picard & Dal Zotto, 2006; Matteo & Dal Zotto, 2015). Increasing the number of vehicles, touch points and the volume of advertising may increase revenues in the short term but it eventually leads to a downward pressure on margins in the medium-long terms (McPhillips & Merlo, 2008).

Similarly, as the attention that consumers can dedicate to media is limited, the direct consequence of media fragmentation, or consumption divergence, is not only the erosion of advertising revenues but also the decline of subscription revenues. Information media have reacted to this effect by cutting budgets and costs. Technological and product convergence can thus have negative effects as it is the case for journalism and journalistic products (Deuze, 2004; Dupagne & Garrison, 2006; Fioretti & Russ-Mohl, 2009; Huang, Davison, Shreve et al., 2006; Huang, Rademakers, Fayemiwo, & Dunlap, 2004; Lawson-Borders, 2006; Meier, 2007). Online information media have a 24-h constantly updated news cycle and require staff to be able to work multiplatform. Further, not only content is constantly updated—thus reducing staff's time to research, report or even think about their work (Jarvis, 2009; Klinenberg, 2005)—but it is also developed across multiple platforms and channels, a phenomenon called multimedia storytelling (Gray, Bounegru & Chambers, 2012; Jenkins, 2001). This inevitably calls for a cultural and organizational change within media firms (Boczkowski, 2005) and increases the level of uncertainty among employees. Within web-driven newsrooms journalists have for instance less time to talk with sources and check facts (Porlezza & Russ-Mohl, 2013; Thurman & Myllylahti, 2009), they will therefore tend to report news that have already been or are being reported by other colleagues in other media in order to reduce uncertainty. As a result information becomes standardized in its format and uniform in its content making it difficult for media firms to distinguish themselves from each other and gain a competitive advantage. Furthermore, multiplatform content production and distribution reflects a process of editorial convergence: a converged newsroom provides content for multiple distribution platforms thanks to the contribution of different, complementary and multi-skilled media professionals. In order to accommodate and take advantage of cross-platform media strategies journalistic and editorial work withing newsrooms needs to be reorganized while staff recruiting, training and development measures become fundamental.

4 Managing the Media Convergence Process: Outline of the Book

So far we have learnt that factors such as digitalization, deregulation and the subsequent changes in user preferences enhance the media convergence process. Depending on the impact of those factors different types of media convergence can be distinguished, even if they may be complementary and share evolution paths.

Indeed, media industry convergence is the result of technological convergence and the subsequent need for media firms—which are characterized by high fixed costs and low marginal costs—to reach economies of scale. As we have seen, technological development has led to an increasing fragmentation of media and audiences, denoting in this case a divergence process, and subsequently to a products and services' price deflation. A viable way to address this inevitable price deflation seems to be the application of competition policy, allowing industries to become more concentrated and so reduce competitive pressures. Considering that the loss of pricing power quickly leads to a decline in industry profitability, the only possible reaction for the media industry has been a horizontal consolidation or vertical integration to gain gatekeeping power (Noam, 2006 and 2009).

Furthermore, both technological and industry convergence leads to experience and social convergence at consumer level, as well as to cultural and editorial convergence within media organizations (See Fig. 1). Due to multimedia and multiplatform content production and distribution environments different professional cultures and tasks are thus melting together. It is therefore clear that, in order to successfully manage media convergence processes, not only technological and media economics aspects but also human resource and organizational development related aspects need to be taken into account.

In this handbook we have tried to touch upon all four main levels of media convergence—technology (e.g., information systems level (Lugmayr, 2013), industry, consumer and organizational level—in order to show its multidisciplinary and multifaceted character (see Fig. 1). As contributions were many and converging around two main areas, we decided to organize them into two volumes, one focusing on the journalism, broadcasting and social media aspects of convergence, and a second one highlighting the perspectives of media firms and users.

4.1 Volume One

The *first volume* of this handbook starts with a closer look at the meaning of media convergence by presenting at the same time three different perspectives on the

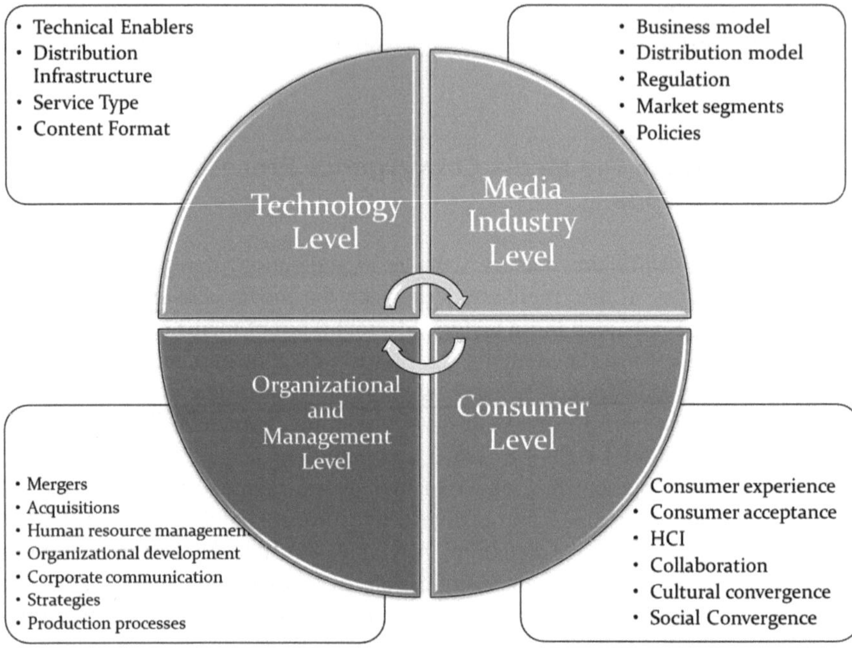

Fig. 1 Impact of media convergence at different levels

concept. As mentioned at the beginning of this introductory chapter, in order to analyze cause and effects of media convergence and thus conceive solutions to current media industry challenges, it is necessary to first understand the multifaceted meaning of media convergence. In the first part Arango-Forero, Roncallo-Dow and Uribe-Jongbloed address the idea of convergence both as an abstract concept and as a word ascribed to specific phenomena. They point out a divergent conceptualization of convergence in the literature so far and the necessity to disentangle the abstract concept from the working definition of convergence which refers to current geo-economic and culturally bound phenomena. The evolution of media convergence from its first stage of merging technologies to its second catalyzing stage is problematized by Marie-José Montpetit in the second chapter. Through a thorough discussion of network and device convergence, she explains how the advent of social connectivity has disrupted the media consumption experience and pushed convergence away from basic functional effects into a richer set of interdependent elements such as hardware, software, content and user interaction. Shantanu Dey concludes the introductory perspectives part on convergence by highlighting that media convergence does not necessarily lead to "power convergence"—i.e., "economic power" in terms of media ownership concentration is not necessarily being imposed on "cultural power". He claims that the existing open architecture of the Internet allows the articulation of divergent powers and proposes an "affirmative public ownership" policy at network infrastructure and applications level in order to maintain the structural openness of the system and enable free speech.

The volume continues by specifically addressing three media fields: journalism, broadcasting and social media. Focusing on convergence challenges in journalism and publishing, **part two** discusses the impact of news media online shift on journalistic quality. Christian Wellbrock argues that this shift is preventing mass media do adequately fulfill their function within democratic societies and suggests that public service broadcasters should be allowed to expand their activities online. On the same line, by discussing the implications of convergence on human resources and labor conditions, according to Spyridou and Veglis emerging convergence skills might reduce professional autonomy and journalistic performance. Kay Hamacher compares blogs and traditional media in order to see if digitalization increases the degree of content diversification, while Veglis, Dimoulas and Kalliris investigate technological issues when implementing cross-media publishing. Pointing out that media publishing channels differ in terms of production and consumption, they propose a modular model to appropriately manage cross-media publishing automation while respecting the peculiarities of the different publishing areas. Moving towards the consumer perspective, Mustic and Plenkovic highlight that the graphic representation of information affect the perception of content and this information credibility. Ebner, Schön and Alimucaj present a new print-on-demand service that can improve the perceived quality of open access online journals by allowing readers to assemble different articles and print their "personal book". Finally, the last paper in this chapter problematizes the fact that, in order for new business models to emerge and for the online shift to have success, media professionals need to reconsider their belief system.

Part three concentrates on broadcasting. In particular, Matteucci discusses intellectual property rights within the context of digital TV convergence in the European Union. He indicates how patent portfolio strategies and control of crucial copyrights become increasingly central for competing in a converging TV sector. Deliyannis looks at presentation requirements for a converged interactive television through some real-life case studies; Tuomi approaches mobile devices as second screen idea to enhance consumers' television experience in our present century, while Ala-Fossi problematizes existing divergences within converging radio and television newsrooms by highlighting the different approach followed by Europe and the USA on one hand and Japan on the other hand.

Social media aspects of convergence are the topic of **part four**. Villi, Matikainen and Khaldarova discuss technology-enabled user distributed content by exploring how the convergence enables a participatory audience to disseminate news media online content within social networks. Their study draws on the analysis of fifteen news media within seven different countries. On the same path Duh, Meznaric and Korosak examine the possibility to use different social media channels for the automatic generation of interactive social media news streams. Zlatanov and Koleva argue that the raising information quantity determines the superiority of technology savvy and communication experienced individuals, as well as their subsequent major influence on information spread within online social networks. The authors point out that technological convergence might thus lead to social divergence. The tension between technological convergence and social divergence is problematized by Damasio, Henriques, Teixeira-Botelho and Dias, too. Their paper concludes the first volume of this handbook and focuses on mobile technology. According to the authors mobile Internet access facilitates more interactive and participative activities on one hand, but it results in network-based individualism on the other hand.

4.2 Volume Two

If the first volume focuses on the impact of digitalization on the media industry and thus on technology convergence, the *second volume* of the handbook considers structural/economic convergence issues as well as cultural convergence. **Part one** deals with the impact of convergence on media organizations and media markets. In the first paper Gershon examines media convergence effects on special design and operations within transnational organizations. Georgiades continues by addressing the critical issue of employees' involvement and highlighting how divergent understandings of concepts such as communication and employee involvement itself may hamper organizational convergence processes. Karmasin, Diehl and Koinig argue that media convergence implies the need for media firms to develop new business models, appropriate organizational structures as well as new managerial competences. Vukanovic concludes the chapter by analyzing how media convergence can create opportunities and shape successful business models.

Chapter two and chapter three of the second volume of the handbook reflect both the impact of changes in user preferences as well as the intertwined issue of cultural change and cultural convergence within the media industry. **Part two** mainly discusses convergence in relation to media production activities. Denward's contribution deals with participation enabling media production. By analyzing the design and implementation phases of a drama production in Sweden, the author approaches various types of convergence—industrial, technology, cultural and social—and the difficulties they created in the production process. Noronha e Sousa, de Lemos Martins and Zagalo investigate the practical implications of producing and consuming transmedia narratives from both the perspective of producers and consumers. Innocenti, Pescatore and Rosati approach narratives, too. By using conceptual tools borrowed from the media, economics and information architecture studies, the authors offer a cross-disciplinary perspective on serial narratives as audiovisual production niche within a converging media environment. Indrek Ibrus applies an evolutionary approach to media change to interpret recent Northern European examples of how audiovisual micro-companies are confronting the "convergence culture" and innovating their output via cross-media strategies and transmedia storytelling. Foster and Brostoff's paper conclude the chapter discussing advances in 3D computer graphics technology and possible convergence enabled synergies between video game and urban design—two fields that share the goal of constructing a cohesive, credible and readable visual narrative about an alternative reality.

Part three focuses on user perspectives and experiences. Alexandre Fleury opens the chapter with an investigation about how convergence between mobile and television technology is experienced in highly connected societies. He explores current practices in Denmark and Japan and identifies a number of future trends for converging media solutions. Reichl continues by analyzing experience from a quality point of view. He addresses issues such as the development of laws as well as models to charge for quality of experience in converging communication systems. Tavares and Schofield deal with multisensory interaction design as critical success factor for a correct user experience of multiple devices, media and platforms. The following paper further addresses the implementation of a multi-modal interface system within various media platforms to enhance the effectiveness of man–machine interaction. McKay points out though that effective media convergence depends on knowledge engineers' understanding of the changeable nature of the human-computer interaction environment. Finally, user experience design is approached from a transmedia storytelling point of view, early in the creative writing process. Ghellal, Wiklund-Engblom, Obal and Morrison define transmedia storytelling as evolving storylines running across multiple platforms. This definition highlights the complexity and importance of transmedia design, including for instance interdisciplinarity, genres and emergent production models.

Last but not least, **part four** presents a few papers discussing future media convergence trends. Grüblbauer and Haric conceptualize how to design and build a convergence application to effectively showcase big amounts of information and data coming from diversified sources. By analyzing existing theory on converging media and media planning, Svahn, Wahlund, Denward, Rademaker and Nilsson present a model for evaluating converging media for advertising purposes. Giesecke analyses the business models of sixteen event ecosystem organizations in four different countries. His conclusions indicate that, in order to achieve convergence, the news media industry needs to learn from the entertainment media industry, build and bridge networks, and ensure affinity between content and advertisements. Finally, Giesecke and Immonen argue that mid-sized mass media companies suffer from a conservative approach to convergence, and encourage mass media company leaders to diverge and focus on business models based on a system approach. The future will show, if convergence will let media environments evolve towards more ubiquitous media or ambient media environments (Lugmayr, 2012).

5 Conclusion

All in all we believe that this handbook provides practitioners and researchers, experts and amateurs with an overview of the most important issues as well as with precious insights concerning media convergence processes. Furthermore, thanks to an in depth content analysis of all contributed chapters at the very end, the handbook contributes to the debate about causes and effects of the media evolution process by highlighting not only general trends in research but also and especially research gaps in the media convergence field.

References

Aitamurto, T., Sirkkunen, E., & Lehtonen, P. (2011). *Trends in data journalism*. Espoo: VTT Next media.

Barwise, P., & Picard, R. G. (2012). *The economics of television in a digital world: What economics tells us for future policy debates*. RISJ Report, Oxford: Reuters Institute, University of Oxford.

Boczkowski, P. (2005). *Digitizing the news, innovation in online newspapers*. Cambridge: MIT Press.

Chan-Olmsted, S. M., & Kang, J. (2003). Theorizing the strategic architecture of a broadband television industry. *Journal of Media Economics, 16*, 3–21.

Dal Zotto, C., & Dowling, M. (2003). Venture Capital Investitionen in Neugründungen der Telekommunikationsbranche: Erfahrungen aus der geplatzten "Blase" in den USA. In F. Habann (Ed.), Innovations management in Medienunternehmen, Gabler Verlag.

Dal Zotto, C., & Picard, R. G. (2015). The dimensions of ownership and control of media, In P. Valcke, M. Sükösd & R. G. Picard (Eds.), Media pluralism: Concepts, risks and global trends, Palgrave McMillan (in Print).

Dal Zotto, C., Dichamp, E., & Sommer, C. (2012). Alte Rezepte und neue Businessmodelle: Die Medienkrise aus der Sicht der Verlagsmanager. In W. Meier, H. Bonfadelli, & J. Trappel (Eds.), *Gehen in den Leuchttürmen die Lichter aus?* Lit Verlag Münster: Was aus den Schweizer Leitmedien wird.

Dal Zotto, C., Galichet, G., & Lugmayr, A. (2011). Managing media convergence: An economic and editorial project. In *European Media Management Conference 2011*. Moscow, Russia: European Media Management Association (EMMA).

Dal Zotto, C., & van Kranenburg, H. (2008). *Management and innovation in the media industry*. Cheltenham: Edward Elgar Publishing.

Deuze, M. (2004). What is multimedia journalism? *Journalism Studies, 5*(2), 139–152.

Deuze, M. (2011). *Managing media work*. Thousand Oaks: Sage.

Dowling, M., Lechner, C., & Thilmann, B. (1998). Convergence—Innovation and change of market structures between television and online services. *Electronic Markets, 8*(4), 31–35.

Dupagne, M., & Garrison, B. (2006). The meaning and influence of convergence: A qualitative case study of newsroom work at the Tampa News Center. *Journalism Studies, 7*(2), 237–255.

Fioretti, N., & Russ-Mohl, S. (Eds.). (2009). *Merging media, converging newsrooms*. Lugano: Giampiero Casagrande editore.

Gray, J., Bounegru, L., & Chambers, L. (2012). *The data journalism handbook*. Sebastopol, CA: O'Reilly.

Greenstein, S., & Kanna, T. (1997). What does industry convergence mean? In D. B. Yoffie (Ed.), *Competing in the age of digital convergence*. Boston: Harvard Business School Press.

Horwitz, R. B. (2005). On media concentration and the diversity question. *The Information Society, 21*, 181–204.

Huang, E., Davison, K., Shreve, S., Davis, T., Bettendorf, E., & Nair, A. (2006). Facing the challenges of convergence: Media professionals' concern of working across media platforms. *Convergence, 12*(1), 83–98.

Huang, E., Rademakers, L., Fayemiwo, M. A., & Dunlap, L. (2004). Converged journalism and quality: A case study of the Tampa Tribune news stories. *Convergence, 10*(4), 73–91.

Jarvis, J. (2009, June 7). Product v. process journalism: The myth of perfection v. beta culture. http://buzzmachine.com/2009/06/07/processjournalism/

Jenkins, H. (2001, June). Convergence? i diverge. *Technology Review*, 93.

Jenkins, H. (2006). *Convergence culture: Where old and new media collide*. New York: New York University Press.

Klinenberg, E. (2005). News production in a digital age. *Annals of the American Academy of Political and Social Science, 597*, 48–64.

Krumsvik, A. H. (2013). Towards a typology of strategies for user-involvement. In M, Friedrichsen & W. Mühl-Benninghaus (Eds.), *Handbook of social media management* (pp. 655–669). Berlin, Heidelberg: Springer.

Lawson-Borders, G. (2006). *Media organizations and convergence: Case studies of media convergence pioneers*. Mahwah, NJ: Lawrence Erlbaum.

Lugmayr, A. (2012). Ambient media culture: What needs to be discussed when defining ambient media from a media cultural viewpoint? *International Journal of Ambient Computing and Intelligence (IJACI), 4*(4), 58–64.

Lugmayr, A. (2013). Brief introduction into information systems and management research in media industries. In *2013 I.E. International Conference on Multimedia and Expo Workshops (ICMEW)*. (pp. 1–6), San Jose, CA, USA.

Lugmayr, A., Niiranen, S., & Kalli, S. (2004). *Digital interactive TV and metadata—Future broadcast multimedia*. New York: Springer.

Lugmayr, A., & Dal Zotto, C. (2013). Convergent divergence—Cross-disciplinary viewpoint on media convergence. In *European Media Management Education Association Conference 2013—Digital Transformations and Transactions in Media Industries*. Bournemouth, UK.

Matteo, S., & Dal Zotto, C. (2015). Native advertising, or how to stretch editorial to sponsored content within a transmedia branding era. In G. Siegert, K. Förster, S. Chan-Olmsted & M. Ots (Eds.), Handbook of media branding, Springer.

Matteo, S., Spolaor, G. & Dal Zotto, C. (2013). Branding with social media at RTS. In M. Friedrichsen & W. Mühl-Benninghaus (Eds.) Handbook of social media management, Springer.

McPhillips, S., & Merlo, O. (2008). Media convergence and the evolving media business model: An overview and strategic opportunities. *The Marketing Review, 8*(3), 237–253.

Meier, K. (2007). Innovations in Central European newsrooms: Overview and case study. *Journalism Practice, 1*(1), 4–19.

Mish, F. C. (Ed.). (1993). *Merriam-webster's collegiate dictionary* (10th ed.). Springfield, MA: Merriam Webster.

Noam, E. (2006). Deregulation and market concentration: An analysis of post-1996 consolidations. *Federal Communications Law Journal, 58*(3), 539–549.

Noam, E. (2009). *Media ownership and concentration in America*. Oxford: Oxford University Press.

Picard, R. G. (2000). Changing business models of online content services: Their implications for multimedia and other content producers. *The International Journal on Media Management, 1*, 31–34.

Picard, R. G. (2009). Media clusters and regional development: Reflections of the significance of location in media production. In I. Bernhard (Ed.), *The geography of innovation and entrepreneurship* (pp. 877–886). Trollhätten: University West.

Picard, R. G. (2010). The future of the news industry. In J. Curran (Ed.), *Media and society*. London: Bloomsbury Academic.

Picard, R., & Dal Zotto, C. (2006). Business models of newspaper publishing companies, in: Where news? The media future research initiative, Ifra Report Nr. 1, May.

Picard, R. G., & van Weezel, A. (2008). Capital and control: Consequences of different forms of newspaper ownership. *The International Journal on Media Management, 10*, 22–31.

Porlezza, C., & Russ-Mohl, S. (2013). Getting the facts straight in a digital era: Journalistic accuracy and trustworthiness. In C. Peters & M. Broersma (Eds.), *Rethinking journalism, trust and participation in a transformed news landscape* (pp. 45–59). New York: Routledge.

Rayport, J. F., & Jaworski, B. J. (2001). *Cases in e-commerce*. Boston: McGraw-Hill/Irwin.

Rheingold, H. (2000). *Smart mobs: The next social revolution* (pp. 157–182). Cambridge, MA: Perseus.
Stöber, R. (2004). What media evolution is. *European Journal of Communication, 19*(4), 483–505.
Tameling, K., & Broersma, M. (2013). De-Converging the newsroom: Strategies for newsroom change and their influence on journalism practice. *International Communication Gazette, 75* (1), 19–34.
Thilmann, B., & Dowling, M. (1999). Convergence and innovation strategy for service provision in emerging web-TV markets. *The International Journal on Media Management, 1*, 4–9.
Thurman, N., & Myllylahti, M. (2009). Taking the paper out of news. *Journalism Studies, 10*(5), 691–708.
Urban, A., & Bodoky, T. (2013). The impact of Facebook on news consumption. In M. Friedrichsen & W. Mühl-Benninghaus (Eds.), *Handbook of social media management*. Heidelberg: Springer.
Valcke, P. (2011). Looking for the user in media pluralism regulation: Unraveling the traditional diversity chain and recent trends of user empowerment in European media regulation. *Journal of Information Policy, 1*, 287–320.
Valcke, P., Picard, R. G., Süksöd, M., & Sanders, J. (2009). Indicators for media pluralism in the member states—Towards a risk based approach, Volume 1, Final Report, Lulu.com
Wirth, M. (2006). Issues in media convergence. In A. Albarran, S. M. Chan-Olmsted, & M. Wirth (Eds.), *Handbook of media management and economics*. Mahwah: Lawrence Erlbaum.
Wirtz, B. (1999). Convergence processes, value constellations and integration strategies in the multimedia business. *The International Journal on Media management, 1*, 14–22.
Wirtz, B. (2011). *Media and internet management*. Wiesbaden: Gabler Verlag.
Wirtz, B., Schilke, O., & Ullrich, S. (2010). Strategic development of business models. Implications of the Web 2.0 for creating value on the Internet. *Long Range Planning, 43* (2–3), 195–215.
Yoffie, D. B. (1997). Introduction: Chess and competing in the age of digital convergence. In D. B. Yoffie (Ed.), *Competing in the age of digital convergence*. Boston: Harvard Business School Press.

Intelligent Networking and the Transnational Corporation: Redefining Business Work Space

Richard A. Gershon

1 Introduction

International business has been transformed by the power of instantaneous communication. The combination of computer and telecommunications has collapsed the time and distance factors that once separated nations, people and business organizations (Castells, 2000; Tapscott, 1996). This has been particularly true for the transnational corporation (TNC) which operates as a nationally based company with overseas operations in two or more countries. Strategic decision-making and the allocation of resources are predicated on economic goals and efficiencies with little regard to national boundaries. The TNC has become a salient feature of our present day global economic landscape (Albarran & Chan-Olmsted, 1998; Compaine & Gomery, 2000; Gershon, 2006).

This chapter will examine the subject of intelligent networking which provides the technologies and electronic pathways that makes international business communication possible. The intelligent network has become the vital nervous system that enables the TNC to be truly global in its design, structure and extended reach. One of the goals of this chapter is to fully understand the ways in which intelligent networks affect the spatial design and operations of the transnational organization. Special attention is given to four work-design principles including: (1) Virtual

Richard A. Gershon, Ph.D. (Ohio University) is Professor and Co-Director of the Telecommunications and Information Management program at Western Michigan University where he teaches courses in Media Management and Telecommunications. Dr. Gershon is the author of *Media, Telecommunications and Business Strategy,* 2nd ed. (2013).

R.A. Gershon, Ph.D. (✉)
Telecommunications and Information Management, School of Communication, Western Michigan University, Kalamazoo, MI 49008, USA
e-mail: Richard.Gershon@wmich.edu

A. Lugmayr, C. Dal Zotto (eds.), *Media Convergence Handbook - Vol. 2*, Media Business and Innovation, DOI 10.1007/978-3-642-54487-3_2

Communication, (2) Information Exchange, (3) Digital Media and Convergence and (4) Decision Support Analysis.

2 The Intelligent Network: Literature Review

Intelligence can be defined as the ability to reason, problem solve, think abstractly, comprehend complex ideas and learn. Halal (1997) describes organizational intelligence as the "capacity of an organization to create knowledge and use it to strategically plan and adapt to its environment." (p. 67). Intelligent networks, therefore, are the systems of communication that organize, transmit and display information with the goal of improving organizational performance. The intelligent networks provide three levels of functionality as illustrated in Fig. 1. They include: (1) Transmission, Display and Storage, (2) Decision Support Analysis and (3) Artificial Intelligence (see Fig. 1).

The first level can be described as *Transmission, Display and Storage (TDS)*. The role of the intelligent network is to provide the proper switching and routing of information between a sender and an intended audience. This can vary in size and complexity from a simple SKYPE video exchange to an international videoconference involving a project team from around the world. In both cases, the goal is to transmit information to an intended audience.

The second level can be described as *Decision Support Analysis*. Here the emphasis is on providing the user with critical information for purposes of information gathering, planning and decision-making (Huber, 1990). The intelligent network is responsible for providing the organization and its users immediate access to a whole host of internal and external data base services that might include investigating infectious diseases (i.e., U.S. Center for Disease Control and Prevention) or pursuing a criminal investigation of a suspected international terrorist (i.e., Interpol, U.S. Department of Homeland Security). Depending on how the information is organized and sorted, there is an abundance of information that can provide the user with critical analysis capability (Gershon, 2017).

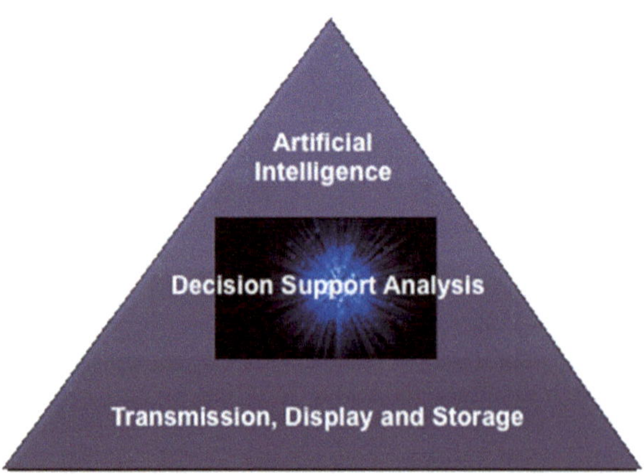

Fig. 1 Intelligent network: three level hierarchy. *Source:* Gershon (2017)

The third level can be described as *Artificial Intelligence (AI)*. The goal of the intelligent network is to make preprogrammed decisions. The network is designed to make recommendations to the user and/or take corrective action based on established algorithms. Once again, examples can vary in size and complexity starting with a proprietary software recommendation system built by electronic commerce companies like Amazon, Netflix and Apple. Such companies make personalized product recommendations (i.e., books, films and music) via their EC websites based on past selections. At a more complex level, automated intelligence refers to preprogrammed decision-making. As an example, modern aviation relies on a AI flight control management system in order to control the aircraft. The AI flight control system can control and automate all phases of a flight operation, including take-off and ascent, flight guidance (autopilot), descent, approach and landing.

2.1 Four Working Assumptions About Intelligent Networks

When engineers discuss the architecture of a network; they are describing how the physical parts of the network are organized, including: (1) Information pathways (network configurations), (2) Terminals (computers, smartphones etc.), (3) Software (applications and protocols) and (4) Data Enhancement Equipment (modems, laser printers, Wifi etc.) First, the intelligent network is not one network, but a series of networks designed to enhance worldwide communication for business and individual users alike (Gershon, 2011; Noam, 2001). Second, what gives the network its unique intelligence are the people and users of the system and the value-added contributions they bring to the system via critical gateway points. Today, the Internet has grown exponentially in size and complexity due to the many contributions of its users ranging from powerful search engines to unique web site design as well as the aggregation of content.

A third assumption is that intelligent networks do not operate in a vacuum. Rather, the use of intelligent networks are part of a greater human communication and organizational decision-making process (Monge & Contractor, 2003). As Berners-Lee (1999) points out, the Internet is as much a social creation as it is a technical one. Nowhere is this more evident than in the creation of social networking sites like *Facebook* which has been described by its founder and CEO, Mark Zuckerberg, as a "mathematical construct that maps the real-life connections between people. Each person is a noderadiating links to other people they know" (Most Innovative Companies, 2007). As friends and acquaintances join Facebook, they become part of a larger social grid that matters to the individual. It creates value to the individual by adding to one's social capital (Ellison, Steinfield, & Lampe, 2007). Finally, as intelligent networks grow and evolve, they often exhibit self-learning qualities in what can be described as *network evolution* (Gershon, 2011; Monge, Heiss, & Magolin, 2008). This is a crucial element in helping to explain what makes an intelligent network, intelligent. It also speaks to the importance of artificial intelligence.

2.2 Network Evolution

The principle of network evolution suggests that a network improves in value over time (Monge et al., 2008). The public switched telephone network, for example, has improved in value from a system once designed for voice communication to one that fully supports a variety of enhanced information services. Similarly, the success of the Internet today is due to a combination of factors, including hypertext linking, improved website design and powerful search engines. Google Inc. is the world's preeminent search engine and was founded by Larry Page and Sergey Brin in 1998. One of Google's stated missions is to organize the world's information. The power and networking capability of the Google search engine has proven highly adaptive and grown exponentially over time. The ever increasing amount of data now generates its own unique networking effect (Auletta, 2009). There is an automatic self-learning feature that is built into the system that facilitates the development of other Google software products (e.g., Google maps, key-word search, etc.). The more people use the Google search engine, the more powerful the network becomes. In sum, the Google network has become greater than the sum of its parts. There is an important self-learning feature that is built into the network.

2.3 Permeability

A second important network principle is permeability which allows information to flow in and out of the network system or organization. Intelligent networks, by definition, presuppose permeable boundaries; that is, structured entry points that allow users to access and contribute to the overall system design. The level of permeability varies according to the openness of the system. The intelligent network must adhere to an internal logic (i.e., system protocols) while having the capacity to grow and develop. This is particularly evident when it comes to the internal logic structure and accessibility of the Internet.

The Permeability Predicament As noted earlier, the Internet has grown in size and complexity due to the many contributions of its users. The principle of permeability is central to this discussion since the Internet must allow easy access points for its users (e.g., personal computers, smart phones, computer tablets). At the same time, permeability also means opening up the system to any number of unwanted influences and outcomes. I call this the permeability predicament. What are some of the unwanted influences that affect network design and critical infrastructure? Such examples might include privacy invasion and network security threats (Gershon, 2011, 2017).

Network Security Threats Both large and small businesses alike are highly dependent on their computing and information storage systems. A network security threat is generally understood to mean unlawful attacks against an intelligent network and the information contained in such networks. Such attacks are directed

Table 1 Select examples of intelligent networking and critical infrastructure

• Banking and financial record keeping
• Infectious Disease and FBI criminal surveillance data bases
• Nuclear reactor and power grid operation and maintenance
• Airport traffic control
• University and student record keeping
• Bridge, tunnel and highway operation and maintenance
• Hospitals and medical record keeping

Source: Gershon (2011)

against critical infrastructure resulting in the destabilization of a network and/or violence against persons and property (Clayton, 2011). The goal is to intimidate an organization in furtherance of a political, military or social objective. Clark (2010) coins the term *cyberwarriors* to describe a new class of military actions taken by a country to penetrate another nation's computer networks for the purpose of intelligence gathering and/or causing severe disruption. Companies like Target, Home Depot and Sony Corporation, to name only a few, have been the target of sophisticated hackers. For such companies, the problem translates into damage to reputation as well as lost productivity and operation time. The data breach that affected Target cost the company an estimated $148 million in lost sales as well as recovery costs. Consider what would happen if the financial record-keeping at American Express was suddenly and irretrievably lost. The enormity of accurately recreating the lost data would fully destabilize the company and have a cascading effect on the world's financial markets. Hence the term, *critical infrastructure*. Select examples of intelligent networking and critical infrastructure can be seen in Table 1.

2.4 The Transnational Corporation

The transnational corporation is a nationally based company with overseas operations in two or more countries. One distinctive feature of the transnational corporation (TNC) is that strategic decision-making and the allocation of resources are based on economic goals and efficiencies with little regard to national boundaries. Through a process of foreign direct investment, the TNC actively promotes the use of advanced media and information technology on a worldwide basis (Behrman & Grosse, 1990; Compaine & Gomery, 2000; Gershon, 2006, 2013).

Most companies do not set out with an established plan for becoming a major international company. Rather, as a company's exports steadily increase, it establishes a foreign office to handle the sales and services of its products. In the beginning stages, the foreign office tends to be flexible and highly independent. As the firm gains experience, it may get involved in other facets of international business such as licensing and manufacturing abroad. Later, as pressures arise from various international operations, the company begins to recognize the need for a more comprehensive global strategy.

2.5 The TNC and Foreign Direct Investment

Foreign Direct Investment (FDI) refers to the ownership of a company in a foreign country. This includes the control of assets. As part of its commitment, the investing company will transfer some of its managerial, financial and technical expertise to the foreign owned company. The decision to engage in FDI is based on the profitability of the market, growth potential, regulatory climate and existing competitive situation. There are four primary reasons that help to explain why a company engages in FDI. They include: (1) Foreign market penetration (2) Proprietary assets and natural resources, (3) Research, production and distribution efficiencies and (4) Overcoming regulatory barriers to entry (Bartlett & Ghoshal, 1998; Behrman & Grosse, 1990).

3 Redefining Business Work Space

Today's knowledge economy involves the full integration of transnational business, nation-states and technologies operating at high speed. It is a global economy that is being driven by free-market capitalism. It stands in marked contrast to many of the basic patterns and assumptions of the industrial age. The once highly centralized business has given way to the transnational organization that operates in multiple countries throughout the world. Instead of time and communication being highly synchronized, today's working professional lives in a digital world of asynchronous and virtual communication that allows for the international collaboration of projects regardless of time zones, geographical borders and physical space (Tapscott, 1996). We have entered the era of global virtual teams where work is produced across multiple time zones and geographic spaces. We now consider four work-design principles that affect the spatial design and operations of the transnational corporation. They include: (1) Virtual Communication, (2) Information Exchange, (3) Digital Media and Convergence, (4) Decision Support Analysis.

4 Virtual Communication

The combination of computer and telecommunications technology has had a major effect on the spatial reorganization of activity for the transnational organization. Time and distance factors have become less important in determining where a company chooses to locate today. The term *virtual communication* can be used to describe the artificial space and network linkages connecting a disparate set of users using both computer mediated and telecommunications technology (Gershon, 2017). The technologies and software applications are varied and can include a combination of wired and wireless telephony, videoconferencing (full motion) and Internet based communication to name only a few. The communication, itself, can be both synchronous (real time) as well as asynchronous (different times).

We begin our discussion by exploring the term *virtual network* and what it means to communicate across time and space without the need for a physical connection. A *virtual private network* (VPN) is a computer network that that uses a public telecommunications infrastructure such as the Internet to provide remote users (or departments) secure access to their organization's network. A VPN can range in size and scale of operation from theTNC that operates on multiple continents to a major medical hospital that must provide secure healthcare information to physicians and other medical professionals located in a variety of clinics and adjoining facilities. The major requirement is the ability to provide immediate and secure information available to the organization and its affiliate sites.

4.1 Global Virtual Teams

International project teams are the key to smart, flexible and cost effective organizations. A global virtual team represents working professionals from a TNC's worldwide operations assembled together for the purpose of working on a common project goal or assignment (Martins, Gibson, & Maynard, 2004; Maznevski & Chudoba, 2000; Lipnack & Stamps, 1997). In such teams, traditional face-to-face communication is partially or fully replaced by electronic forms of communication, including videoconferencing, desktop video, (SKYPE, Facetime etc.), web-based shared documents (Google Docs, Dropbox etc.) and electronic mail. The TNC's use of global virtual teams as part of a larger effort to share international expertise across the entire organization. The global virtual team offers up certain distinct advantages, including collaborative research and design work, reduced travel costs etc. (DeSanctis, Poole & Dickerson, 2000).

Advancements in communication technology and intelligent networking have elevated the principle of teamwork to a whole new level in terms of collaborative effort. At the same time, the use of global virtual teams brings with it a unique set of challenges. Foremost, are issues pertaining to trust involving differences of culture, geographic dislocation, complex problem solving and the effective collaboration of ideas. Specifically, how does one creatively engage a group of people that one has never physically met and trusting that everyone is equal to the task? (Ramirez & Zhang, 2007; Evaristo, 2003; Jarvenpaa, Knoll, & Leidner, 1998; Potter & Balthazard, 2002). The global virtual team presents both opportunities and challenges in terms of utilizing the principles of virtual communication in tandem with intelligent networks.

4.2 Videoconferencing: Tools and Applications

One of the standard tools for the global virtual team is the videoconference. The videoconference provides an electronic meeting format using video images and audio sound. People are typically linked together at two or more locations via a telephone codec or satellite. The videoconference permits the exchange of information, data and/or drawings between two or more people at separate locations.

The organizational videoconference has become a standard feature of transnational communication and serves as an alternative to face-to-face meetings. This can include both to point-to-multipoint videoconferences using structured conference meeting rooms as well as Internet-based conferencing.

Videoconferencing offers the transnational organization some distinct advantages, including (1) information exchange, (2) increased productivity, (3) cost savings and (4) employee training. First and foremost, the TNC videoconference is about information exchange. This gives the large scale organization a level of productivity and efficiency that would otherwise be unavailable. The TNC videoconference can link together marketing and engineering teams and, thereby, eliminate the need for travel, hotel and lodging expenses associated with travel. It also contributes to increased productivity since key management people are in the office and not in transit. Videoconferencing companies like Cisco and Polycom have tried to simplify the conferencing process by introducing easy meeting set ups, as well as making the design of the conferencing rooms (including table shape, wall colors and back drop) look as though all participating members are seated in the same room. This stands in marked contrast to older videoconferencing systems that required dedicated and expensive room-based facilities that were difficult to set up (Passmore, 2007).

Alternatively, an Internet-based videoconference can provide a low cost way to electronically link members of the transnational organization. Today, nearly all desktop, laptop and tablet computers are equipped with a webcam and microphone. Software applications like *SKYPE*, *Facetime* and *Go to Meeting* make international project meetings and consultations a fairly routine exercise. In addition, a prearranged webinar can provide transnational employees with a timely method for obtaining educational training and information updates. Typically, the webinar is scheduled ahead of time and involves an information specialist as well as written materials as part of a larger information package.

5 Information Exchange

Intelligent networks are at the heart of international business finance. The transfer of goods and services is premised on information exchange; that is, electronic communication that executes and verifies that a transaction has occurred. A credit card transaction provides a good illustration of intelligent networking and the principle of information exchange. A successful credit card transaction sets into motion a three-way information exchange between the consumer, the retailer/store and credit card company. A related term is *exchange efficiency* which is an important concept found in management theory. It has to do with creating the optimum conditions through which a consumer can obtain a product or service. Traditional examples of exchange efficiency can be seen with speed lanes in a supermarket, thereby, allowing customers to move quickly through the checkout line. Similarly, the principle of exchange efficiency can be seen with companies who specialize in E-Commerce shopping such as *Amazon.com, Zappos and Booking.com* to name only a few (Gershon, 2017).

5.1 Electronic Commerce

Today, electronic commerce has taken the principle of exchange efficiency to a whole new level in terms of retail trade. E-commerce represents the ability to buy and sell products and services electronically via the Internet. They include business-to-consumer (B2C), business-to-business (B2B), and consumer-to-consumer (C2C). Of particular note, is B2C electronic commerce which involves selling products and services directly to consumers via the Internet. B2C comes in two general formats, including traditional retailers (e.g., Sears, Target, etc.) as well as those companies whose primary business model depends on the Internet as the basis for retail trade (e.g., Amazon.com, Apple and Netflix). B2C electronic commerce has fundamentally changed how retail trade is conducted in terms of information gathering, production and distribution. It has created an altogether new business model that maximizes the potential for instantaneous communication to a world-wide customer base. This is particularly evident in the field of music sales and distribution. The speed and efficiency of producing Internet delivered music using MP3 file-sharing software, has fundamentally changed the cost structure of music recording and distribution on a worldwide basis. Specifically, the combination of the Apple iPod and iTunes media store created the first sustainable music EC business model of its kind (Gershon, 2013). It has redefined the way music and video entertainment is distributed to the consumer and given new meaning to the term exchange efficiency.

5.2 Supply Chain Management and Business Process

Supply chain management (SCM) is a complex business model that considers the entire set of linking steps necessary to produce and deliver a product to the end consumer. SCM philosophy is grounded in the belief that everyone involved in the supply chain is both a supplier and customer and requires access to timely, up-to-date information. The goal is to optimize organizational efficiency by meeting the needs of both suppliers and customers. SCM forces companies to move away from an organizational structure designed around functional silos toward one designed around the end-to-end flow of business processes. A well-designed SCM system requires the ability to give real time information to an extended network of suppliers, manufacturers, distributors and retailers (Tarn, Razi, Yen, & Xu, 2002). SCM makes it possible for companies like Dell Computers to engage in just-in-time manufacturing and Amazon.com to offer direct-to-home delivery of goods and services.

5.3 Just-in-Time Manufacturing

Most companies have access to excellent hardware and software capability that enables them to operate in a global business environment. The distinguishing factor often centers on speed and turn around time. Faster product cycles and the ability to train and produce worldwide production teams have transnationalized the manufacturing and distribution process. It is the ability to apply time-based competitive strategies at the global level that enables the TNC to manage inventories across borders. At the heart of time base competitiveness is just-in-time manufacturing which allows a company to meet an order in the least amount of time. Just-in-time manufacturing relies on the use of supply chain management and enterprise resource planning systems (ERP) for the purpose of tracking customer orders. ERP tends to focus on internal business processes within the boundaries of a single organization (Zheng, Yen, & Tarn, 2000).

Dell Computers and Just-in-Time Manufacturing Dell Computers was established by Michael Dell in 1984 and has grown to become one of the world's preeminent manufacturers of desktop and laptop computers. By 1996, Internet sales had taken off and the company realized that computer savvy shoppers preferred the convenience of custom ordering what they wanted directly from Dell and having it delivered to their door. During this time, Dell had become master innovators involving two important business processes. The first process was customization using a just-in-time manufacturing capability. Dell built computers to customer order and specification, thereby, eliminating excess inventory and the need for storage. Because of Dell's build-to-order philosophy, Dell has evolved a highly sophisticated manufacturing and logistics capability. The second important process was direct-to-consumer sales delivery thus avoiding costly investment in retail store infrastructure. It was a process model that other computer manufacturers would later adopt (Fields, 2006).

6 Digital Media and Convergence

Digital media is at the heart of today's communication revolution. Digital media represents the artistic convergence of various kinds of hardware and software design elements to create entirely new forms of communication expression (Gershon, 2017). From electronic commerce (Amazon.com) to music and video streaming (iTunes, Netflix), digital media has transformed the business of retail selling and personal lifestyle. Digital technology improves the quality and efficiency of switching, routing and storing of information (Wirtz, 2011; Wirth, 2006). It increases the potential for manipulation and transformation of data. Digital technology makes it possible to achieve convergence between different electronic media forms, including voice, data and video communication.

Electronic Commerce and Amazon.com Amazon.com is a transnational media and electronic commerce company headquartered in Seattle, Washington. Company founder, Jeff Bezos, incorporated the company in July 1994. Today, Amazon.com is the largest EC retailer in the world. Amazon.com started with online books given the large world-wide demand for literature, the low price points for books, as well as the large number of titles available in print. In its formative years, Amazon focused on business-to-customer (B-to-C) electronic commerce. The challenge was to become more fully diversified in terms of product and service offerings. In time, they incorporated customer reviews and leveraged such information as a way to sell more products and services as well as improve the customer experience. In addition, Amazon has greatly expanded its third-party marketplace, where merchants worldwide can set up their own virtual stores on Amazon.com and sell their products alongside Amazon's—all the while leveraging Amazon's large customer base and credit-card-processing services (i.e., exchange efficiency). Both retailers and individual sellers utilize the Amazon.com platform to sell goods. Large retailers like Nordstrom and Target use Amazon.com to sell their products in addition to their own web sites.

Amazon Kindle In November 2007, Amazon launched its Amazon Kindle; an E-book reader. Today, most analysts agree that Amazon probably sold its original Kindle hardware at breakeven or a small loss to subsidize media sales. With the original Kindle, Amazon pioneered the sale of digital books, and as a result owns over 90 % of their distribution. By July 2010, Amazon E-book sales for its Kindle reader outnumbered sales of hardcover books for the first time ever. In September 2011, Amazon announced its entry into the computer tablet market with the introduction of its Kindle Fire, which runs on the Android operating system. The Amazon Kindle is much more than an E-reader. It represents the foundation for an entire media ecosystem (Matthews, 2012). Specifically, the Kindle Fire is a computer tablet, media store, a platform for digital media sales as well as a publishing imprint. The goal is to make all of Amazon's media offerings part of the Kindle digital media ecosystem. This includes: (1) Digital books, (2) Mp3 music and software products, (3) Amazon Instant Video (videostreaming), (4) Software apps. and (5) Advertising. The Amazon Kindle is a decade-long investment in a media planning, product design, development and distribution.

7 Decision Support Analysis

Decision theory is concerned with identifying the values, risks and uncertainties associated with important decisions. The goal is to find the best tools, methodologies and software to help people and organizations make better decisions. The most systematic and comprehensive software tools developed in this way are called decision support systems (DSS). A properly designed DSS is an interactive software-based system intended to help decision makers compile useful information from a combination of business intelligence documents, including relational data sources, financial performance data and personal/experiential knowledge in the field (Russell & Norvig, 2003).

7.1 Decision Theory and Artificial Intelligence

Artificial intelligence (AI) is the study and design of intelligent agents or networks. AI is closely tied to the study of decision theory in mathematics and computer science. The goal of artificial intelligence is to develop new approaches to reasoning and problem solving. What all AI systems share in common is the ability to reason, problem solve and take corrective action based on preprogrammed assumptions and information inputs (Kurzweil, 1990). There are two distinguishing features that characterize all AI systems. First, the AI system must have the ability to scan its external surroundings. Second, the AI system must have the ability to evaluate a situation and initiate an appropriate decision/response. This is referred to as adaptation (Holland, 1995). The decision must be rationale (Russell & Norvig, 2003). In practical terms, AI systems can be both simple and complex in design. The movie rental service, *Netflix*, for example, exhibits simple AI features to the extent that it knows how to create recommended film viewing lists based on user inputs as well as a preprogrammed algorithmic based rating system. Alternatively, an automobile collision avoidance system is designed to react to situations that humans can not or choose not to, due to driver error. An automobile collision avoidance system (ACAS), will automatically initiate stability control; including the use of anti-lock braking and sensing systems to determine the optimal requirements to support driver safety and prevent accidents. What is important to remember is the degree to which the AI system makes preprogrammed choices on behalf of the user. In principle, the AI system can make faster calculations and decisions than humans involving high speeds as well as reacting to unexpected changes in the external environment.

7.2 Intelligent Agents

From Douglas Engelbart's original prototype of the computer mouse to present day voice recognition systems, one of the promising areas of human/computer interface design and artificial intelligence can be seen in the area of intelligent (or software) agents. An intelligent agent (IA) is a software program that organizes information in support of personal and professional decision-making (Kurzweil, 2000). Think of the IA as a virtual secretary whose job is to maintain the user's calendar, organize appointments, prioritize incoming information and scan relevant websites for important news and information items. All of the essential software elements for IA presently exists. What remains is the task of making them fully integrated.

In keeping with the principle of network evolution, the IA will use knowledge and past experience as the basis for growth and improved decision-making. This could include everything from a simple Amazon product alert to aiding in a complex research project using voice activated information exchange. Each successive generation in voice recognition software is better able to adapt to the user by keying on select words and phrases and storing it in its internal memory. Apple and Google's voice activated search application software that responds to verbal queries is merely the tip of the iceberg. In practical terms, this is just the first step in combining artificial intelligence and voice activated information exchange.

8 Discussion

The chapter looked at several ways in which the intelligent network affects the spatial design and operations of the transnational organization. Special attention was given to four work-design principles that affect the spatial design and operations of the transnational corporation. They include: (1) Virtual Communication, (2) Information exchange, (3) Convergence and Digital Media and (4) Decision Support Analysis. A central argument is that the intelligent network is not one network, but a series of networks designed to enhance worldwide communication for business and individual users alike. What gives the intelligent network its unique intelligence are the people and users of the system and the value-added contributions they bring to the system via critical gateway points.

Intelligent networks, by definition, presuppose permeable boundaries; that is, structured entry points that allow users to access and contribute to the overall system design. The same gateway points also means opening up the system to any number of unwanted influences and outcomes. Such unwanted influences and outcomes can include network security threats, privacy invasion, and copyright infringement of posted Internet material. We refer to this as the permeability predicament. Providing structured gateway points to the network is at the heart of making the network qualitatively better and more efficient. The downside risk is giving users with bad intentions access to the same network on-ramps.

The stakes become that much higher when dealing with cyberterrorism and critical infrastructure systems.

8.1 Redefining Business Work Space

Social/scientific terms like *knowledge economy*, *information society*, or *digital age* do not lend themselves to precise definition or meaning. What is beyond dispute, is the role of intelligent networks in helping to advance strategic decision making in support of international business communication. The knowledge economy involves the full integration of transnational business, nation-states and technologies operating at high speed. The once highly centralized business has given way to the transnational organization that operates in multiple countries throughout the world. Instead of time and communication being highly synchronized, today's working professional lives in a digital world of global virtual teams, videoconferencing and cloud computing that allows for the international collaboration of projects regardless of time zones, geographical borders and physical space. Voice, data and video communication speak the common language of digital communication. Information is digitally organized; reduced to 1's and 0's on computers while racing at high speed across a variety of transmission pathways.

Nowhere is this more evident than in the Internet itself. The Internet has become steadily woven into all aspects of work and leisure. The intelligent network has become the vital nervous system that enables the transnational organization to stay virtually connected. It has become the all important network engine that drives globalization forward.

References

Albarran, A., & Chan-Olmsted, S. (Eds.). (1998). *Global media economics*. Ames, IA: Iowa State Press.
Auletta, K. (2009). *Googled: The end of the world as we know it*. New York: Penguin Press.
Bartlett, C., & Ghoshal, S. (1998). *Managing across borders: The transnational solution*. Boston, MA: Harvard Business School Press.
Behrman, J., & Grosse, R. E. (1990). *International business and governments: Issues and institutions*. Columbia, SC: University of South Carolina Press.
Berners-Lee, T. (1999). *Weaving the web*. New York: Harper Collins.
Castells, M. (2000). *The rise of the networked society, the information age: Economy, society and culture* (2nd ed.). Oxford: Blackwell.
Clark, R. (2010). *Cyber war*. New York: Harper-Collins.
Clayton, M. (2011, March 7). The new cyber arms race. *Christian Science Monitor*, 27–33.
Compaine, B., & Gomery, D. (2000). *Who owns the media?* (3rd ed.). Mahwah, NJ: Lawrence Erlbaum.
DeSanctis, G., Poole, M., & Dickerson, G. (2000). Teams and technology: Interactions over time. In M. Neale & T. Griffin (Eds.), *Research on groups and teams* (pp. 1–27). Greenwich, CT: JAI Press.
Ellison, N., Steinfield, C., & Lampe, C. (2007). The benefits of Facebook friends: Social capital and college students' use of online social network sites. *Journal of Computer Mediated Communication*, 12(4), http://jcmc.indiana.edu/vol12/issue4/
Evaristo, J. (2003). The management of distributed projects across cultures. *Journal of Global Information Management, 11*(4), 58–70.
Fields, G. (2006). Innovation, time and territory: Space and the business organization of Dell Computer. *Economic Geography, 82*(2), 119–147.
Gershon, R. (2006). Issues in transnational media management. In A. Albarran, M. Wirth, & S. Chan-Olmsted (Eds.), *Handbook of media management and economics* (pp. 203–228). Mahwah, NJ: Lawrence Erlbaum & Associates.
Gershon, R. (2011). *Intelligent networks and international business communication: A systems theory interpretation* (Media markets monographs). Pamplona: Universidad de Navarra Press.
Gershon, R. (2013). *Media, telecommunications and business strategy* (2nd ed.). New York, NY: Routledge.
Gershon, R. (2017). *Digital media and innovation*. Los Angeles, CA: Sage.
Halal, W. (1997). Organizational intelligence: What is it and how can managers use it? In J. Cortada & J. Woods. (Eds.), *The knowledge management yearbook 1999–2000* (pp. 66–69). Burlington, MA: Butterworth-Heinemann.
Holland, J. (1995). *Hidden order: How adaptation builds complexity*. New York: Helix Books.
Huber, G. (1990). A theory of the effects of advanced information technologies on organizational design, intelligence and decision-making. *Academy of Management Reviews, 15*, 195–204.
Jarvenpaa, S., Knoll, K., & Leidner, D. (1998). Is anybody out there? antecedents of trust in global teams. *Journal of Management Information Systems, 14*(4), 29–64.
Kurzweil, R. (1990). *The age of intelligent machines*. Boston, MA: MIT Press.
Kurzweil, R. (2000). *The age of spiritual machines: When computers exceed human intelligence*. New York: Penguin.

Lipnack, J., & Stamps, J. (1997). *Virtual teams: Reaching across space, time and organizations with technology*. New York: Wiley.
Martins, L., Gibson, L., & Maynard, M. (2004). Virtual teams: What do we know and where do we go from here? *Journal of Management, 30*(6), 805–835.
Matthews, C. (2012, July 16). Will Amazon take over the world? *Time*. http://business.time.com/2012/07/16/will-amazon-take-over-the-world/
Maznevski, M., & Chudoba, K. (2000). Bridging space over time: Global virtual team dynamics and effectiveness. *Organization Science, 11*(5), 473–492.
Monge, P., & Contractor, N. (2003). *Theories of communication networks*. New York: Oxford Press.
Monge, P., Heiss, B., & Magolin, D. (2008). Communication network evolution in organizational communities. *Communication Theory, 18*(4), 449–477.
Most Innovative Companies. (2007, May 14). *Business Week*, 60.
Noam, E. M. (2001). *Interconnecting the network of networks*. Cambridge, MA: MIT Press.
Passmore, D. (2007). Videoconferencing opportunities and challenges. *Business Communication Review, 37*(2), 19–21.
Potter, R., & Balthazard, P. (2002). Understanding human interactions and performance in the virtual team. *Journal of Information Technology Theory and Application, 4*, 1–23.
Ramirez, A., & Zhang, S. (2007). When online meets offline: The effect of modality switching on relational communication. *Communication Monographs, 74*(3), 287–310.
Russell, S., & Norvig, P. (2003). *Artificial intelligence: A modern approach* (2nd ed.). New York: Prentice-Hall.
Tapscott. (1996). *The digital economy: Promise and peril in the age of networked intelligence*. New York: McGraw-Hill.
Tarn, M., Razi, M., Yen, D., & Xu, Z. (2002). Linking ERP and SCM systems. *International Journal of Manufacturing Technology and Management, 4*(5), 420–439.
Wirth, M. (2006). Issues in media convergence. In A. Albarran, M. Wirth, & S. Chan-Olmsted (Eds.), *Handbook of media management and economics* (pp. 445–462). Mahwah, NJ: Lawrence Erlbaum & Associates.
Wirtz, B. (2011). *Media and internet management*. Wiesbaden: Springer Gabler.
Zheng, S., Yen, D., & Tarn, J. M. (2000). The new spectrum of the cross-enterprise solution: The integration of supply chain management and enterprise resource planning systems. *The Journal of Computer Information Systems, 41*(2), 84–93.

The Effects of Information to Gain Employee Involvement and Achieve Organizational Convergence in a Media Setting

Stavros Georgiades

1 Introduction and Problem Discussion

The purpose of this study is to increase our understanding of how the way managers seeking to achieve organizational convergence think about gaining employee involvement by closing a gap in the current literature. Specifically, we believe that scholars have failed to adequately study how managers view employee involvement. This lacuna is noteworthy given an apparent consensus about the value of employee involvement to successfully achieve organizational convergence.

There is consensus among academics that to achieve organizational convergence via employee involvement successfully, managers often need to gain the support of employees (Cobb, Folger, & Wooten, 1995; Delaney & Sockell, 1990; Morgan & Zeffane, 2003; Tesluk, Vance, & Mathieu, 1999). To do so, they are advised to use practices that empower employees via both the provision of employee development (Block, 1987; Conger & Kanungo, 1988; Detert & Burris, 2007; Rusaw, 2000; Tangirala & Ramanujam, 2008) and operational freedom (Bandura, 1986; Conger, 1986; Nord, Rosenblatt, & Rogers, 1993; Tierney, 1999; Zhang & Bartol, 2010). Also they should use several motivational methods (Black & Margulies, 1989; Ketokivi & Castaner, 2004; Neubert & Cady, 2001; Schwochau & Delaney, 1997) to ensure employees are motivated to deal with problems that arise without delay (Clayton & Gregory, 2000; Gill, 1996; Lewis, Schmisseur, Stephens, & Weir, 2006; McHugh, 1997).

Despite Lewin's (1943) pronouncement about the practicality of a good theory, if managers and academics think about employee involvement to achieve organizational convergence differently, even if an underlying theory is correct, then managers might not use it. To see how wide the gap might be, we begin with a

S. Georgiades (✉)
School of Economic Sciences and Administration, Frederick University, Limassol, Cyprus
e-mail: bus.gs@frederick.ac.cy

A. Lugmayr, C. Dal Zotto (eds.), *Media Convergence Handbook - Vol. 2*, Media Business and Innovation, DOI 10.1007/978-3-642-54487-3_3

brief overview of significant portions of related academic work. Then, we present the results of an empirical study designed to learn how at least some media managers think about introducing employee involvement to achieve organizational convergence. One caveat before we begin—we are academics, consequently our understanding of how media managers think about employee involvement was developed from the perspective of prior academic research.

2 Literature Review

Research stressing the importance of gaining employee involvement to achieve organizational convergence has a long history. Existent research on involvement has appeared under a variety of labels, including participation and empowerment.

2.1 History of Academic Interest

Much of the interest in topics such as participative management is grounded in the 1947 work of Kurt Lewin (Maynard, Mathieu, Gilson, O'Boyle, & Cigularov, 2012). Some of Lewin's most well-known inquiry dealt with the use of participation to reduce resistance to change at the Harwood Manufacturing Corporation, where participation entailed involving workers in small discussion groups. The results were published in the now-famous paper by Coch and French (1948). This lineage apparently led many academics to view group discussions as the major step in implementing participative management and achieve organizational convergence. Even though one of the co-authors, Coch, was a personnel manager, this study gave us little insight about how the managers in general conceptualized employee involvement.

The emphasis on groups was evident in a second early stream of research, the contingency theory of leadership advanced by Vroom (1973). Vroom and Yetton proposed a normative theory to guide leaders choices concerning the degree to which groups should be involved in decisions. As with the Lewin and Coch & French works, this research did not investigate how the managers viewed the involvement process.

Since this early research that was the foundation for interest in employee involvement, more recent work has increased our understanding of employee involvement in efforts to achieve organizational convergence. Much of this work concerned communication.

2.2 Importance of Communication in Involvement and Convergence

The way communication can help gain employee involvement to achieve organizational convergence has been widely recognized. Lundberg's (1990) theoretical work explained how different types of communication can aid employee involvement. Further, Klein (1996) provided evidence concerning the way the communication process can help employees. He found that: (1) feedback and adjustments, (2) strategies that encourage disclosure of problems, and (3) discussion of solutions, was helpful.

In a similar vein, Gill (1996) observed that recent company history, trust factors, and communication methods could introduce barriers when aiming to achieve employee involvement. Introduction of a communication exercise prior to implementing involvement contributed to success in its implementation. Similarly, Tierney (1999) found that the quality of supervisor/employee relationships affected the degree to which an organization's climate was conducive to employee involvement leading towards organizational convergence.

2.3 Overview of History

The longstanding interest in employee involvement has produced a rather strong consensus in the academic literature that employee participation, communication, empowerment, etc. can contribute to successful employee involvement and organizational convergence.

However, it appears to us that there is a significant omission in this stream of research. There has been little study of how practicing managers view employee involvement. This omission is particularly troublesome given the wide variety of labels academics have used to refer to employee involvement. Still, some of the more recent inquiry has provided some clues about the managers' views.

2.4 Some Clues Concerning Managers' Conceptualizations of Employee Involvement

Several scholars that have dealt directly with how managers view the role of employee involvement have provided some clues. First, practicing managers may not perceive terms that academics use to refer to employee involvement, particularly empowerment, as applicable to organizational convergence.

Second, managers consider the provision of information in achieving employee involvement and organizational convergence to be important.

2.5 Managers May Not Perceive Concept of Empowerment as Applicable to Convergence

Edwards and Collinson's (2002) work provided some more direct information about managers' conceptions of employee involvement efforts. They treated employee involvement under the heading of empowerment. Their findings supported Hales (2000) analysis that the empowerment concept was elastic.

Managers had difficulty defining the term, but whatever it was it was difficult to implement and the term was . . .not a word used at local level (p. 287). Managers viewed the idea of empowerment as inapplicable to organizational convergence because it meant giving workers a wide range of freedom. "Managers were much happier with terms such as involvement" (p. 288). They concluded that, in practice, managers "rarely attempt the far-reaching transformations that the langue of empowerment" (p. 293) suggests.

2.6 Information Provision in Employee Involvement and Convergence

The value of the type of information management provides employees with to achieve their involvement and attain organizational convergence has been widely recognized. According to this research stream the information provided needs to be (a) relevant, essential and specific to a particular operation/decision, (b) adequate, structured to separate all its components and accurate, (c) controllable and not imposed without the employees influence, (d) timely so that it can be of any use to employees, and (e) relate to the employees' area of operations (Jackson-Cox, McQueeney, & Thirkell, 1987; Lewis et al., 2006; Morgan & Zeffane, 2003; Purdy, 1993; Wall & Lisheron, 1977; Zhang & Bartol, 2010).

In a similar vein, several studies emphasized the organizational arrangements that can be provided to achieve employee appreciation of the information. Academics have observed that regular expressions of confidence in employees, formal technical training, assistance provided to ensure employees can understand how the information has been prepared, management control over the amount of information provided, and methods to identify and rectify related employee problems can help employees appreciate and use the information provided by management (Clayton & Gregory, 2000; Conger, 1986; Conger & Kanungo, 1988; Detert & Burris, 2007; Ketokivi & Castaner, 2004; Klein, 1996; Nord et al., 1993; Rusaw, 2000; Tangirala & Ramanujam, 2008).

2.7 Overview

Based on the literature, there is a good reason to believe that the provision of information and related organizational arrangements to employees by management can lead towards their involvement that can help achieve organizational convergence.

There has been little study of how media managers conceptualize employee involvement via the provision of information and related organizational arrangements, but the available clues suggest that inquiry is needed. The study reported below was motivated by our desire to satisfy this need. We conducted an inductive case study.

3 Methodology and Approach

3.1 Setting and the Change

The study was conducted in a Greek-speaking, medium-sized media organization that recently implemented a major change—the Personal Development Program (PDP). The program was introduced because due to the convergence of the European Union markets, this Greek-speaking organization had to operate in a very competitive media environment.

The PDP was intended to reduce costs and increase both the quantity and the quality of the programs produced. It was one of several programs the organization had introduced aimed at improving personnel quality to achieve organizational convergence.

The PDP included the introduction of new technologies, accompanied by special personnel educational programs aiming to improve employee skills and knowledge related to the new technologies. It also provided specific organizational arrangements that management hoped would achieve employee involvement through which employees could influence and contribute toward organizational decisions and operations in an effort to achieve organizational convergence.

Gaining employee involvement via the implementation of the PDP was intended to ensure that the opinions of the employees were taken into account and that they influenced both the organizational decisions and work level operations. However, management did not want to make joint decisions with employees. They only aimed for employees to have an input in the decisions the managers made.

To achieve the union's agreement and ensure all employees were thoroughly informed about the PDP, the Board held meetings with the employee trade union to explain the aims of the PDP and how it was expected to operate. Besides the information and explanations received by their union, employees could also communicate with the management staff of their department for any further clarifications or questions focusing on specific departmental matters.

The program also included training. During the second 3 months of the program all organizational staff went through some Internet training conducted by external specialists in order to ensure they understood and were able to utilize the new facilities at work. By the middle of the year all new facilities were in place and all staff members were trained to use them.

3.2 Research Design

Following Yin (2003), we employed a case-study design. The design permitted within study replication by treating a series of interviews as a series of experiments with each case serving to confirm or disconfirm the inferences drawn from the others. Data were collected primarily through 22 in-depth interviews conducted over a period of 5 months with the departmental manager (DM) of the technical department and the TM, the middle managers (MMS) and the working unit leaders

(WULS) of the three departmental areas of the radio and music department (Talk Area First Radio, Music Area First Radio, Third Radio).

Two sets of interviews were conducted: (1) an initial interview with the manager of the technical department, and (2) semi-structured interviews with the seven managers of the radio and music department. Immediately after the interview, facts and impressions were cross-checked. Several rules were followed. First, the 24-h rule required that detailed interview notes be completed within one day of the interview. Second, at the time of the interview, all data, regardless of apparent importance were included. In addition, secondary sources were consulted. Data were collected from internal documents, including the organization's annual summary and the annual published radio and music departmental review.

Initial interviews centered on departments' operations and structures and on the implementation of the PDP. The second and third set of interviews were more structured and concerned the provision of information and related organizational arrangements management provided employees with to achieve their involvement. 13 matters were discussed in total (See Appendix).

3.2.1 Semi-Structured Interviews of Seven Managers of the R&M Department

After the initial interview with the manager of the technical department, semi-structured interviews with all managers of the radio and music department were conducted.

In addition, questions were directed at obtaining factual accounts of what informants did or observed others doing.

3.3 Data Analysis

The search for consistent perspectives was assisted by selecting pairs of managers (MMS and WULS) within each of the three different departmental areas of the radio and music department, and listing similarities and differences between each pair.

Managers' thoughts about involvement were classified under three general headings: (1) Information Characteristics, (2) Empowerment Practices (3) Organizational Arrangements.

4 Results

4.1 Information Characteristics

While considering the information characteristics, the media managers thought that providing employees with information that relates to their area of operation, is relevant to a particular operation/decision (essential and specific), adequate (structured to separate all its components and accurate), controllable and timely, encourages and enables them to get involved.

4.1.1 Talk Area

The Talk Area Working Unit Leader (WUL) claimed that the information provided to employees had to relate to their area of operations for them to "become interested and able to understand and thus utilize it to get involved and to contribute". He added that "this way employees can influence matters of their area and consequently feel that they have the necessary job independence". According to the Talk Area Middle Manager (MM) employees had a limited amount of time available to appreciate the information provided by management because they were very busy working on many different jobs, and also spent a lot of time working out of office. Consequently "management needs to provide them with only the essential information".

According to the Talk Area WUL the information provided needed also to be structured to separate all its components so that management could ensure that employees were not overloaded with unnecessary information. The two Talk Area managers considered that the information provided should not be imposed without the employees influence, because otherwise "employees feel that they lose their job independence and consequently do not want to use the information to get involved and to contribute. Job independence is a particularly important matter because employees do not trust management to be completely independent".

Finally, the Talk Area MM explained that employees aimed to produce programs/activities of high quality, and this could be achieved when the information used in the programs/activities was prompt and continuously updated in order to be accurate. Consequently the issue of information timeliness was major.

4.1.2 Third Radio Station

The Third Radio station case also indicates the linkage between several information characteristics and the achievement of employee involvement.

For example, the Third Radio Station WUL believed that the information provided had to relate to the employees' area of operations because via their involvement employees believed that they could assist management make decisions that could improve the quality of their area's operations and as a result benefit themselves via both job promotions and the provision of better working conditions. The two Third Radio Station managers also added that the information had to be relevant and specific towards a particular operation/decision in order for employees not to be overloaded with unnecessary information.

4.1.3 Music Area

Several different issues were also mentioned by the Music Area managers linking the information characteristics to the achievement of employee involvement.

More specifically, the two managers considered necessary that the information provided to employees be relevant, that is essential and specific to a particular decision/operation. They explained that because their area's operations had either changed or become more sophisticated after the internet implementation, when the information provided by management was not relevant to the particular operation/ decision it was difficult for employees to appreciate and use it to get involved and to contribute.

Also, because most of the information was collected and transmitted online, management was tempted to provide employees with too much information. Management thus had to chose and transmit only the essential information because otherwise employees would not be able to focus on the specific operation/decision and their contributions would be vague as a result. Information according to the two managers needed also to be "relevant to the operations of more advanced organizations of the same industry so that employees are aware and comprehend the developments that are already taking place in the more advanced organizations, and as a result their contributions take these changes into account".

The MM also considered necessary the information provided was structured to separate all its components and accurate so that employees could appreciate and use during the decision making process, and that the manager's job was assisted by the fact that a lot of the information relating to the music area operations was produced, collected and stored online. The WUL noted that "the information provided to employees has to be under the employees control to also alleviate employee worries that they can be responsible for decisions which are out of their control".

The WUL considered necessary management provided employees with timely information "because the aim of the music area is to produce programs and activities of a current nature, for different groups of people whose tastes and needs change continuously".

4.2 Empowerment Practices

In relation to empowerment practices, the media managers thought that they can assist employees understand and appreciate the information provided by regularly expressing confidence to them, clearly defining their responsibilities, and involving them with its production.

4.2.1 Expressions of Confidence

The managers think that expressions of confidence do not always have a positive effect on the employee appreciation of the information provided.

According to the DM, expressing confidence to employees can assist them appreciate the information provided and that this can be done by "highlighting cases in which employees use the information to argue a particular matter or to draw their own conclusion and contribute". The Music Area WUL added that "expressing confidence to employees is necessary because the information provided has been more advanced and technical after the internet implementation and as a result more difficult for employees to understand and appreciate". In addition, the Music Area MM noted that "expressing confidence is crucial because employees are usually sentimental and easily affected and influenced based on other people's comments".

In contrast, the Third Radio Station MM considered that there is "no need to regularly express confidence because this will probably make them think that it is very difficult to appreciate the information provided by management. When employees have problems they need the provision of specific management assistance that can help them overcome their problems without delay and not words of confidence".

4.2.2 Definition of Employee Responsibilities

The data collected illustrate a linkage between the definition of employee responsibilities in relation to the information provided and employee appreciation of that information. The managers in all three areas think that "management needs to clearly define employee responsibilities in relation to the information provided so that employees only have to deal with the specific information assigned to each one. The definition of responsibilities must be done in cooperation with employees because they are very close and thus familiar with each other's abilities, knowledge and expertise". The Talk Area WUL added that "as a result employees feel satisfied and comfortable with the information assigned to each one of them and consequently appreciate and utilize it to contribute".

4.2.3 Involvement with the Production of the Information

The managers of all areas pointed out that employees were also involved with the production of the assigned information. The DM noted that this "helps employees attain a good understanding and enables them to use the information to comprehend situations, get involved and contribute".

For example in the music area, according to the MM, "the continuous changes occurring in relation to its operations and activities mainly caused by the internet implementation, resulted in continuous changes in relation to the information used in the decision making process. Consequently, because employees need a lot of explanations and clarifications in order to reach a satisfactory level of understanding, it is also necessary to involve them with the production of the assigned information in order to obtain the necessary insights which will enable them to appreciate and utilize the information to contribute towards the management decisions".

Also, according to the MM of the Talk Area "by involving employees with the production of the information they have the necessary assurance about the independence of the information used to make decisions".

4.3 Organizational Arrangements

Media managers commented extensively, on how to assist employees appreciate the information provided and encourage them to use it to get involved by providing them with technical training, ensuring each employee receives all assigned information and by utilizing a system of feedback and adjustments.

4.3.1 Training

All three areas illustrate a linkage between the provision of training sessions and employee appreciation of the information provided. The Music Area MM noted that "training sessions are necessary for the employees because some of the information has become very technical, and thus difficult for employees to understand and appreciate, due to the fact that some of the activities have become more sophisticated after the internet implementation".

The Third Radio station WUL pointed out that "the discussions with management after the end of the training sessions are considered very useful by employees because they have the opportunity to clarify all outstanding matters, including simple matters they would be embarrassed to mention in front of their area's staff".

According to the Talk Area MM "explanations in relation to the information provided are also available online for those employees who cannot take part in some of the training sessions because of their work duties".

4.3.2 Receipt of all Assigned Information

The managers also noted the importance of ensuring all employees receive all assigned information. According to the DM this should be done based on each employee's experience, knowledge and expertise.

The managers of the Talk Area pointed out that "it is very important for us to ensure that each employee receives and appreciates all information he/she is assigned to deal with. Employees can then get together, discuss and appreciate all the information and related matters concerning a particular operation or decision in total, then as a team be in a position to get involved during the discussions with management and contribute towards the final management decisions".

The MM of the Third Area and the Music Area WUL added that "management uses the internet to provide some of the assigned information and via regular face to face discussions ensures the correct information is received by all employees".

4.3.3 Feedback and Adjustments

Finally, the data illustrate the importance of utilizing a system of feedback and adjustments. For example, the Third Radio station MM noted that "we set up a mechanism to ensure all problems and misunderstandings the employees face, in relation to the information provided, reach management as soon as possible in order to be solved without delay. Otherwise they become an obstacle to employee understanding, because employees are influenced negatively and feel that they can not deal with the information, and thus get involved and contribute".

The WUL of the Talk Area added that "WULS are assigned to view employee problems daily, both online and via personal contact. If problems are not minor they can consider solutions together with the MM, and in case of a difference in opinion discuss also with the DM. Solutions are then presented and explained to employees".

Similarly, the Talk Area managers noted that "we inform employees during the provision of information that they must communicate all problems instantaneously to us in order to be tackled immediately".

5 Viewpoint on Convergence

A major purpose of this chapter was to learn how closely the way media managers seeking to achieve organizational convergence think about gaining employee involvement via the provision of information and related organizational arrangements, and compare them to the academic theories mentioned earlier on.

When the comments are viewed together, an interesting pattern can be seen.

First, the media managers shared the academic literature's emphasis on the importance of the provision of information to employees to achieve their involvement and attain organizational convergence, and that this information needs to have certain characteristics for employees to be encouraged and enabled to get involved in the decision making process. In addition, the media managers suggested that they need to provide employees with the necessary organizational arrangements in order to assist them comprehend, interpret and appreciate the information provided.

Importantly, adoption of the media managers' perspective helped us to see some interesting nuances in how managers thought about how to gain this involvement via the provision of information and related assistance, leading to organizational convergence. Two things were noteworthy and both related to employee trust. First, their idea on involvement via the provision of controllable information places emphasis on job independence which deals with the issue of employee trust towards management. The lack of employee trust in a media organization appears to be a major issue which can be considered further by future studies. Second, it was the proactive stance they felt they needed to take to ensure employees consider the information independent. In particular, they saw the involvement of employees with the production of the information provided as highly important. Also, they felt they needed to clearly define employee responsibilities in relation to the information provided in cooperation with employees so that they feel they can trust and can consequently use it to get involved.

In addition, there were a number of things the media managers believed they needed to do to achieve organizational convergence via employee involvement. Consistent with much of the academic literature many of these were quite pragmatic. The provision of a specific type of information was clearly the most salient. Other pragmatic-centered matters included: ensuring all employees receive all assigned information, providing training and the importance of immediately attending to problems.

Whereas much of the academic literature also stresses the idea of communication via the provision of specific information, our results point to the need to recognize the somewhat obvious idea that not all calls for communication are necessarily calls for the same thing. Communication focused on ensuring employee job independence is not likely to be a call for the dialogue many academics seem to have in mind.

Of course, these specific thoughts may well be idiosyncratic to the media setting and thus can not be generalized, at least at this point.

On the other hand, what may be generalized and serve as a stimulus for future research is the findings that these managers thought of rather specific things that they needed to do at least in this media context, to gain the desired employee involvement and achieve organizational convergence. Thus, the results are a first step in addressing the lacuna in the academic literature we pointed out at the start.

While transmitting specific information seemed to be a major issue, when the data were considered through an empowerment lens, some limited themes consistent with the emancipatory spirit appeared. For example, media managers emphasized the expressions of confidence, definition of employee responsibilities in relation to the information provided in cooperation with employees and the employee involvement with the production of the information.

Whereas much of the academic literature stresses expressions of confidence via words of encouragement, our results point to the need to recognize the different effects on different types of media employees. When employees seem to be sentimental, they can be positively affected. However, there are situations when employees are negatively affected, feeling that the reason why management regularly express confidence to them is because they do not consider them capable of dealing with the information provided.

This study also adds insights relating to the organizational arrangements provided by management towards employees like the provision of training, all assigned information and a system of feedback and adjustments. These arrangements point towards a very interesting direction that could be advanced by further studies, that of the importance of informal face to face interactions on a personal level.

5.1 Limitations

It is important to recognize some limitations of this study. In interpreting these results, it must be recognized that the study dealt with one particular type of change in one setting and included only a few managers. Thus while the findings suggest the need for future research, they must be interpreted cautiously.

6 Conclusions

This chapter addresses managers' conceptions of employee involvement in a rapidly-changing media environment. The results are derived from a single case and thus are, at best, suggestive. The study arose from concern that little was known about how media managers view employee involvement to attain organizational convergence. Although a case study of this sort cannot provide generalized results, it did yield some important insights into an underdeveloped topic. Specifically, the study of involvement (and the various headings related to it) can be advanced by study of how managers view it.

Matters of information provision and related assistance are likely to be central, leading towards some pragmatic matters apt to be salient. Interestingly, whereas the academic literature frequently suggests that the provision of a specific type of information assists employees to appreciate and use it to get involved and achieve organizational convergence, the media managers in this study viewed the provision of this type of information as a means to achieve employee trust.

Moreover, consistent with some previous findings, it appears that media managers believe that the provision of several organizational arrangements can help gain employee involvement and achieve organizational convergence. However, the managers seem to highlight the importance of informal face to face interactions on a personal level with employees rather than formal organizational programs.

Of special note is that while both managers and academics emphasize the same word—communication—as a benefit of employee involvement to achieve organizational convergence, they are not necessarily discussing the same thing. Study of how media managers think about communication revealed what may be an important difference. Media managers may be focusing on ensuring employee job independence. Academics may be more apt to concentrate on dialogue. Future inquiry into this subtle dissimilarity that might make all the difference would seem to be a good direction.

Appendix

List of Matters discussed during the interviews

Description of the Department

1. Operations and Structure
2. Implementation of the PDP

Characteristics of the information provided

In order to achieve employee appreciation the information provided by management to employees needs to

3. relate to the employees' area of operations
4. be relevant, essential and specific to a particular operation/decision
5. be adequate, structured to separate all its components and accurate
6. be controllable and not imposed without the employees influence
7. be timely so that it can be of any use to employees

Organizational Arrangements in relation to the information provided

In order to assist employees appreciate and use the information to get involved and contribute and thus achieve organizational convergence management needs to

8. provide employees with technical training
9. regularly express confidence in employees
10. clearly define employee responsibilities in relation to the information provided
11. promote and encourage feedback and adjustments

12. explain to employees the way the information each one is assigned to deal with is produced
13. ensure that each employee receives all the information assigned by management to deal with

References

Bandura, A. (1986). *Social foundations of thought and action: A social cognitive view*. Englewood Cliffs, NJ: Prentice-Hall.

Black, S., & Margulies, N. (1989). An ideological perspective on participation: A case for integration. *Journal of Organizational Change Management, 2*(1), 13–35.

Block, P. (1987). *The empowered manager*. San Francisco: Jossey–Bass.

Clayton, J., & Gregory, W. J. (2000). Reflections on critical systems thinking and the management of change in rule-bound systems. *Journal of Organizational Change Management, 13*, 140–161.

Cobb, A. T., Folger, R., & Wooten, K. (1995). The role justice plays in organizational change. *Public Administration Quarterly, 19*(2), 135–152.

Coch, L., & French, J. R. P., Jr. (1948). Overcoming resistance to change. *Human Relations, 1*, 512–532.

Conger, J. A. (1986). *Empowering leadership,* Working Paper. Montreal: McGill University.

Conger, J. A., & Kanungo, R. N. (1988). The empowerment process: Integrating theory and practice. *Academy of Management Review, 13*, 471–482.

Delaney, J. T., & Sockell, D. (1990). Employee involvement programs, unionization and organizational flexibility. *Academy of Management Best Paper Proceedings,* 1990, (pp 264–269).

Detert, R. J., & Burris, R. E. (2007). Leadership behavior and employee voice: Is the door really open? *Academy of Management Journal, 50*, 869–884.

Edwards, P., & Collinson, M. (2002). Empowerment and managerial labor strategies: Pragmatism regained. *Work and Occupations, 29*(3), 272–299.

Gill, J. (1996). Communication—Is it really that simple? An analysis of a communication exercise in a case study. *Personnel Review, 25*, 23–37.

Hales, C. (2000). Management and empowerment programmes. *Work, Employment and Society, 14*(3), 501–519.

Jackson-Cox, J., McQueeney, J., & Thirkell, J. E. M. (1987). *Strategies issues and events in industrial relations: disclosure of information in context*. London: Routledge and Kegan Paul.

Ketokivi, M., & Castaner, X. (2004). Strategic planning as an integrating device. *Administrative Science Quarterly, 49*, 337–365.

Klein, S. M. (1996). A management communication strategy for change. *Journal of Organizational Change Management, 9*, 32–46.

Lewin, K. (1943). Psychology and the process of group living. *Journal of Social Psychology, 17*, 113–131.

Lewis, K. L., Schmisseur, M. A., Stephens, K. K., & Weir, K. E. (2006). Advice on communicating during organizational change. *Journal of Business Communication, 43*, 113–137.

Lundberg, C. C. (1990). Towards mapping the communication targets of organisational change. *Journal of Organizational Change Management, 3*(3), 6–14.

Maynard, M. T., Mathieu, J. E., Gilson, L. L., O'Boyle, E. H., Jr., & Cigularov, K. P. (2012). Drivers and outcomes of team psychological empowerment: A meta-analytic review and model test. *Organizational Psychology Review, 3*(2), 101–137.

McHugh, M. (1997). The stress factor: Another item for the change management agenda? *Journal of Organizational Change Management, 10*, 345–362.

Morgan, D. E., & Zeffane, R. (2003). Employee involvement, organizational change and trust in management. *International Journal of Human Resource Management, 14*, 55–75.

Neubert, M. J., & Cady, S. H. (2001). Program commitment: A multi-study longitudinal field investigation of its impact and antecedents. *Personnel Psychology, 54*, 421–449.

Nord, W. R., Rosenblatt, Z., & Rogers, K. (1993). Toward a political framework for flexible management of decline. *Organization Science, 4*, 76–91.

Purdy, D. E. (1993). Accounting controls, change and the perceptions of managers: A longitudinal study of ward units in a teaching hospital. *Financial Accountability and Management, 9*, 45–66.

Rusaw, C. A. (2000). Uncovering training resistance—A critical theory perspective. *Journal of Organizational Change Management, 13*, 249–263.

Schwochau, S., & Delaney, J. (1997). Employee participation and assessments of support for organizational policy changes. *Journal of Labor Research, 18*, 379–402.

Tangirala, S., & Ramanujam, R. (2008). Exploring nonlinearity in employee voice: The effect of personal control and organizational identification. *Academy of Management Journal, 51*, 1189–1203.

Tesluk, P. E., Vance, R. J., & Mathieu, J. E. (1999). Examining employee involvement in the context of participative work environments. *Group and Organization Management, 24*, 271–300.

Tierney, P. (1999). Work relations as a precursor to a psychological climate for change. *Journal of Organizational Change Management, 12*(2), 120–135.

Vroom, V. H. (1973). A new look at managerial decision making. *Organizational Dynamics, 1*(4), 66–80.

Wall, T., & Lisheron, J. A. (1977). *Worker participation*. London: McGraw-Hill.

Yin, R. K. (2003). *Case study research—Design and methods* (3rd ed.). Thousand Oaks, CA: Sage.

Zhang, X., & Bartol, M. K. (2010). Linking empowering leadership and employee creativity: The influence of psychological empowerment, intrinsic motivation and creative process engagement. *Academy of Management Journal, 53*, 107–128.

Convergent Business Environments: Debating the Need for New Business Models, Organizational Structures and Management respectively Employee Competencies

Matthias Karmasin, Sandra Diehl, and Isabell Koinig

1 Introduction and Problem Discussion: Convergence and Management Interdependencies

Even though postulated as a new concept, convergence has been around for much longer than anticipated and has been applied to the field of media and business studies approximately 30 years ago for the very first time (Micó, Masip, & Barbosa, 2009: 123). Nowadays, it is perceived as "a multidimensional process that, facilitated by the generalized implantation of digital telecommunications technologies, affects the technological, business, professional and publishing environment of the [. . .] media. This promotes an integration of previously dispersed tools, spaces, working methods and languages, with the result that journalists [organizations, as well as management and staff] produce content that is distributed through numerous platforms, using the languages corresponding to each of these" (Salaverría, García Avilés, & Masip, 2008).

In our article we define convergence as a complex phenomenon that alludes to the merging and overlapping of (media) services, features and functionalities that once belonged to different sectors/industries, taking a variety of forms: convergence of media, businesses, technologies and content (Diehl, Karmasin, Leopold, & Koinig, 2013; Jenkins, 2006).

M. Karmasin (✉) • S. Diehl • I. Koinig
Department of Media and Communications, Alpen-Adria-Universitaet, Klagenfurt, Austria
e-mail: matthias.karmasin@aau.at; sandra.diehl@aau.at; isabelle.koinig@aau.at

A. Lugmayr, C. Dal Zotto (eds.), *Media Convergence Handbook - Vol. 2*, Media Business and Innovation, DOI 10.1007/978-3-642-54487-3_4

- In the case of convergence of media, formerly distinct products and services melt into one another, bringing about new technological innovations in form of Internet TV, Internet telephony or mobile Internet.
- Convergence of businesses alludes to firms or companies crossing their original areas of economic operation and expanding into new fields of expertise. This is usually achieved by means of collaboration or cooperation, which varies in terms of objectives, time exposure, and intensiveness.
- Convergence of technologies has established itself as the most common example, easing working relationships and changing personal interactions sustainably. Two prominent cases, Unified Messaging (UM) and Unified Communications (UC), have particularly had an impact on the professional domain, with the work routine becoming less stationary and more flexible.
- Convergence of content is especially vital to media enterprises which can use different channels when targeting consumers and recipients alike. This means that content is not media-bound anymore, but needs to be adapted to changing environments of production and consumption (Karmasin & Winter, 2000).

As indicated above, convergence has had a drastic impact on the media and communication landscape, "changing the way we create, consume, learn and interact with each other" (Jenkins, 2006). Media convergence is perceived as a "process whereby new technologies are accommodated by existing media and communication industries and cultures", further referring to an "adaptation, merging together, and transitioning process" (Dwyer, 2010: 2). According to McPhillips and Merlo (2008), three main areas of effect need to be differentiated, which concern the sector's major agents: media owners, advertisers as well as consumers, who are given "the ability [. . .] to obtain multiple services on a single platform or device or obtain any given service on multiple platforms or devices" (Ofcom, 2008: 1).

Closely related to business practices is management. "Management is conceptualized as a business administration discipline that identifies and describes strategic and operational phenomena and problems in the leadership of media enterprises. At the same time, it is an applied science that is intended to provide assistance to the business practice regarding the leadership of media enterprises. [It] covers all the goal oriented activities of planning, organization and control within the framework of the creation and distribution processes for information or entertainment content in media enterprises" (Wirtz, 2011: 5–15).

Conditioned by shifts in media and technological environments, also the craft of managing companies is subject to change. Existing managerial tasks and competencies are only applicable to a varying degree anymore, and need to be expanded as well as adapted to contemporary standards. While traditional management functions centered on Planning, Organizing, Staffing, Directing/Leading and Controlling (Isaacs & McAllister, 2005; Koontz & O'Donnell, 1955; Weihrich & Koontz, 1993), this categorization proves to be insufficient in a converged surrounding. In addition, several tasks (such as Coordination and Decision-Making) need to be taken into account, which are, however, not listed separately. The same applies to so-called "patterns" involving oral communication, networking and workplace routines (Schreyögg & Koch, 2007).

While convergence facilitates change and substantially alters existing management practices, it "is a new media ideology too: that is, a way of thinking that facilitates the operation of neoliberal global markets" (Dwyer, 2010: 2f.), often demanding businesses to rethink their established operations and strategies to fit the requirements of present-day standards. Technical innovations and technological advancements in particular drive and facilitate those far-reaching changes, leading to new working environments as to production, distribution and consumption. Thereby, sectorial or industrial convergence is to be understood as "a 'blurring' of boundaries between industries, induced by converging value propositions, technologies and markets that lead to the emergence of inter-industry segments" (Bröring, Coultier, & Leker, 2006: 487). Yet, "[c]onvergence industries and convergence products are not mere results of radically novel technologies; most importantly, they support innovative business models [and organizational structures]" (Schwarz & Gustafsson, 2013).

2 Present and Future Trends: A review

Numerous trends stir up the media industry and have a bearing on the industry itself as well as on the managerial profession and the tasks associated therewith. These include, amongst others, technological and communication trends, as well as organizational and business trends.

Technological trends are triggered by rapidly growing technical/technological infrastructures and especially concern the media and communication industries. They are brought about by increasing broadband Internet usage rates and a larger dissemination of mobile communication devices (International Delphi Study, 2009). These two aspects are of utmost importance in those media environments characterized by convergence and are dramatically transforming workplace conditions. Web 2.0 technologies have become an immanent part of contemporary business practices, revolutionizing as well as improving professional communication and collaboration (Andriole, 2010: 67).

Communication trends enable the application of new technologies across varying fields and involve the areas of customer relationship management or knowledge management, indicating that they are useful means for both internal and external communication practices (Andriole, 2010: 67). Moreover, network communication is on the rise and presupposes a connected approach to communication. Those innovations have also led communication to increase to previously unknown dimensions, with more individualistic and interactive approaches dominating (Schmutzer, 2010). When comparing old and new media, the two forms differ in that new media is technology immanent, meaning resting on technology, as it is the case with software products (Schelhowe, 2007). Hence, users are increasingly asked to become media literate (Roth-Ebner, 2012; Zorn, 2011). Media literacy is already a prerequisite in today's labor market and, according to Schachtner (2010), consists of five dimensions: an instrumental dimension (the suitable application of new devices and technologies), an adoptive dimension (the capable usage of tools and services), a reflexive dimension (the conscious reflection on media content and processes), a contrastive dimension (an incorporation of technologies into daily routines), and a communicative, transcultural as well as co-operative dimension. The last aspect must not be left out of sight for it considers that businesses are often leaving their original fields of operation in order to operate globally, also investing in joint-projects.

Organizational trends change the way information is processed and handled within a company. As convergence has led to an information overload that is hard to tackle, organizational structures are affected by those changes and need to be adapted accordingly. With new media channels surfacing, information cannot only be distributed multiple times but can also be used on multiple platforms. As a consequence, staff members need to develop new ways of dealing with the pieces of data they are presented with. In this context, it is inevitable to filter irrelevant from relevant information in order to reduce the overall amount of information (Zorn, 2011). Moreover, collaboration is doomed to change, as project-oriented work gains in importance, which is characterized as temporary and task-oriented. This development also changes organizational structures drastically, which then have to be dismantled and restructured. Instead of hierarchical structures, either network structures or resilient structures are introduced. Network structures allow for faster communication but, at the same time, bear the danger of self-amplification. Resilient structures, which present a merger between a hierarchical and network system approach, are especially suited for present-day standards, allowing for fast reaction and flexibility in cases of change (Hernstein Institute for Management and Leadership, 2012b; Zukunftsinstitut, 2012).

Business trends need to be addressed in the age of convergence as well since industry boundaries blur or are increasingly dismantled, with firms extending their operations to prior unknown domains. These chances demand businesses to be flexible and adapt quickly to alterations in their immediate environments, which can be either of environmental, social, ecological or technical nature. One way of meeting today's requirements can be induced in form of change management. The term reckons "the coordination of a structure period of transition from situation A to situation B in order to achieve lasting change within an organization" (BNET Business Dictionary). It presents a beneficial and suitable means of bridging the transition from traditional to more open and flexible office structures, such as the divided workplace, consisting of both a home and a mobile office (International Delphi Study, 2009).

3 Methodology and Approach

In order to fully grasp the concept of convergence, the following paragraphs will outline how business environments have changed, conditioned by far reaching technical, technological as well as industrial alterations in contemporary business settings. In there, modifications to existing business models and organizational structures will be thematized, while also demonstrating to which extent present-day professional standards call for either a renewal or dismantlement of existing modes and means of operation. By means of examples, the article tries to both draw some conclusions and derive at implications of how established organizational trends and managerial roles can be copied from some top-notch/successful enterprises since they are well-suited for implementation in related industries.

3.1 The Effects of Convergence on Business Models

As already mentioned above, previously established industry boundaries increasingly blur in the digital age and, as a consequence, firms are put under pressure when facing (and successfully managing) new challenges brought about by the convergence of industries (Hacklin, 2007; Hacklin, Klang, & Baschera, 2013; Hacklin, Marxt, & Fahrni, 2010; Lei, 2000). As "traditional models tend to induce managers on the lockout for changes to stick within their own particular industry or sector boundaries" (Hacklin et al., 2013), new, open and flexible models are demanded.

Following Chesborough and Rosenbloom (2002), business models fulfill numerous functions and serve as a way of articulating a value proposition, identifying a market segment, specifying the revenue generation, defining the value chain's structure, detailing revenue mechanisms, estimating a cost structure, describing a firm's position within a network and formulating a competitive strategy.

Hacklin et al. (2013) recommend that, in order for firms to remain competitive, they are required to adapt their business models accordingly to dramatically changing environments. For this purpose, the authors have developed several archetypical business models that are meant to provide companies with basic guidelines as how to proceed in times of uncertainty and disillusion. Hence, an established model (Johnson, Christensen, & Kagermann, 2008) is re-modeled, allowing for capturing and embracing convergence not only at the very start but also at its core.

"[W]hen industries grow together, it makes little sense to try to compete on existing knowledge, technologies, products, or services only. To be successful even after [...] convergence has become established, a firm needs to fundamentally rethink the four basic elements of its business model, that is, the customer value proposition, the profit formula, the key resources and processes. This requires the ability—and willingness—to think beyond the existing boundaries of one's own industry or sector, allowing future differentiation to be achieved on the basis of the business model, rather than the positioning within the old industry" (Hacklin et al., 2013).

In order to grasp the extensive transformation in the media industry and related sectors, Hacklin et al. (2013) develop some recommendations or organizational archetypes in response to the above-mentioned trends. Generally, they can take three distinct forms: (1) brokering between industries, (2) opening-up the ecosystem, and (3) attacking head-on:

(1) *Brokering between industries* presupposes an early identification of trends that, for instance, enable collaborations and cooperation between different industrial

sectors and ultimately expands a firm's existing product portfolio. These often involve third-party engagements, which accumulate some indirect sales revenue for the original company and enable them to improve as well as expand the knowledge in their original area of expertise as well as in the newly accessed field. Enterprises allocated in the information and technology field are especially prone to engage in such co-operations with telecommunication firms in order to establish themselves in the area of Unified Communications, for instance.

(2) *Opening-up the ecosystem* requires a company to leave its original field of operation in order to appeal to customer bases from different markets and industrial segments, creating a novel customer proposition. This "deliberate and targeted orchestration of ecosystems" (Hacklin et al., 2013) enables an existing business model's extension and by taking advantage of network effects and a fast-working production pipeline, consumers can be appealed by use of intensive marketing activities. Here, for example, telecommunication companies could opt to enter the entertainment market in order to offer their customers with a broader selection of content.

(3) *Attacking head-on* is a strategy brought about by increasing competition, particularly in the information and communication technology (ICT) sector. Due to the availability of almost identical services and products, an extensive as well as intensive customer relationship management becomes a core feature in building and maintaining consumer loyalty, which, in turn, is essential in tying customers to a business and its associated/connected product range. By offering complementary products, firms can supply their existing customer base and extend their former business operations at the same time, using their original infrastructure to their benefit (Hacklin et al., 2013). A very prominent example of companies attacking heads on is the on-going battle between Apple and Samsung as the companies are matching for the leadership position in the mobile communications market.

Due to numerous and lasting changes, some of these commonly used business models are not appropriate for application anymore, at least not on their own. They require a rethinking as to their functionalities and demand organizational alterations to take place at the company's core, mandating a re-definition of organizational structures.

3.2 The Effects of Convergence on Organizational Structures

Organizational structures are defined by means of three basic components, namely (formal) reporting relationships, groupings/departments as well as system designs (Daft, 2001: 86). Thereby, the last aspect is of vital importance as differences become striking in terms of the designs' objective/goal orientation: while a traditional organizational structure emphasizes vertical communication and control to achieve the highest efficiency possible, the learning organization—a more contemporary approach—centers on horizontal communication and coordination instead

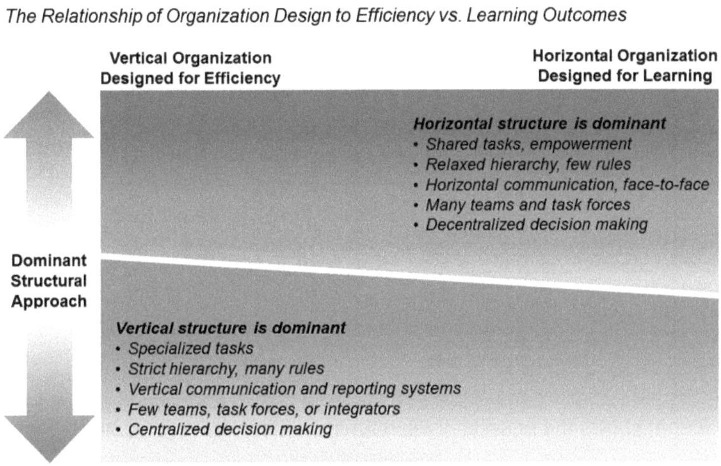

Fig. 1 Organizational design and its outcomes (adapted from Daft, 2001)

(see Fig. 1; Daft, 2001: 86). Below, the different organizational forms and their manifestations will be discussed in more detail.

(1) *Traditional Organization:* In this model, tasks originate in the top hierarchical sections, are passed down towards the bottom and with the intention to ensure the overall controllability of processes. Thereby, top-level executives have to be informed about all processes executed underneath their supervision and enforce rules and plans in hierarchical order, using predominantly formal communication and information systems, such as reports or written orders/statements (Daft, 2001; Galbraith, 1973; Galbraith, 1977). In terms of its usability, this form is quite common in family or owner-run businesses.
(2) *Learning Organization:* Whereas communication in traditional structures is very rigid and strict, the learning organization allows for communication to flow more freely both within departments and amongst employees. Coordination often requires additional mechanisms that are not directly (aka visually) integrated into the firm's structure itself, such as (cross-functional) information systems, task forces or teams (Brown, 1999; Cronin, 1997; Daft, 2001; Galbraith, 1973). An enterprise utilizing this structural form is Unilever, which has started to implement flexible open-space working environments which enable new team and project constellations and foster mutual learning and exchange.

As to organizations overall design structures, several different types with varying focal points need to be differentiated (Fig. 2):

(1) *Functional structure:* This structural type follows a "design that groups people on basis of their common expertise and experience or because they use the same resources" (Jones, 2004: 160). In organizations organized along a functional

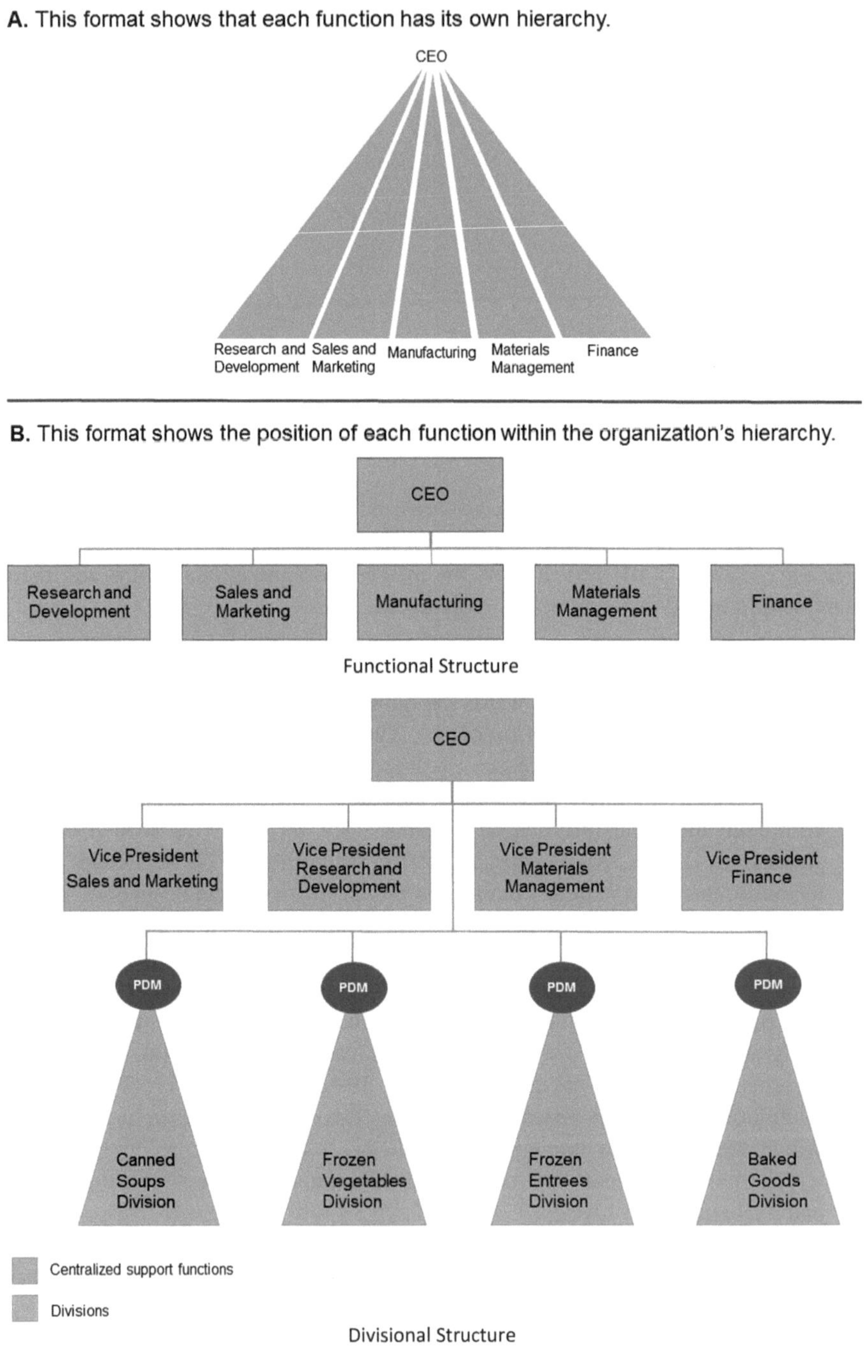

Fig. 2 (continued)

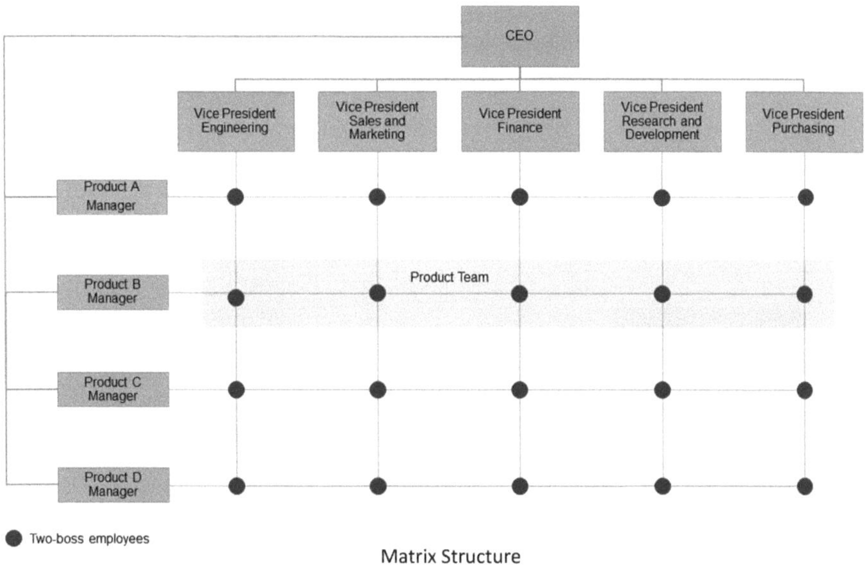

Fig. 2 Overview on organizational designs (adapted from Jones, 2004)

structure, activities are functionally grouped and located in a top-to-bottom hierarchy. As such, a particular area know-how is consolidated and especially benefits the achievement of organizational goals as well as the firm's overall efficacy. Some advantages of this structure are the potential usage of economies of scale, the development of in-depth skills and the completion of functional goals, especially when only centering on a few selected products (Daft, 2001; Duncan, 1979; Randolph & Dess, 1984). Potential disadvantages consist in slow adoption processes, response times and innovations, together with restricted views as to company objectives, poor coordination and an overwhelmed board, who is supervising and approving every decision (Daft, 2001; Duncan, 1979). A company applying this type of structure is, for instance, Microsoft or Fuji TV (Jones, 2004) (Fig. 3).

(2) *Divisional structure:* This approach has to be perceived as "a structure in which functions are grouped according to the specific demands of products, markets, or customers" (Jones, 2004: 167). The divisional structure, as opposed to the functional structure, is at times also labeled a product or strategic business unit structure, alluding to its organization in terms of business, product or task segments. Hereby, organizational design is closely related to the expected output and intends to maximize flexibility as well as change/adaptation of the departments involved. At the same time, decision making is not a centralized task anymore as authority is directed towards lower hierarchical levels (Daft, 2001; Duncan, 1979). As already indicated before, this concept's advantages involve fast adaptation, high levels of customer satisfaction due to multiple contact points, increased coordination, potential regional/local adaptations and

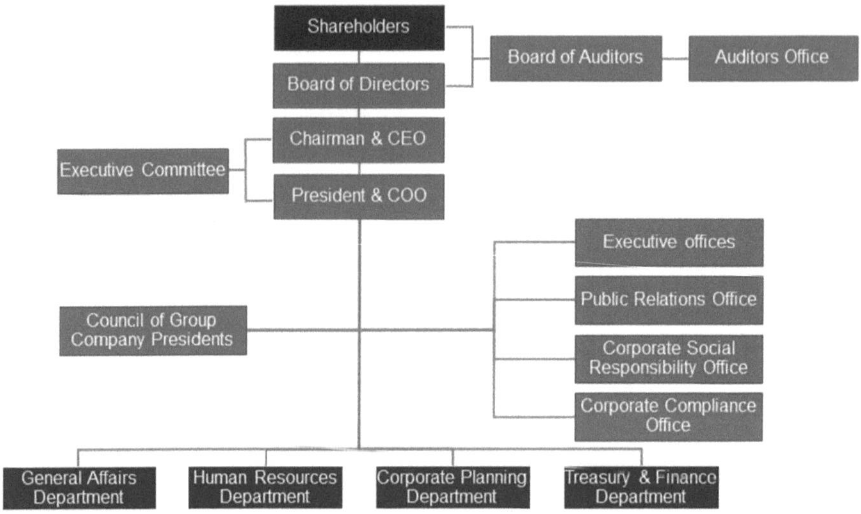

Fig. 3 Fuji's organizational chart (adapted from http://www.fujimediahd.co.jp)

decentralized decision processes. In terms of company size, it is best suited for large organizations with a substantial product variety. Yet, some disadvantages must not be left out of sight: economies of scale and in-depth competencies are eliminated, coordination across product lines suffers significantly, and a standardization of product lines is almost completely impossible (Duncan, 1979: 431). For example, General Motors, Cadillac and Disney have been successfully applying this design for years (Jones, 2004) (Fig. 4).

(3) *Matrix structure:* The matrix presents a structure designed to cope with more complex organizational environments that "can be used when both technical expertise and product innovation and change are important for meeting organizational goals" (Daft, 2001: 103). Following a rather horizontal linkage, this design manages to combine functional characteristics with (product) divisional aspects, granting each party an equal amount of authority and keeping hierarchies very flat (Jones, 2004: 183). In order for the matrix organization to work properly, several conditions need to be met: shared resources across product lines, (environment) pressure to produce in-depth knowledge as well as new products on a regular basis; and (environmental) complexities and uncertainties (Davis & Lawrence, 1977).

A matrix organization emphasizes/stresses the positive side and potentials of conflict, fostering constructive exchanges between different parties. Thereby, it intends to achieve not only an integration of competencies, but also effects of synergy, requiring the ability and willingness for both conflicts and compromises (Jones, 2004). Moreover, this structural approach is especially recommended to

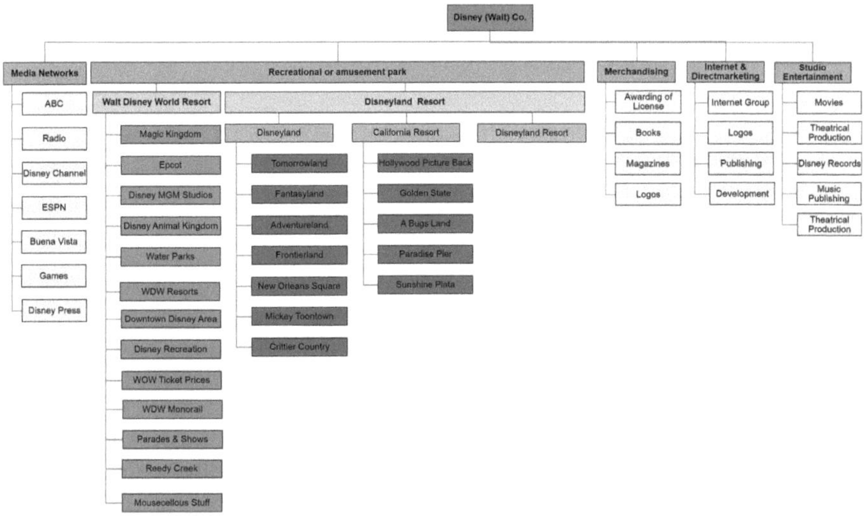

Fig. 4 Disney's organizational chart (adapted from www.disney.com)

organizations with a dual focus (on both product and functional objectives) and is meant to pay tribute to fast-changing environments. Besides those two advantages, it is also ideal to deal with complex decisions in uncertain areas of operation and fits the requirements of medium sized businesses with diverse product ranges. Yet, some weaknesses must not be overlooked for it is subject to a dual chain of command, demands intensive training of staff members, and is also very time-intense. Moreover, it presupposes employees' cooperation plus collegiality and takes a lot of effort to achieve a power balance (Daft, 2001; Duncan, 1979).

3.3 New Trends in a Convergent Work Environment

Regardless of the fact that those organizational models have proven to be successful over the past decades, where they have managed to stand the winds of both time and change, some concepts can be regarded as insufficient against the background of an ever- and faster-changing business environment. Hence, organizational mechanisms are subject to modification and at times even significant alteration. In the following, two major trends—hybrid organizations as well as team- and project-work—will be briefly introduced.

3.3.1 Hybrid Organizations

With the intention of taking specific strategic needs into consideration, hybrid and resilient organizational designs were developed. This model merges elements of the abovementioned structural approaches (functional and divisional) and utilizes them to their fullest in varying business environments (Daft, 2001) (Fig. 5).

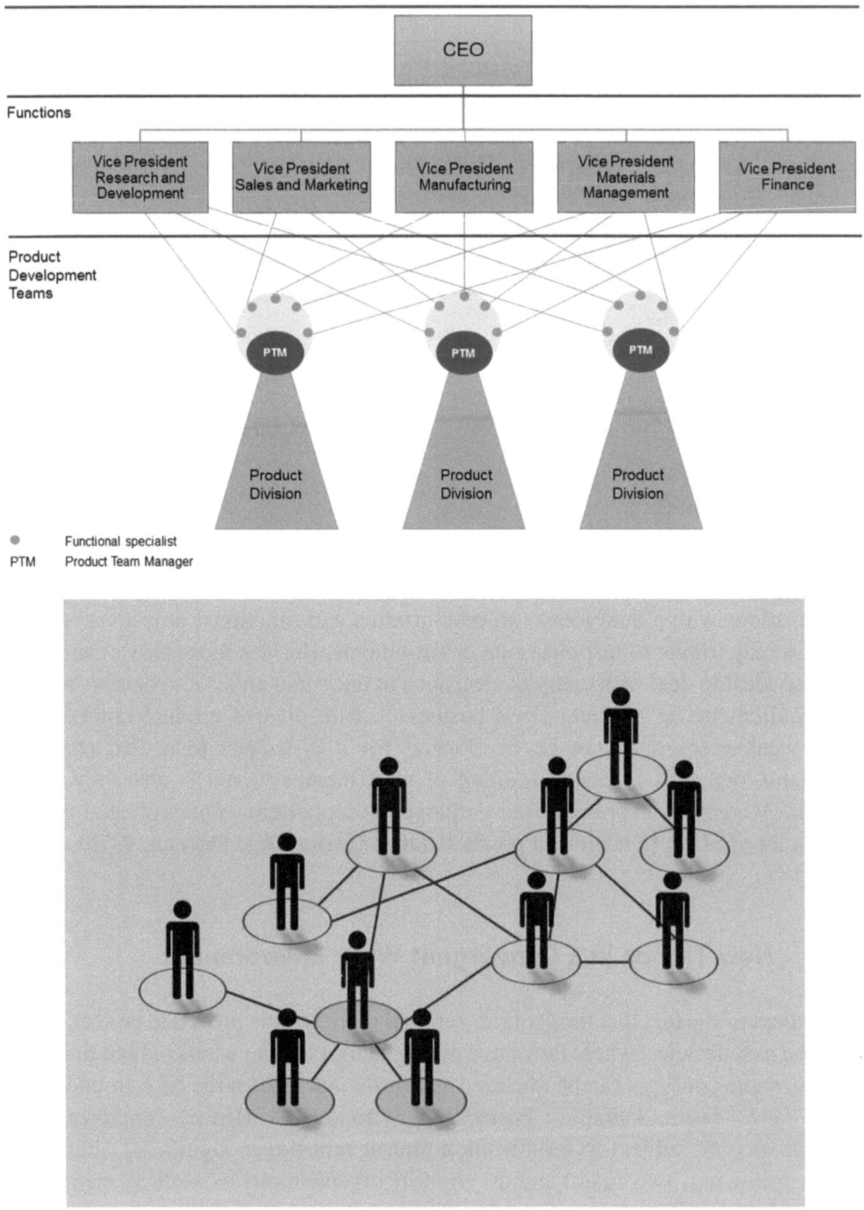

Fig. 5 Examples of hybrid organizational structures (adapted from Jones, 2004; Zukunftsinstitut, 2012)

Thereby, two major forms can be distinguished: a Network structure and a Boundaryless structure (Jones, 2004).

- A Network Structure is defined as "a cluster of different organizations whose actions are coordinated by contracts and agreements rather than through a formal hierarchy of authority" (Miles & Snow, 1992). Hereby, one organization is in the lead and enters a contract of this form to increase its effectiveness, often by means of outsourcing particular value creation processes (Jones, 2004). This leads to a reduction in production costs, fast adaptations in case of environmental changes and a quick replacement of network partners if they fail to fulfill their obligations. Yet, major drawbacks concern a considerable amount of mutual adjustment, which is difficult to achieve, coordination and control problems together with potential trust issues (Bettis, Bradley, & Hamel, 1992; Jones, 2004; Snow, Miles, & Coleman, 1992).
- Quite similar to the network organization is the Boundaryless Structure, which is "composed of people who are linked by computers, faxes, computer-aided design systems, and video teleconferencing, and who may rarely or ever see one another face to face" (Jones, 2004: 189; Fulk & Desanctis, 1995). Thus, it presents the archetype of a convergent workplace setting that is based on both technological devices and project or team work; additionally, high degrees of flexibility are achieved, while costs are reduced at the same time (Jones, 2004).

3.3.2 Increasing Team and Project Work

In convergent times and workplace environments, team work is gaining in importance, whereby a group of people presents the strongest and most powerful horizontal linkage mechanism and needs to be perceived as an enterprise's permanent task force. Usually, teams are installed as facilitators of department work and cross-functional solutions (Daft, 2001: 92), meeting the complexities of today's fragmented work processes (Clegg, Kornberger, & Pitsis, 2008: 92). They present a viable asset as teams are not only more flexible in terms of decision making, but also adapt faster and achieve a higher performance than ordinary work structures would (Clegg et al., 2008). Thereby, different types of teams can be formed to ease organizational work: Advice and Involvement Teams (Quality Control Circles or Management Decision Making Committees), Production and Service Teams (Consulting Teams or Assembly Teams), Project and Development Teams (Software Development Teams or Research Teams) and/or Action and Negotiation Teams (Trade Union Negotiation Teams; West, 2008). Due to numerous team roles [for instance, see Belbin (1993) or (2000)], individual members are constantly called upon to reflect on their own plus their team's decisions, fostering individual growth as well as group cohesiveness (Clegg et al., 2008; West, 2008).

Closely related to the previous discussion is project oriented work, which can take different forms, such as teams, core teams and workshops or conferences, with each type fulfilling different functions and purposes. While teams are mostly composed of members from different hierarchical levels and different fields of operation (Schneider and Barsoux, 2003), core teams are established within teams themselves (divided according to tasks or functions; Kutschker & Schmid, 2008). Workshops or conferences, by contrast, present events where staff members gather in order to engage in an active exchange concerning innovative processes, presentations as well as reflections respectively (Kutschker & Schmid, 2008).

4 The Effects of Convergence on Managers and Staff: New Competencies and Skills in the Workplace

In order to meet contemporary requirements, changes in the workplace cannot remain unanswered. Rather it is up to businesses to develop and utilize appropriate ways of dealing with altered circumstances to ensure to not lose their "license to operate", while managing to remain competitive at the same time. Technological advances and industrial alterations have affected organizations at their core and mandate adaptations. These changes do not only concern established business models and organizational structures, but also managerial and employee competencies. In the workplace setting, present challenges must not be overcome by sticking to traditional competencies and skills, but these are in need to be expanded. Increasingly, workers have to broaden their horizons. For this purpose, the original managerial roles developed by Mintzberg (1980: 92; see Fig. 6) need to be amplified as some central aspects have not found consideration until now.

Convergence can be made responsible for the emergence of three new roles, which need to be added to the illustration below, namely Technological Scouts, Media Literacy Practitioners and Change Managers (see Diehl et al., 2013).

- *Technological Scouts'* major task lies in identifying potentials for innovation in numerous and diverse areas. Initially, they have to detect a need for change, followed by raising awareness amongst the workforce before ultimately

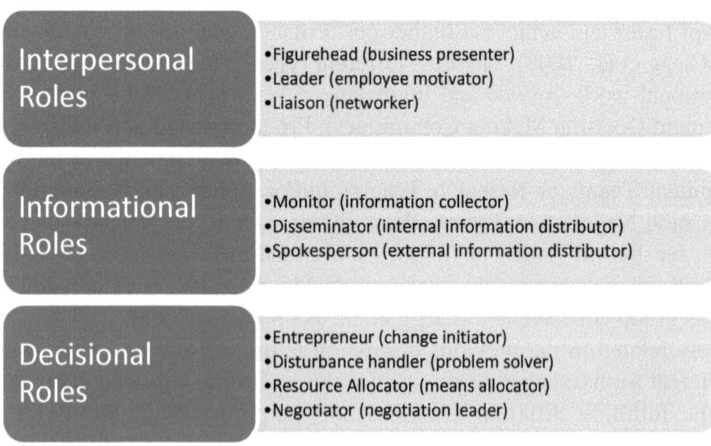

Fig. 6 Traditional management roles (adapted from Mintzberg, 1980)

stimulating the implementation of new technologies. Additionally, their duty involves accompanying, monitoring and facilitating the transition process (Rohrbeck, Heuer, & Arnold, 2006).

- *Media Literacy Practitioners (MLPs)* are the managerial roles in charge of fostering technological and content-related media usage. They somehow present a mix between Technological Scouts and Change Mangers and are responsible for accomplishing an awareness for new media services and devices, which is characterized not only by technological standards but also boundaries (Schachtner, 2010). Therefore, it is up to people holding this position to manage a challenge characteristic of convergent times: the challenge of coming up with appropriate ways of coping with convergent media platforms and environments. For instance, MLPs have to secure that shareholders are informed about companies' operations; yet, they have to filter out appropriate ways of communicating this information (Zorn, 2011).
- *Change Managers'* duties are amongst the most complex, as they operate at almost all levels of the managerial process. They are assigned "a key role in operating the centralized part of the business change process, including assessing new Requests for Change (RFCs), allocating Impact Assessors, coordinating Impact Assessment Review Meetings and keeping Quality Officers informed" (Quality Management Department, 2011). In addition, both the allocation of resources and development of new products are part of the Change Manager's responsibilities (Diehl et al., 2013).

The roles above-introduced are meant to complement the five original management functions, to which interpersonal, informal and decisional roles are assigned (Schreyögg & Koch, 2007). If, however, the complexity of present-day business relations is meant to be grasped, the three roles need to be added, whereby they are assigned to more than one management functions simultaneously (Diehl et al., 2013; see Fig. 7).

In the era of convergence, management roles are also predicted to change dramatically as "leadership has become an unnecessarily complex, confusing, and contradictory domain of interest" (Clegg et al., 2008: 128). Thus, present leadership and managerial roles are said to be transformed into "coaches" and "mentors", who leave their hierarchical positions behind to actively engage with their employees on all hierarchical levels (Schreyögg & Koch, 2007: 275f.). As the terms conjecture, managerial tasks are twofold and involve coaching as well as mentoring. While the prior alludes to strengthening an employee's knowledge and skill to improve overall job outcomes, the latter refers to managers passing on their very own expertise and knowledge, with the ultimate goal of developing qualified workers or even a protégé for the future (Clegg et al., 2008). Following Dubrin (2005), successful mentors and coaches must possess the following core competencies: trust building, empathy, active listening, influence tactics, set goals, monitor performance, feedback, encourage positive actions and discourage negative actions.

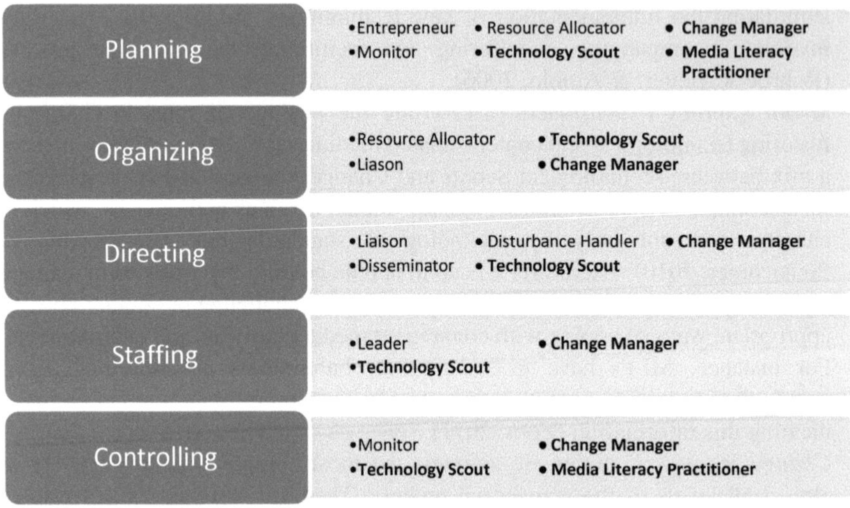

Fig. 7 Management roles on different managerial levels (see Diehl et al., 2013; Schreyögg & Koch, 2007)

5 Conclusion and Outlook

The chapter's main focus has been put on convergence, with the issue of divergence taking a backseat. "The old idea of convergence was that all devices would converge into one central device that did everything for you; [however, what] we are seeing now is the hardware diverging while the content converges" (Cheskin Research, 2002). In detail, this means that while content is used on multiple platforms and multiple times and, thus, converges, it takes more devices to display it, meaning that technology diverges. Moreover, interactivity enables divergence as consumers are eagerly trying to contribute by creating content that can at best be described as fragmented, shattered across diverse platforms (Appelgren, 2004). At the same time, skills in the workplace are mostly subject to converge, explaining the focus taken in this chapter.

The question we asked in the beginning of our chapter was: "Is there a demand for new business models, organizational structures and management/employee competencies?" and can clearly be answered with yes. This is the case as "[d] esigning organizational structure is becoming an increasingly complex management activity in today's changing world" (Jones, 2004: 190); moreover, future trends in the field of convergence have led to significant changes in business practices, organizational designs and managers' professional profiles. In these changed settings, classical and traditional management competencies proof to be insufficient when it comes to dealing with new and technologically advanced workplace situations. As a consequence, these skills need to be adapted, extended or even overcome—meaning they need to be substituted with new competencies that fit the present and future requirements of a mediatized and converged work environment. In our article we outlined some possibilities and solutions as to how existing business models, organizational structures and management competencies can be adapted and modified with regard to both current and future needs.

The trends discussed before do not present any utopic and farfetched scenarios but are supported by future development analyses (Hernstein Institute for Management and Leadership, 2012a), which have already started to thematize and address some of the roles and competencies attributed to convergence given above (see Fig. 3), such as the emerging complexity of the managerial position, the rising importance of teamwork as well as the dismantlement of traditional organizational hierarchies (Hernstein Institute for Management and Leadership, 2012a).

Organizations fit for the media- and information society are mediated and convergent organizations. Not exclusively, but first of all, they can be found in the media industry. The future challenges indicated beforehand illustrate the need for a convergent organization and merging management competencies, which means that companies do not only have to understand the changes in society, consumer behavior and business models, but they also have to be able to react appropriately and proactively.

Yet, the above discussion is far from being complete. Since changes are often unpredicted, surprising and faster than ever anticipated this article does not present permanent but only temporary solutions which are subject to change with new (technological/technical) innovations and altered work-routines and skills. So the debate is far from being over.

Of course, there are some limitations to this present discussion. For instance, future research could conduct a real-life case study contrasting two managerial profiles and business models taken from media companies in competing industries. Moreover, since only few booklets address the issue of convergence (e.g., *The Global Skills Convergence*; KPMG International, 2008), especially in detail, a next step could involve the development of a specialized guide that is meant to aid professionals in successfully managing their enterprises and staff in an ever-changing environment.

References

Andriole, S. J. (2010). Business impact of web 2.0 technologies. *Communications of the ACM, 53* (12), 67–79.

Appelgren, E. (2004). *Convergence and divergence in media: Different perspectives*. Paper presented at the 8th ICCC International Conference on Electronic Publishing. Brazil, June 2004.

Belbin, R. M. (1993). *Team roles at work*. Oxford: Butterworth-Heinemann.

Belbin, R. M. (2000). *Beyond the team*. Oxford: Butterworth-Heinemann.

Bettis, R. A., Bradley, S. P., & Hamel, G. (1992). Outsourcing and industrial decline. *Academy of Management Executive, 6*(1), 7–22.

Bröring, S., Coultier, L. M., & Leker, J. (2006). The front end of innovation in an era of industry convergence: Evidence from neutraceuticals and functional foods. *R&D Management, 36*(5), 487–498.

Brown, E. (1999, May 24). 9 ways to win on the web. *Fortune, 139*, 112–125.

Chesborough, H., & Rosenbloom, R. S. (2002). The role of the business model in capturing value from innovation: Evidence from Xerox corporation's technology spin-off companies. *Industrial and Corporate Change, 11*(3), 529.

Cheskin Research. (2002). Designing digital experiences for youth. Market insights series, Fall 8–9.

Clegg, S., Kornberger, M., & Pitsis, T. (2008). *Managing and organizations: An introduction to theory and practice*. London: Sage.

Cronin, M. J. (1997, August 18). Intranets reach the factory floor. *Fortune*, 208.

Daft, R. L. (2001). *Organization theory and design* (7th ed.). Cincinnati: South Western College Publishing.

Davis, S. M., & Lawrence, P. R. (1977). *Matrix*. Reading, MA: Addison-Wesley.

Diehl, S., Karmasin, M., Leopold, A., & Koinig, I. (2013). New competencies for the future: How changes and trends in media convergence demand new skills from the workforce. In S. Diehl & M. Karmasin (Eds.), *Media and convergence management* (pp. 353–376). Berlin: Springer.

Dubrin, A. J. (2005). *Coaching and mentoring skills*. Englewood Cliffs, NJ: Pearson Prentice Hall.

Duncan, R. (1979). What is the right organization structure? *Organizational Dynamics, 7*(3), 59–80.

Dwyer, T. (2010). *Media convergence*. Maidenhead: McGraw Hill.

Fulk, J., & Desanctis, G. (1995). Electronic communication and changing organizational forms. *Organizational Science, 6*, 337–349.

Galbraith, J. R. (1973). *Designing complex organizations*. Reading, MA: Addison-Wesley.

Galbraith, J. R. (1977). *Organization design*. Reading, MA: Addison-Wesley.

Hacklin, F. (2007). *Management of convergence in innovation*. Wiesbaden: Springer.

Hacklin, F., Klang, D., & Baschera, P. (2013). Managing the convergence of industries: Archetypes for successful business models. In S. Diehl & M. Karmasin (Eds.), *Media and convergence management*. Berlin: Springer. in press.

Hacklin, F., Marxt, C., & Fahrni, F. (2010). An evolutionary perspective on convergence: Inducing a stage model of inter-industry innovation. *International Journal of Technology Management, 49*(1–3), 220–249.

Hernstein Institute for Management and Leadership. (2012a). *Rolle von Führungskräften in Unternehmen*. Retrieved July 24, 2012, from http://www.hernstein.at/Wissenswert/Hernstein-Management-Report/Aktuelle-Reports/Rolle-von-Fuehrungskraeften-in-Unternehmen/

Hernstein Institute for Management and Leadership. (2012b). *Organisationsstrukturen*. Retrieved July 24, 2012, from http://www.hernstein.at/Wissenswert/Hernstein-Management-Report/Aktuelle-Reports/Organisationsstrukturen/

International Delphi Study. (2009). *Zukunft und Zukunftsfähigkeit der Informations- und Kommunikationstechnologien und Medien*. Retrieved May 5, 2012, from http://www.tns-infratest.com/presse/pdf/Zukunft_IKT/Zukunft_und_Zukunftsfaehigkeit_der_IKT_2009.pdf

Isaacs, S., & McAllister, J. (2005). *Management skills*. Retrieved February 10, 2013, from http://www.ca.uky.edu/agc/pubs/id/id108/10.pdf

Jenkins, H. (2006). *Convergence culture*. New York: New York University Press.

Johnson, M., Christensen, C., & Kagermann, H. (2008, December). Reinventing your business model. *Harvard Business Review*. Retrieved February 10, 2013, from http://www.ondernemenzondergrenzen.nl/uploaded_files/6_reinventing_your_business_model.pdf

Jones, G. R. (2004). *Organizational theory, design and change: Text and cases* (4th ed.). Upper Saddle River, NJ: Pearson-Prentice Hall.

Karmasin, M., & Winter, C. (Eds.). (2000). *Grundlagen des Medienmanagements*. Munich: UTB.

Koontz, H., & O'Donnell, C. (1955). *Principles of management: An analysis of management functions*. New York: McGraw-Hill.
KPMG International. (2008). *The global skills convergence: Issues and ideas for the management of an international workforce*. Retrieved August 20, 2013, from http://www.areadevelopment.com/article_pdf/id86996_GlobalSkills2.pdf
Kutschker, M., & Schmid, S. (2008). *Internationales management*. München: Oldenbourg.
Lei, D. T. (2000). Industry evolution and competence development: The imperatives of technological convergence. *International Journal of Technology Management, 19*(7–8), 699–738.
McPhillips, S., & Merlo, O. (2008). Media convergence and the evolving media business model: An overview and strategic opportunities. *The Marketing Review, 8*(3), 237–253.
Micó, J., Masip, P., & Barbosa, S. (2009). Models of business convergence in the information industry: A mapping of cases in Brazil and Spain. *Brazilian Journalism Research, 5*(1), 123–140.
Miles, R. E., & Snow, C. C. (1992). Causes of failure in network organizations. *California Management Review, 34*, 53–72.
Mintzberg, H. (1980). *The nature of managerial work*. Englewood Cliffs, NJ: Prentice-Hall.
Ofcom. (2008, December). *The international communications market 2008: Convergence*. London: Ofcom.
Quality Management Department. (2011). *Quality management system manual: Office for harmonization in the internal market*. Retrieved May 03, 2012, from http://oami.europa.eu/ows/rw/resource/documents/OHIM/serviceCharter/quality_management_system_manual_en.pdf
Randolph, W. A., & Dess, G. G. (1984). The congruence perspective of organization design: A conceptual model and multivariate research approach. *Academy of Management Review, 9*, 114–127.
Rohrbeck, R., Heuer, J., & Arnold, H. (2006). *The technology radar. A tool of technology intelligence and innovation strategy*. Retrieved May 03, 2012, from http://www.rene-rohrbeck.de/documents/Rohrbeck_Heuer_Arnold_%282006%29_Technology-Radar_Paper.pdf
Roth-Ebner, C. (2012). *Medienkompetenz und Genderkompetenz*. Retrieved July 14, 2012, from http://www.medienimpulse.at/articles/view/352
Salaverría, R., García Avilés, J. A., & Masip, P. (2008). Media convergence as a research concept: A proposal for its theoretical and operational definition. In *Second ECREA Conference*, Barcelona.
Schachtner, C. (2010). *Thesen zum Thema Feminismus im Web 2.0. Im Spannungsverhältnis zwischen Öffentlichkeit und Privatheit*. Retrieved June 14, 2012, from http://www.uni-klu.ac.at/cschacht/Vortrag_Feminismus_im_Web_2.0.pdf
Schelhowe, H. (2007). *Technologie, Imagination und Lernen: Grundlagen für Bildungsprozesse mit Digitalen Medien*. Münster: Waxmann.
Schmutzer, T. (2010). *Security in unified communications umgebungen*. Retrieved February 13, 2013, from http://www.hmp-consulting.com/fileadmin/user_upload/HMP_UC_Securityaspekte_2010_Auszug_TS.pdf
Schneider, S. C., & Barsoux, J.-L. (2003). *Managing across cultures* (2nd ed.). New York: Prentice Hall.
Schreyögg, G., & Koch, J. (2007). *Grundlagen des Managements: Basiswissen für Studium und Praxis*. Wiesbaden: Gabler.
Schwarz, E., & Gustafsson, V. (2013). Business modelling and convergence. In S. Diehl & M. Karmasin (Eds.), *Media and convergence management* (pp. 9–23). Berlin: Springer.
Snow, C. C., Miles, R. E., & Coleman, H. J., Jr. (1992). Managing 21st century network organizations. *Organizational Dynamics, Winter*, 5–20.
Weihrich, H., & Koontz, H. (1993). *Management: A global perspective*. New York: McGraw-Hill.
West, M. A. (2008). Team performance. In S. R. Clegg & J. Bailey (Eds.), *The sage international encyclopedia of organizational studies*. London: Sage.

Wirtz, B. (2011). *Media and internet management*. Wiesbaden: Gabler.
Zorn, I. (2011). Medienkompetenz und Medienbildung mit Fokus auf Digitale Medien. In H. Moser, P. Grell, & H. Niesyto (Eds.), *Medienbildung und Medienkompetenz* (pp. 175–209). München: Kopaed.
Zukunftsinstitut .(2012). *Work: Design. Die Zukunft der Arbeit gestalten*. Kelkheim: Zukunftsinstitut GmbH.

Converging Technologies and Diverging Market Trends of Internet/Web and Traditional Media

Zvezdan Vukanovic

1 Introduction: The Need for the Application of New Media Business Models and the Decrease of the Economic and Social Influence of Traditional Media

One of the challenges of studying new media business models in the age of media convergence is that the concept is so multifaceted and broad that it has multiple meanings. As a result, the academic and scholarly literature in this area is diverse and remains under-researched, under-explored and under-developed from both a theoretical and an empirical perspective. This article reviews scholarly studies that identify the range of strategic options available for sustainable business models in new media industry.

Identification of sustainable and hyper-competitive new media business models is an urgent priority as continuing decline in audiences and collapse of traditional/old media organizations pose a major threat to media, democracy, ICT and telecommunications industry, with scholars agreeing that further erosion of media industry also have major implications for the advertising industry and a wide range of content producers.

Referred to in the industry as 'audience fragmentation' or 'disaggregation', this breakdown of large mass audiences of mass media is resulting in both advertising volume and rates falling within the dominant commercial media business model (Macnamara, 2010b). As Henry Jenkins warns, 'monolithic blocks of eyeballs are gone' (2006:66). Notwithstanding, John Pavlik concludes in his 2008 text *Media in the Digital Age* that 'few media organizations have settled on a viable long-term strategy for making money in a sustainable fashion' (2008:173). As a result of this lack of foresight, John Pavlik points out that media organizations—particularly news companies and departments—have not invested sufficiently in research and

Z. Vukanovic (✉)
University of Donja Gorica, Podgorica, Montenegro
e-mail: zvezdanv@t-com.me

A. Lugmayr, C. Dal Zotto (eds.), *Media Convergence Handbook - Vol. 2*, Media Business and Innovation, DOI 10.1007/978-3-642-54487-3_5

development to expand or update their product line over recent decades (2008: 20). Pavlik estimates that many news media have invested less than 1 % of their operating budgets in R&D to develop new products and new business models. The cost of failing to recognize the potential and public demand for new forms of content and distribution methods has been that media organizations have not developed new products tailored to the Web 2.0 and Web 3.0 environment of social media and social networks or the changing media and ICT economy (Macnamara, 2010a, b).

The lack of efficient development of new business models caused the decrease of the economic and social influence of traditional media (print, radio and TV). It is particularly evident in the statistical data which show that radio needed 38 years in order to reach 50 million users. TV needed 13 years to reach the same number of users, Internet 4 years, IPOD—3 years; while Facebook added 100 million users in less than 5 months (between April 8, 2009 and September 15, 2009). Moreover, iPod application downloads hit one billion in 9 month. On the other hand, global internet companies such as Google in 2012 have increased the value of their brand for 26 %.

Simultaneously, all the trend lines went downwards for the newspaper business. Global newspaper advertising revenues fell −17 % in 2009; North American newspapers lost a quarter of their advertising revenues. Ad spending was also down in Western Europe −13.7 %, Central and Eastern Europe −18.7 %, Asia −9.6 %, Latin America −2.9 % and was stable in the Middle East and Africa. Between 2004 and 2009, the US newspaper industry lost 34 % of its readers; the UK industry lost 22 %.

The research of the Newspaper Association of America shows that daily newspaper print ad has been constantly decreasing since 2005. Furthermore, in the U.S., *The Wall Street Journal* is the only newspapers in 2010 to gain in circulation among the top 25 newspapers. The importance as well as the market expansion of digital media is evident in the bookseller Barnes and Noble assertion that the company in 2010 sells more digital books than physical books on its Web site. Accordingly, Forrester Research expects U.S. e-book sales to total $2.8 billion in 2015, up from nearly $1 billion in 2010. The research firm projects the number of e-readers and tablets in the U.S. will soar from more than 15 million in 2010 to nearly 60 million in 2015.

2 Literature Review and Discussion on Recent Models of Convergence in Media Research

The concept of convergence is frequently used both in the academic field and within the media industry to denote the ongoing restructuring of media companies as well as to describe the latest developments in media forms, distribution, and consumption (Appelgren, 2004). However, there is currently no generally accepted definition of the concept. Depending on the context, the meaning and connotations vary. It is generally accepted among media business scholars that convergence denotes the actual process toward a more efficient management of the media value chain. The use of the concept has therefore developed from being mainly connected with digitalization in media technology to also include elements of integration, combination, competition and divergence. This paper supports Ester Appelgren's (2004) perspective. It suggests that convergence should be seen as an ongoing process of media and media industry development that is dependent on and in continuous interplay with a contrasting and complementary process of media divergence.

Jenkins (2001) divides convergence into five areas: technological, economic, social or organic, cultural and global convergence. Technological convergence is the digitalization of all media content, economic convergence deals with the integration of the entertainment industry and the social or organic version of the process handles the consumers. According to Jenkins, cultural convergence is the explosion of new forms of creativity at the intersections of various media technologies, industries and consumers. Finally, global convergence is the cultural hybridity that results from the international circulation of media content. This definition is in line with the notion that convergence is an ongoing process, occurring at various intersections between media technologies, industries, content and audiences; it is not an end state (Jenkins, 2001). The effects of the process of convergence are visible, measurable and possible to detect, while the actual process might not be (Appelgren, 2004).

Lawson-Borders (2003) suggests another model of convergence, where the starting point is that convergence is a concept as well as a process. Lawson-Borders has identified seven observations. of convergence all beginning with the letter c: Communication, commitment, cooperation, compensation, culture, competition, and customer. These seven areas partly overlap and can serve as a guideline for best practices to expound on convergence both as a concept and a process Lawson-Borders (2003).

In addition, Lawson-Borders (2003) believes that for convergence to succeed, media firms must:

(a) Engage in high quality communication about what the organization is trying to accomplish;
(b) Be committed to incorporating convergence into their organizational mission and philosophy;
(c) Promote cooperation among everyone involved in the journalistic process "to share stories and ideas;"
(d) Revise compensation plans to fairly compensate multimedia journalists for taking on the new roles and responsibilities required by convergence;
(e) Facilitate the blending of different cultures in the newsroom (i.e., print, radio, television, and online) (see also Killebrew, 2003);

(f) Develop strategies and alliances capable of allowing media firms to successfully compete in local markets and globally; and
(g) Develop convergence strategies capable of serving evolving consumer needs in a dynamic and increasingly competitive/challenging marketplace (pp. 94–96).

Furthermore, Lee (2003) describes four categories and eight levels of digital convergence:

1. Data convergence (Media convergence and Domain convergence)
2. Structural convergence (Architecture convergence and Infrastructure convergence)
3. Application convergence (Platform convergence and Device convergence)
4. Industrial convergence (Intra-industry convergence and Inter-industry convergence).

Dennis (2003, p. 7) identified four stages of communication industry convergence: "incremental awakening"—the 1980s, "early adoption"—early to mid-1990s, "uncritical acceptance"—late 1990s, and "presumptions of failure"—early 2000s. Greenstein and Khanna (1997, pp. 203–204) define convergence in terms of substitutes and complements: "Two products converge in substitutes when users consider either product interchangeable with the other. . . . Two products converge in complements when the products work better together than separately or when they work better together now than they worked together formerly." Allison, DeSonne, Rutenbeck, and Yadon (2002, p. 61) consider convergence as a "business trend where previously separate industries … are converging through megamergers, buyouts, partnerships and strategic alliances."

3 Driving Forces of Media Convergence

As the concept of media convergence appears to be a multifaceted process there are apparently many driving forces behind convergence and the increased interest in the concept (Wirth, 2003). The most dominant driving forces include, but are not limited to:

(a) Technological innovation, including the rise of the Internet and the digital revolution;
(b) Deregulation/liberalization and globalization, including passage of the Telecommunications Act of 1996, formation of the European Union and the privatization of telecommunications and media around the world;
(c) Changing consumer tastes and increased consumer affluence;
(d) Technological standardization;
(e) The search for synergy (i.e., $1 + 1 = 3$);

(f) Increasing global competition (which has resulted in high levels of merger and acquisition activity among media and telecommunication companies around the world); and
(g) Repurposing of old media content for distribution via various forms of new media (Wirth, 2003).

4 Review of Literature in New Media Business Models

This section provides a summary classification of the new media business models scholarly studies. Accordingly, the most important business models that are analyzed in the book chapter include:

Tipping point strategy—Gladwell (2002); Hoegg, Martignoni, Meckel, & Stanoevska-Slabeva (2006).
Crowd sourcing strategy—Dokoupil & Wu (2010); Edwards (2009).
Mesh companies strategy—Gansky (2010).
Micropayment and, nicheization of media market—Anderson (2006); Jaring Matinmikko & Abrahamsson (2006); Mings & White (2000); Graybeal & Lee Hayes (2011); Ryu & Feick (2007); Clemons (2009); Zeng & Reinartz (2003); Foremski (2009); Tam & Ho (2007); Ahonen (2010).
User-generated content—Daugherty, Eastin, & Bright (2008); Schaedel & Clement (2010); Chevalier & Mayzlin (2006); Godes & Mayzlin (2004); Li & Hitt (2008); Ransbotham, Kane, & Lurie (2012); Snuderl (2008); Cattuto, Loreto, & Pietronero (2007); Golder & Huberman (2006); Marlow, Naaman, Davis, & Boyd (2006); Ames & Naaman (2007); Nov & Ye (2010); Qualman (2012); Koh, Kim, Butler & Bock (2007).
Content re-purposing, cross-media content and global convergence—Vizjak & Ringlstetter (2003); Bakos & Brynjolfsson (2000); Doyle (2002).
Experience economy—Pine & Gilmore (1999); Manovich (2012).

5 Research Methodology Approach

This study is based on an analytical and empirical research methodology. The author generated relevant data on old and new/digital media corporations from the Ycharts.com commercial database. The database contains financial information on leading international corporations across all industries.

Furthermore, it contains corporation-level financial data in a standardized financial format including financial ratios and business activities for up to 10 years. For building the dataset, the author analyzed the following micro-economic and financial indicators: enterprise value, market capitalization, annual revenue, annual net income, debt to equity ratio, return on assets, return on equity) of the leading global old and new media corporation.

The analyzed data prove the author's main hypothesis in which most successful global new media corporations maintain their competitiveness over old media corporations applying the following business models: tipping point strategy, Crowd sourcing strategy, Mesh Companies Strategy, Micropayment, nicheization of media market, User-generated content, content re-purposing, cross-media content and global convergence, and Experience Economy.

6 Empirical Case Studies of Leading Global New and Old Media Corporations

The empirical case study of fourteen global leaders in the new and old media business industry reveals that that the new media (internet and web) companies are definite winners in four out of seven micro-economic categories. New media corporations are dominant within the financial scores of net income, market capitalization, debt to equity ratio and return on assets while old media corporations dominate the aspects of revenue, enterprise value and return on equity. Among new media corporations, Google is the winner in three categories: net income, market capitalization and enterprise value; Baidu dominates two financial indicators: return on equity and return on assets. Linkedin and Yahoo lead the category of debt to equity ratio and Amazon is well positioned in the sector of revenue.

It is necessary to point out that debt to equity ratio of new media corporations is five times smaller as opposed to old media corporations. Also, the net income of new media corporations is twice as high compared to old media corporations. In addition, Google's annual net income and Amazon's annual revenue have been the highest recorded in the last 5 years.

The financial and business dominance of new media corporations is further reinforced by the fact that on the Millward Brown's list of top 100 most valuable brands in 2013 new media corporations feature six corporate entities (Google, Facebook, Yahoo, Baidu, Amazon and eBay) and old media companies feature only The Walt Disney Company. Google is positioned as the second most valuable global brand with the brand equity worth of $113.071 billion. In addition, Amazon and eBay have seen respectively a 34 and 40 % rise in brand value compared with last year.

Among old media corporations, Comcast Corporation is the winner in three financial categories: annual revenue, annual net income and enterprise value. The Walt Disney Company dominates sectors of market capitalization and debt to equity ratio, while Time Warner and Viacom lead in the segments of return on equity and media business are content distributors—Google and Comcast. In the Tables 1 and 2, the author provides a more detailed analysis of the most important financial indicators, metrics and scores.

Table 1 Analysis of financial indicators of the leading global new media corporations.

Name of the company	Enterprise value ($)	Market capitalization ($)	Annual revenue ($)	Annual net income ($)	Debt to equity ratio	Return on assets (%)	Return on equity (%)
Google	227.98 billion	271.59 billion	53.50 billion	11.19 billion	0.0681	12.62	16.38
Amazon	63.98 billion	112.81 billion	63.9 billion	−87 million	0.3605	−0.35	−1.12
eBay	62.96 billion	67.86 billion	14.5 billion	2.71 billion	0.2139	8.28	13.65
Facebook	58.74 billion	66.87 billion	5.08 billion	53 million	0.1276	0.44	0.54
Baidu	26.53 billion	29.69 billion	3.82 billion	1.69 billion	0.4177	29.48	46.19
Yahoo	23.88 billion	26.85 billion	4.90 billion	4.04 billion	0.00	24.21	29.11
LinkedIn Corporation	20.31 billion	21.05 billion	972.31 million	21.61 million	0.00	1.94	2.84
Netflix	11.81 billion	12.13 billion	3.76 billion	24.42 million	0.8611	0.64	3.37
Total	542.86 billion	608.85 billion	150.57 billion	65.53 billion			
Average	67.85 billion	76.10 billion	18.82 billion	8.19 billion	0.25	9.65	13.87

Source: Ycharts.com, May 1, 2013

Table 2 Analysis of financial indicators of the leading global old media corporations

Name of the company	Enterprise value ($)	Market capitalization ($)	Annual revenue ($)	Annual net income ($)	Debt to equity ratio	Return on assets (%)	Return on Equity (%)
Comcast Corporation	137.49 billion	110.52 billion	62.57 billion	6.203 billion	0.8197	3.87	12.88
The Walt Disney Company	130.05 billion	113.45 billion	42.84 billion	5.60 billion	0.4254	7.37	14.28
News Corporation	81.89 billion	72.39 billion	7.41 billion	3.72 billion	0.5846	7.49	16.82
Time Warner	73.29 billion	54.55 billion	28.73 billion	3.01 billion	0.6651	4.49	10.13
Time Warner Cable	50.85 billion	27.58 billion	21.73 billion	2.17 billion	3.823	4.38	29.62
Viacom	39.89 billion	32.18 billion	13.25 billion	2.23 billion	1.171	10.05	29.53
Total	513.45 billion	410.67 billion	163.288 billion	22.96 billion	–	–	–
Average	85.57 billion	68.445 billion	27.21 billion	3.82 billion	1.248	6.27	18.87

Source: Ycharts.com, May 1, 2013

7 New Media Business Models

After a detailed analysis of the major financial scores and metrics of new and old media corporations, the author identified seven business strategies that are the characteristics of the most successful new media corporations such as Google and Baidu. As both corporations are content distributors, it is clear that technological and economic aspects of distribution, access, usability and perceptive focus on demand and consumer needs are crucial in establishing effective and sustainable media business strategy. Each of these business strategies is respectively discussed and analyzed in the following sections.

7.1 Tipping Point Strategy

It is important to point out that the efficient usage of the long tail economic strategy leads to the reaching of the Tipping Point. Tipping points are "the levels at which the momentum for change becomes unstoppable" (Gladwell, 2002). Furthermore, the term is represented as "the precise moment of critical mass, the threshold, the boiling point when a trend becomes a trend" (Gladwell, 2002). In economics, the tipping point represents the point at which a dominant technology or player defines the standard for an industry-resulting in "winner-take-all" economies of scale and scope.

An excellent example of the application of the tipping point strategy in social media is the launch of Facebook. Since its inception, it has positioned itself as leader of interactive, participant-based online Web 2.0 media that creates value from the sharing of information between participants (Hoegg et al., 2006). Between August 2008 and September 2011, the number of Facebook users increased eight times (from 100 to 800 million). If Facebook were a country it would be the world's third largest between India and the United States. In addition, the revenue of the Facebook company increased from 52 million dollars in 2006 to two billion dollars in 2010. Based on traffic data from Alexa and Google Trends in June 2011, Facebook was the most popular social network in 119 out of 134 countries. More than 50 % of active users log on to Facebook in any given day.

7.2 Crowd Sourcing Strategy

In order to expand markets, social media such as Facebook uses Crowd sourcing strategy. Such strategy functions as reward programs and is only likely to grow more important, especially as the Web reaches into corners of the world where it never benefited from the frisson of a social movement (Dokoupil & Wu, 2010). In 2009, Google successfully launched the Kiswahili Wikipedia Challenge to grow the number of Swahili-language Wikipedia entries in parts of Eastern Africa by tying them to the chance to win modems, cell phones, and a laptop (Dokoupil & Wu, 2010). In this new world of social networks, the blogosphere, online communities, the ever-growing notion of crowd sourcing ("collective wisdom"), factual information of the masses provides the "true statements and facts" by testing a wide range of users with vastly different opinions (Edwards, 2009).

7.3 Mesh Companies Strategy

Unlike the traditional businesses which follow a simple formula of creating a product or service, selling it and collecting money, in the past few years, a fundamentally innovative business model has taken root-one in which consumers have more choices, more tools, more information, and more peer-to-peer power. Organizations that use social media, wireless networks, and data crunched from every available source to provide people with goods and services at the exact moment they need them, without the burden and expense of owning them outright are called "Mesh companies" (Gansky, 2010). This strategy can be profitable as it creates trusted brands and build strong communities by helping customers buy less but use more products and services. Mesh strategy if successfully aligned with the peer-to-peer power of social media networks can inspire customers in a highly competitive world where access trumps ownership (Gansky, 2010).

7.4 Micropayment and, Nicheization of Media Market

Industry and market structure of the social media industry will be more niche-oriented. If the twentieth century was about hits, the 21st will be equally about niches (Anderson, 2006). On demand media and particularly VoD—Video on Demand, will considerably gain more importance. As such, long tail economics will become more prevalent in capturing the fragmented media market. In terms of advertising and marketing revenue, it is advisable to point out that online and interactive advertising as well as micropayment strategies will be increasingly important. Micropayment will provide potential consumers with immediate transaction processing and will increase VoD—Video-on-demand and PPV—Payment-Per View models. It is argued that micropayment is in the process of becoming the web's new currency and will be especially useful in purchasing electronic books, online articles, music, video and film files. The case of micropayment strategy is additionally supported by the exponential growth of the Internet during the past decade. Thus, between 2000 and 2010 the number of internet users worldwide increased for 445 %.

Micropayment strategy is widely becoming an alternative to subscriptions as it moves content creators closer to consumers. The competitive advantage of micropayments can potentially provide consumers with a payment model in which content can be unbundled and further sold via B2C channel. On the other hand, cloud computing will be especially important in terms of B2B marketing as many international companies will hire another firms to manage their data via the Internet in private spaces, rather than those companies using their own servers, in an effort to gain storage. The increasing development of social media, web, personal computing devices (PCs, mobile phones and portable media players) made possible the wide dissemination of various online contents over the consumer-to-consumer (C2C) channel.

To date, no micropayment standards have been established, and interoperability between micropayment systems has not been solved (Jaring et al., 2006). Difficult usability, high registered customer acquisition costs, lack of universal acceptability, and lax security in traditional micropayment systems have been cited as reasons for a pure play micropayment model's lack of widespread success (Jaring et al., 2006; Mings & White, 2000). Also, there are currently few, if any, online payment solutions that can support transactions in the range of a few dollars or even cents (Tam & Ho, 2007).

Rather than a pure play micropayment model, the authors argue for a "Modified News Micropayment Model" that is constituted and contained by four primary drivers that make the idea of micropayments a feasible and attractive idea for news industries in the Social Web environment: a microearn component, socialization/sharing, local focus, and a centralized banking system (Graybeal & Lee Hayes, 2011).

7.4.1 The Micro-Earn Component

The Modified News Micropayment Model suggests that the ability to microearn is the most critical missing component of a sustainable model Graybeal & Lee Hayes (2011). A microearn system could function much like a referral rewards program: Users can earn points for disseminating news, information, and online content to friends and followers. Taking an equity view of social exchange, Ryu and Feick (2007) found that rewarding the recommender, regardless of the size of the reward, increases the likelihood of referral reducing consumer feelings of inequity in the exchange relation. Further, the likelihood of referral to weak ties (casual acquaintances) significantly increases when extrinsic rewards are given (Graybeal & Lee Hayes, 2011). Microearn enhances the value of shared content because it is disseminated, distributed, and discussed in social circles. The socialization of news (the next driver, discussed later) increases the social value of the content and also allows for a monetary reward for the dissemination of news. Microearn functions much like a rewards program, where users earn points for disseminating news, information, and online content to friends and followers.

7.4.2 The Socialization/Sharing

This system argues that social networking sites capitalize on their logistical streams of social networks social capital to disseminate valued information to trusted peers for peer review only. Therefore, the social aspect of payment for Web content is also extremely vital. Clemons (2009) argued that, although traditional media, specifically newspapers, have the capacity to create unique valuable content, they lack the ability to share it. Online, traditional media also lack logistical streams for distribution that are integral to any business model. In the Social Web, these logistical streams allow for value creation through facilitating interaction and sharing. Without these streams, it is difficult, if not impossible, to generate a critical mass of users, which Zeng and Reinartz (2003) showed to be a crucial revenue driver in business models for the Social Web. The best examples the Socialization/ Sharing function include "Re-tweet" feature on Twitter, "Share" function on Facebook® and Blogroll, links on blog sites.

7.4.3 The Local Focus

News sites retain local pricing decisions that will attract local audiences with its focus on local content. The function of the local focus includes a hyper-local blogs. Foremski (2009) called on newspapers to focus on original content, which people are more likely to pay for because they cannot get it anywhere else. He also said newspapers should focus on hyper-local coverage, where they "own" their regional beat.

7.4.4 The Centralized Banking System

This system allows universal currency exchange so that users can swap "currencies" from different platforms and trade in for cash. It includes Google TM Checkout and PayPal TM. As Tam and Ho (2007) noted, "it is important to establish an economy-wide micropayment infrastructure to settle very small transactions online" (p. 146). As society increasingly moves from a cash-based currency to digital currency, such a system becomes vital (Ahonen, 2010).

7.5 User-Generated Content as a Promoter of Collaborative Information Services

User generated content is characterized as 'Conversational Media', as opposed to the 'Packaged Goods Media' of the past century. The former is a two-way process in contrast to the one-way distribution of the latter. Conversational or two-way media is a key characteristic of so-called Web 2.0 which encourages the publishing of one's own content and commenting on other people's. UGC can be twofold and include both personal and collaborative publishing. The personal publishing consists of weblog, podcast, photo, whereas the collaborative publishing consists of the internet forum wiki. Thus, consumer becomes Prosumer—both producer and consumer of information goods. The proliferation of UGC has made a strong impact on consumers, media suppliers, and marketing professionals while necessitating research in order to understand both the short and long-term implications of this media content (Daugherty et al., 2008).

One of the main competitive advantages of the conversational media is that within the UGC, all digital media technologies are included, such as question-answer databases, digital video, blogging, podcasting, mobile phone photography and wikis. In addition to these technologies, user generated content may also employ a combination of open source, free software, and flexible licensing or related agreements to further reduce the barriers to collaboration, skill-building and discovery. As the consumption, creation, and distribution of UGC continues to evolve, content aggregation tools and Web 2.0 applications built on Really Simple Syndication (RSS) technology will become more usable and accessible to consumers, helping create a manageable information space that is both customized and relevant (Daugherty et al., 2008).

Each step of the traditional value chain of media production—from concepts, know-how, and technology to content production, packaging, marketing and distribution—has a user-generated equivalent (Schaedel & Clement, 2010). This strategy allows social media to considerably increase market share and generate exponential returns for consumers and businesses. Those returns could vary for media businesses from sales, brand awareness, and customer service. A subset of this is that in the future, we will no longer search for products and services; rather they will find us via social media. Due to the speed in which social media enables communication, word of mouth now becomes world of mouth (Qualman, 2012). Therefore it is not surprising that there is considerable interest in the value of user generated content and its antecedents. Research shows that product reviews, for instance, influence consumer search and product choice, enhance sales forecast quality, affect product sales, and drive viewership (Chevalier & Mayzlin, 2006, Godes & Mayzlin, 2004, Li & Hitt, 2008).

There are three important hypothetical findings that define the network characteristics and the value of collaborative user-generated content:

Hypothesis 1. The market value of collaborative user generated content has a curvilinear (inverted U) relationship with the number of contributors to it.

Hypothesis 2. The market value of collaborative user generated content will be positively related to its embeddedness in the content-contributor network.

Hypothesis 3. The impact of (a) the number of contributors and (b) embeddedness on the market value of collaborative user-generated content declines with content age (Ransbotham, Kane and Lurie, 2012).

Christodoulides, Jevons and Bonhomme (2012) in their research "Memo to Marketers: Quantitative Evidence for Change How User-Generated Content Really Affects Brands", maintain that consumer perceptions of co-creation, community, and self-concept have a positive impact on UGC involvement that, in turn, positively affects consumer-based brand equity. A brand with stronger brand equity is likely to lead a more involving user-generated campaign through enhanced perceptions of co-creation, community, and empowerment.

User-generated content is a part of the development of collaborative information services and the usage of folksonomies. Folksonomies represents a collection of tags. The term folksonomy is a portmanteau of the words folk (or folks) and taxonomy that specifically refers to subject indexing systems created within Internet communities (Snuderl, 2008). Folksonomy has little to do with taxonomy—the latter refers to an ontological, hierarchical way of categorizing, while folksonomy establishes categories (each tag is a category) that are theoretically "equal" to each other (Snuderl, 2008). Folksonomies turn the classification system from criteria-centric into a resource-centric approach (Peters, 2009: 3).

On the other hand, Tags are a "bottom-up" type of classification, compared to hierarchies, which are "top-down" (Snuderl, 2008). Tags are keywords, entered as additional metadata to each uploaded file—words that describe the content according to author's opinion and experiences (Snuderl, 2008). So tagging is a method of categorizing information in a collaborative and decentralized way. Tagging, or using keywords to add metadata to shared content, is gaining much popularity in recent years. (Cattuto et al., 2007; Golder & Huberman, 2006; Marlow et al., 2006). Tags are used to annotate various types of content, including images, videos, bookmarks, and blogs, through web-based systems such as Flickr, YouTube, del.icio.us, and Technorati. The popularity of tagging is attributed, at least in part, to the benefits users gain from effective sharing and from organization of very large amounts of information (Ames & Naaman, 2007; Cattuto et al., 2007). Due to the fact that user participation is critical to the sustainability of content sharing communities, a collaborative tagging system cannot succeed without higher level of user contribution (Nov & Ye, 2010; Koh, Kim, Butler & Bock, 2007).

7.6 Content Re-purposing, Cross-Media Content and Global Convergence

Content re-purposing is particularly important because in the future, only media companies focusing on selling content and services in maximum quantities will manage to maintain a profitable position in this highly volatile market (Vizjak & Ringlstetter, 2003: 17). Moreover, the strategic management of cross-media content and platform is important because of two dominant reasons: (1) It increases the number of media distribution platforms and services, and (2) it diversifies a firms' corporate portfolios while reducing financial risk in highly volatile global markets.

The concept of cross-media content will integrate both the hypermedia and multimedia models. Cross-media and on-demand content offer the enormous content base (linear and nonlinear) as a part of web and social media content. In addition, on-demand web and social media services are able to promote premium, niche, and user generated content. As such, innovative services are based on convergent technological architecture (Bakos & Brynjolfsson, 2000). Due to the faster product life cycles, volatile markets, and increased competition, future cross-media services will be more interactive, dynamic, enhanced, and flexible. This enhanced technological and content integration will more efficiently stimulate the economies of aggregation that, in turn, will bring value added services to the media business and industry.

Moreover, globalization and convergence have created additional possibilities and incentives to repackage or to repurpose media content into as many different formats as is technically and commercially feasible (books, magazine serializations, television programs and formats, videos, etc.) and to sell those products through as many distribution channels, outlets, or windows in as many geographic markets and to as many paying consumers as possible (Doyle, 2002:

22). Accordingly, repurposing represents the joint emphasis of media firms on both the content and distribution.

7.7 Experience Economy

Due to the exponential increase of the web and internet media, it is necessary today to reorganize the media economy and business to deal with a new level of human needs, expectations and experiences. The aestheticization of hardware, software design and user interfaces that gradually took place throughout the industry in the decade following Joseph Pine and James H. Gilmore's book Experience Economy: Work is Theatre and Every Business a Stage (1999) fits very well with the idea of the "experience economy". In the age of social media, internet and mobile TV interaction with information devices became a designed experience (Manovich, 2012). In Manovich's opinion:

> ... we can say that the three stages in the development of user interfaces—command-line interfaces of the 1970s (Unix), graphical user interfaces of the 1980s and 1990s (Mac OS), and the new sensual, highly aestheticized interfaces of the post-OS X era—can be correlated to the three stages of consumer economy as a whole: goods, services, and experiences. Command-line interfaces "deliver the goods": that is, they focus on pure functionality and utility. GUIs, in turn, add "service" to interfaces. And at the next stage, interfaces become "experiences". The concept of the experience economy works particularly well to explain how the physical interaction with technology objects—as opposed to their physical forms and screen interfaces—turned into a stage for delivering rich sensorial, immersive, visual, tactile and three-dimensional experiences (Manovich, 2012).

Therefore, the "experience economy" delivers a dynamic, decentralized, non/-linear, on-demand, interactive, immersive mode and habit of consumers' self-direction. In addition, consumers' experiences become a new source of value creation.

8 Main Paradigm Shifts in New/Social Media Over Old/Traditional Media

Although, both the old/traditional and new/social media can reach small or large audiences, there are many fundamental differences in terms of the competitive advantage in distribution, production, technology, market targeting that favor new/social media over old/traditional media. In the Tables 3 and 4, these marking differences are exposed in order to more effectively outline the major conceptual differences between new and old media.

Table 3 Main paradigm shifts in new/social media over old/traditional media

Old/traditional media	New/social, Web and UGC media
Industrial media dominantly produced by large multinational corporations	Personal media primarily produced by internet users
Top-down content production	Bottom-up content production
Centralized framework for organization, production, and dissemination of media One to many content distribution	Decentralized (network and on-demand) based media Many to many content distribution
Linear, One-way media communication	Interactive and immersive media communication
Reaching the audience	Connecting the audience
Passive users—Users as Recipients	Active users—Users as participants
Static media	Mobile media
Economies of scale	Economies of scope (Long tail Economics)
One-sided platform distribution	More diversified multi-platform (hypermedia and multimedia) distribution, less hierarchical, and distinguished by multiple points of production and utility
Less available and accessible to the public, distribution costs and viewing is more expensive	Generally available and accessible to the public at little or no cost
The time lag between communications produced by industrial media can be long (days, weeks, or even months)	Capable of virtually instantaneous responses; only the participants determine any delay in response
Once created content, it cannot be altered (once a magazine article is printed and distributed changes cannot be made to that same article)	Easily altered content by almost instantaneously editing and writing comments
Less creative content creation	More creative content creation
Storage capacity for media content is relatively low	Storage capacity for media content is very high Acts as an online database
Low level of content categorization and sharing	High level of content categorization, annotation and sharing: Widgets, collaborative tagging, social classification, social indexing, and social tagging, folksonomy
Less peer-to-peer power Publisher-centric	More peer-to-peer power User-centric model UGC—User generated content
Analogue	Digital media Digital convergence Mobile and wireless media Ambient media Augmented media Widget(ized) media Tagged media

Table 4 Main paradigm shifts in new/social media over old/traditional media

Two-dimensional media	3D media
Traditional market targeting (B2C and B2B marketing)	Better and more efficient market and consumer marketing (B2C and C2C) Nicheization Social network and online communities
Web 1.0 and web 2.0	Web 3.0 (semantic web) and Web 4.0 (symbiotic web)
Value chain	Value network
Collaborative consumption	Collaborative creation
Producer	Produser
Broadcasting	Narrowcasting, microcasting and egocasting
Interactive media	Immersive media
Consumerism	Prosumerism
Top-down organizational structure	Bottom up organizational structure
Upstream supply chain (push marketing, low-cost producers)	Downstream supply chain (customization, targetization, high margins)
One to many distribution	Many to many distribution
Symmetric information flow	Asymmetric information flow
First build a marketplace, than a community.	First build a community, than a marketplace.
Attention span is longer	Attention span is shorter
Owning the accessed content	Sharing the accessed content
Searching the data	Searching the metadata
Hardware based media	Software based (cloud) media
Demand is the king	Choice is the king
Industrial, tangible economy	Information, network, intangible, experience economy
Connect individual with the information/content/product	Share content and experience among groups
Information based service	Conversation/Communication based service
Partial information access	24/7 information access
Place bounded media	Space bounded media
Individual/one screen media	Multi-screen media
Value is contained in transaction	Value is contained in relationship
Information based service	Conversation/Communication based service
Usage-based pricing	Access-based pricing

9 A Paradigm Shift in Social Media Content Production

When Web 2.0 applications emerged in 2005–2006, cultural theorist Henry Jenkins (2006: 24) was one of the first to notice a definite paradigm shift in the way social media content is produced and circulated: 'Audiences, empowered by these new technologies, occupying a space at the intersection between old and new media, are demanding the right to participate within the culture.' The result, according to Jenkins, was a participatory culture which increasingly demanded room for ordinary citizens to wield media technologies—technologies that were once the privilege of capital-intensive industries—to express themselves and distribute those creations as they seem fit (Dijck, 2011). When 'old media' still reigned, media recipients had little direct power to shape the media content and faced enormous barriers to enter the marketplace, whereas 'the new digital environment expands the scope and reach of consumer activities' (Jenkins, 2006, 215). The technological opportunities seized by grassroots movements and individuals increase their creativity and provide a diverse palette of voices (Deuze, 2007).

Moreover, with the emergence of Web 2.0 applications, most prominently UGC-platforms, the qualification of 'user' has gradually entered the common parlance of media theorists (Livingstone, 2004). Users are generally referred to as active Internet-contributors, who put in a 'certain amount of creative effort' which is 'created outside of professional routines and platforms' (Dijck, 2011). Since the 1980s, the term 'prosumer' has been deployed by various academics to denote how user's agency hovers between the bipolar categories of producer versus consumer, and of professional versus consumer. New hybrid terms such as 'produser' and 'co-creator' have meanwhile entered academic discourse to accentuate user's increased production prowess (Bruns, 2007).

The ubiquity of Web 2.0 services has transformed the landscape of online content consumption (Szabo & Huberman, 2010). With the Web, content producers can reach an audience in numbers inconceivable through conventional channels. Examples of services that have made the exchange between producer and consumer possible on a global scale include video, photo, music sharing, blogs, wikis, social bookmarking, collaborative portals, and news aggregators, whereby content is submitted, perused, rated, and discussed by the user community. Portals often rank and categorize content based on past popularity and user appeal, especially for aggregators, where the "wisdom of the crowd" provides collaborative filtering to select submissions favored by as many visitors as possible. Over the last few years, the Web 2.0, now uniformly tagged as social media, has fundamentally shifted towards user-driven technologies such as blogs, social networks and video-sharing platforms (Smith, 2009).

Social media focus on both global and personal topics demonstrating how the future of content will be increasingly bottom up and consumer driven (Smith, 2009). Characteristics of user generated reviews and reviewers can affect ecommerce demand; feedback in blogs can affect the firms' pricing policies and the nature of competition; the attributes of user-generated search queries can affect the performance of search engine advertising, and the content of customer support dialogues can affect product design (Ghose & Ipeirotis, 2009).

In order to become flexible, adaptive, immediate and accessible social media have to develop personalized, immersive, customized, innovative, engaging and user-friendly applications and, services that can be easily accessed as well as shared. The strategic shift of media business moves toward Internet of Smart Things, Web 3.0 and Web 4.0, cloud media, personalized, ubiquitous, software based, on-demand, wearable and database generated media and distributor of aggregated content (widgetization of media), Flattening of distribution chain, content aggregators and multiplatform distribution. Content and multiplatform distribution aggregators are the winners in the digital future as the availability and the internet speed significantly reduces the cost of media content as well as distribution.

Moreover, in the near future, contextual and behavioral micro targeting in advertising will be more prevalently supported by geospatial tagging, location-based marketing in which social interaction becomes a value. New media has to offer at the same time personal and intimate as well as multifaced experiences firstly attempting to build a community, than a marketplace.

Also, media consumption is not becoming exclusively about demand, but it is also becoming about choice that represents a prospective lock in and barrier entry into a new media ecosystem. Miniaturization in media production and ubiquitous access will inherently favor usage of social media via mobile phones.

10 The Business and Technological Impact of Media Convergence/Divergence

One important point that needs to be mentioned and analyzed is that media convergence refers to an evolutionary process, not an endpoint. It is not simply a technological shift, but it alters the relationship between existing technologies, industries, markets, genres, and audiences (Jenkins, 2006). Furthermore, convergence alters the logic by which media industries operate and by media consumers process news and influence public opinion.

The new media audience is being more divergent due to the increase of media production and its content. The media market is increasingly fragmented and users' taste is more versatile than ever. Different services and applications on the social networks create their own terminal and multiplatform ecosystem that is becoming increasingly unreachable to consumers unless they pay for premium services/applications. What we are now seeing is the distribution platforms converging while the content diverges.

However, a positive side of digital convergence is that it leads to a democratization of content because of the development of web 2.0; where users generate and upload content for a public access (Diehl & Karmatin, 2013). On the other hand, media convergence represents a risk for content producers and distribution operators since most of the new media companies fear a fragmentation or erosion of their markets. Valerie Feldman in her monographic publication 'Leveraging Mobile Media: Cross Media Strategy and Innovation Policy for Mobile Media Communication' further substantiate the competitive and technological advantage of media divergence over media convergence by stating that:

> Multiple utilization of content in the divergent media is one possible leverage for media companies to raise revenue potentials from existing media content and establish multiple revenue streams. The precondition is the production of platform—neutral content that enables repurposing of content according to the syntax specifications of different distribution platforms. The profitability in the media divergent production and distribution is achieved as the content becomes disaggregated and re-bundled according to the characteristics of the medium.

Thus, the development of multiple utilization of content decreases the technological and economical importance of media convergence. The proliferation of channels and the increasingly ubiquitous nature of computing and communications rather contributes to media divergence. Even on the device level, the plethora of specific devices does not suggest convergence, either, albeit digitization enforces technological convergence to some extent (Goldhammer, 2005). Yet, consumers' demand for context specificity as well as parallel media usage at the intersection of various media access modes, devices and contents rather suggest increase in media divergence (Cole, 2004).

Neverthless, Enlund and Lindskog (2000) describe how the range of information from a consumer perspective has widened, as content now is available in many more media channels than before. In this manner, interactivity and online media encourage divergence, but at the same time the technology behind the service and the production work flow, prior to distributing and broadcasting in the different channels, are converging (Appelgren, 2004).

Another important characteristic that further favourizes the media divergence over media convergence is the fact that in the digital media distribution channels and platforms the importance of intermediaries is largely minimized. However, one of the few media industry sectors that need media convergence in order to distribute efficiently its media content is IPTV. The reason for increasing interest in media convergence from the IPTV industry viewpoint consists in the fact that IPTV channels are dominantly distributed to prospective subscribers via telecom multiplatforms. Despite its reliance on media convergence the industrial sector of IPTV has achieved limited commercial success as presently only 6 % of the global TV viewers are payed IPTV subscribers. Moreover, the global future of IPTV market appears to be relatively uncertain as major international consulting and telecom agencies project that until 2018 the number of payed IPTV subscribers will increase approximately just 19 million per year.

It is advisable to point out that media divergence is particularly profitable if the media company decides to use the 'cloud' vendor lock-in. The main competitive advantage of the 'cloud' vendor lock-in is that it makes a customer dependent on a content producer/service/application or distribution channel/platform. This is explained as that typical customers are unable to use another vendor without substantial switching costs or inconvenience. This is predominantly the case when there is a lack of compatibility or, interoperability between content producer/service/application and distribution channel/platform.

11 Major Research Agenda

It seems clear that media entrepreneurs will continue to pursue various types of convergence-based business strategies. Thus, as a result of significant shifts in marketplace realities, convergence-based impacts and strategies are likely to become increasingly important elements of the studies conducted by media economics and management scholars as well as industry practitioners. This means that media researchers and executives need to continue to expand and improve their work in this area. Specific suggestions for improvement in the use of media convergence can be summarized as the following:

- Researchers need to do more in the way of empirical research in this area. Scholars need to design sophisticated empirical studies capable of quantitatively measuring and testing convergence-based theories and impacts. In particular, the author recommends that: (1) researchers adopt more sophisticated empirical methods to address empirical questions; (2) efforts be made to overcome the disciplinary fragmentation that afflicts the larger field of media economics (Fu and Wildman, 2008);
- As scholars plan future work focused on identifying the impact of media convergence, they need to pose and attempt to answer analytical, as opposed to descriptive, research questions (Wirth, 2003).

Some possible research questions scholars might utilize as a basis for future empirical research in this area include:

- What are the main differences between old and new media consumption patterns?
- How have various types of convergence affected old media/new media/telecommunication company performance (e.g., usability, consumer and choice demand)?
- How has convergence affected the availability of substitutes and complements within the media/telecommunication marketplace?
- How have the marketing strategies of media/telecommunication firms been affected by convergence?
- How have the value chains of old media/new media/telecommunication companies been affected by media convergence?
- How has the value creation process of old media/new media/telecommunication companies been affected by media convergence?
- How have the corporate financial strategies pursued by media/telecommunication firms been affected by convergence?
- How have the globalization merger and acquisition strategies pursued by media/telecommunication firms been affected by convergence?

In sum, convergence-based studies of media and telecommunications are still in an early stage of development (Wirth, 2003). As a result, there are a wide array of possible studies and research directions available for scholars to pursue (Wirth, 2003). One of the major challenges faced by researchers as they conduct research in this area is to clearly define what they mean by *convergence*, and to then operationalize and measure convergence in social media so that they can assess its impact on the phenomenon under study. This will be efficiently done if researchers attempt to improve the Social Media Analytics and Measurement of ROI and examine its major parameters such as: Unique visitors, Member registrations, Interaction rate, Member Engagement rate (share files/documents), Product and Service Feedback, Social media follower, Website visitor, Transparency and participation, and comments per post.

12 Conclusion

The successful social media corporations will have to act more as corporate planners, as well as 'cloud', 'on-demand' and 'ubiquitous' content and distributor disaggregates, than traditional content and advertising providers. Moreover, with all these changes, media will need to accommodate various consumer lifestyles. In an increasingly global and mobile digital media landscape, it is easier than ever to reach a large audience, but it is harder than ever to effectively connect with it. The old media traditional preoccupation was to reach the audience, however, in the age of digital media globalization, new media companies have a twofold task to reach and connect the audience.

In summary, the second decade of the twenty-first century digital media is apparently becoming increasingly interactive, mobile, immersive, and ubiquitous. Furthermore, the future of the media appears to be specifically oriented towards the establishment of, networkes, 3D, on-demand, broadband and unicast as well as multimedia and hypermedia models of distribution, communication and content creation. Therefore, it is crucial that profitable digital media companies realize that media divergence can successfully perform as vendor lock-in, top-down corporate process and a bottom-up consumer-driven process.

References

Ahonen, T. T. (2010). End of cash? First blog in a series examining the pending doom of minted coins and printed banknotes [Web log post, October 11, 2010]. Retrieved December 20, 2010, from http://communitiesdominate.blogs.com/brands/2010/10/end-of-cash-first-blog-in-a-seriesexamining-the-pending-doom-of-minted-coins-and-printed-banknotes.html

Allison, A. W., III, DeSonne, M. L., Rutenbeck, J., & Yadon, R. E. (2002). *Tech terms* (2nd ed.). Washington, DC: National Association of Broadcasters.

Ames, M., & Naaman, M. (2007). Why we tag: Motivations for annotation in mobile and online media. *Proceedings of the ACM SIGCHI Conference on Human Factors in Computing Systems*. San Jose, CA.

Anderson, C. (2006). The Long Tail: Why the Future of Business is Selling Less of More. Hyperion

Appelgren, E. (2004). Convergence and divergence in media: Different perspectives. *8th ICCC International Conference on Electronic Publishing Brasilia* (pp. 237–248). DF, Brazil, June.

Bakos, Y., & Brynjolfsson, E. (2000). Bundling and competition on the internet: Aggregation strategies for information goods. *Marketing Science, 19*(1), 63–82.

Bruns, A. (2007). Produsage: Towards a broader framework for user-led content creation. *Proceedings of the 6th ACM SIGCHI Conference on Creativity & Cognition*, Washington, DC, USA—June 13–15, 2007, (consulted April 23, 2007) http://snurb.info/node/720

Cattuto, C., Loreto, V., & Pietronero, L. (2007). Semiotic dynamics and collaborative tagging. *Proceedings of the National Academy of Sciences, 104*(5), 1461.

Chevalier, J. A., & Mayzlin, D. (2006). The effect of word of mouth on sales: Online book reviews. *Journal of Marketing Research, 43*(3), 345–354.

Christodoulides, G., Jevons, C., & Bonhomme, J. (2012). Memo to marketers: Quantitative evidence for change how user-generated content really affects brands. *Journal of Advertising Research, 52*, 53–64.

Clemons, E. K. (2009). Business models for monetizing internet applications and web sites: Experience, theory, and predictions. *Journal of Management Information Systems, 26*(2), 15–41.

Cole, J. (2004). Multitasking bei der Internetnutzung. In A. Zerdick, A. Picot, K. Schrape, J. C. Burgelman, R. Silverstone, V. Feldmann, D. K. Heger, & C. Wolff (Eds.), *E-merging media. Kommunikation und Medienwirtschaft der Zukunft* (pp. 82–83). Heidelberg: Springer.

Daugherty, T., Eastin, M. S., & Bright, L. (2008). Exploring consumer motivations for creating user-generated content. *Journal of Interactive Advertising, 8*(2 Spring), 1–24.

Dennis, E. E. (2003). Prospects for a big idea—Is there a future for convergence? *International Journal on Media Management, 5*(1), 7–11.

Deuze, M. (2007). Convergence culture in the creative industries. *International Journal of Cultural Studies, 10*(2), 243–263.

Diehl, S., & Karmatin, M. (2013). *Media and convergence management*. Berlin: Springer-Verlag.

Dijck, J. (2011). You as in "You Tube": Defining user agency in social media platforms. In Z. Vukanovic & P. Faustino (Eds.), *Managing media economy, media content and technology in the age of digital convergence*. Lisbon: Media XXI.

Dokoupil, T., & Wu, A. (2010). Take this blog and shove it! *Newsweek, 156*(7) 42. Retrieved August 16, 2010.

Doyle, G. (2002). *Understanding media economics*. London: SAGE Publications.

Edwards, T. (2009). User-generated content. *MultiLingual, 20*(3), 18–19. Number 103.

Enlund N., & Lindskog T. (2000). Nya redaktionella processer vid flerkanalspublicering. På väg mot medievärlden 2020. In H. Hvitfelt, & G. Nygren (Eds.), Sweden: Studentlitteratur.

Foremski, T. (2009, February 17). Some ideas on reinventing newspapers. *Silicon Valley Watcher*. Retrieved November 20, 2009, from http://www.siliconvalleywatcher.com/mt/archives/2009/02/some_ideas_on_r.php

Fu, W. W., & Wildman, S. (2008). Economic research on asian media markets and industries: A critical reflection. *Asian Journal of Communication, 18*(2), 92–101.

Gansky, L. (2010). *The mesh: Why the future of business is sharing*. New York: Penguin Group.

Ghose, A., & Ipeirotis, P. (2009). The econo mining project at NYU: Studying the economic value of user-generated content on the internet. *Journal of Revenue and Pricing Management, 8*(2/3), 241–246.

Gladwell, M. (2002). *The tipping point: How little things can make a big difference*. New York: Brown & Company.

Godes, D., & Mayzlin, D. (2004). Using online conversations to measure word-of-mouth communication. *Marketing Research, 23*(4), 545–560.

Golder, S., & Huberman, B. (2006). Usage patterns of collaborative tagging systems. *Journal of Information Science, 32*(2), 198–208.
Goldhammer, K. (2005). On the myth of convergence. In J. Groebel, E. M. Noam, & V. Feldmann (Eds.), *Mobile media. Content and services for wireless communications*, Mahwah, NJ: Lawrence Erlbaum Associates.
Graybeal, G. M., & Lee Hayes, J. (2011). A modified news micropayment model for newspapers on the social web. *The International Journal on Media Management, 13*, 129–148.
Greenstein, S., & Khanna, T. (1997). What does industry convergence mean? In D. B. Yoffie (Ed.), *Competing in the age of digital convergence* (pp. 201–226). Boston: Harvard Business School Press.
Hoegg, R., Martignoni, R., Meckel, M., & Stanoevska-Slabeva, K. (2006). *Overview of business models for web 2.0 communities*. From Proceedings of GeNeMe—Gemeinschaften in Neuen Medien, Technische Universität Dresden.
Jaring, P., Matinmikko, T., & Abrahamsson, P. (2006). Micropayment business in Finland—Forming the basis for development of micropayment methods and business. In M. Avital & K. Crowston (Eds.), *Proceedings of the Helsinki Mobility Roundtable. Sprouts: Working papers on information systems (Vol. 6)*. Amsterdam: Sprouts Alliance. Retrieved November 1, 2010, from http://sprouts.aisnet.org/6-40
Jenkins, H. (2001). Convergence? i diverge. *Technology Review, 104*(5), 93.
Jenkins, H. (2006). *Convergence culture. Where old and new media collide*. Cambridge: MIT Press.
Killebrew, K. C. (2003). Culture, creativity and convergence: Managing journalists in a changing information workplace. *International Journal on Media Management, 5*, 39–46.
Koh, J., Kim, Y., Butler, B., & Bock, G. (2007). Encouraging participation in virtual communities. *Communications of the ACM, 50*(2), 68–73.
Lawson-Borders, G. (2003). Integrating new media and old media: Seven observations of convergence as a strategy for best practices in media organizations. JMM, *The International Journal on Media Management*, *5*(II), 91–99.
Lee, W. C. L. (2003, June). *Clash of the titans: Impact of convergence and divergence on digital media*. Unpublished doctoral thesis, Massachusetts Institute of Technology, Cambridge, MA.
Li, X., & Hitt, L. M. (2008). Self-selection and information role of online product reviews. *Information Systems Research, 19*(4), 456–474.
Livingstone, S. (2004). The challenge of changing audiences. *European Journal of Communication, 19*(1), 75–86.
Macnamara, J. (2010a). *The 21st century media (R)evolution: Emergent communication practices*. New York: Peter Lang.
Macnamara, J. (2010b). Remodelling media: The urgent search for new media business models. *Media International Australia, 137*, 20–35.
Manovich, L. (2012). The back of our devices looks better than the front of anyone else's: On apple and interface design. In P. Snickars & P. Vonderau (Eds.), *Moving data: The iPhone and the future of media*. New York: Columbia University Press.
Marlow, C., Naaman, M., Davis, M. and Boyd, D. (2006). Tagging paper, taxonomy, Flickr. *Proceedings of the 17th ACM Conference on Hypertext and Hypermedia*. Odense, Denmark.
Mings, S. M., & White, P. B. (2000). Profiting from online news: The search for viable business models. In B. Kahn & H. R. Varian (Eds.), *Internet publishing and beyond: The economics of digital information and intellectual property* (pp. 62–96). Cambridge, MA: MIT Press.
Nov, O., & Ye, C. (2010). Why do people tag?: Motivations for photo tagging. *Communications of the ACM, 53*(7), 128–131.
Pavlik, J. (2008). *Media in the digital age*. New York: Columbia University Press.
Peters, I. (2009). *Folksonomies*. Berlin: De Gruyter Saur.
Pine, J., & Gilmore, J. H. (1999). *Experience economy: Work is theatre and every business a stage*. Boston, MA: Harvard Business School.

Qualman, E. (2012). *Socialnomics: How social media transforms the way we live and do business*. New York: Wiley.
Ransbotham, S., Kane, G. C., & Lurie, N. H. (2012). Network characteristics and the value of collaborative user-generated content. *Marketing Science, 31*(3), 387–405.
Ryu, G., & Feick, L. (2007). A penny for your thoughts: Referral rewards programs and referral likelihood. *Journal of Marketing, 71*, 84–94.
Schaedel, U., & Clement, M. (2010). Managing the online crowd: Motivations for engagement in user-generated content. *Journal of Media Business Studies, 7*(3), 17–36.
Smith, T. (2009). The social media revolution. *International Journal of Market Research, 51*(4), 559–561.
Snuderl, K. (2008). Tagging: Can user-generated content improve our services? statistical. *Journal of the IAOS, 25*, 125–132. IOS Press.
Szabo, G., & Huberman, B. A. (2010). Predicting the popularity of online content. *Communications of the ACM, 53*(8), 80–88.
Tam, K. Y., & Ho, S. Y. (2007). A smart card based Internet micropayment infrastructure: Technical development and user adoption. *Journal of Organizational Computing and Electronic Commerce, 17*, 145–173.
Vizjak, A., & Ringlstetter, M. (2003). *Media management: Leveraging content for profitable growth*. New York: Springer.
Wirth, M. O. (2003, July 31). *New media strategy: Convergence-based driving forces & challenges*. Paper presented at the annual convention of the Association for Education in Journalism and Mass Communication, Kansas City, KS.
Zeng, M., & Reinartz, W. (2003). Beyond online search: The road to profitability. *California Management Review, 45*, 107–130.

Part II

Media Production and Convergence

Designing, Implementing and Producing for Participation: Media Convergence in Practice

Marie Denward

1 Introduction

> One day a few years ago a young woman disappears from her wedding without a trace. The case gets attention when a drama series based on the event is promoted. A group of people claims that the woman became a victim of a conspiracy, and when the husband tries to prove them wrong there are no traces of her having existed at all. What is the truth? Is it a game or is it a true story? The mystery is a fact! What is the truth about Marika?
>
> SVT opens up a web office with the mission to seek out the truth, engages the audience in the quest. /.. / As the team starts to investigate the matter, traces and evidence of Marika's existence begin to pop up. The question the team tries to answer is: Who lies behind her disappearance and why does someone try to hide her traces?. . .
>
> The treatment: 11

In Sweden around 20,000 people have disappeared since the mid 1960s. And year-to-date, many of them have actually still not been found. The lines above depict the base story of an unusual and somewhat crazy hunt for a disappeared person that took place in Sweden during 2007. Thousands of people followed the hunt online in forums and chatrooms, and searched webpages for traces. Some of them also scouted streets and woods all over the country, guided by the disappeared woman's friend. The hunt was a hybrid media production created by Swedish Television (SVT), the largest public service television broadcaster in Sweden, and the small experimental games producer, The company P[1] (P). 'Participation drama' was the format name given indicating that it was more of a story than a game. A drama series was created around the disappearance and debate programs dealing with the issue were set up. Background facts of the production were interwoven with made up fictional content, communicating a dispersed narrative

[1] This is how the company spelled its name at the time.

M. Denward (✉)
Interactive Institute Swedish ICT, Box 1197, 164 26 Kista, Sweden
e-mail: marie.denward@tii.se

A. Lugmayr, C. Dal Zotto (eds.), *Media Convergence Handbook - Vol. 2*, Media Business and Innovation, DOI 10.1007/978-3-642-54487-3_6

using various platforms such as television, internet, mobile technology and game-like online and physical world activities. Social media like *YouTube*, *Flickr* and *Google Maps* were used too. It was claimed that it was a real investigation, and that the disappeared woman existed. At least this was what some participants thought. All was set up to help find out *if* she had existed at all, and if so, trying to find her whereabouts.

Up to that date the production did not resemble anything ever experienced in Swedish media. The aim was to create a (pilot) format that would invite a mass audience to more than ordinary TV spectatorship. Inspired from a range of emerging games, the idea was to support a variety of engagement levels, based on the consumer's choice of frequency, depth and immersion. It had no clear (game) conditions regarding winning or losing. Instead it offered collaborative activities like puzzle solving, deciphering, and missions to be corporally performed/enacted in physical world settings. Together producers and participants created a coherent game world. Even if it in different aspects did not reach the types of activities and the size of audience aimed for, it got international acknowledgement for its novelty. In April 2008 it won an International Interactive Emmy Award for best television service.

1.1 Problem Discussion

Sanningen om Marika (hereafter called *SOM*) was a significant example of media convergence in that many types of convergences took place: industrial, technical, cultural and social. It offered opportunities to identify, examine and analyze priorities and challenges in the making, and the various difficulties the convergences created in the (blurred) processes of production and consumption. A second reason why it was an interesting study object was that the producers were an unusual constellation. Never before had SVT collaborated with a similar games startup, and, as it would show, never before had SVT given a contracted company such loose reins during production. Thirdly, production processes of games—how they are produced, work processes and practices of design and implementation, is an under-researched area. So when I got the offer from the IPerG[2] project leader to take part, it was a singular opportunity to conduct research with my own research focus which I could not resist.

I was able to study the production from idea to implementation, what it consisted of and how the parts worked together. I could study the teams' cultures of production, and examine their practices, and whether and how the media logic of the two

[2] IPerG was a European Union-funded project (FP6—004457) run between 2004 and 2008. The aim was to create entirely new game experiences, tightly interwoven with everyday life through objects, devices and people that surround people and places. Through an exploratory approach several showcase games were created called "pervasive games". New technologies and methods were explored for the creation of novel and compelling forms of content. http://iperg.sics.se/index.php Accessed 28th Jan 2013.

companies differed. Did it have any impact on the collaboration and outcome? The new forms of audience interaction aimed for implicated that the two medium specific types of storytelling for games and television drama would be merged, and form a novel format. How would that work? Lastly, I was able to study the changing relationship of producers and consumers. As an observing participant-researcher, I observed participants, their actions and the interplay with the producers. As I will show, some aspects of media convergence were fruitful, while others did not really succeed.

2 Literature Review

Media convergence is a concept with many connotations and definitions. This is partly depending on the current state of development, but is also due to the fact that different actors use the concept for different reasons. For my own empirically grounded observations of *SOM* the work of Henry Jenkins (2004, 2006) and Jenkins and Deuze (2008) forms an important base. Jenkins embraces not only the common technological and/or industrial perspectives, but shows that convergence is a multi-layered and complex phenomenon that also includes social and cultural processes. The following section provides both the theoretical underpinnings and the backdrop for the analysis of the hybrid media production *SOM*.

2.1 Industrial Convergence

The last 40 years of technological, economic, cultural and political shifts have affected media industries. Digital technology, the internet, inexpensive computers and increasing bandwidth have pushed changes and boundaries between them. The strategies have been to integrate both vertically and horizontally to handle the technological shifts as well as the economical and cultural ones that followed. The television industry faces increasingly unforeseeable changes in audience preferences and is confronted with the consumers-producer shift which leads to a balance between being flexible and innovative in production on the one hand, and keeping steady routine within parts of the more or less reliable structure of the creative process, on the other (Deuze, 2007: 191). The games industry has developed fast and become a global industry. With games, the consumer pattern has changed towards interactive-participant-co-producer. Meanwhile, growth towards an increasing merge with other cultural industries like film, television and advertising takes place (Deuze, 2007; Johns, 2006). Compared with the broadcast industry's hierarchical and bureaucratic features, the games industry features a kind of democracy. The culture of small development studios is often described as anti-authoritarian and rebellious (Kline, Dyer-Witheford, & De Peuter, 2003).

2.2 Technological Convergence: Multi Media Platforming

The internet with its almost infinite information-providing capacity enables enhanced audience communication, relations and services. For broadcasters, the strategy has been to migrate towards a diversified (digital) multi-platform resulting in new forms of content, multiple layers and interactivity (Roscoe, 2004). Changing consumer patterns have imposed on this as well. The usage of the internet has also made the number and types of new game concepts grow, as well as the number of new marketing and new collaboration platforms. The broadcast industry grows in the direction of multiplayer online games and increasingly relies on online business models (Küng, Picard, & Towse, 2008: 89).

2.3 Convergence of Consumer: Producer

For decades the internet and digital technological tools have enabled consumers to develop and spread personal ideas, news and content. Individuals, groups and networked communities co-produce different cultural works. Companies also embed interactivity and related services into their products since it has shown to have a growing value for digital advertising and a way to better target consumers and manage subscriptions. Traditional TV operators' models of interaction, however, have been characterised by closed technical/multi media platforms and content formats that have limited viewers' freedom and participation. This may look paradoxical. Industries seem to gain and loose power at the same time as media control seems to be concentrated and dispersed (Jenkins & Deuze, 2008: 7). Similarly consumers gain and lose control. The process of media convergence can be described as twofold, taking place both top-down and bottom-up: sometimes the processes are reinforcing each other, and sometimes they are in conflict (Jenkins, 2006).

Multiple forms of media formats and technical platforms are today used in media productions in which every part and technology in itself holds distinctive contribution to a consumer's understanding. This 'aesthetic', transmedia storytelling, arises as an effect of media convergence (Jenkins, 2006). Dena (2008) points out that games like alternate reality games, contain another form of participatory culture. Here the primary producer has designed 'gaps' to enable participation for massive global audiences. With these tiers, different audience segments are targeted, and provided with separate content to facilitate and differentiate the experiences. Through filling in these intentionally (or unintentionally) left gaps the audience may co-create and develop new content. This way a smaller (game) audience segment with specific skills create content that becomes the main consumption part for a large 'non-playing' (television) audience. In this respect these new forms of games may offer new degrees of freedom and participation for media consumers, whether they are television viewers or game participants.

2.4 Convergence of Games, Play and the Ordinary

Contemporary games and play are not only leisure activities but grow into other social domains like work and education. Cultural productions like films and books not only illustrate the ambiguity of real and fictive, but the growth of the ludic in society, a kind of maturation of the gamer generations in contemporary society (Stenros, Montola, Waern, Jonsson et al., 2007). This development, coupled with the cultural awareness and struggle over public space, results in the growth of a particular type of games.

Such games, pervasive games, form a broad category and are played in public spaces and urban areas (Montola, Stenros, & Waern, 2009).[3] The word pervasive 'having the quality or tendency to pervade or permeate'[4] denotes the expansion 'outside' of the agreed on limits of playing. Outsiders are involved and intertwined with participants' everyday life and thus flavour it. Game actions can be difficult to separate from non-game ones (Montola, 2005), shaping ambiguous and unique 'brink' experiences (Poremba, 2007). Since they are neither just game experiences nor normal everyday experiences they always work at the borders in terms of both the nature of play and its contextualisation in culture. The game world is brought into the workplace, school building, or the home and grows online into chat rooms and forums (Montola et al., 2009). Game-related activities can take place with bystanders participating both in their ordinary life as well as in online spaces. Commonly, pervasive games are designed not to involve non-participants in ways that are ethically infringing, dangerous and/or risky.

2.5 Convergence of Cultural Commodities

Previously cultural commodities could easily be identified and interpreted through their common features and content. They were tied to a specific medium. This is no longer a fact. Types, formats and genres merge and form new types of entertainment. Games, previously produced for one single media, are nowadays developed for new platforms and TV programs which contain contests and game activities. Online viewing and co-authorship in programming are other examples.

There are important distinctions between storytelling in television and games, like the ways the content is organised and how it is meant to be consumed. In literature, film and television the dominant user function is interpretive (Aarseth, 1997: 62–65). In games, configuration—the process of selecting content and

[3] As Montola et al. state this new family of games has been called many names like 'adaptronic games, alternate reality games, ambient games, appropriative games, augmented reality games, brink games, context aware games, crossmedia games, geogames, hybrid games, immersive games, invasive games, location-based games, locative games, massive games, mixed reality games, mobile games, pervasive games, reality games, supergames, total games, transreality games, ubiquitous games, urban games, and so on' (2009: xix–xx).

[4] http://www.thefreedictionary.com/pervasive, accessed on 6th February 2013.

putting together what is going to be experienced—is the main player activity (Eskelinen, 2001). The concepts of text and story also express the differences. The term 'ergodic literature' denotes a text that demands some effort to be 'read' (Aarseth, 1997). A computer game consists of a network of choices, in which each choice has a number of outcomes. Each choice will affect the particular story (outcome) to be experienced. A characteristic for interpretive practices (i.e. reading, viewing) is the need to read (more or less) the entire text. In (computer) games, you often don't come across every possible combinatory event. Thus, the entire network of choices and outcomes is the text, but a player can never see (experience) the (entire) text. Consequently, games that are not settled, but open-ended, do not take the player's activity and game contribution into consideration. However, in live action role-playing where collective story building is the main activity, the player's agency and actions are emphasised and thus a crucial contribution to the game.

2.6 Convergence of Fiction and Facts

The previously rather stable notions of fiction and fact in media and cultural expressions seem to be continuously breaking up. Genres, the common agreement between consumers and producers, have since long helped consumers read/interpret the cultural product at hand telling them if the production was a fiction (i.e. a fabrication of events and people) or factual product (i.e. referring to the reality).

Expressing realism in media has grown in art, film and various mass media in recent decades and is reflected in new genres like talkshows, docusoaps and reality television (Sandbye, 2001). Today's perception of reality is a kind of 'performative realism' (Knudsen & Thomsen, 2002) where reality and the authentic are staged, created and constructed. However, new genres like docudrama, docusoaps and dramadocs show that nuances have appeared. Realism in program trailers, listing information, titles of program and social gossip also influence. Television genres in motion, using authenticity tools and methods of television, obstruct the viewers' interpretation of the show as being either factual or fiction. It may seem controversial and could be questioned ethically, but it becomes even more problematic when the audience is expected to engage, and interact, in the production.

3 Methodology and Approach

This chapter is based on the research I undertook of the *SOM* production during the phases of design, implementation and production of the interactive participatory parts. I here present my main results.[5]

[5] The reader who wants to dig deeper into my analysis will find it in Pretend it is real!: Convergence culture in practice (Denward, 2011).

I followed the production team for 9 months in 2007, joining the team when it was about to start the creation of the interactive participatory parts. An ethnographic approach would provide my research scope—to learn from the teams' own perspectives and within their own contexts (O'Reilly, 2005, 2009; Spradley, 1980). Everyday practice of observing and categorizing allows development of a consolidated, systematic study of structures, cultures, people and phenomena (de Certeau, 1984). I used a wide array of ethnographic methods applicable both online and in the physical world (Hine, 2000; O'Reilly, 2005, 2009). I conducted participant observation and interviews, and I spent long periods with the production teams, observing meetings, interviewing the designers and game masters recurrently. I collected a variety of other data such as (design) documents, sketches and pictures. I attended game events and met with, and observed, participants, both online in forums and chat rooms and in different physical locations. At the end I conducted in-depth participant interviews and a participant web survey. The broad scope implied the use of an iterative-inductive approach (O'Reilly, 2005) where I started early on to analyze data and turned back to search new data, while writing up.

4 Analysis: The Truth About Marika

The analysis is divided into three parts. The first sections deals with the background and early ideas of the production and the final outcome (4.1–4.3). Secondly follows the production process and the collaboration (4.4), and thirdly the reception in general (4.5). The chapter ends with a conclusion of the analysis.

4.1 Background and Ideas

At this time (2007), the company P was a small newly founded pervasive entertainment company consisting of a group of very engaged individuals, all members of the Nordic live action role-playing (larp) subculture. They were experienced in organizing larps and other entertainment productions built on participation.[6] Their goal was to commercialize their knowledge and skills. P also had strong ideological political ideas: pervasive entertainment had powers that could transform individuals and society. SVT, on the other hand, is the oldest and largest national television broadcaster, responsible for the national production of public service television programming and broadcast activities. SVT had to deal with similar problems as other European public service broadcasters (Enli, 2008)—loosing viewers due to 'competition, channel proliferation, digitalization, digital media and technology'. Their analyses showed that TV viewing had declined in favor of

[6] See www.thecompanyp.se

computer gaming and social media offered on digital platforms. A design document expressed the agreed quantitatively goal:

- Viewers, who watch and possibly follow the drama series (600,000 persons)
- Engaged, who visit SVT.se and follow the debate (60,000 persons)
- Participants, who take on missions in the game and that may report (6000 persons)
- Co-creators, who propel and report back (600 persons)

However, qualitatively slightly different pictures surfaced. SVT wanted to reach new and younger audience groups, explore new production methods built on well-known ones. P's main qualitative goal was to create a reality game where reality outside the game started to intervene with the game[7] and to create a transmedia narrative production layered with game components that fully invited to participation.

Moreover, the companies' inspiration also differed and imposed on the expectations on the future production. P's inspiration came from alternate reality games (ARG) and the Nordic style of role-playing games (larps) that use real-world history and events, and are played out in the physical—pervasive games. Both producers (game masters) and participants take part and influence the storyline development in these games tailored for around ten to a few hundreds of participants. The SVT team's inspiration emanated from an ARG that was combined with a television production.[8] It was largely based on fictional content, a clue hunt that took part mostly online and where the game masters were hidden for the millions of participants that it was tailored for.

4.2 Production Overview

SOM can be defined as a game of progression—a strong story containing a series of sequentially presented challenges (Juul, 2003), and aimed to be performed only once. Meticulously the designers puzzled a lot of design elements together in order to craft a game full of rich engagement, entertainment, and problem solving opportunities. It was made up of a distributed narrative coupled with story fragments that the participants, through play, would piece together to form a collective story (Montola et al., 2009). In order to create a solid and unified story universe, the parts had to be spliced together and run simultaneously on different platforms. A substantial number of the story elements were scripted in advance, ready to be implemented in an on-the-fly production manner, if participants took the bait. *SOM* was designed to create an experience in which participants, stepwise, would uncover layers of truth to reveal a 'true story' behind a televised drama series

[7] Interviews P producer, CEO and Creative Director, May 2007.

[8] The ReGenisis ARG was produced by The Movie Network.

based on 'real events'. It also contained a TV debate and several media/game activities. SVT was the responsible publisher and P was commissioned to design, implement and manage the (street) game parts.

SOM was a game-mastered game (Fine, 1983; Mackay, 2001), meaning that it was not entirely run by the players (it would require that they adhered to pre-written instructions), and neither was it controlled by an automatic game engine (as in a computer game). The production team monitored and steered the game and remained hidden from the participants. This way several team members could be given roles in the game (called The True Story). Pervasive games, that are not automated, are experienced as less predictable and feel more realistic (Jonsson & Waern, 2008) and can dynamically adapt to unexpected participants initiatives. These roles also allow for the on-the-fly addition of content during runtime (McGonigal, 2003) and sustain a responsive game world (Montola et al., 2009). Additionally, this allowed the organizers to fix bugs and other game technology. It also made the game extremely realistic. The game masters' characters and names were closely based on the real people playing them and lay very close to what was actually going on outside the game. The P producer put it[9]:

> The True Story was built on many facts from the real lives of the people playing characters in it, and during the game runtime their actions were both adapted for the story and the story adapted to explain their actions. Every piece of truth that somehow fitted the game narrative and was not too personal to be brought into The True Story was generally used in it, to strengthen the illusion that The True Story was real.

To be able to create a reality game with an indistinguishable borderline between fiction and facts, the production team went to great length to uphold the illusion. This conscious obfuscation and ambiguity continued throughout the production and was employed in all media channels. Many production parts were posing as fictional and factual at the same time.

4.2.1 Storylines

The communicated fiction was a TV drama (hereon called The Drama) that told a story about a woman, Marika, gone missing on her wedding night. Her husband Andreas and childhood friend Janna started to search for her. Janna had a theory of the disappearance as being part of a conspiracy. Trails pointed towards an underground organization into which she might have been pulled. The clues were uncovered bit by bit, and Marika was ultimately found. The drama had its own storyline and had some significant disparities from the 'true' background story as told in the game. According to the game activities, the woman had stopped looking for her disappeared friend after an encounter with a security company that threatened to kill her and had stolen her computer with all her research. The posters, that she still had the nerve to put up, were later found by SVT who planned to create a TV series about missing people. After finding out about SVT's plans the woman

[9] Nordgren, A. (unpublished work) Tools for Runtime Game Mastering of Large-Scale Pervasive Games. Stories and Lessons from The Truth of Marika.

started touring Sweden, communicating that the upcoming TV series was based on her true story, and that her friend had really disappeared. People signed up on the website conspirare.se to take part in the search. SVT publicly announced they would search too, and create a debate program to explore the issue, as an answer to the allegations that they were exploiting a personal tragedy. The story continued with the woman getting in contact with the secret organization and a week before the TV series started, the forum members uncovered the secret society's website. Both story layers were fictional, although The True Story (the game activities) had elements of facts woven into the narrative.

4.3 Production Parts

In this section the most important parts are described. Table 1 showing the main production parts with their function, content form and media use can be found at the end of the section. Table 2 deals with the types of activities and engagement levels that were offered to participants.

The TV drama series was called *Sanningen om Marika*.[10] Its function was to tell the backstory and to work as the (main) game entrance. One of the many planted game entrances and clues could be found via the URL conspirare.se appearing in the first episode. The theme of an upcoming episode was followed up in the game

Table 1 Main production parts

Component	Function	Media used	Production form	Content
Drama series	Backstory Game entrance	Broadcast Streaming	Pre recorded	Five TV programs
Conspirare.se	Hub of game activities	Internet	Continuously edited, real time and asynchronous monitoring	Blog Forum Chat
Debate series	Splicing game and drama Game input	Broadcast Streaming	Pre recorded	Five TV programs
Svt.se/marika	Production website	Internet	Partly edited through production, asynchronous monitoring	Production facts Eva's video blog Editor's 'chat' SVT inbox
Ordo Serpentis	Street game	Physical world Internet	Game-mastered real time and asynchronously	Street game Game report system on web

[10] It was aired on five consecutive Sundays evenings on channel SVT2, starting end of October 2007. Each 45-min-long sequence was followed by a 15-min debate show.

Table 2 Participant activities and engagement

Component	Activity	Engagement level
Drama series	TV viewing	Low
Conspirare. se	Reading, following story development, communicating, collaborating, playing, socializing, chatting	Medium-High
Debate series	TV viewing	Low
Svt.se/ marika	Reading, following story development, communicating	Low-Medium
Ordo Serpentis	Physical world/embodied game activities	High

activities. It was also meant to work as a standalone TV drama production. According to the trailer, the drama was based on the fact that 6000 Swedes disappear annually. This contextualization was confirming its factional status. The formal aspects, such as genre features or the marketing, contradicted this set up.

The website *Conspirare* (www.conspirare.se) worked as the hub of the game activities and as a communication platform. The site looked like a contemporary community website comprising a forum, a chat, and a text and video blog. Apart from freely accessing the front page with its recent blog entries, the participants had to register to get access to all forum posts, to contribute with material and participate in the chat. The motives behind crafting a forum were to enable steering the game and monitoring all contributions. It also made the game world dynamic and responsive, and as such more authentic. The updates were almost daily during the intense game runtime. The chat offered live communication and as such carried a significant social and practical function. This way P could control the players in the street game, instead of letting the players themselves take command, as is common in ARG (Martin, Thomson, & Hatfield, 2006). Participants (and game masters) socialized, discussed clues, and planned real-world actions to be performed. All communication on *Conspirare* took place in character with no chance to meta-discuss, something that had a significant implication of the fact-fiction blur. The organizers worked hard to moderate and control it real-time, to uphold the reality illusion, which did not always work. Partly this was due to television viewers going online wanting to know what it was all about asking meta-questions. Closing the chat for outside participants and only giving *Conspirare* members log in access solved it. On occasions members were banned from the chat discussions.

The svt.se/marika web was fairly static and offered only limited asynchronous interaction. This was due to a decision that put P in command of the interactive parts since they were the game experts. However, the website held some importance with features like a video blog, a 'chat', an archive, an application for uploading user-generated material and a picture quiz tool. On the site the television drama, debate, and actors were promoted. Eva's video blog handled the SVT investigation of the game events. It complemented the one on *Conspirare*. The online tool, SVT inboxen (Eng. The SVT inbox), enabled the submission of video snippets, photos and sound tracks that were then compiled in the archive. Participants could phone or email the editorial office and also register to get SMS or email notifications when a drama episode was due to start. Another offer was to 'chat' with the editorial staff in 'Redaktionschatten' (Eng. Editorial chat). This was in turn moderated, rendering the communication with participants selective and asynchronous. A picture quiz would offer television viewers a lightweight form of involvement, as an option to the more demanding streetgame's physical play and the online oriented *Conspirare* engagement. However, in reality the active participants were the ones that solved the riddles. The last feature was the then recently implemented online service, SVT Play, offering on-demand streaming of the TV series.

Several different marketing activities were conducted, from viral marketing to more or less conventional strategies. The main intention was to try to create media buzz to kick off the game with a snowball effect. Due to different reasons several of the efforts failed—there was no media buzz—an issue that was on the team's table during the entire production.

In the current affairs debate program, following the drama, a serious editorial staff, led by a credible host examined the truth-value of the drama's claims about the missing people. The original idea to connect the drama and game activities through audience interaction in the debate program was thoroughly changed, and created tensions, challenges and negotiation between the two teams. The SVT director used his artistic creativity and individual influence and decision on the form of the program, to craft a 'quality' product according to common broadcast production demands (Küng-Shankleman, 2000). The result was a novel product, merging a fiction story into a factual programming format, however it was without the planned (live) viewer interaction.

The game activities were a group of more or less game-structured experiences: the street game *Ordo Serpentis*, the picture game *Spektaklet*, the *Entropia* Universe parts, and all the pre-game activities that promoted it. These activities started months ahead of the airing of the drama. The early on small, but engaged, participant group grew slowly with the initial viral marketing activities. The more intense parts ran in parallel with the TV drama, and directly influenced the content of the debate. The game ended simultaneously as the TV series. The game involved the participants in a dramatic story with experiences such as treasure hunts, meet-ups with other participants, being chased by an evil guard company, and receiving strange missions from a mysterious secret society. The tasks were to gather information on the disappeared woman. Traces were spread on all platforms. Participants uploaded documentation of completed missions (video and photos) to *Flickr*, *YouTube* or to *Conspirare* and svt.se/marika. The producers then selected and reused the material. Participants accepting obfuscations between facts and fiction were offered immersive play in the ordinary, with real/physical people in real/physical locations.

4.4 The Collaboration and the Production Process

4.4.1 Division of Labor and Daily Work

Early plans roughly formulated the work division, utilizing each company's particular specialty and skills. But this changed. SVT produced the drama series, studio debates, trailers and other marketing activities and the svt.se/marika webpage. A web editorial team worked with and implemented SVT's interactive components and carried out some game mastering. P, subcontracted to design and produce the game parts, staged the real life installations, crafted the online puzzles and produced props, sounds, choreography and subliminal images for the key scenes of the drama series as well as edited the scripts for the series and the debate. They developed the technology platform[11] that enabled participation on all available media—websites, forum, chat, mobile application, game—and the orchestration of the larp parts.[12] Two groups of seasoned live action role-players, situated in Stockholm and Gothenburg, volunteered to plant clues and props.

The key team members cooperated primarily through weekly phone meetings and a project web portal, due to their geographical distance (400 km). Apart from differing in their physical environment, size and history, their societal, institutional and organizational contexts also differed. This was acknowledged and there was a strong motivation to understand each other's cultures and languages.[13] The features of non-profit production culture were strong at P (Svahn, Kullgard et al., 2006), as was the culture of a games developer; a democratic and anti-bureaucratic way of doing things, with expectations to work long hours and with gross engagement (Kline et al., 2003). This clashed with SVT's hierarchical and bureaucratic broadcasting production culture coupled with the involvement of many departments and regulated working hours. Some of the reasons why the teams managed collaboration probably lay in their akin personal engagement and the similarity in the dynamic and informal nature of their project-based employment.

4.4.2 Opposing Cultures of Production

Television programming and role-playing games diverge highly in their practical and technical production processes. A broadcaster operates in terms of a one-way communication process and creates a show to be transmitted to an audience (Lowe & Bardoel, 2008). The (mass) audience is commonly given limited influence and interaction. Viewer ratings measure success. A director uses a written manuscript containing a set narrative and plot, to implement the show with employed actors. The result is a controlled, one-directional and easily broadcast product. So, at SVT, the production followed ordinary broadcast production processes, divided into parts

[11] This was carried out with two research institutes, the Swedish Institute of Computer Science and The Interactive Institute.

[12] This game mastering tool, called Games Creator, was part of the mentioned IPERG research project.

[13] This was expressed in several interviews by different team members.

and teams depending on the type of content. But the features were also controversial and unique[14] like when high-level managers took part in the spontaneous decision-making that was usually carried out at lower levels,[15] and that different departments worked tightly together.

P had a different work process. The team worked satisfactorily in iterations, using a moderated software development method, to conduct the design and implementation process of the game. However, during 'crunchtime' game features were cut out to manage implementing the most crucial ones. The team had weekly, and during runtime daily, follow-up planning meetings. The design document was viewed as a dynamic piece, and numerous iterations and game test performed gradually. They knew that success demanded a flexible design with open-ended game elements and open-minded game masters and aimed for a planned but open-ended, interactive and multi-directional experience, played out online and in the physical. In role-playing productions all participants are players and viewed as co-creators. Participants and game masters hold equal positions (Montola, 2005) and are expected to contribute to each other's experiences. This form of multi-directional production process, like other types of interactive narratives (Rilstone, 2000), or ergodic literature (Aarseth, 1997), represents a blur of the common distinction between producer and consumer.

By applying a multi-platform approach SVT aimed at audience interaction. It was a clear goal, pushed internally. It enabled SVT to maintain full control, as the 'user generated' material could be selected, edited and broadcast back. P, in turn, aimed to foster (their type of) participatory culture. The differences in attitudes and approaches were observed in the wordings in blogs, the choice of pictures and photos, and the selection of communication tools such as in the chatting tools and the usage of *Flickr*, *YouTube* and *Google Maps* (at *Conspirare*).

4.4.3 Impacts

In some ways one can argue that the novel production with its planned convergence succeeded. In other perspectives, the two different organizations with their differing organizational cultures, cultures of production, and medium specific logics were not easily overcome. Tensions and impacts appeared both in the final production and as collaborative tensions, both internally at SVT and between the teams. Within SVT the production pulled organizational changes like on how to deal with economical and production organizational who-is-doing-what questions (i.e. marketing, promotion snippets, replay decisions). The impacts could also be read in how the teams chose to design specific features. The svt.se/marika content largely consisted of the TV medium's specific one-way directed communication, relying to a lesser extent on features of asynchronous interaction. At *Conspirare*, on the other hand, the games responsiveness was crucial for success, and thus much effort put into it.

[14] Several SVT team workers confirmed this during interviews.

[15] Interviews with SVT management.

This resulted in low activity on the SVT parts. The forum at conspirare.se, on the other hand, grew quickly containing vivid discussions and a stream of participant material. An argument in favor of the interaction at svt.se was that the modes were supposed to demand less engagement to be consumed (i.e. tiered participation). Other likely explanations were that the SVT inbox application was a prototype and technologically deficient. SVT policies also imposed. For example, the SVT workers had regulated working hours, but P workers accepted more hours of workload. Extensive working hours during larp productions is a well-known fact.[16] All in all, the SVT management had to deal with several virgin areas like the external moderation coupled with the rather controversial themes and content of *SOM*. Other issues were the subcontract with the content producer P, that implemented and managed (SVT) production parts during transmission, as well as the implicit game nature and the production's blur of facts and fiction.

4.5 The Reception

For a transmedia production, like *SOM*, the interest from the mainstream audiences is important as it determines the business model's bottom line, affecting the engagement from the most ardent fans (Davis, 2012). Thus, the number of viewers affects the number of highly engaged participants. According to SVT statistics the *SOM* TV drama had 350,000 viewers week one, a number that decreased to 210,000 the second week and decreased even more the following weeks.[17] In total 92,000 people watched the streamed *SOM* material[18] (drama, debate and Eva's video blogs), with a weekly rate around 8400 people. According to P's statistics conspirare.se had around 47,000 unique visitors.[19] The forum got 490 registered members and almost 5800 unique visitors. The *Conspirare* chat had 100,000 posts. The *Ordo Serpentis* game got 751 registered members. This can be compared with the statistics of *ReGenisis*.[20] A fair assumption is that the 385 survey respondents were among the active participants. Most of them had followed the production for at least 1 or 2 weeks and the most popular activities were to watch the drama and debate and/or browse the websites (70 % active at *Conspirare* at least once a week). Research show that only a small minority of viewers are likely to fully engage as

[16] Interviews with P workers.

[17] This could be compared with the common reach of 600,000 viewers for similar SVT drama productions. Interview with SVT production manager, December 2007.

[18] SVT did not carry our regular measuring until January 2009, then viewed as a stable activity. To compare, in 2011 a viewing rate of 10 % was considered high. Interview M Sterner, SVT Analyst, 6th February 2011.

[19] The average time on site was around 5 min where 31 % visited 3 pages or more, and less than 3 % visited 15 pages or more. Sandberg, C (unpublished work) Truly a lot of lies.

[20] Dena (2009: 241) describes how 10 % of the audience of the TV drama visited the website and 10 % of these visitors signed up for accompanying game, and that 10 % of these players became hardcore players.

fans (Evans, 2011) and as highly engaged collaborate, moderate discussions, and lead development (Davis, 2012). According to the survey, it is also likely that the production attracted a fairly gender-balanced audience and had a geographical reach nation wise. Collaboration seemed more popular than competition, which is typical for ARG (Martin et al., 2006). Larpers were overrepresented among the respondents, probably a result of P's promotion in their own larper network.

The survey shows that all parts had to be read, or participated in, in order to grasp the (entire) production. *Conspirare* and the TV programs were considered core parts, although the svt.se/marika site did not spur the same engagement. Even if tiers of engagement cannot be traced in the data the participants' interpretational stances may say a bit more. The majority were TV viewers perceiving *SOM* as a conventional TV drama. Another group seemed to investigate further online, concluding it was a game, and showing up at public events. The third group, the players of the game, consisted partly of people understanding they were playing a game. The other part seemed to shift from being unaware to becoming aware of the gameness. There is reason, however, to believe that some still kept the belief that they were taking part in a proper search for a missing person. The survey also shows that the street game (i.e. *OS*) did not work as expected. Possible explanations are that the debate was not linked to the game as intended, the content too occult and that the missions were perceived as scary, making even the seasoned gamers hesitate. Also, the difficult game design task, to get people to rise from their TV sofas to engage in real-world enacted/corporeally mission, did not succeed. One week was probably also a too short timeframe to manage to gather the player groups, find time to perform and report missions. The elaborate narrative and complexity of the production apparently posed challenges to the audience, which is often the case (Davis, 2012) and casual viewers may lose interest if the production is too complex for casual consumption (Dena, 2009; Jenkins, 2006; Leiter, 2011).

5 Conclusion and Viewpoints on Convergence

This book chapter presents an in-depth analysis of a transmedia storytelling production *Sanningen om Marika* (Eng. The Truth About Marika). As such, transmedia production strategies are today common among television and film industries as a means to promote their narrative content and attracting larger audiences. However, a number of 'pure' transmedia productions have shown difficulties to create the aimed for synergy between different types of media (Bourdaa, 2011). The legal and economic vagueness has also been difficult to deal with. *Sanningen om Marika* was no exception.

The producers, Swedish Television (SVT) and the games company, The company P (P), jointly created what they called a 'participation drama' The aim from start was to build on one central story that extended narratively over several media (pure transmedia), including pervasive games activities such as alternate realty gaming and live action role-playing, Nordic style. Multiple processes, or types of convergences took place on various levels and places in the two companies' collaboration process. The ambition, to consciously form a hybrid media production offering novel forms of audience interaction thus merging their expertise in broadcasting and games development, would show to be a difficult task.

As an example of industrial convergence, it was novel and unusual for Swedish circumstances. Collaboration was problematic, where cultures of production and features such as size, history, corporate culture, ideology and sub-cultures imposed. The process of product merging created tensions although the companies acknowledged their differing cultures of production. Less easy to merge was the different views of what quality was and how to achieve it. Both were driven by their own media industrial logics followed by different viewpoints on what and how to produce, and for whom. Being a bureaucratic company and with complex government mandates, the collaboration gave rise to internal tensions and communication issues at SVT, and forced a more pragmatic decision-making. Later a new improved organization was launched. Inter-organizational cultures that cut across the corporate culture at SVT also existed, creating tensions between the more bureaucratic corporate culture, and the small project-based team's culture. The different corporate culture at P featured a similar culture to other game developers with their rebellious anti-authoritarian attitude of operating. At P the collaborative authorship and a common political and ideological mission made them solve internal conflicts differently than at SVT.

The merging of the two different types of storytelling in the design of audience participation is an example of the mentioned synergy difficulties for pure transmedia products. The two media's specific features and elements and the quality demands coupled with their interaction models opposed, as described above. The TV audience activity is mainly interpretation, with a low ambiguity factor. This makes it opposite to a (digital) game, which is an ergodic text (i.e. interactive) and an effort is demanded to experience it. The user has to configure game elements to create a game experience, watching is not enough. These different demands could not be merged fully as planned in the design of audience interaction in some of the production parts, and thus partly failed.

Other 'convergences' appearing were the reality-fiction content blur and the reality-games blur. Elaborate transmedia storyworlds, like *Sanningen om Marika*, must be designed to appeal to, and be accessible to, casual media consumers in addition to more highly engaged audiences (Davis, 2012). In many respects this was not the case. The story world was very multifaceted and complex, and as if this was not enough, two opposing pervasive games design elements were added. Even if alternate reality games blur reality and fiction, an online space allowing for meta-discussions about the game is often added to support the notion that you are actually playing a game. However, the Nordic style live role-playing did not allow for such a space, and thus not set up. A consumer non-familiar with these types of games could not detect the gameness. And casual consumers lost interest early on due to these complexities and thus the lack to casually follow the production. Furthermore, the producer-consumer agreement of TV genres implicates that SVT should have been clear about the fictional nature of it, but this was not the case. The producers' unclear communication divided the consumers. Some probably chose not to participate, because they did not grasp the blur and for others (the aware/hardcore players) this feature disturbed their participation through the (unaware) participants trying to unveil that it was all set up, and not (a) real (hunt). The producers solved it by controlling participation and interaction in various ways, strategies content owners may use to control audience expression to prevent behavior that would damage the brand (Scott, 1996).

The motive to craft audience participation/participatory culture also parted the collaborators, an institutional strategy versus a potential for activism. Their attitude to the audience was similar though. As many individuals as possible were to be as active as possible, believing that participation is a basic consumer characteristic (Carpentier, 2009). Generally P, as interaction specialists, criticized conventional media for not being participatory, arguing that live action role-playing empower the individuals and promote personal development. Still P wanted to keep the audience under control to fulfill their aims, see previous paragraph. This was obvious in the *Conspirare* chat. Overall, participants had no chance to question the frames of the production. The many qualitative web survey comments can be interpreted as if there was a wish, and need, to question it.

Finally, transmedia storytelling products as a result of media convergence are still an experimental phenomenon. The use and ownership-taking of new technologies in production practices have enabled the phenomenon to develop, as has the high level of competition in the audiovisual sector and the variety of consumer and fan practices. The book chapter has highlighted one product example showing that the process of consciously striving for/towards convergence in crafting pure transmedia is complex and not an easy task. Producers have to take into consideration the continuously changing consumer behaviors and new modes, kinds, degrees and dynamics of audience engagement. How to effectively accommodate business logic with audience logic and aesthetic logic in a variety of offerings is a central transmedia design challenge (Davis, 2012). Developing audience information systems using multiple sources of data and deep analytical capabilities would be an important instrument in enabling the measurement of the effectiveness of transmedia products, campaigns and integrated market campaigns as well as to provide greater insights into the value to the firm of the individual audience member [Breur (2011), Gunzerath (2012), and Napoli (2011) in Davis (2012)].

References

Aarseth, E. (1997). *Cybertext: Perspectives on ergodic literature*. Baltimore: Johns Hopkins University Press.

Bourdaa, M. (2011). Quality Television : Construction and de-construction of seriality. In *Previously On. Interdisciplinary Studies in TV series in the Third Golden Age of Television* (pp. 33–44). Sevilla: Frame.

Breur, T. (2011). Data analysis across various media: Data fusion, direct marketing, clickstream data and social media. *Journal of Direct, Data and Digital Marketing Practice, 13*(2), 95–105.

Carpentier, N. (2009). Participation is not enough: The conditions of possibility of mediated participatory practices. *European Journal of Communication, 24*(4), 407–420.
Davis, C. H. (2012). Audience value and transmedia products. In *International Symposium on Media Innovations, Oslo.*
de Certeau, M. D. (1984). *The practice of everyday life*. Berkeley, CA: University of California Press.
Dena, C. (2008). Emerging participatory culture practices: Player-created tiers in alternate reality games. *Convergence, 14*(1), 41–57.
Dena, C. (2009). *Transmedia Practice: Theorising the practice of expressing a fictional world across distinct media and environment*. Unpublished doctoral dissertation, University of Sydney.
Denward, M. (2011). *Pretend that it is real!: Convergence culture in practice*. Doctoral thesis, Malmö University, Holmbergs, Malmö.
Deuze, M. (2007). *Media work*. Cambridge: Polity.
Enli, S. (2008). Redefining public service broadcasting: Multi-platform participation. *Convergence, 14*(1), 105–120.
Eskelinen, M. (2001). The gaming situation. *Game Studies, 1*, 1.
Evans, E. (2011). *Transmedia television. Audiences, new media and daily life*. New York: Routledge.
Fine, G. A. (1983). *Shared fantasy: Role-playing games as social worlds*. Chicago: Univ. of Chicago.
Gunzerath, D. (2012). Current trends in U.S. media measurement methods. *International Journal on Media Management, 14*, 99–106.
Hine, C. (2000). *Virtual ethnography*. London: Sage.
Jenkins, H. (2004). The cultural logic of media convergence. *The International Journal of Cultural Studies, 7*(33), 33–34.
Jenkins, H. (2006). *Convergence culture: Where old and new media collide*. New York: New York University Press.
Jenkins, H., & Deuze, M. (2008). *Convergence Culture Convergence, 14*(1), 5–12.
Johns, J. (2006). Video game production networks: value capture, power relations and embeddedness. *Journal of Economic Geography, 6*(2), 151–180.
Jonsson, S., & Waern, A. (2008). The art of game-mastering pervasive games. In *International Conference on Advances in Computer Entertainment Technology*. Yokohama, Japan: ACM.
Juul, J. (2003). *Half-real: Video games between real rules and fictional worlds*. Doctoral Thesis, IT University of Copenhagen.
Kline, S., Dyer-Witheford, N., & De Peuter, G. (2003). *Digital play: The interaction of technology, culture and marketing*. Montréal: McGill-Queen's University Press.
Knudsen, B. T., & Thomsen, B. M. (2002). Indledning. In B. T. Knudsen & B. M. Thomsen (Eds.), *Virkelighedshunger: nyrealismen i visuel optik: antologi*. København: Tiderne Skifter.
Küng, L., Picard, R. G., & Towse, R. (2008). *The internet and the mass media*. Los Angeles: Sage.
Küng-Shankleman, L. (2000). *Inside the BBC and CNN: Managing media organisations*. London: Routledge.
Leiter, D. S. (2011). *Detecting audience design: The rhetoric of mystery and transmedia storytelling in the case of Harper's Island*. Unpublished doctoral dissertation, Purdue University, West Lafayette, IN.
Lowe, G., & Bardoel, J. (2008). From public service media to public service media the core challenge. In G. Lowe & J. Bardoel (Eds.), *From public service media to public service media RIPE@2007*. Göteborg: Nordicom, Göteborg University.
Mackay, D. (2001). *The fantasy role-playing game: A new performing art*. Jefferson, NC: McFarland.
Martin, A., Thompson, B., & Hatfield, T. (Eds.). (2006). *2006 alternate reality games white paper*. International Game Developers Association IGDA.

McGonigal, J. (2003). A real little game: The performance of belief in pervasive play. Level up. *Digital games research conference proceedings, DiGRA 2003*. Utrecht Universiteit, Utrecht.

Montola, M. (2005). *Exploring the edge of the magic circle: Defining pervasive games*. DAC 2005, IT University of Copenhagen.

Montola, M., Stenros, J., & Waern, A. (2009). *Pervasive games, theory and design*. Burlington: Morgan Kaufmann.

Napoli, P. (2011). *Audience evolution. New technologies and the transformation of media audiences*. New York: Columbia University Press.

O'Reilly, K. (2005). *Ethnographic methods*. London: Routledge.

O'Reilly, K. (2009). *Key concepts in ethnography*. Los Angeles: Sage.

Poremba, C. (2007). *Critical potential on the brink of the magic circle*. DiGRA 2007 Situated Play conference, University of Tokyo, Tokyo.

Rilstone, A. (2000). Role-playing games: An overview. The Oracle, Essays.

Roscoe, J. (2004). Multi-platform event television: Reconceptualizing our relationship with television. *The Communication Review, 7*(4), 363–369.

Sandbye, M. (2001). *Mindesmærker: tid og erindring i fotografiet*. København: Politisk revy.

Scott, G. G. (1996). *Can we talk?: The power and influence of talk shows*. New York: Insight.

Spradley, J. P. (1980). *Participant observation holt*. New York: Rinehart and Winston.

Stenros, J., Montola, M., Waern, A., & Jonsson, S. (2007). Play it for real: Sustained seamless life/game merger in momentum. In *DiGRA 2007 Situated Play conference*. Tokyo: University of Tokyo.

Svahn, M., Kullgard, P., Waern, A., Holopainen, J., Koivisto, E., Stenros, J. et al. (2006). *Business opportunities and business structures, Appendix 5* (IPERG Report D4.5). Event arrangements in volunteer organisations.

Transmedia Storytelling: The Roles and Stakes of the Different Participants in the Process of a Convergent Story, in Divergent Media and Artefacts

M.N. Sousa, M.L. Martins, and N. Zagalo

1 Introduction and Problem Discussion

Transmedia Storytelling has lately become a buzzword, massively used by academics, professionals, and consumers alike to describe a novel phenomenon promoted by digital media. Though this is not altogether a new practice, it was primarily identified in the 1990s, by authors in different areas, such as media economy, video games and television studies. It was not, however, until Henry Jenkins's coining (2003) that authors were brought together and started to circumscribe an autonomous concept, and an independent field of inquiry. As a diffuse field, and a diffuse practice, and because it embraces several media (media studies have tended to be single-media), there are still serious gaps in the understanding of what transmedia storytelling is, which extends to non-academic discourse.

Scolari (2009) speaks of an apparent "conceptual chaos", surrounding the terminology. Many authors use Jenkins's definition, but adopt different terms to refer to the same phenomenon; others use the same term to refer to different practices, either different types of transmedia storytelling, or other media convergence dynamics, cross-fertilizations between media, or intertextual practices such as adaptation or the serialization of stories (Sousa, 2011). In non-academic discourse, the term seems to be used rather indiscriminately whenever several media are used in the same media project, which, at this day and age, is almost always the case.

But this indefinition originates and runs alongside other problems. Though it might not be entirely novel, transmedia-related projects have sprung from

M.N. Sousa (✉) • M.L. Martins
Communication and Society Research Centre, University of Minho, Braga, Portugal
e-mail: martanoronhasousa@gmail.com; moisesm@ics.uminho.pt

N. Zagalo
Communication and Society Research Centre & Engage Lab, University of Minho, Braga, Portugal
e-mail: nzagalo@ics.uminho.pt

A. Lugmayr, C. Dal Zotto (eds.), *Media Convergence Handbook - Vol. 2*, Media Business and Innovation, DOI 10.1007/978-3-642-54487-3_7

everywhere like mushrooms in the past few years, not always identified as that, not always produced by the media industries or distributed through the usual channels, with a growing audience intervention. Formats are increasingly diverse and fluid. The media landscape is changing very fast and very radically, and we do not yet fully grasp its potential and limitations. The new convergence culture, as Jenkins (2006) puts it, involves a change both in media production and in media consumption, stories along with every other type of media content. As such, the relationship between producers and consumers is also changing dramatically.

As a result, the different stakeholders in the matter are struggling to find the best way to adjust their practices, in order to adapt to the new challenges. Media companies, authors, and the public (which is much more empowered than ever before), have their own particular interests in the matter, and they are not always compatible. Consumers are becoming more intervening and demanding; media producers have to adapt, even though, whatever they do, the outcomes are often unpredictable and not always the most desirable. As Jenkins (2006: 11) points out, "we are in an age of media transition, one marked by tactical decisions and unintended consequences, mixed signals and competing interests, and most of all, unclear directions and unpredictable outcomes".

But we cannot settle for chaos. What are the stakes for each of these groups? How do their stakes diverge and converge, and in what way do they influence how transmedia stories are being produced, distributed and consumed? Is transmedia storytelling simply a way for media conglomerates to cash out on their resources, or are there real creative possibilities? Media companies have to earn their keeping. How can they guarantee profit, in a context of free flow of contents, decreasing consumer loyalty and generalized public participation? Is consumer participation actually threatening the media industry? On the creative side, a lot of practices are being called transmedia, but are they actually *transmedia storytelling*? Could this be a good opportunity for content creators or is authorship getting lost in the collaborative and participative process? Is the public able and willing to actively participate in the transmedia process?

This paper springs from questions already raised by scholars and practitioners, and aims to further investigate the practical implications of the power struggle between the major stakeholders in the process, and the results of that struggle on how transmedia stories are being told and experienced. We do this by observing the phenomenon through the perspectives of the different stakeholders, describing their interests and roles in shaping the transmedia process. Our thesis is that a better understanding of what transmedia storytelling is, or could be, and a more "transmedia storytelling approach", could lead to a more effective and productive convergence, and a harmonization of the stakeholders' conflicting interests.

2 Literature State-of-the-Art of Transmedia Storytelling and the New Media Culture

Before we take a closer look at these questions, it is vital to understand the development they have already had and frame the context in which they spring. We begin by clarifying the theoretical concept of *transmedia storytelling*, and then review the main literature on the characteristics of this new communication model, this new culture so deeply influenced by digital media.

In Jenkins's (2007) view, "Transmedia storytelling represents a process where integral elements of a fiction get dispersed systematically across multiple delivery channels for the purpose of creating a unified and coordinated entertainment experience". Though it is often summoned by different terms, such as cross-media, multi-channel or multi-platform storytelling or entertainment, it basically refers to "a particular narrative structure that expands through both different languages (verbal, iconic, etc.) and media (cinema, comics, television, video games, etc.)" (Scolari, 2009: 587), in order to create a larger story, or story world (Dena 2009), where "the whole is greater than the sum of its parts" (Phillips, 2012: xi).

As such, this larger story becomes so rich and compelling, and the story world so complex, that it "cannot be fully explored or exhausted within a single work or even a single medium" (Jenkins, 2006: 114). As Bernardo (2011) puts it, the story content is "platform and media agnostic", because it can be told through many—possibly all—media platforms, even though, of course, it must then be materialized and adapted to each medium's specificities (Bolin, 2007: 243).

Many different media may be used, not only traditional media, such as books, comics, films, and television fiction shows, but also more recent ones, such as narrative games, blogs, websites, or social media. These franchises might even include theme parks, mobile phone text messages, and the staging of live events. In its ideal form, each medium tells the part of the story that is most suited to its specific semiotic determinations (Jenkins, 2006).

In one approach, maintained by Jenkins and his followers, each extension should be "self-contained", so the public can enjoy it even if they have not experienced the other extensions, while at the same time completing the story (Jenkins, 2006). Each extension can "be experienced separately and still be enjoyable", but at the same time become "part of a single unified storytelling experience" (Long, 2007: 15). Other authors (e.g. Dena, 2009), however, consider that extensions need not be "self-contained", the consumer having to experience all of them so as to make sense of the whole story. Some authors include both possibilities (Long, 2007; Phillips, 2012). In any case, each transmedia extension becomes a different "point of entry" into the story as a whole (Jenkins, 2006), because it "relates a different aspect of the story or relates it in a different manner" (Miller, 2008: 150).

Another characteristic of transmedia storytelling is that at least part of the story is interactive, so people can participate in it (Miller, 2008: 150). In most cases, we only have interaction, such as the one enabled by browsing a website or playing a videogame. In a growing number of cases, however, "fan speculations and elaborations also expand the world in a variety of directions" (Jenkins, 2006: 114), since they are given the chance to actively participate, either by choosing from a set of given possibilities or by pitching their own original ideas as the story is being developed. In some cases, as those described by Bernardo (2011), fiction is extended through social media and cell phone communications so that the public can interact and get responded to as if the characters were real-living people.

One of the reasons why this practice is so difficult to define, possibly the strongest one, is the seemingly endless mutability of the new media landscape. This communicative paradigm was promoted by the advent of new media and their connection through the Internet, due to the compatibility of their format: the digital format. The question with digital data is that it "can be stored easily, accessed quickly, and transferred among a great variety of devices. It can also be readily reassembled in an almost infinite number of ways, and thus it becomes a viable form of content for interactivity" (Miller, 2008: 4). As such, it also promotes the design of innovative and creative formats, both narrative and non-narrative.

New media, however, despite all apocalyptic forecasts, did not replace traditional media, which, though mostly digitized and available on different "delivery technologies" (Jenkins, 2006), did not disappear. For that reason, we have now more media than we have ever had in the past, old and new, existing side by side.

Also, both new and old media have very different specifications: they allow us to do many different (and very different) things, since their technology is becoming more complex. Being digital, the content in one medium can be easily accessed in other devices: in a simple device, such as a mobile phone or tablet, we can access the internet, watch films, read books, listen to music, play games, take pictures, etc. (Bolin, 2007; Jenkins, 2006). As a result, the traditional boundaries between media are becoming more and more blurred: "it becomes increasingly difficult to make distinctions between different media technologies, as they adopt functions and forms from each other" (Bolin, 2007: 237).

Furthermore, the creative tools and massive distribution systems now available make it possible for anyone not only to easily access most media messages, anywhere in the world, but to create their own messages as well, and spread them massively, through growing networks of people connected through the Internet (Shirky, 2008; Sousa, Zagalo, & Martins, 2012).

Though media technologies are at the centre of these discussions, Jenkins (2006: 15/6) believes that this is not fundamentally a technological, but rather a cultural, shift. "Convergence", as he calls this phenomenon,

> ... alters the relationship between existing technologies, industries, markets, genres, and audiences. Convergence alters the logic by which media industries operate and by which media consumers process news and entertainment. Keep this in mind: convergence refers to a process, not an endpoint.

In that sense, the dynamics between the stakeholders are constantly changing. The public is becoming a definitely active participant: being now able to massively distribute their own contents and opinions, and even to make the industry's contents circulate (almost) freely (even if not always legally), they can exert a much greater power in what is being said and done in the media sphere. Media companies and creators were not used to this, and they are having to adapt, though most of the time, they are just guessing to what they are adapting, because no one is able to predict what the public will do next (Bolin, 2007; Jenkins, 2006).

So, if media and media uses change, so do narratives. New media are "transforming the way that we communicate with each other and how we tell, deliver and share stories. We're beginning to see the emergence of new forms of storytelling inconceivable before the Internet" (Rosenthal, 2011: XIII). According to Murray (1997: 9/10), the computer promises "to reshape the spectrum of narrative expression, not by replacing the novel or the movie but by continuing their timeless bardic work within another framework". But, in Phillips's (2012: 9) view, these emerging kinds of storytelling "are more than just "a book you read on your Kindle" or "a movie you watch on your iPad"." It means more than putting old contents in new delivery technologies, or using several media alongside each other. It means creating new forms of storytelling, and new ways to consume and participate in the stories, that imply new ways to produce and finance stories.

3 Methodology and Approach

The question now is to understand how these changes affect the participation of each stakeholder in the process of a transmedia story. Our approach begins by evaluating the particular points of view of each partaker separately. We do so by applying a deductive methodology: based on data found in other scholarly works, in layman and professional accounts in the media, and in deductive reflections on observed empirical dynamics, we try to infer logically what their actual role and interests are, and how much they converge or diverge. The main interlocutors we focus on are the media industry, big and small, the creators who design and materialize the stories, and the public (or better, publics, since audience members are not all alike). This is presented in the next part of the chapter.

In Sect. 5, we then put the several views in perspective, so as to devise a possible solution to harmonize those divergent interests.

4 Role and Stakes of the Different Partakers in the Transmedia Process: A Deductive Survey

4.1 The Media Industry

To the media industry, this has not been an easy change, because they have had to adapt to totally different market logics, technologies and distribution systems, to the public's changing demands, and to the need to diversify their offer. The biggest change probably resides in the empowerment of the public, which is taking an increasing control of the media (Jenkins, 2006; Rosenthal, 2011).

Some companies are more open to innovation and to the public's inputs, being able to recognize the advantages of having the public not only give their intake on the changes, but also actually drive the process of convergence (Jenkins, 2006; Phillips, 2012). In this model, they have a constant barometer of the public's approval, which allows them to adapt more rapidly and invest only on the products they know will be well received. However, these market dynamics are very unpredictable, and thus hard to manage. As the audience gains power over contents, these become harder to control and capitalize, and, being free to choose from a wide span of media, the audience becomes a "moving target" (Bolin, 2007: 246).

Furthermore, the digitization of media technologies has led to a dramatic decrease in production and distribution costs, which is an obvious economic benefit for the industry; but, at the same time, this also enabled the consumers to have easy access to these tools. Consumers are now able "to archive, annotate, appropriate, and recirculate media content in powerful new ways" (Jenkins, 2006: 17/8). Their resources might not compare to the resources of the big industry (as much as their skills might not compare to those of the professionals), but now the public can intervene in a sphere that used to be exclusive to the media industry.

This participation often collides with the need to protect intellectual property. Production may be cheaper, but it is also harder to capitalize, because a lot of contents are being shared for free, without due financial retribution to the producers and often without their agreement. When companies work on these new models, they have to devise new ways to finance their activity. And this is not impossible: the owners of companies that work on this new model, such as Apple, Amazon, or Facebook, are millionaires. Parallel products and advertising, often conquered from traditional media, are two of those strategies. They work because audiences are increasingly dependent on the means of consumption (Bolin, 2007: 242), being predisposed to decide what to consume by influence of these new platforms, and to consume the new, parallel products they have to offer. Furthermore, if one medium or one media product runs onto the public spotlights, gains are possibly massive. Of course the industry has to develop a great deal of creativity and improvisational skills to be able to quickly respond, since public adhesion is fleeting. Company policies and procedures have to be restructured often and rapidly, and the managerial models have to be driven by consumption rather than production. Not all companies are ready to take the leap; many still prefer to adopt aggressive copyright-protection measures instead of embracing change.

According to Jenkins, the balance has not yet been established: "Sometimes, corporate and grassroots convergence reinforce each other, creating closer, more rewarding relations between media producers and consumers. Sometimes, these two forces are at war" (Jenkins, 2006: 18).

In any case, the new economic model, namely the part of it that concerns industry ownership, has been favourable to transmedia storytelling. In the past, book publishing, newspaper, cinema, television, videogames, or radio industries were made of independent, specialized companies, but they have tended to integrate into large conglomerates (Jenkins, 2006). This monopolist tendency has been a cause for alarm to many media professionals and thinkers, because it might threaten diversity, independence and creative freedom. On the other hand, this has also favoured the achievement of synergies: since copyright is held by the same label, a story can be passed on freely from one media company to the next. Experts in different media can cooperate more closely, integrating their different resources and know-how to create transmedia texts (Bolin, 2007; Jenkins, 2006). In this perspective, transmedia storytelling "makes economic sense" (Scolari, 2009: 589).

These kinds of synergies have long since been identified and taken profit of: in the cinema industry, for instance, blockbusters have tended to be surrounded by an extensive "commercial paratext", composed of toys, games, merchandising, books, music, etc., the story becoming a powerful commercial brand that stimulates further consumption (Stam, 2005: 28). Television, book publishing and games industries have, for decades now, used the same strategy. These "paratexts" attract the attention of the consumers and allow them a continued contact with the story, even after they have left the theatre, closed the book, or shut down the computer. These are, of course, also ways to cash out on the notoriety of the main narrative artefact: stories such as *Lord of the Rings*, *The Matrix* and *Harry Potter* become "heavyweight narrative brands" (Scolari, 2009: 590), which tend to constantly surpass the boundaries of one single medium and be recycled and reshaped to fit other media. These brands, ever extendible (note the overwhelming number of extensions to the *Star Wars* franchise), are aimed "to create a symbolic universe endowed with meaning" (Scolari, 2009: 599), which appeals, seemingly irresistibly, to fan communities and individuals. The industry has long since recognized this marketing break and turned fiction stories and characters into high-rate commodities (Bolin, 2007).

Actually, the first authors to refer to *transmedia*, Kinder and Kearny, used the term to express a "primarily promotional practice involving merchandising, adaptations, sequels and franchising", not necessarily dealing with digital media, but with commercial practices involving different media (Evans, 2011: 21). And this is an undeniably good way to make money. Media industries embrace these franchises because they consolidate consumer loyalty in an age of market fragmentation (Jenkins, 2006: 243). In this logic, creating licensed goods and narrative spin-offs is an excellent way to capitalize upon previous work.

The transmedia storytelling logic in particular makes the consumer dependent on the story, whatever platform or format, whether narrative or not, it is presented on. It stimulates the public's curiosity to learn more about the story and track down the characters' adventures wherever they are available, thus being led to consume more, because each extension provides new and diverse information.

However, many of the fore-mentioned franchises and spin-offs, though often called thus, are not properly transmedia storytelling: could it be that the public notices the difference and responds negatively? We will come back to this later.

Another question is that Transmedia Storytelling demands multidisciplinary teams: being spread through different media, you need different specialists, with skills in different media, to collaboratively create the franchise. Media conglomeration obviously favours these exchanges. However, Jenkins (2006: 107) believes that there is often aggressive competition, rather than collaboration within media conglomerates: "While the technological infrastructure is ready, the economic prospects sweet, and the audiences primed, the media industries haven't done a very good job of collaborating to produce compelling transmedia experiences."

Of course, things are changing very rapidly. Big companies are starting to accept collaboration in new projects more easily, because the market demands it (Phillips, 2012). Also, the new generations of professionals were already born in this new paradigm, and are thus used to a closer social connectivity and collaboration, which is the hallmark of this new era. Though business structures at the higher level are often still working on the old logic, the professionals getting the work done will increasingly press old mentalities to break.

New projects and new business models, by small, independent companies or even individual creators, are springing everywhere. Many tend to escape the mainstream radar, maybe received well, but by a very limited audience; some, however, have had significant success. Such is the case of *The Blair Witch Project*, created by three students that proudly stated their limited budget but still made millions. Of course, small companies do not have the resources to compete continually with big conglomerates; however, because they are small, they are not only more prone to synergies with other companies, big and small, and easier to sustain if results are not significant, but also willing to try innovative projects that are cheaper, but still possibly very appealing to this new type of audience. And this often indeed challenges the big industry. All it takes to succeed is a good story.

4.2 The Creators: Artists and Storytellers

Transmedia Storytelling has strong advantages, not just for media companies looking for profit, but also for the authors, since it presents undeniable creative, artistic possibilities (e.g. Miller, 2008; Rosenthal, 2011). Typical franchisation of stories involves the creation of new narratives, and thus demands creative work, but all the authors have to do is reshape the same story into other formats. Transmedia stories, however, demand the creation of new content for each extension; each of them needs to add something new to the story, and that *something new* has to be intertwined with the other parts of the story, so that they all make sense and form a larger whole. They obviously demand a harder creative work to coordinate the story and create new nuances that may be explored in new extensions.

Therefore, transmedia may be a lot more than a mere marketing strategy, than the mere spreading of a story through a commercial franchise. It means more work, and for different professionals, specialized in different media. In order to unfold across multiple media and artefacts stories need to be extremely rich, deep or complex: transmedia storytelling has a "unique ability (. . .) to import a rich dimensionality to a property and to tell a story in a deeper and more lifelike, immersive way than could be possible via a single medium" (Miller, 2008: 153). Of course, this is not new: Tolkien's *The Lord of the Rings* is possibly one of the best examples of a story so rich that it could easily generate a large number of extensions, in different formats. Today, with the ease to create more effective synergies, this can be more easily accomplished. And, being necessary to coordinate the extensions of a story from within, so that every artefact becomes an integrated part of a larger puzzle, we agree with Jenkins (2006: 115) when he says that art direction is replacing production and marketing departments at the centre of franchise conception.

The problem here lies in the fact that, often, big companies with good resources still settle for spin-offs that simply repeat the same story, and often not very skilfully; though transposed into another medium, they present no novelty, they are not true *transmedia storytelling*. In this context, the authors' creative efforts are highly restrained, especially if they are urged to work fast to comply with release deadlines and deterred from making significant changes or appropriations.

Another obstacle is the need for collaboration between professionals trained in different media and art forms. Experts in different areas, with differently formatted views on storytelling, can sustain conflicting perspectives on how the story should evolve. Each author is experienced in telling stories in a specific medium, characterized by a specific semiotic language, and can find it difficult to agree with other experts' views, as well as to create stories that can be effective in other media. Training courses still focus mostly on specific media. But the reverse is also true: each extension is to be presented in a specific medium, with specific determinations, in which not all professionals are skilled. So more effective interdisciplinarity and collaboration are highly necessarily, and are still to be attained.

A specific type of transmedia projects presents good advantages for authors, though; it includes frequent (sometimes daily) updates of the story, on web platforms such as blogs and social media. One example is *Sofia's Diary*, a Portuguese project by BeActive, later on internationalized. Before they created the television show, and the books series, they began with a website, where fans could not only read Sofia's daily updates on her diary, but also post messages and subscribe to receive mobile phone text messages with the latest updates (Bernardo, 2011). Well, this implies the constant need for creative work, because writers have to design new contents continuously. Also, that creation has to be conciliated with the audience's suggestions, which is good for authors, on the one hand, because they can take fresh new ideas and make their work grow from there; they also have a constant feedback on what they are doing, so they can fit it to the audience's demands, and, as such, create much more compelling and sellable products.

On the other hand, however, exposure to fans' opinions might also be cruel. During the development of the *Lost* and *The Matrix* sagas, for instance, fan communities were formed on the web, to share their views and theories on each story and how it would end. In *Lost*, a fan's idea at one time anticipated something the authors meant to do, thus ruining the surprise and dissuading them; in the case of *The Matrix* some fan theories have been reputed as better than what was actually done with the story (Jenkins, 2006: 96). There is even a fan fiction website, *howitshouldhaveended.com*, where fans propose different—often parodic—endings to popular films. Storytellers have to consider the chance that their work might disappoint the fans. And these fans will publicly assert their discontent, which could be embarrassing for creators and bad for sales.

Moreover, audience participation can be difficult because, in an honest attitude, all suggestions, even incoherent and uninteresting ones, have to be considered and weighed and decided upon. If the author has a highly formed, original idea, taking too much input from the fans might make the story stray into something totally different, and not always more interesting. Either for that reason or for demagogic purposes, participation is usually somewhat illusory: though it is asserted, the public's inputs are highly restrained, or manipulated into what the authors want the story to be, or simply disregarded.

In any case, the author's personal creative effort might be overshadowed and restrained by the public participation, as well as by the collaboration with other authors. A participative, collaborative process might not comply with personal goals, glory, or copyright claims. Questions of authorship in this new age are very pungent and there is still very little conversation on the matter, mostly, possibly, because the answers are extremely complex and hard to devise with any certainty.

4.3 The Public (or Publics)

Audiences are changing dramatically; as we have stated, they are becoming increasingly intervening, demanding, and active. The truth is they were never totally passive, since the reception of any message demands the use of intellectual operations—possibly different ones, for each media, but nonetheless necessary in any of them. A story is always presented in pieces, with gaps that must be filled, between a series of interrelated events and characters that must be organized so as to create a logical diegesis and fictional world, and to clarify the cause-effect relations between them (e.g. Cook & Bernink, 1999: 322).

However, in the old days, the consumers' activity was placed mostly in their own minds, or on the feedback they passed on to their personal, private and thus limited, relations. Today, however, digital technologies have enabled a radical change, best described by Jenkins (2006: 18/9):

> If old consumers were assumed to be passive, the new consumers are active. If old consumers were predictable and stayed where you told them to stay, then new consumers are migratory, showing a declining loyalty to networks or media. If old consumers were isolated individuals, the new consumers are more socially connected. If the work of media consumers was once silent and invisible, the new consumers are now noisy and public.

Consumers are now willing to share their opinions, and also participate, both collaborating with media companies on the production of artefacts, and producing their own. The cheap and easy access to multiple creative technologies enables them to produce their own materials, with no need for highly specialized training or big budgets (Sousa et al., 2012). Therefore, they are able to compete, when competent enough, with industries and trained professionals. Often, high budgets do not correspond to high quality stories and, although indie projects do not have the significant marketing apparatus of the big industries, the Internet and social media allow them to gather an unprecedented attention and engagement from the public whenever they can accomplish a good story.

Stories are a particularly keen context for this type of production because they stimulate the consumers' imagination, and inspire them to produce continuations for the adventures they are not willing to abandon after they consume their formal, commercial manifestations. Stories that are able to enter people's hearts and imagination become myths, being endlessly recycled and appropriated (e.g. Campbell, 1949), and inspire the public to use new media technologies to do that (Jenkins, 2006: 131). Even though some of these things already happened before, the Internet allowed fan participation to be visible on a wide, global level.

Furthermore, participation is not primarily individual, as reception has been in the past: it has become collective (Shirky, 2008). Hard-core fans used to be many, but they were disconnected, and sometimes even frowned upon; now, they are able to find each other and gather around communities with the same interests that they find on the Internet, thus working together to push things to their advantage. Their power, both because they are many and because they are more visible, is increasing, and pressing established institutions, dominant in the past, to reconsider their behaviours and strategies (e.g. Jenkins, Ford, & Green, 2013; Shirky, 2008).

However, this new approach to media and fiction consumption is not evenly spread through all audience members. If media consumers were always somewhat different in the past, convergence has brought about what Bolin (2007: 241) calls "user divergence". There are still different types of users, but now they are more different from each other. Many still prefer the old media model, because they simply are not willing to spend so much time and effort in media surfing and participation (Evans, 2008; Jenkins, 2006). Also, not all have the same abilities to participate. The number of people who still cannot at all access new technologies is highly decreasing, but they are still plenty, around the world. Many others do have that access, but do not have the skills to fully understand media messages and participate in their exchange, production, and discussion (Jenkins, 2006).

The ones pioneering convergence are typically young people that were already born in a world packed with digital media—the "digital natives" (Prenski, 2001). Though they still constitute a narrow group now, they will carry their habits along as they grow older, and that means more people will consume media in this way. Also, being young, they are entering the media sphere primarily through entertainment, namely through games and fiction (Jenkins, 2006). These audiences are "actively engaged in the stories that they love, sometimes as collaborators in terms of co-creation and fan fiction or avidly spreading the word as evangelical marketers and distributors" (Rosenthal, 2011: XIII). Considering how likely it is that this market segment will grow in the future, this will be a vast opportunity for media producers willing to engage with transmedia and other new forms of storytelling.

Another feature of this "new" public is that they are more dispersed along media (e.g. Bolin, 2007). Since there are more media than ever, the public has to choose some media in favour of others, and though they might not totally abandon any of them, they certainly have to divide their time. That choice, though it may seem that way at times, is not random: they search for those media that provide the contents they prefer, and when they enjoy a specific story, they expect to be given more, and on formats that are stimulating and satisfy their need for novelty. They expect, not to say demand, media industries to provide that (Rosenthal, 2011).

The fact that they are more dispersed means they are willing to surf the media in search of what they want. The skills they need to follow the flow of the stories are being developed on the way (Scolari, 2009: 589), so this is also a pedagogical exercise. These new consumers are becoming "information hunters and gatherers, taking pleasure in tracking down character backgrounds and plot points and making connections between different texts within the same franchise" (Jenkins, 2003). And they are expecting to find these flows and connections; if they do not, they feel discouraged, and seek other media products that indeed provide that dynamic.

5 Overcoming Divergence in Transmedia Practices: A Proposal

We argue here that one source of divergence, which may be delaying the progress of the transmedia practice and theory, is the fact that, in many cases, so-called transmedia franchises are not properly *transmedia storytelling,* as we understand it here. One example is Tim Burton's version of *Alice in Wonderland*, released by Walt Disney Pictures: a novel, a videogame, an illustrated book, and a website were released, among other products like toys and merchandising, but none of these artefacts actually adds any new information to the film. Each of them makes perfect sense within the whole, but they do not complete each other, expand the story in any way, or create a larger world. Hence, though some might use this fresh, new buzz term to describe the franchise, this is not *transmedia storytelling*.

All media enunciations are, in some way, derivative, in the sense that they are built upon what has come before, upon previous texts: this is the principle of intertextuality (Kristeva, 1981). Some enunciations, however, are more derivative that others, as are adaptations, franchises, serial fiction and transmedia storytelling, because they derive directly from other narratives, presenting the same characters and world, and some of the same actions. As such, adaptations, sequels, prequels, and other highly intertextual practices work on the same basic principle as transmedia: when consumers love or are in some way stirred by a story, or a fictional character or world, they enjoy coming back to it, reliving it somehow.

As far as adaptations go, Hutcheon (2006: 4) believes that part of the pleasure of experiencing them "comes simply from repetition with variation, from the comfort of ritual combined with the piquancy of surprise". Some adaptations, though, are better than others. They *can* be creative (or faithful, whatever criteria is used), and interesting, and inspire the public's appraisal, or simply more of the same, and even be fiercely criticized at times. In any case, revenues are typically high, because the story itself draws the public into consumption. And box-office figures show how profitable adaptations are. However, commercially produced, derivative franchises often called transmedia, not being particularly piquant, interesting, or innovative, might represent a disappointment to fans. Redundancy between artefacts can be tiresome and deterring because there is no surprise or novelty; redundancy "burns up fan interest and causes franchises to fail" (Jenkins, 2003). The public, who expects more, and is better informed than ever before, might recognise the merely commercial strategy (as they often do), and, being able to make themselves heard, reproach it massively, thus influencing others against it.

In contrast, and that is our thesis, if franchises would better incorporate the transmedia storytelling principles, they might become more appealing to the public. As we have said, there are two sorts of transmedia stories: one includes more finished works, which are complementary but for the most part self-contained and do not usually include public participation, though they do include interaction. *The Matrix* is one of those cases. The attraction here is that the fan who experiences several of the extensions, or all of them, will get a different, richer, and more comprehensive experience out of the story; in a way, in searching for and uncovering the secrets behind it, they live their own little adventure. This corresponds to a greater immersion in the story and a more vivid experience of it, which can be highly rewarding for fans and is not accomplished by other types of franchises.

Other transmedia projects, on the other hand, do include participation: the public is allowed to pitch original ideas for the story's continuance (as in *Sofia's Diary*), solve mysteries (*Perplex City*), vote for several main decisions in the story, participate in it as an actor or inspire a character (*Axe Anarchy*), or post messages on social media and *get answered*, through other posts, emails or text messages (*Dawsons' Desktop*). In this kind of projects, the transmedia approach is much more powerful, because it engages the users as if the characters were real-life people and they were part of the adventure (Bernardo, 2011; Miller, 2008; Phillips, 2012).

The case of *Perplex City*, provided by Phillips (2012), can exemplify the full power of this approach: being used to "talking" frequently with the main character, Anna Heath, to getting her feedback, and helping her solve the mystery of a theft, audience members felt as if she was a real person. So much so that, when, in the story, she was sent to an ambush and killed, the fans felt her (fictional) death as strongly as if they had known her in real life. They sent condolence emails to the other characters and thought up a way to honour Anna's memory in real life, folding 333 origami cranes and personally delivering them to the producer Mind Candy's office in London. *Perplex City* was not a pure narrative, but a game, an ARG (Alternative Reality Game); however, these games function in the same way as this type of transmedia storytelling. As such, they have the same power:

> Any single-medium work can in theory make an audience laugh or cry. But make an audience feel directly involved in the events in a story? (...) This is the power of transmedia. (...) And it's not just you—it's a joyful collaboration, with hundreds or even thousands of individuals fabricating a common fiction together. (Phillips, 2012: 4/5)

Chasing after clues to unravel more information on the characters, world and story is gratifying and challenging in itself. But if the public understands that they have an actual say on how the story evolves and is given the opportunity to do so, it will become involved in the story in a totally different way. Consumers will think this story to be their own, and thus consume more, and on top of it all promote it believingly. Though this is still a narrow market segment, they gather globally and their number is rising, as more and more people are growing into this new way of consuming and participating in media experiences and stories.

5.1 The Proposal Viewed Under Each Stakeholder's Perspective

If transmedia storytelling principles are more often properly used, as we propose here, this will have consequences for each partaker in the process. On the perspective of the media industry, with transmedia, instead of one artefact, companies can produce several: the more products you create, the more you can sell, the more money you can make. Of course, many of these products have lower prices, some of them are even free (as are Facebook profiles and blogs), and audiences might not be interested in all of them, but they are also much easier and cheaper to create. Under the principle of synergy, when you create something based on what you have done before, it will be less laborious, since you already have something to build on. Also, a website, blog or social media profile update, though it is cost-less, does not take a lot of highly specialized technicians, as those you need to produce, say, a feature film, but only a few scriptwriters.

When companies are conglomerated, and partnerships and synergies exist, a competing or conflicting attitude will render difficult, if not impossible, the effort to create a coordinated story between the different artefacts, because communication between co-creators will fail. And other, sometimes very small, independent companies, or professionals, willing to unite, effectively manage to maximize efforts and resources, and to devise successful transmedia stories. Many recent transmedia projects are independent projects, because they are easier and cheaper to make and risk is minimum, since they have little to lose. In synergies, partners work side-by-side, and not one under the other. Independence frees the authors from corporate demands and creative restraints and they still get their name onto the spotlights. This model is proving itself to work, and all odds point to an increase in this tendency. If big companies are able to follow, in the spirit of making true transmedia stories (whichever the type), rather than often disappointing, more-of-the-same adaptations and franchises, they will have an edge on the future.

Transmedia is also a good way to attract to different targets (Scolari, 2009). Since there is a growing "user divergence" (Bolin, 2007), the different formats in the transmedia whole can attract to different market segments, traditionally not interested in mainstream media. Those audience members who enjoy surfing the media to complete a wider, more complex story can do so, but those who prefer the old paradigm can still consume the self-contained artefacts as they did in the past, finding the story in whatever format they prefer. Many did not watch the *The Matrix* animes, read the comics, or play the games, but they still enjoyed the films (even if they may not have fully understood them); furthermore, the comics, anime and videogames fans had the chance to access the story in their favourite formats. This is also more democratic: people with different preferences, skills, and literacy levels can all find the type of stories that better suit their needs.

So, if there are such advantages to transmedia storytelling, in not pursuing it, media companies are missing out on a very profitable opportunity. The age of convergence is settled, and there is no turning back. It is not possible to put a stop to market dynamics where things are endlessly shared, and where the public demands to participate in the creative process. Trying to do so is like wanting to stop an avalanche with a stop sign: it will be violently run over. Contents flow, whether copyright holders like it or not. As we have come to think, today, they *need* to flow: the top-selling films, the most watched videos on Youtube, the hottest books and videogames are the ones people are talking about on social networks, and yes, sometimes sharing between themselves (not always legally). But, as Jenkins et al. (2013: 1) say: "if it doesn't spread, it's dead". Though some extensions of the transmedia franchises are free, they stimulate a buzz around the story that, in our view, is essential today for the massive acceptance and consumption of any media product. It is thus not in the media industry's best interest to stop that buzz.

Surely, the industry needs to make money; they need to survive somehow, and that should not be shameful to say. But even when contents get pirated, they also in a way addict the public to the story, and make way for new products; people *will* buy them because the story will have entered their hearts and minds. If, on top of that, fans are given the chance to participate, they will take the stories as their own and become their privileged ambassadors.

Since they are increasingly connected and organized into communities, they will have an important part in attracting still more public and in inspiring still more fans. The *Harry Potter*, *Millenium,* and *Fifty Shades of Grey* sagas may be said to have had success because the readers liked them so much that they inspired other people to read them; from there grew the (very profitable) idea to adapt the books to films. Had it been tried, a transmedia approach would have been economically viable. This mouth-to-mouth publicity is proven to be ever more effective than advertising, and, on top of it all, it is free. There are new models to be explored and capitalized upon. So the industry might as well embrace it. And, if some companies do not, the public will not hesitate to turn to those who do.

Undoubtedly, media conglomerates can be seen as a threat and thus feared: being bigger and stronger, they might (think they) have less need to accommodate the needs of the public and of the creative professionals, and they keep pressing their products to the market with massive amounts of advertising, with which indie projects will never be able to compete. On the other hand, independent projects are a possibility for anyone who does not share the corporative views, and these projects are so abundant that they do not allow big companies to monopolize the market. They are a force big corporations will somehow have to reckon with.

The solution to fighting anti-democratic pressures and corporative manipulation, we believe, must be based on media—or better, on transmedia—literacy. Being able not just to read and write but also to create and share their opinions and contents with others, consumers will be more empowered, and more able to participate on their own matters and interests, and thus become full citizens on their own right (Jenkins, 2006).

As far as creators go, has we have seen, they have everything to gain from embracing transmedia storytelling. Of course not all narratives in the future need to be transmedia, but the target audience is growing, and so are the demands for new transmedia stories. If seen through the right perspective, co-creation and participation can actually generate more diverse, creative and original artefacts: in the *Sofia's Diary* saga, Bernardo (2011: XXI) assumed that

> Instead of just shaping the concept in a room with writers, we were shaping the concept every day with the help of the growing number of fans. This daily process helped us to shape and develop storylines and, as a team, get us into sync with what the audience liked and disliked about what we were doing.

In the *The Matrix* project, the Wachowskis' strategy was to call renowned artists in other creative areas (videogames, comics, and anime), already with a cult audience behind them, brief them upfront about what they absolutely had to respect in the story, and then give them total creative freedom to build the transmedia paratexts to the films. As a result, the artists felt free and inspired, and a horde of fans, that were not typical film fans but followed their idols wherever they went, was conquered to the saga.

Of course, it takes a lot of work to coordinate this collaboration. Each of the contributions has to make sense in the context of the whole story, and, in order to plant clues and informations that are complementary in different artefacts, there must be a general conceptual direction to coordinate them, as happened with the Wachowskis' project. Within a large, co-authoring team, not every idea is good, but having many ideas rather than few allows the authors to choose the best ones, and these will probably be better than anything any single author could have devised alone. All it takes is a good direction so everything fits together.

Authorship must, of course, be shared. But in the case of the Wachowskis, their talent was not overshadowed, but rather fortified by the fact that they surrounded themselves with worthy, competent and already famous artists in other areas and media. Artists willing to innovate and dare (at least those who show real talent) will still be recognized for their work. In Bernardo's (2011: XIX) view, "in a media world with an excess of stories, if you do something new, you can still capture an audience". Many consumers are actually more and more prone to following a limited set of stories more deeply, instead of many momentary and isolated narratives (Jenkins, 2003). And this is happening precisely because, and only if, stories are compelling, and rich, and stimulate the search for further information.

In the case of adaptations, they demand less creative work, because no new story is needed. Historically, we know that adaptation practices have always been seen with suspicion, but they have also always been profitable. However, it is plausible to say that the public might be even more enthusiastic about products that actually add something to the story, instead of simply reproducing it in another format. Plus, the more intertwined the different artefacts are, the more the public will become "hooked" and feel the need to know (i.e., consume) the different extensions in order to grasp their full meaning and live the full experience.

6 Conclusion

The truth is people have always loved stories. This is a natural human activity, a natural human need. Stories have always been told, ever since the dawn of Humankind, in every human community (Barthes, 1966; Ryan, 2004). Through stories, humans learn to know the world and other people around them, and how to deal with the human condition and issues such as mortality, morality, and identity (Campbell, 1949); they also enable us to "explore alternate realities and expand our mental horizon beyond the physical, actual world—towards the worlds of dreams, phantasms, fantasy, possibilities, and counterfactuality" (Ryan, 2004: 2/3).

We, as well as Phillips (2012: 6), believe that transmedia storytelling is actually feeding "a core hunger of their truest fans: to have more, richer, deeper stories. Fans who love your creation are going to want to see more of it. They want to be a part of it. Transmedia (...) is the way to give them what they want". The public, the fans in particular, wish to immerse themselves in the fantasy world, because this represents a more enjoyable experience than the mere, passive consumption we were used to (Bernardo, 2011: XIX). Connections to other artefacts make the audience crave for more information, and search for it, and build communities so they can share their knowledge. As a result, the ancient need to live out our fantasy may even be being intensified by the participatory, immersive possibilities of digital media (Murray, 1997: 98).

Creators are still learning how to create good transmedia stories, whatever *good* means, to each different type of consumer/participant. They are still learning how to intertwine the partial stories in each artefact, so they can compose that larger story and fictional world, and stimulate the search for the other extensions. There are no guaranteed formulas yet (as there are in the cinema, for instance, with classic Hollywood films). Producers have to readjust heir way of thinking to the new ways of media technology and consumption, which are, most often, different from those they knew as they were growing up.

Academics and critics must also help in this process, not by closing themselves in nostalgic or catastrophic views of digital change, but by stepping out into the world with curious eyes and getting to know how this fresh, new model is actually functioning. Furthermore, it is vital that they start thinking in trans-media rather single-media terms, since these approaches, most common in the academia and training disciplines, are too narrow to fully account for transmedia phenomena.

Most importantly, we must take a step beyond established beliefs and learn how to know the public. It has taken charge of things. Anyone with a stake in creating positive, constructive, and successful stories has to get to know how they work so as to provide creators and the public with the most challenging options they can. And, of course, audience members are also still learning, and trying to figure out what works better for them, and how to use the amazing new tools at their service. And this is not just important to create better narrative experiences. The shared participation and the convergence of efforts between all the partakers in the process may be a way to create a better, more informed, skilled, participative, equalitarian and creative society.

References

Barthes, R. (1966). Análise estrutural da narrativa. In R. Barthes et al. (Eds.) (1982), *Communications* (pp. 19–60). Rio de Janeiro: Editora Vozes.

Bernardo, N. (2011). *The producer's guide to transmedia. How to develop, fund, produce and distribute compelling stories across multiple platforms*. Lisbon: beActiv Books.

Bolin, G. (2007). Media technologies, transmedia storytelling and commodification. In T. Storsul & D. Stuedahl (Eds.), *Ambivalence towards convergence: Digitalization and media change*. Gothenburg: Nordicom.

Campbell, J. (1949). *The hero with a thousand faces*. New Jersey: Princeton University Press.

Cook, P., & Bernink, M. (Eds.). (1999). *The cinema book*. London: BFI Publishing.

Dena, C. (2009). Transmedia practice: Theorizing the practice of expressing a fictional world across distinct media and environments. PhD Dissertation, University of Sydney, New South Wales, Australia.

Evans, E. (2008). Character, audience agency and transmedia drama. *Media Cult Soc, 30*(2), 197–213. doi:10.1177/0163443707086861.

Evans, E. (2011). *Transmedia television. Audiences, new media, and daily life*. New York: Routledge.

Hutcheon, L. (2006). *A theory of adaptation*. New York: Routledge.

Jenkins, H. (2003). Transmedia storytelling. Moving characters from books to films to video games can make them stronger and more compelling. *Technology Review*. Accessed June 15, 2011, from http://www.technologyreview.com/biomedicine/13052/page1/

Jenkins, H. (2006). *Convergence culture: Where old and new media collide*. New York: New York University Press.

Jenkins, H. (2007). Transmedia storytelling 101. Accessed January 11, 2011, from http://henryjenkins.org/2007/03/transmedia_storytelling_101.html

Jenkins, H., Ford, S., & Green, J. (2013). *Spreadable media: Creating value and meaning in a networked culture*. New York: New York University Press.

Kristeva, J. (1981). *Desire in language. A semiotic approach to literature and art*. Oxford: Basil Blackwell.

Long, G. (2007). *Transmedia storytelling. Business, aesthetics and production at the Jim Henson company*. Dissertation, Massachusetts Institute of Technology.

Miller, C. H. (2008). *Digital storytelling: a creator's guide to interactive entertainment*. Burlington: Focal Press.

Murray, J. (1997). *Hamlet on the holodeck. The future of narrative in cyberspace*. New York: Free Press.

Phillips, A. (2012). *A creator's guide to transmedia storytelling. How to captivate and engage audiences across multiple platforms*. New York: McGraw-Hill.

Prenski, M. (2001). Digital natives, digital immigrants part 1. *On the Horizon, 9*(5), 1–6. doi:10.1108/10748120110424816.

Rosenthal, L. (2011). Foreword. In N. Bernardo (Ed.), *The producer's guide to transmedia. How to develop, fund, produce and distribute compelling stories across multiple platforms*. Lisbon: beActiv Books.

Ryan, M. L. (Ed.). (2004). *Narrative across media. The languages of storytelling*. Lincoln, NE: University of Nebraska Press.

Scolari, C. A. (2009). Transmedia storytelling: Implicit consumers, narrative worlds and branding in contemporary media production. *International Journal of Communication, 3*(4), 586–606.

Shirky, C. (2008). *Here comes everybody. How change happens when people come together.* London: Penguin.

Sousa, M. N. (2011). Adaptação versus transmediação: O impacto dos meios digitais na propagação de histórias. In *Proceedings VII SOPCOM Congress, Porto University, Portugal, 15–16 December 2011* (pp. 3445–3459). Accessed May 05, 2012, from http://sopcom2011.up.pt/media/SOPCOM_2011_Atas.pdf

Sousa, M. N., Zagalo, N., & Martins, M. L. (2012). "Eu também posso propagar histórias" A adaptação e as narrativas transmediáticas na era da participação. *Com Soc, 22*, 167–183.

Stam, R. (2005). Introduction. In R. Stam & A. Raengo (Eds.), *Literature and film. A guide to the theory and practice of film adaptation.* Oxford: Blackwell.

Converging Universes and Media Niches in Serial Narratives: An Approach Through Information Architecture

V. Innocenti, G. Pescatore, and L. Rosati

1 Introduction

Our paper deals with models of convergence pertaining to audiovisual media production, with specific reference to serial narratives, since their peculiar features (expansion and persistence over space and time) allow a privileged point of view on convergence phenomena. The paper will adopt an interdisciplinary approach, using conceptual tools from several disciplinary fields, namely film and media studies, as well as media economics and information architecture. Our ultimate purpose is to offer a cross-disciplinary perspective that can be profit from the combination of all these fields. In order to do so, we will start our paper by focusing on some preliminary notions and relevant changes concerning TV series production in recent years. What is most evident in this study area is that we can no longer view TV series as just "a text", but instead need to study and analyze TV series as narrative ecosystems, that is as a composite environment that can integrate energy flows originated by users, technologies and media items.

We will focus on three different production strategies (*derivation, cross-media, ecosystem*), mainly concentrating on the ecosystemic structure of contemporary TV series. In order to overcome the limits of traditional methodology in audiovisual materials analysis, we will apply information architecture and cross-channel user experience principles to the media field, adopting an overall systemic perspective, and focusing on the bias between *local* experience (the single product) and *global* design (the whole ecosystem) and vice-versa.

V. Innocenti (✉) • G. Pescatore
University of Bologna, Bologna, Italy
e-mail: veronica.innocenti@unibo.it; guglielmo.pescatore@unibo.it

L. Rosati
University for Foreigners of Perugia, Perugia, Italy
e-mail: luca.rosati@unistrapg.it

A. Lugmayr, C. Dal Zotto (eds.), *Media Convergence Handbook - Vol. 2*, Media Business and Innovation, DOI 10.1007/978-3-642-54487-3_8

We argue that the convergent development of certain media productions involves both a project (planning) as well as a practice (made by desire and use), and that these two polarities contribute to define a complex and multifaceted phenomenon. We will focus on the misalignments between project and practice, as well as on possible forms of conflict and adverse reactions to planning. We account for divergence models that lead to the building of "media niches", which often remain external or peripheral to the circulation and delivery of media content—for instance, the case of fans using and re-appropriating content linked to specific gender practices and identities).

2 State of the Art: Serial Narrative in the Contemporary Media Landscape

TV industry, especially in the US, has recently started relying a lot on serial narratives that have enjoyed great success and have attracted considerable critical attention from scholars of audiovisual media (see, among many others, Gwenllian-Jones & Pearson, 2004; Innocenti & Pescatore, 2008; Jenkins, 2006a, 2006b; Mittell, 2009, 2012). Many scholars have actually concentrated on the important experimental work that TV series have done on narrative structures and staging, as well as on the complex mechanisms of their production and promotion, and the impact of the response these products generate in their viewers (Curtin & Shattuc, 2009; Grasso & Scaglioni, 2009; Hammond & Mazdon, 2005; Jenkins, 2013; Scaglioni, 2006).

The modern scene of American TV series is the result of a long process of regulation and competition, a story that has shaped the way these shows are produced, and had a strong influence on the need to create recognizable and distinctive brands for products, as well as for networks and TV channels. Nowadays, the proliferation of distribution channels and media conglomerates (Hesmondhalgh, 2007; Tungate, 2004), along with the convergence among TV, Internet and other entertainment devices, such as videogame consoles, tablet pcs, smart phones and, of course, computers, has led to the decline of the traditional TV system based on just three or four big networks. Basically, we are witnessing what Lynn Spiegel referred to as "the phase after TV" (Spiegel, 2004: 2). In this phase, we as members of the audience are facing two major changes: a constant growth in technological convergence and a constant increase in cultural convergence (Jenkins, 2006a), as well as a significant improvement in the quality of the TV production, often condensed in the label *Quality Television* (Jancovic & Lyons, 2013; Leverette, Lott, & Buckley, 2008; McCabe & Akass, 2011; Thompson, 1997, 2003).

Within the macro-category of the TV series, we tend to differentiate products into two primary models:

1. series consisting of episodes with a self-contained narrative; the episodes privilege the development of an **anthology plot**, or episodic plot, where the events

being narrated always arrive at a conclusion at the end of the episode. The most common forms of this type are the so-called "anthology series", with episodes that are completely independent of one another and with non-recurring characters (for instance *The Twilight Zone*, 1959–1964, resumed between 1985 and 1989 and later between 2002 and 2003) and sitcoms (such as *Friends*, 1994–2004 or *Modern Family*, since 2009).
2. series consisting of installments characterized by suspended or open endings; the events narrated in a single installment never arrive at a conclusion, and we need to follow the story through many episodes in order to see how it ends. This model privileges a **running plot**, that is, a plot that is carried on for a long time and for many episodes (at least 4–5, in some cases for an entire season). The most evident example is *soap operas*, which mainly come from Anglo-Saxon context and are characterized by a story that does not offer any resolution nor come to an end (for example *The Bold and the Beautiful*, since 1987).

Since the mid-eighties, however, TV series have undergone a process of "serialization". That meant that, although the classic structure of a series with self-contained episodes has not yet disappeared completely, [1]the narrative formulas have gone through a process of mutation and hybridization, and many TV series have become serialized, developing a mixed structure influenced by soap-operas. In the current "serialized series" (Innocenti, 2007; Mattessi, 2003; Newman, 2006), the episodes usually retain a high degree of autonomy, so there is always a central story that comes to end within the episode (anthology plot), but there is also a framework that continues for several episodes (running plot). This adds an element of temporal progression and narrative opening, which was missing in traditional formulas. Therefore, contemporary TV avoids the risk of narrative atrophy by creating a diegetic world where things constantly change at all levels—characters, scenes, and narrative technique, an approach that is much appreciated praised by fans (Innocenti & Pescatore, 2008).

3 Approach and Methodology: The Role of Information Architecture

When watching a TV series, apparently we get caught up with narrative developments involving a group of recurring characters; we focus on their stories and relationships and finally surrender to the pleasure of the story. This is no different than what happens with any form of storytelling, be it literature, cinema or television. On closer inspection, however, TV series show a number of anomalies as to their narrative progression. They are in fact "abnormal" elements, exceeding in terms of both time and space. Some series lasted many years, such as *ER* (15 seasons, 1994–2009), *24* (8 seasons, 2001–2010), *Doctor Who* (26 continuous

[1] See for instance such productions as *Cold Case* (2003–2010) or *Law & Order* (1990–2010).

seasons between 1963 and 1989, then the series resumed in 2005 and is still in production). Other TV series, however, have branched out across different media, following the model of "trans-media storytelling" suggested by Henry Jenkins (2003, 2006a), i.e. to generate different products for different media spaces, such as novelizations, comic books, video games, fan guides, trading cards and mobisodes. This ability to overflow into media space is the result of a mutation, both formal and epistemological, of narrative structures. It is clear that what we are dealing with today are no longer "unique" objects, as in self-contained, finished texts, but long-term, trans-media, convergent phenomena. Rather than just "a story", contemporary TV series provide viewers with an entry point to a whole universe. This is the case with *Heroes* (2006–2010) and ARG (Alternate Reality Game[2]) *Heroes 360° Experience*, or with *Lost Experience*—an ARG played by fans during the second season of *Lost* (2004–2010) in the UK and in the interval between the end of the second season and the beginning of the third in the U.S. These are clear examples of extensions of the series narrative across several digital, but also physical, platforms.

Such high degree of narrative and production complexity as has been reached by current TV series makes them hard to investigate with the sole help of traditional tools of analysis, such as semiotics or narratology. These methodologies actually treat the TV series as a "text", that is, as an object with a clear beginning and ending, clear boundaries a target and a direction, albeit rhizomatic and manifold. But things have changed, and today a TV series is the result of the union of several "texts" that have no definite beginning or ending. In fact, the development of a series is the result of the combination of different narrative solutions, offered by different forms of narration and designed for different media. Thus, it is necessary to broaden our perspective and hybridize our approach to media studies with disciplines that have an intrinsic ability to embrace complex and multifaceted phenomena. Information architecture is one of these.

Although if the label "information architecture" originally refererd to the practice of designing websites (Rosenfeld & Morville, 1998), now it describes the "structural design of every shared information space", physical, digital or procedural (Rosenfeld & Morville, 2006: 4). In other words, it refers to the act of organizing items in an environment in such a way as to create a specific experience.

Shifting the focus from the notion of text to that of information environment, information architecture enables to better understand complex phenomena that span several media and artifacts—such as serialization and trans-media. Indeed, the roots of information architecture lie in principles that are largely independent of the support they apply to (papery, audio-visual, digital etc.—Resmini & Rosati, 2011).

[2] Alternate Reality Games (ARG) are games that combine the Internet and the real world. They are developed with the help of several Web tools (blogs, email etc.) and confront the player with a mysterious story, full of clues that point to the real world. See Montola, Stenros, and Waern (2009).

In this perspective, information architecture can be viewed as a deep, invisible layer, whose key function is to connect the parts of a complex system. This deep layer takes different forms depending on the medium, via set of "execution / translation" rules. The impact of information architecture is relevant insofar as its invisible layer is essential in order to keep the system in a state of balance, which is also the result of the system being harmonious and organized, open, interconnected, persistent, and resilient (as we will see in the next paragraph). Many of the tasks we perform everyday require that we move between different media, and between the digital and the physical world. Some of these tasks can also be linked to the consumption of entertainment products, TV series in particular (Innocenti & Pescatore, 2012).

4 Narration in a Convergent Environment

Watching a TV series is a composite, diversified experience that often leads members of the audience to move back and forth between the physical and digital sphere; this generates participation and stimulates further consumption. It is more like living in the world of the series itself, rather than simply following a storyline. The narrative no longer has a single core, from which the plot radiates, but tends to develop along different lines.

It is therefore necessary to broaden our perspective, shifting the focus from individual items to clusters of items. This shift is a consequence of the fact that today physical and digital environments are designed, created and experienced as a seamless whole that uses old as well as new media (Resmini & Rosati, 2009). This is a key feature in the general evolution of design, from the design of artefacts to that of processes, or experiences (Sterling, 2005), which cross physical and digital, offline and online space. This development has of course influenced the designing of narrations and of storytelling experiences, bringing about a significant change from serial narratives to what we call narrative ecosystems.

The different ways of designing information content (Giovanazzi & Mascaro, 2009) make it possible for us to rethink TV series design and production in the contemporary mediascape, outlining three production strategies that can be associated with as many different moments in the history of TV serial narratives, as well as with three peculiar design paradigms:

1. derivation
2. parallel or cross-media
3. ecosystem.

The first wave of recent TV series production belonged to the design mode of *derivation*. We are referring to the series from the mid-1980s that introduced multi-linear narratives. The multi-strand narrative structure allows for an easy scalability of content, whereby storylines can be broken down to modular packages. Therefore, thanks to the process of digitization, they are distributed in various media formats other than the one they were conceived for. See for instance *ER*, or *Buffy the Vampire Slayer* (1997–2003), from which video games and novelizations were generated in the wake of the success of the series.

The *parallel design mode* or *cross-media* design mode is typical of the era of convergence. The designing of audio-visual products starts from a narrative mould, and subsequently develops into a more complex project through the use of several media. This narrative mould generates ad hoc objects, available for all target audiences. Nevertheless, this way of conceiving audiovisual cultural products is much more frequent in film production rather than in TV production. A representative example are the Matrix[3] trilogy and its parallel products, the short animation film collection *The Animatrix* (2003), two video games (*Enter The Matrix*, 2003 and *The Matrix: Path of Neo*, 2005), comic books, and so on. Parallel production is also a typical feature of the relationship between cinema and video games. Because different sectors require different timings for production, it is necessary to harmonize production timing in order to ensure that the release dates of individual objects belonging to the same story are in sync. To keep this kind of tool standing, substantial *consistency* across different products for different platforms and production models is required. There has to be effective integration among all the products in order to avoid loss of information, as well as information redundancy. In this second case, however, we are still dealing with a "finished" narrative universe presented in the form of multiple texts linked by a common origin.

However, as mentioned before, contemporary TV series are no longer simple textual objects, but are instead the result of an *ecosystemic design*, in which a general model is developed in advance, as an evolutionary system with a high degree of consistency among all its components. See for instance *Lost*, whose pilot episode cost 10 million dollars and was shot in 3 months from the production's green light. According to Pearson (2009: 240), *Lost* can be placed in a phase of transition between the relatively stable conditions of the post-network era and the increasingly unstable situation of the post-TV era at the beginning of the twenty-first century, when the progressive convergence of channels and platforms, and the fragmentation of audiences, started to give a new and peculiar shape to the medium. In other words, *Lost* represents a vertical integration between the production of narrative content and its distribution within a media conglomerate (such is Disney-ABC).

Besides, rather than taking advantage of product placement, *Lost* worked in a more unusual way. It tended to erase space/time divergences, extending the narrative elements not only across different media channels, but also into the spaces and experience of everyday life (Johnsons, 2009: 34). This means that *Lost* does not fill its narrative universe with physical merchandise. On the contrary, it is *Lost* itself that becomes the product to be incorporated in real life. All the viral marketing strategies implemented for *Lost* are mainly aimed at bringing the world of

[3] *The Matrix*, 1999; *Matrix Reloaded* and *Matrix Revolutions*, 2003, all directed by Andy and Larry Wachowski.

production and the fictional universe in the daily life of the user, overcoming the boundaries between physical, digital and fictional space, and enhancing the experience of the TV show as a "whole" that is made available to the audience in several different forms.

5 Beyond the Text: The Narrative Ecosystems

We already pointed out that contemporary TV series are often the result of an ecosystemic design which, from an economical point of view, exploits a key concept/narrative in different environments, thereby reaching bigger and broader audiences. These products have an intrinsic modular structure, can be serialized and used in different occasions and contexts, often according to customized schedules. They are also "pervasive" and enable viewers who share a common interest in them to join together in communities. These new forms of narration have no centre and are characterized by an increasing interactivity, which leaves more and more space and autonomy to active users. As we already pointed out, we privilege a change of perspective from the idea of "text" or "narration" to that of "narrative ecosystems", i.e. systems that have the following features:

They are **open systems**, inhabited by stories and characters that change over space and time. Ecosystems are also **interconnected structures**. In terms of relations among different media, they are configured through mechanisms of remediation/relocation[4] (Bolter & Grusin, 2000; Casetti, 2011) as well as, at the level of narrative elements, through the mechanism of convergence, which can take the form of:

- sequel and prequel, for instance *Ashes to Ashes*, 2008–2010, is the sequel to the British TV series *Life on Mars* 2006–2007; or *Bates Motel* (2013), a TV series to be regarded as the prequel to the classic Hitchcock movie *Psycho*;
- reboot,[5] the classic example here is *Battlestar Galactica* (1978–1979), then the series resumed in 1980 as *Galactica 1980* and was finally recast as *Battlestar Galactica* (2004–2009);
- spin-off, i.e. a derivative work developed from an already existing one, in which a character often migrates from one production to another. For instance, *Angel* (1999–2004) was a spin off of the popular teen drama *Buffy the Vampire Slayer* (1997–2003);
- fictional crossover, i.e. the introduction of one or more fictional characters, settings or narrative universes into the context of another story, thereby creating

[4] Remediation is defined as the representation of a medium in another medium, or the use of some typical features of a medium inside another. Relocation, as described by Francesco Casetti, "is the movement that allows the media to work in new environments and new platforms by bringing back to life the experience it was connected with" (Casetti, 2011).

[5] "In serial fiction, to reboot means to discard all established continuity in a series and start over from the beginning" (Wikipedia Contributors, 2013).

a link between two different products. This happens for instance when a character moves between different shows (as in *Friends* and *Mad About You*), which creates a link between the two shows with the purpose of capturing and preserving the audience's attention.

Precisely because of the complex (rhizomatic, unpredictable) nature and behaviour of the contemporary serial narrative, we propose an alternative approach through the principles of information architecture. These principles are independent of the media—and hence transversal (see above par. *Complexity and information architecture*). This makes them a valid tool to explain the behaviour of non-discrete, cross-media phenomena. Here we will focus especially on the principles of correlation, consistency and, overall, resilience.

The ability of a system to suggest relevant connections among pieces of information, as well as among services or goods, and across different media and channels—in order to help users achieve explicit goals or stimulate latent needs, is a significant feature of narrative ecosystems, and is called principle of *correlation* in information architecture.

Ecosystems tend to reach, and keep, a certain balance over time, arranging a coherent, persistent universe that continues outside the confined space of the screen, whilst evolving according to unpredictable lines of development. A system is in equilibrium when it is consistent and resilient. According to information architecture heuristics, *consistency* is the ability of a system to suit the purposes, contexts, and people it is designed for (*internal* consistency), and to maintain the same logic and recognisability across different media, environments/spaces, and times in which it acts (*external* consistency). External consistency is the most important quality for our discussion.

Resilience is the ability of a system to shape and adapt itself to different users, needs, and experiential strategies. A successful, resilient environment implements its information architecture as a dynamic process, in which people are active players and constitute an integral component of the design. In addition, resilience indicates the capability of the system to react to change (overall, radical or unexpected changes) and recover its equilibrium.[6]

Historically, TV series have been built on long/very long narratives, and have shown a high degree of consistency and persistence. They are also resilient in that they can survive perturbation, both external (such as changes in programming slots, a decline in rating, changes in the composition of audiences[7] or exceptional events,

---

[6] In biology and ecology, "resilience" is the ability of an ecosystem to maintain or restore the equilibrium condition during or after a perturbative event. For a deeper exploration of this notion see Zolli and Healy (2013).

[7] These are events resulting answers from the narrative universe; for example, with the insertion of new ad hoc characters, designed to recapture the attention of a lost audience sector. This was the case of the sudden introduction of the character of Dawn in *Buffy the Vampire Slayer*, specifically created to recapture the teenage audience, which was moving away from the series as a result of the aging of its stars.

such as the writers' strike that lasted 100 days between 2007 and 2008), and internal—such as radical changes in the cast, actors leaving the production,[8] or spoiling (Jenkins, 2006a).

Narrative ecosystems are non-procedural systems, which means that they are not determined by a syntagmatic sequence of functions, but by declarative elements that describe environments, characters and relationships, turning the narrative material into a universe explored by the viewer, where experience can be randomly reconfigured. The procedural narrative structure exists only at a local level, in the single episode, which, although complex, is usually directed towards its end and is constrained by its duration. As part of an ecosystem, the weekly episode is instead just one of several potential entry points for the viewer to get caught in the narrative. Access to the serial narrative universe does not have to necessarily happen through the weekly episode aired on TV, but might instead be prompted by other narratives, linked to the TV series and supported by other media such as viral videos, mobisodes or comic books.

Finally, drawing on the model of natural ecosystems, we could say that narrative ecosystems are also formed by an **abiotic** component and a **biotic** component. The abiotic component is the media context, while the biotic component coincides with the narrative structures. The narrative material is a living material, a vital subject that undergoes processes of competition, adaptation, change, and modification. On the contrary, the abiotic component is provided by the mediascape, in which the structures of the series are inserted, and is less easy to be modified, since the mediascape is shaped by the economic, and cultural, structure of a specific country. For instance, screenwriters and showrunners, apparently the creators of a TV series, do not have total control over serial products that evolve according to unpredictable schemes that often cannot be anticipated when the product is designed and put into production. Once again, *Lost* is a good example of a narrative ecosystem that possesses all the features we described above. *Lost* is in fact much more than just "a story". It is an interactive narration that takes advantage of several tools in order to attract more viewers. According to former ABC prime time president Steve McPherson, *Lost* is an all-encompassing experience for the very dedicated fans who want to extend their experience with the show far beyond the weekly viewing.

In order for a narrative ecosystem to function, it has to be sufficiently resilient to tolerate potentially infinite propagations, interpolations and spin-offs. Here lies the dialectical relationship so typical of complex systems, between order and disorder, balance and chaos. And, precisely in the perspective of the convergence-divergence mechanism, such a dialectical relationship plays a crucial role. In other words, it reflects the manifold tensions between project and practice, internal and external processes, constraints and concessions, as we are about to see.

---

[8] The classic example comes from the world of soap operas, where the same character is played by different actors, without this leading to major problems for viewers. Take, for example, the character of Thorne in *The Bold and the Beautiful*, who has been played by three different actors since 1987.

6 Misalignments Between Production and Use

An interesting phenomenon in TV serial narratives is the misalignment between the complex design and organization of the TV series and the actual use that viewers make of these media objects. Although this kind of production is the result of a very accurate design, it is often received by audiences in unpredictable and irregular ways, which go much deeper than the merely entertaining purpose they seem to be produced for.

We have seen how contemporary media production involves an accurate project planning (see paragraph *Narration in a convergent environment*). Now we will see how media objects, especially TV serial narratives, also require practice and use. Production/enjoyment are two polarities that need to be read with the help of new theoretical tools, such as the ones we propose, derived from information architecture. In a convergent mediascape, we can think about convergence as the cultural convergence well described by Jenkins (2006a), as well as the technological convergence generated by the intensive process of digitization which, in the last decades, has completely reshaped the production/enjoyment paradigm. Apart from convergent phenomena, we can focus on the divergent phenomena that are nonetheless relevant for a better description of the complex setting of audio-visual production. Media texts have been regarded as products designed for a specific medium, but today the concept of narrative ecosystems demonstrates that they are in fact produced for active audiences who can access these contents through a range of different devices. Thus, often times it does not matter if the entry point is the weekly episode on TV or an ancillary product. Moreover, since we are no longer dealing with single media texts/products, but narrative universes that take different forms on different cross-media channels, media industry investments are increasingly concentrating on audiences rather than on products, as the growth in marketing expenses for audio-visual media productions demonstrates (Bolin, 2007).

The success and effectiveness of market penetration of these products is not evaluated solely in terms of audience measurement and ratings, but also and especially on the basis of their ability to elicit reactions in the audience, and to prompt and encourage their creative criticism, as shown by the failure of the U.S. series *Firefly* (one season in 2002), created by Joss Whedon. This series was closed very early, even before the entire season was aired. However, this caused the most regular viewers to buy a page of *Variety*, and urge a protest campaign against the UPN network, in an attempt to save their favourite show.

In creating their own version of the narrative universe of the series they are fond of, fans use writing as a tool to express the emotional response that the show generates in them, they comment on what happens in the program, discuss characters and events, and analyse the plot and its narrative twists. Some of the stories written by fans, for example, have the purpose of filling in the gaps left by official screenplays. In addition, these parallel narrations often enrich the characters with new emotional nuances or tend to establish romantic relationships between characters that are only hinted at in the show.

An interesting example of misalignment between the decisions of the show production and the destiny of a serial product is the teen drama *Buffy the Vampire Slayer*, when its producers decided to end the show with the last episode of season 7 (2003). The series, which was quite popular, (especially among academics, who have written several books, essays and papers on this show) had been created by writer and director Joss Whedon based on his 1993 movie of the same name. Aired 1997–2003, the series attracted a very loyal audience and was the object of convergent modes of production, through the creation of a spin off (*Angel*) as well as over sixty novels, five videogames, merchandising and a series of comic books. The Buffy comics were issued in between episodes of the series, but more interestingly they continued to be released, with the help of Whedon himself and of many other TV series screenwriters, after the television series ended. Issues 1–63 followed the storyline while the series was still on air, and none of them were written by Whedon himself. Starting in 2007, a new series of Buffy comics has been produced, as a canonical continuation of the television series, and as such it is regarded as the never shot, never aired *Buffy the Vampire Slayer* Season 8.

The case of *Buffy* is an example of an interesting way of recasting the relationship between product and audience, achieving a delicate balance between reality and fiction. Cult TV (Gwenllian-Jones & Pearson, 2004; Scaglioni, 2006) does not integrate reality in fiction, but instead tends to export the fictional world to the behaviour, habits, and sphere of experience of its audiences.

Another remarkable example that involved a community of viewers/fans, which was established many years ago and spread all over the world is *Star Trek: New Voyages*. Created by James Cawley and Jack Marshall in 2003 and set in the Star Trek universe, distributed exclusively as an Internet download, the series is meant to be a continuation, written, directed and produced by non professional fans, of the original *Star Trek*, starring Captain Kirk and the legendary Mr. Spock. Set in the fourth year of travel of the Enterprise, the series aims at completing the initial draft of Gene Roddenberry, who had planned to shoot five seasons, while only three were in fact produced and aired. Although the rights of the *Star Trek* franchise belong to CBS, no action was taken by the network to block the production and its circulation. *Star Trek: New Voyages* was tolerated to some degree, and also enjoyed the support of Gene Roddenberry's son as consultant, and saw the participation of several actors from the original cast (George Takei and Walter Koenig, respectively Sulu and Chekov).

To sum up, relating to a TV serial narrative means creating not only a recognizable emotional space, in which the product can live for a long time, but also adjusting to the interferences and frictions between the fictional world and reality, thus allowing viewers to use cult shows as a mode of interpretation that promotes knowledge and social life.

7 Divergence and Media Niches

Now that we have seen some examples of misalignment in TV serial narratives, we will try to focus on how this divergence paradigm can lead to the formation of "media niches" which, although remaining external or peripheral to the traditional transmission and delivery of media content, are nevertheless extremely relevant in the process of circulating media items. We are thinking, for instance, of fans who are willing to contribute to the circulation of media products (Appelgren, 2004) through the use and re-appropriation of media content linked to specific gender practices and identities.

Writing fan fiction or creating fan series is a common practice among the viewers of TV serial narratives. Audiences are no longer a formless mass of couch potatoes, but are made up of active viewers, who often provide critical feedback about the things they watch. As we have seen at the end of the previous paragraph, fans feel intimately connected to the narrative ecosystems they are interested in; they feel that they belong to this ecosystem, and that they can provide feedback over the ecosystem's life span. The production of written or audio-visual fiction authored by fans allows participants to play with the characters and situations, thus becoming an active part of the narrative universe.

This process can be positive for the media industry, because it can enhance the audience's awareness about an audio-visual production, as well as contribute to strengthen a community of fans that can guarantee the success of the product for quite a long time. On the other hand, beyond a certain limit this practice of appropriation and divergent use by the audience risks becoming unproductive for the media industry, especially when it is limited to very small communities that foster underground cult practices that don't become popular.

A complementary phenomenon to that of fan activity is called slash fiction, i.e. stories in which the names of the main characters involved are marked with the "/" symbol to unite them, denoting the presence of explicit sexual content in the plot. The fan feedback/production comes in different formats: written productions (short stories, screenplays, poems) that coexist with fan art (graphics, posters, photomontages), songfic (fans create songs on the basis of existing songs, and change the lyrics by inserting references to the world of the show they admire), the phenomenon of cosplaying (a sort of performance in which the participants wear costumes and accessories, embodying a specific character of a fictional universe) and, finally, the production of fan videos, in the form of mash-up (derivative works that combine multiple video sources), remix (alternative versions) and vidding (fan practice of creating music videos from several media sources).

In particular, we can find quite a few subgenres in the stories written by fans, including slash fiction and everything that has to do with the sexual sphere, but also what is called "hurt-comfort", that is, the story of a character imagined to be sick, injured or tortured, who is healed thanks to the care of another character, creating deep intimacy that often leads to unusual romantic plots.

Basically, what happens in all these situations is that a slice of the narrative ecosystem is appropriated by the audience, who uses it in an unpredictable way. More than just trans-media extensions, these parallel productions should be viewed as misappropriation, because they belong to the sphere of subtraction. The accurate planning of plot and characters by the official production clashes against the often-unexpected use that viewers make of them. Although even the official production is the result of negotiation between the demands of production and consumption, there is a certain amount of unpredictability in the reactions/actions of the audiences that can seriously affect the balance of a narrative ecosystem.

Divergence and media niches are the consequence of a complementary evolution of the circulation paradigm in today's mediascape. The traditional broadcasting system is based on a hierarchical relationship between production and audiences: the production exercises control over consumption timing and channels, via the programme schedule or the release window system. The convergence phenomenon deconstructs this hierarchy: the digitalization process and the transition from a one-to-many to a one-to-one relational model brings about a strong decentralization and a transfer of control over consumption timing to the audiences.

The implementation of the release window system generates divergence phenomena and media niches. In the case of free-to-air TV, it is often the production that embraces a convergent model, for instance by making available through web streaming contents that have just been aired. In these cases, possible misalignments between production and use may be managed by treating them as elements that strengthen the cross-media strategy. On the contrary, pay TV adopts a strategy of product exploitation which, by reinforcing control over content access, creates niches of users. But then again, the choice of the media niche can be a market strategy when value is attached to the media product itself and not to the experience of the ecosystem.

8 Conclusions

First of all, we highlighted the limits of the traditional approaches of narratology and semiotics when studying the phenomenon of convergence/divergence in the serial narrative. These methodologies adopt a discrete item as their object of analysis, a text which has clear boundaries, while the serialization phenomenon generates a complex environment, which can be more aptly defined as an ecosystem (in the meaning given to the term by biology and natural sciences). In this perspective, each single element of a series work functions as an interface, an entry point to the whole, complex environment it is included in.

As a consequence of this, it is necessary to hybridize the media study field with other approaches that can more adequately face complex phenomena, and provide a more accurate analysis of environments and ecosystems, instead of just isolating artefacts. One of these approaches might be information architecture. It is not, strictly speaking, a discipline, but a field resulting from the interaction of several branches of research, old and new, such as architecture, librarianship, information science, and human-computer interaction. Its basic principles are largely independent of the information support, and focus on the notions of environment, interaction and experience. Hence, this cross-disciplinary field can be a valuable alternative, which can help overcome the limits of traditional approaches.

So far, we have tried to apply some key notions in the heuristics of information architecture to serial narratives. In particular, we have emphasized the principle of resilience (see par. *The narrative ecosystems*), that is, the ability of the system to adapt to change, and at the same time react in such a way as to recover its balance. Apparently, this heuristic approach turns out to be a valid tool in understanding the bias between convergence and divergence.

When the divergence dynamics described above (see par. *Divergence and media niches*) becomes so extreme as to take away portions of the whole, then the ecosystem breaks down: despite its resilience, it is no longer able to restore the balance, because what takes place is not an alteration of the system (however severe), but a severance in its elements.

Resilience provides us with an essential reading key, not only in understanding the design of convergent systems but also in defining their economic value. Value attribution shifts from the single artefact (a single product) to the whole (the entire ecosystem). This in turn means that it is preferable to sacrifice local details and local precision for a better global experience than vice versa, as *local imprecision* might result in *global precision* (Resmini & Rosati, 2011: 45). "Precision" is a catchall label that describes a plurality of concepts. In our case, it is synonym with "control": tighter local control might result in reduced global stability. In other words, a convergent strategy has to embrace openness, imprecision, and a certain amount of instability—that is to say, resilience[9]—in order to be successful. The experience of this kind of ecosystems is a process, a transit across time and space, so that the more the system allows both users and producers to shape and reshape the ecosystem itself in a potentially infinite series of variations, the more it will be capable to grow and spread.

The top-down, hierarchical models of the broadcasting industry are therefore inadequate to explain the media ecosystems, precisely because they treat each channel and consumption behaviour separately. Since ecosystems are not merely a sum of items (products, channels, distribution times), both the design and the value processes should embrace a global perspective.

To sum up, we can evaluate a convergent strategy, and measure its potential success, using the parameters of pervasive information architecture: resilience (overall), correlation and consistency (see par. *The narrative ecosystems*). In this

[9] Rosen (1999) offers an interesting, relational view on complexity. A system is "simple if all its models are simulable. A system that is not simple, and that accordingly must have a nonsimulable model, is complex". In other words, Rosen links his notion of complexity to modeling: a simple system can be simulated and predicted; a complex system cannot. "When a single dynamical description is capable of successfully modeling a system, then the behaviors of that system will, by definition, always be correctly predicted. Hence, such a system will not have any 'complexity' in the sense above, in that there will exist no unexpected or unanticipated behavior".

perspective, the concept of niche may be expanded, to include any process of resilience reduction operated by users through appropriation processes, or by the production itself, in order to over-control some items in the ecosystem.

In conclusion, phenomena that are very similar to serial narrative and media convergence are now taking place in many other fields, such as industrial design, services design and marketing—where such labels as cross-channel design or cross-channel experience are becoming increasingly popular.[10] It is therefore desirable to promote a joint study of these phenomena, carried out at the intersection of these different fields of analysis: we need meta-models capable of explaining a convergence process that (albeit with different nuances) has a transversal influence on many of our everyday experiences. This is another point of view on the convergence/divergence issue: while media and processes converge, disciplines are still too loosely interlaced to comprehend the more tightly interconnected cases.

References

Ambient Media Association (AMEA). Accessed February 16, 2013, from http://www.ambientmediaassociation.org

Appelgren, E. (2004). Convergence and divergence in media: Different perspectives. In: J. Engelen, S. Costa, & A. C. Moreira (Eds.), *Building Digital Bridges: Linking Cultures, Commerce and Science. Proceedings of the 8th ICCC/IFIP International Conference on Electronic Publishing*, Brasília 23–26 June.

Bates, M. (2002). Toward an integrated model of information seeking and searching. Keynote Address, *Fourth International Conference on Information Needs, Seeking and Use in Different Contexts*. Lisbon. Accessed February 10, 2013, from http://www.gseis.ucla.edu/faculty/bates/articles/info_SeekSearch-i-030329.html

Bolin, G. (2007). Media technologies, transmedia storytelling and commodification. In T. Storsul & D. Stuedahl (Eds.), *Ambivalence towards convergence. Digitalization and media change*. Göteborgs: Nordicom.

Bolter, J. D., & Grusin, R. (2000). *Remediation. Understanding new media*. Cambridge, MA: The MIT Press.

[10] As Donald Norman (2009) claims "[a] product is actually a service. Although the designer, manufacturer, distributor, and seller may think it is a product, to the buyer, it offers a valuable service. [. . .] Amazon's Kindle is [an] example of superb systems thinking. This is Amazon.com's ePaper-based book reader. Now, there are competing products on the market which offer superior features. Amazon wins, however, because of its systems thinking. No computer is necessary for most transactions. When the Kindle arrives, it is preloaded with the books that were ordered. Moreover, it can work instantly. Even more important, Amazon thought through the entire system, from discovering a book to loading it onto the Kindle. [. . .] No product is an island. A product is more than the product. It is a cohesive, integrated set of experiences. Think through all of the stages of a product or service—from initial intentions through final reflections, from first usage to help, service, and maintenance. Make them all work together seamlessly. That's systems thinking". In the information science field, both Marcia Bates (2002) and Gary Marchionini (2012) advocate the need for a shift of perspective, from pure disciplines to connective ones, what they call meta- or cross-boundary disciplines. They have the ability to grow up through synthesis and coordination, and to work across our physical and digital lives.

Casetti, F. (2011). Cinema lost and found: Trajectories of relocation. *Screening the Past* 32 (Screen Attachments: New practices of viewing moving images). Accessed February 9, 2013, from http://www.screeningthepast.com/2011/11/cinema-lost-and-found-trajectories-of-relocation/
Curtin, M., & Shattuc, J. (2009). *The American television industry*. London: British Film Institute.
Giovanazzi, L., & Mascaro, L. (2009). Architettura informativa ed ecosistemi (Information architecture and ecosystems). Speech at *3rd Italian Information Architecture Summit* (Forlì, Italy). Accessed February 9, 2013, from http://www.sketchin.ch/it/blog/conferenze-workshop/iasummit-ecosistemi-e-adattabilita.html
Grasso, A., & Scaglioni, M. (2009). *Arredo di serie I mondi possibili della serialità televisiva Americana*. Milano: V&P.
Gwenllian-Jones, S., & Pearson, R. E. (2004). *Cult television*. Minneapolis, MN: University of Minnesota Press.
Hammond, M., & Mazdon, L. (2005). *The contemporary television series*. Edinburgh: Edinburgh University Press.
Hesmondhalgh, D. (2007). *The cultural industries*. London: Sage.
Innocenti, V. (2007). Buffy the Vampire Slayer e la serializzazione della serie. In B. Maio (Ed.), *Buffy the Vampire Slayer. Legittimare la Cacciatrice*. Roma: Bulzoni.
Innocenti, V., & Pescatore, G. (2008). *Le nuove forme della serialità televisiva. Storia, linguaggio, temi* (The New Shapes of TV Series. History, Language, Topics). Bologna: Archetipolibri.
Innocenti V., & Pescatore G. (2012). Information architecture in contemporary television series. *Journal of Information Architecture, 4*(1–2). Accessed September 2, 2013, from http://journalofia.org/volume4/issue2/05-pescatore/
Jancovic, M., & Lyons, J. (Eds.). (2013). *Quality popular television*. London: British Film Institute.
Jenkins, H. (2003). *Transmedia storytelling*. Technology Review. January 15. Accessed February 10, 2013, from http://www.technologyreview.com/news/401760/transmedia-storytelling/
Jenkins, H. (2006a). *Convergence culture. Where old and new media collide*. New York: New York University Press.
Jenkins, H. (2006b). *Fans, bloggers, and gamers. Exploring participatory culture*. NewYork: New York University Press.
Jenkins, H. (2013). *Textual poachers: Television fans and participatory culture*. Updated Twentieth Anniversary Edition. New York: Routledge.
Jenkins, H., Ford, S., & Green, J. (2013). *Spreadable media: Creating value and meaning in a networked culture*. New York: New York University Press.
Johnsons, D. (2009). The fictional institution of lost: World building, reality and the economic possibilities of narrative divergence. In R. Pearson (Ed.), *Reading lost*. New York: Tauris.
Leverette, M., Lott, B., & Buckley, C. (Eds.). (2008). *It's not TV. Watching HBO in post-television era*. New York: Routledge.
Lugmayr, A., Stockleben, B., Risse, T., & Kaario, J. (2012). Editorial: Re-thinking the future of semantic ambient media. *Multimedia Tools and Applications, 58*, 289–292. doi:10.1007/s11042-011-0899-6.
Lugmayr, A., Zou, Y., Stockleben, B., Lindfors, K., & Melakoski, C. (2012). Categorization of ambient media projects on their business models, innovativeness, and characteristics—Evaluation of Nokia Ubimedia MindTrek award projects of 2010. *Multimedia Tools and Applications, 66*(1), 33–57. doi:10.1007/s11042-012-1143-8.
Marchionini, G. (2012). Bridges, linchpins and membranes: From I to we. *Bulletin of the American Society for Information Science and Technology*. December/January. Accessed February 10, 2013, from http://www.asis.org/Bulletin/Dec-11/DecJan12_Marchionini.html
Mattessi, P. (2003). Buffy and the serialization of the series. *Metro Magazine, 137*, 76–77.
McCabe, J., & Akass, K. (Eds.). (2011). *Quality Tv. Contemporary American television and beyond*. London: I. B. Tauris.
Mittell, J. (2009). *Television and American culture*. NewYork: Oxford University Press.

Mittell, J. (2012). *Complex TV: The Poetics of Contemporary Television Storytelling*. Pre-publication edition. MediaCommons Press.

Montola, M., Stenros, J., & Waern, A. (2009). *Pervasive games: Theory and design*. San Francisco: Morgan Kaufmann.

Newman, M. Z. (2006). From beats to arcs: Toward a poetics of television narrative. *The Velvet Light Trap, 58*, 16–28.

Norman, D. (2009). Systems thinking: A product is more than the product. *Interactions, 16*(5). Accessed February 4, 2013, from http://www.jnd.org/dn.mss/systems_thinking_a_.html

Pearson, R. (Ed.). (2009). *Reading lost: perspectives on a hit television show*. London: I. B. Tauris.

Resmini, A., & Rosati, L. (2009). Information Architecture for Ubiquitous Ecologies. *MEDES'09 The International Conference on Management of Emergent Digital EcoSystems*, Lyon, October 27–30. Accessed February 4, 2013, fromhttp://andrearesmini.com/blog/ia-for-ubiquitous-ecologies

Resmini, A., & Rosati, L. (2011). *Pervasive information architecture. Designing cross-channel user experiences*. Burlington: Morgan Kaufman.

Retail Touchpoints. (2011). *Shopper preferences shift as customers expect improvements in retailers' cross-channel capabilities*. Accessed February 4, 2013, fromhttp://www.retailtouchpoints.com/resource-center/51-white-papers/1028-shopper-preferences-shift-as-customer-expect-improvements-in-retailers-cross-channel-capabilities

Rosen, R. (1999). *Essays on life itself*. New York: Columbia University Press.

Rosenfeld, L., & Morville, P. (1998). *Information architecture for the world wide web*. Brooklyn, NY: O'Reilly Media.

Rosenfeld, L., & Morville, P. (2006). *Information architecture for the world wide web* (3rd ed.). Brooklyn, NY: O'Reilly Media.

Scaglioni, M. (2006). *TV di culto. La serialità televisiva americana e il suo fandom* (Cult TV. The American Television Series and its Fandom). Milano: V&P

Sepinwall, A. (2013). *The revolution was televised: The cops, crooks, slingers, and slayers who changed TV drama forever*. New York: Touchstone.

Spiegel, L. (2004). Introduction. In L. Spiegel & J. Olsson (Eds.), *Television after TV essays on a medium in transition*. Durham, NC: Dukes University Press.

Sterling, B. (2005). *Shaping things*. Cambridge, MA: The MIT Press.

Thompson, R. J. (1997). *Television's second golden age. From hill street blues to ER*. Syracuse, NY: Syracuse University Press.

Thompson, K. (2003). *Storytelling in film and television*. Cambridge, MA: Harvard University Press.

Thompson, R. J. (2007). Preface. In J. McCabe & K. Akass (Eds.), *Quality TV. Contemporary American television and beyond*. London and New York: I. B. Tauris.

Tungate, M. (2004). *Media monoliths. How great media brands thrive and survive*. London: Kogan Page.

Wikipedia Contributors. (2013). Reboot (fiction). Wikipedia (English). Accessed February 9, 2013, from http://en.wikipedia.org/wiki/Reboot_(fiction

Zolli, A., & Healy, A. M. (2013). *Resilience. The science of why things bounce back*. New York: Random House.

Micro-studios Meet Convergence Culture: Crossmedia, Clustering, Dialogues, Auto-communication

Indrek Ibrus

1 Introduction: Convergences, Divergences and Their Co-evolution

Convergence is a concept with many implications, as has been well demonstrated by Fagerjord and Storsul (2007). They suggested that the term should be understood, therefore, as a useful rhetorical device that denotes the complexities of the modern media evolution. However, in order to start interpreting these ongoing dynamic and evolutionary processes, it is perhaps justifiable to start out by briefly reviewing some of these underlying 'dimensions' of modern media convergence. Fagerjord and Storsul distinguished between 'network convergence', 'terminal convergence', 'service convergence', 'rhetorical convergence', 'market convergence', and 'regulatory convergence'. Also 'industry convergence' is often referred to in the dedicated academic discourse. All these dimensions of convergence processes could be understood as being autonomous, but only to an extent. They are also interdependent, mutually conditioning and co-evolving, however, in complex and non-linear ways. In order to interpret the theme of this book, of how convergence is related to the processes of divergence or emergence in media evolution, let me sketch some of these interdependencies here.

Network convergence normally refers to the phenomenon that once contents are digitised any network could effectively transmit any sort of information and content, which in turn makes it easy to harmonise the technical communication protocols of these platforms, eventually enabling the integration of these networks and their

The preparation of the article was supported by the European Union through its European Regional Development Fund (Centre of Excellence in Cultural Theory) as well as by research grant ERMOS79 financed by the Estonian Science Foundation and co-funded by Marie Curie Actions.

I. Ibrus, Ph.D. (✉)
Tallinn University, Tallinn, Estonia
e-mail: Indrek.ibrus@gmail.com

A. Lugmayr, C. Dal Zotto (eds.), *Media Convergence Handbook - Vol. 2*, Media Business and Innovation, DOI 10.1007/978-3-642-54487-3_9

seamless co-operation. The Web was once a 'desktop' platform but now, after decades of work on fixed-mobile network convergence, is effectively a ubiquitous platform, able to take many shapes and forms on a variety of access platforms. This means that fixed and mobile networks have converged, facilitating in turn, for instance, the process of *terminal convergence*. This of course refers to the fact that many of these 'access devices' are in parallel, also increasingly multifunctional, integrating many of the functionalities formerly associated with autonomous dedicated devices, such as: photo cameras, video recorders, type writers, phones, calculators, walkmans, books, etc. The fact that modern tablet computers or smartphones can carry out all these functionalities is no news, but such a convergence process is not uni-directional. It is remarkable how the many new applications, originally developed for mobile devices, are later also deployed on 'previous' platforms, such as: 'app stores', touch screens or specific web browser applications. So, the 'old terminals' are convergent too. However, the differences between these terminals, in their forms, shapes and affordances, have remained largely intact, referring to the fact that the differences in 'use cases' also continue motivating 'terminal divergence'.

However, the fact that functionalities converge or that the same functionalities can be found in a variety of media access devices, refers to the fact that the related *service provisions are converging* too. Similarly to the Web, the formerly device-specific services such as telephony, television, radio are now device-agnostic and ubiquitous. What is more, these different services themselves are also converging, in the wake of 'social TV', for instance, we experience the emergence of a convergent format. Relatedly, as the services converge, the providers of these integrated services also converge and grow in size. That is, we recognise a movement towards increasing *media concentration* globally, whereas it is, increasingly, Facebook, Apple, Google and the like that provide, or at least mediate for the global population, the majority of their media content. This phenomenon is in turn immediately related to *market convergence*, as the services are integrated, so are also the markets for these services. For instance, there is increasingly less of exchange relationships relating exclusively to either mobile or desktop webs (except the 'advertising networks' that are mostly under the helm of the same global players—i.e., Google's AdMob and AdWords). Also, the phenomenon known as *industry convergence* is closely related to these developments. This refers to the already described tendency towards media concentration or the formation of increasingly large crossmedia conglomerations, or simply to increasing cooperation between institutions of formerly rather distant sub-sectors. This means that the subsector-specific value chains are getting more complex and tend to cross the conventional industry borders. As a result, it is increasingly difficult for outside observers to distinguish between the formerly separate industries, for instance, between the telecoms, online communication services and television as in the case of 'social TV'.

Last but not least, *rhetorical convergence* results from the technical capability of combining the existing media forms of various modalities. The creative repurposing of representational conventions from a variety of media and genres and their creative recombining into innovative *convergent genres* results, when these gain popularity, in effect in the emergence of new genres and forms; that is, in the *divergence* of media forms. This recognition indicates how, in the case of media forms, the processes of convergence, divergence and emergence form a somewhat paradoxical trinity that is, however, central to the modern evolutionary dynamics of media. For instance, when the genre of action films and the genre of first-person shooter games convergence the outcome may be a divergent new genre (be it 'interactive action film' or 'filmic videogame'), in effect an emergent new media form.

I will return to this discussion in few pages, but here I would first like to point out once again the interdependencies between these various 'dimensions' of media convergence processes. As networks are made to technically converge, institutions also react by starting to provide new combined services, which in turn gradually results in increasingly integrated exchange relationships and value chains—i.e. market convergence eventually affects the formation of new sorts of institutions, either in the form of crossmedia conglomerations or industry meta-bodies that start 'covering' and integrating the previously autonomous industries or sub-sectors. This, either bottom-up or top-down integration normally results in various standardisation protocols that in turn could be seen to contribute to the 'convergence in practice' and also in meta-discourses that codify both the practices as well as the many codes of conduct which define the industry. It can be suggested that once such codification takes place, it could motivate further investments into network interfacing standards or into the development of converged service provisions, etc. However, what this article will also discuss and propose is that convergence cannot be ultimate or absolute. There may be resistances or refusals to converge by some of the sub-systems and there could be divergences conditioned by the need for difference experienced both by industries, as well as by consumers. The paradoxical parallelity in movements towards both, convergence and divergence, will be the focus of this chapter.

2 Literature Review on 'Convergence Paradox': Conditioned by 'Dialogues' and 'Auto-communication'

Codification that takes place on a variety of 'levels' takes us to the concept of auto-communication, originally suggested by Russian-Estonian semiotician Juri Lotman (1990). The term denotes communication from and to oneself, where the self-communicating entity can be both an individual and a larger social structure. As several authors have elaborated (Broms & Gahmberg, 1983; Christensen, 1997; Morsing, 2006; Steedman, 2006) in the modern-day context all kinds of communications (such as strategic plans, corporate reports, marketing communications, press releases) that organised bodies or systemic structures might produce could eventually start working auto-communicatively. Even if the communicative act was originally not meant for internal use but for the outside audience, once the message feeds back to its authoring structure the auto-communicative effect has taken place. Hence, as it has been argued (Christensen, 1997; Morsing, 2006), the auto-communication is not primarily oriented toward sending and receiveing messages, but towards the production and celebration of metatexts on the identity and nature of the communicating structure. It could be suggested that such 'celebrations' may often, especially if the communicating system is convergent (i.e., it is nascent, not yet established) contribute towards 'confirming' the system towards its self-creation by standardisation and institutionalisation. I have suggested elsewhere (Ibrus, 2010, 2013b) that also in the technoculture the above described various 'dimensions' of convergence processes could be understood as contributing to the auto-communicative process that results in the emergence of a qualitatively new system—the one that is then converged. That is, in addition to various layers of metatexts that codify the distinctions of the new media system, also the 'system' itself—the harmonised networks, their communication protocols, their integrated services, the new universalised business models, shared exchange standards and codes of conduct for the industry, these all participate in generating self-oriented aggregations of communicative acts and in this way in constituting and fixing the new system—as for instance, the 'mobile web' (see Ibrus, 2010, 2013a, 2013b).

However, what about divergence? How is it conditioned? For interpreting this from the perspective of evolutionary approaches to media change, let me briefly look at Lotman's 'semiotics of culture'. Lotman stressed that 'well-defined and functionally unambiguous systems' never exist in isolation. Instead, they acquire their role and meaning when perceived as one segment of the continuum of multifaceted, multi-leveled and variegated semiotic formations; that is, when being immersed in semiotic space (Lotman, 1990). To interpret the dynamics within this space, Lotman proposed his concept of 'semiosphere'. Originally coined by Lotman in an analogy with Vladmir Vernadsky's 'biosphere' and 'noosphere' and Bakhtin's logosphere (Mandelker, 1994), it refers to an abstract 'semiotic continuum' that is inherently heterogeneous and enclosed in itself, that functions as a self-referential system (Nöth, 2006), but is also in constant interaction with other similar structures (Kotov, 2002).

The analytic function of this concept is to facilitate the analysis of the complex relationships between a culture, its different sub-components and its semiotic environment. Relying on the organicist philosophical strategy, it presumes and analyses isomorphic relations between all the structures and levels of a semiosphere (Alexandrov, 2000; Mandelker, 1994). For instance, if we relate a single website as an inherently heterogeneous but bounded textual entity to the whole 'web-culture' as another textual entirety, and then to the whole global 'sphere' of human culture, their differences should be understood only as quantitative—one 'level' cannot exist without the other and one cannot be interpreted without knowledge of the other. Still, as Nöth (2006) explained, Lotman's hierarchy of levels makes up a system of relational stratifications in a way that higher levels are always conceived as semiotic spaces with more dimensions in relation to the spaces of the lower levels that they embrace. But the dichotomy of super-systems/sub-systems is not linear or simple. Instead, systems of different levels intertwine and relate to one another in complex ways. Super-systems embrace numerous smaller ones. For instance, a carefully designed crossmedia strategy as a textual system could consist of a variety of media types and genres [print media, TV, Internet, mobile apps, videogames, alternate reality games (ARGs), etc.]. At the same time smaller systems could also be perceived as parts of several bigger ones—a website could be seen as part of an internationally recognised webmedia genre, or again as a component of a carefully orchestrated transmedia narrative, or be seen as an entity of a distinct national media system. Therefore, the evolution of various groupings of websites as distinct textual domains is also conditioned by the semi-autonomous operations of all these super-systems. What is more, as such the 'websites', as well as most individual media platforms, constitute dialogic spaces where the various super-systems meet, partly converge and, therefore, are conditioned to co-evolve.

All in all, Lotman argues (1990) that the entire space of a semiosphere is transected by boundaries between different levels, sub-sphericules and texts. These boundaries, as they separate, although they also connect and translate (Lotman, 1990), are inseparable from the term 'individuality' (Torop, 1999). Individuality is seen as the outcome of the autopoietic process where a cultural system, for instance 'television', identifies itself and its boundaries in space and/or time. It is the self-defined continuum inside the self-generated boundaries that thereafter become the mechanisms of translation—as identifying oneself presumes the realization that between 'own domain' and 'alien domain' differences exist and that the alien domain ('Theydom', as Hartley explicates it—see Hartley, 1996) then needs to be understood and translated. It is the coexistence of the infinite number of such sub-systems of culture, with their asymmetrically different languages, discourses and identities that, despite their differences, forces them into dialogues.

The suggestion here is that the ability to observe the 'Theydom' and potentially to understand and talk to 'Others' also brings a potential for a dialogue and, subsequently, the emergence of new systems. Luhmann explained that 'every social contact is understood as a system, up to and including, society as the inclusion of all possible contacts' (Luhmann, 1995). Therefore, when contact between two participants is established, there takes place a dialogue, then there is a possibility for autopoietic closure and an emergent social system.

Regarding the emergence of various new media out of the dialogic contacts among the variety of societal sub-systems, of institutions and organisations, I start with Krippendorff (1995) who has shown how the vocabulary of the modern design discourse stems from several sources—the arts, engineering, ergonomics, advertising, popular culture, software manufacturing, and so on. Similarly, any new medium could be understood as a convergent domain that has taken shape in dialogues among a variety of social systems and institutions. Related to this, at the stage when a medium is still young, there are not many new and medium-specific norms for the texts as well as for their production practices, since much is simply inherited from 'parent domains'. However, the new 'emancipated' practices and medium-specific conventions are expected to evolve as part of the new system's auto-communicative process.

In this context I propose that the meta-discourse that tries to impose norms for a design from a perspective of a certain social system is also, for the most part, the same discourse that auto-communicatively articulates the identity of the same social system. Krippendorff (1995) has established three ways of how design discourse should be instituting its recurrent practices:

(i) enabling social organizations to thrive on controlling the technical means of (re)producing and disseminating the discourse—not only its textual matter and its community, but, most importantly, its very own organizational forms (social autopoiesis),
(ii) legitimizing its procedures, methods, theories, schools of thought, and criteria through the very acts of making them selectively available, especially to members of its discourse community who may turn the benefits of participation into loyalties to particular organizations operating within that discourse, and by
(iii) applying its axioms relative to which a discourse (its textual matter, conversations, and organizations) can achieve a certain autonomy, coherence, and direction.

Krippendorff also argues that discourse 'surfaces in textual matter' which is continuously (re)read, (re)written, (re)produced, (re)searched, (re)articulated, elaborated or rejected. 'A community continually (re)generates its textual matter and acquires the character of a dynamically connected diversity' (Krippendorff, 1995, 2008). I propose that this principle of 'dynamically connected diversity' is beneficial for interpreting cultural evolution as it helps us to understand the dynamic processes of emergence at the borderlines of the existing social structures, discourses and languages. Both the emergent, as well as older, systems of media production/consumption are inherently heterogeneous in terms of their textual or discursive constellations as they are also in constant dialogues (in multitudes) with the outside (since in different ways they are parts of a variety of super-systems) and hence the auto-communicative function only reacts to all this diverse dynamic by making suitable connections and rearticulating the functions and identity of the own-domain. For instance, the cultural domain of film production is extremely rich in terms of its inherent subsystems (different professions, genres, approaches, national cinematic cultures, etc.) and all these separately as well as in aggregate are in multiple dialogues with their environment—other arts, technology providers of various kinds, distributors, policy makers, audiences, etc. This means much of complexity in terms of the reoccurring need to auto-communicatively re-articulate what the 'film art' may actually constitute in the context of changing technological, social and cultural contexts.

Another suggestion that can be made based on Lotman's semiotics of culture is that the evolutionary dynamics of culture is paradoxical for being bidirectional—it evidences both centrifugal and centripetal forces, which play themselves out on various, coexisting layers (Schönle & Shine, 2006). That is, it is the auto-communicative functioning of larger systems that is having a centripetal effect—i.e., homogenizing the cultural space, creating unities between its smaller sub-systems. And it is the dialogues between these smaller systems and their parallel acts of auto-communication that is, in turn, conditioning the emergence of difference within the larger space of culture, facilitating plurality and dynamics, constituting, in effect, the 'centrifugal' force of cultural change. It is important to understand that both of these forces are mutually conditioned, the need for creating unities is effected by growing entropy, while the act to create difference is motivated by the need for meaningful experiences, for creating new and socially relevant experiences and meanings in the too universalized culture. It is a significant paradox of media evolution, that the convergence of various media conventions in new media applications does not only mean apocalyptic one-way flows into semantic implosions as Baudrillard (1983) suggested, but also the evolution of new languages, new borders, new differentiations and functionalities, new discontinuities within new continuities.

3 Methods and Studies

The broad analysis in the rest of this chapter will rely on a series of studies conducted by myself and my colleagues during the course of 2011–2013. The first of the studies investigated small crossmedia projects funded by the First Motion consortium in the Baltic Sea region. The consortium consisted of, among others, various national or regional film funds, film and media schools and regional development authorities from seven countries[1] around the Baltic Sea. The rationale of First Motion came from the fact that the audiovisual industry in the Baltic Sea Region (as elsewhere) consists mainly of small and medium sized enterprises (SMEs) with limited capacities for innovation by means of their own resources. At the same time, as discussed above, multiplatform publishing has emerged and offers opportunities for innovation. Hence, First Motion was initiated to investigate and work on the issue from different angles: researching new markets, clusters and value chains; funding experimental new projects; developing distribution platforms and higher education programmes; making policy suggestions. This chapter will rely on some of these collectively produced studies on the AV-industry, clustering in various cities and regions around the Baltic Sea (see Filmby Alexandra, & High Tech Accelerator Innovation Center University of Lodz Foundation, 2011). However, one of the main activities of First Motion was to fund experimental crossmedia productions by SMEs from the participating countries. In the course of 2 years (2010–2011) 16 projects were funded. Of all the 16 projects, I interviewed people involved with six of them. The interviews were paralleled with close textual analyses of their productions together with occasional documentary analysis, if additional documentation or background material about these projects was available. The interviews with projects' representatives (occasionally individuals, some group interviews, in several cases interviews repeated over time) were transcribed. Transcriptions were analysed by means of robust discourse analysis, a form of grounded theory development that focused on the above-described conceptual phenomena—constituting an experimental form of 'semiospheric discourse analysis'.

[1] Denmark, Estonia, Germany, Latvia, Norway, Poland, Sweden.

The second study this chapter discusses was conducted in 2012 and focused on the clustering of the AV-industry in Tallinn, Estonia and specifically investigated, among other things, what are the potentials of the AV-industry to start cooperating and eventually to converge with ICT-industries in order to innovate, to develop new business models and to start outputting content on a variety of digital platforms with innovative affordances? In terms of the empirical study (see Tafel-Viia, Lassur, Ibrus, Tafel, & Terk, 2012) we first interviewed the representatives of 19 enterprises of the local film industry, including representatives from studios, post-production companies and distributors. The interviews were made with either the owners of the companies or with their senior executives. Some of the interviews had only one respondent whereas others had more than one, for instance, two owners of the company. All of the approached companies were generally regarded as rather central to the operations of the local film industry. The interviews were transcribed and analysed using a grounded theory method. The second phase of the same study was a confirmative focus group with the select group that included, not only the previously interviewed filmmakers, but also a few additional industry insiders. The function of the focus group was to reflect further on the proposed findings, to extend and confirm them.

4 Crossmedia Emergence and Effects

In the conceptual context described above the emergence of the crossmedia or transmedia phenomena is not insignificant. Because they mark both of the forces of evolutionary dynamics of culture described by Lotman, the centrifugal as well as the centripetal development. *Crossmedia* is normally understood as intellectual property, service, story or experience that is distributed across multiple media platforms using a variety of media forms. Such distribution results often from strategic endeavours by media companies. In effect, crossmedia strategies result from what are known as 'diversification strategies' in media economics (Chan-Olmsted & Chang, 2003, 2006); that is, attempts by rights' holders to creatively adapt their property for a variety of media platforms, in order to either obtain a higher margin from that property, or strengthen it via cross-promotion among platforms. In the era of networked media and user participation, such crossmedia strategies have, of course, evolved and become ever more complex. In addition, the emergence of crossmedia is conditioned not by strategies by media companies, but by media users that adapt, modify and further develop content and publish and disseminate their work it various ways, especially by posting on the Internet. Returning to the second, closely related concept, *transmedia storytelling* is a technique of telling a single story across multiple platforms and formats, including modern interactive technologies that, in turn, enable user participation and contributions to the story (Jenkins, 2006).

Both of these phenomena facilitate connection making and unities (convergence) in the culture as well as disconnections and pluralities—i.e. autonomy for culture's many subsystems (divergence). A transmedia narrative, as for instance in the case of the notorious science fiction franchise Fringe, consists of the TV-series (the narrative 'mothership'), series of fictional and meta-textual websites, three collections of mini-comics, a five-part series of video-comics, an alternate reality game (ARG), records and merchandise of a fictional band, and a fictional character that 'appears' occasionally on other U.S. television shows or real world entertainment events (see Belsunces Gonçalves, 2012). As already discussed above, a narrative has the potential to integrate these different media outlets and platforms and make them co-evolutionary to an extent. The more there are such narratives or crossmedia strategies nurtured by media industries or their (active) audiences/users, the more these media platforms are made to be interdependent in their operations and development, resulting in similarities and continuities between them. Taken that such multiplatform strategies are increasingly prevalent for modern media practice, it implies that certain evolution of 'unities' may be characteristic to convergence culture. On the other hand, the individual media that may be included into various transmedia narratives could be presumed, following Lotman's theorisations, to continue their autonomous evolution, to be 'path dependent' (David, 2000; Garrouste & Ioannides, 2000; Ibrus, 2008) on the textual history and affordances of the particular medium, platform, modality or a genre. Not to mention that the institutional structures involved with the production of the particular medium or genre, could be presumed to be path-dependent as well—i.e. operating auto-communicatively, in that reinforcing the medium-specific practices of production and hence also the institutional and textual divide between the various media, platforms, genres, etc. As such all crossmedia or transmedia phenomena consist effectively of distinctions—of either slight modifications of the same text for different media or of compilations of principally different texts that are still intertextually connected and as such are in complementary, but dialogical relationship. These distinctions, either slight or principal, effect difference, that is, new meaning, into the cultural space. These distinctions are mostly motivated by specific affordances of media platforms or by specific characteristics of their dominant user groups (together with their 'needs for meanings' and interpretative limits). The resulting dynamics effects plurality—that is, plurality in culture is one of the essential effects of crossmedia strategies. Therefore, paradoxically not only is homogenisation among the effects of 'crossmedia emergence' in contemporary media culture, but there is also the emergence of difference and pluralisation. In other words, following Lotman's original argument, the modern 'convergence culture' is a new intensified mode of culture's rather universal paradox of evolution—it striving in parallel towards multiplication of forms and meanings (divergence) as well as towards their homogenisation (convergence).

5 Autonomy, Attention Economy and Political Economy of Media Clusters

The balancing between convergence and divergence takes us to the question of autonomy of different media; to what extent can it be sustained, where are the limits and what are the degrees of freedom? Our study of Estonia's film industry's clustering tendencies (Tafel-Viia et al., 2012) included a scenario of a new inter-industry cluster combining film production with the ICT industry. Although some fractions of the film industry were fascinated about the scenario there were also others that were not and were concerned about the possible loss of identity and 'quality'. Now, realising that the expected outcomes of the inter-industry cluster scenario were also to be various innovative crossmedia solutions and strategies, we need to recognise how crossmedia could be interpreted as threatening the autonomy of a media domain; film medium in the particular case. If crossmedia is the prevailing trend in the television-content industries, as has been demonstrated by many (Bennet, Strange, Kerr, & Medrado, 2012; Doyle, 2010; Erdal, 2009; Evans, 2011; Perryman, 2008; Suárez Candel, 2012)—to the extent that documentary films are rarely commissioned without crossmedia extensions—this is often experienced as de-powering by documentary filmmakers. They want to concentrate on their own art—documentary film—and not, for example, to develop in parallel clever social media marketing strategies. We need to recognise that from the perspective of filmmakers, crossmedia is often perceived as a challenge that is not well received; it is seen to undermine an established and distinctive art form and a profession and can be interpreted, therefore, as a social, if not a political issue.

Related to this, we should also interpret crossmedia as a concern in terms of its political economy. The question is, to what extent is the crossmedia consolidation and associated control by the dominant industry fractions problematic as it may undermine the public value of media systems? It is important to recall that 'crossmedia' is often also used to refer to concentrated and oligopolistic media markets where the connections between media outlets and platforms may be strictly controlled by only a few large enterprises (e.g. Congdon, Graham, Green, & Robinson, 1995). Such concentration of control may be suspected, however, to contribute towards effecting homogeneity across media, especially towards standardising the ways in which content is translated between platforms and adapted for the various perceived affordances, use cases, etc. As the Internet as a marketplace is effectively global, also the corporate authorities that may dominate the codifying practices of such standardisation are now increasingly less local. Globally dominant players such as Google, Apple or Facebook develop specific crossmedia package solutions that combine, for instance, TV or film viewing experiences with social networking, online shopping or targeted social advertising. These globally universal solutions may emerge as de facto standards for crossmedia content presentation. In this way, the evolutionary dynamics of crossmedia forms may be historically distinct for it being a manifestly global phenomenon, facilitated by the few globally dominant industry players.

But the challenge of oligopolistic markets occurs not only with global providers of communications infrastructures and media platforms (such as Google, Apple or Facebook), but also regarding content provision. Terranova (2012) has accused the 'attention economy' theory (for instance Goldhaber, 1997, 2006) for re-economizing the potentially productive 'attention labour' and presenting the related problematics solely from the perspective of seller/provider of corporate commodities and problematising the 'non-perfect' behaviours of customers. Still, when questioning if the content provision is rich enough in specific markets, we should also ask if the structuring of the market is such that it can secure the diverse offering, i.e., if there is enough content producers of various sizes with adequate access to market. Or again in other words, if the majority of creative industries companies are in most countries rather small, then what are their opportunities to access and participate in the marketplace? In the Schumpeterian 'market first' tradition of evolutionary economics, it is perceived, of course, that the core source of innovations is the 'creative entrepreneur' and his/her institutional frame, the producing firm. As has also often been demonstrated within innovation studies (Tether, 1998; Tether, Smith, & Thwaites, 1997), the stronger and bigger the firm, the more it has resources to systematically invest and innovate, resulting in notably more innovations being generated by sizable companies. Still, as it has also been demonstrated, it is the relationships between the large enterprises and clusters "around them" that generally consist of multitudes of independent micro-companies of different kinds (Britton, 2007) that could be understood to contribute to the evolution of an healthy innovation system. As suggested by Potts and Keane (2011), in the case of creative industries such clusters have a potential to constitute the groundwork for a diverse and dynamic cultural milieu that provides the broader innovation system with much needed circulation of alternate ideas and reflections. As demonstrated by Bennet et al. (2012), in the UK, as an instance, there has evolved a 'compact' between the large Public Service Broadcasters (PSBs) and the independent sector that is built on the balancing of economic rewards with a cultural commitment to the purposes, characteristics and production modes of the PSBs found across the independent sector. This compact has made the UK's content unique and exportable. As Bennet and his colleagues emphasise, this is especially true for crossmedia content; PSB investment in multiplatform commissioning has helped to make the UK a world-leader in interactive, multiplatform content. What is more, multiplatform production has created unique forms of public service value that harness the power of audience participation and empower audiences to 'make a difference' in their personal, social and public lives. In terms of industry evolution, such PSB investment has incubated and grown digital agencies and divisions within television content producers, providing scope for new business and production models to emerge.

However, next to this rather positive example we also have an expectation that the network era and new 'social network markets' (Potts, Cunningham, Hartley, & Ormerod, 2008) could offer even better opportunities to independent producers and would undermine such 'compacts' as instances of oligopolistic media markets. In parallel to that, there is conversely still the 'attention economy' claim that should make us to question if there is any 'attention left' for the small independent players in the marketplace? Or is it that the heavily saturated online content markets are dominated by the few large multiplatform players with big enough (cross-) marketing muscle?

The latter possibility refers a finding of my study on small independent film studios that had undertaken multiplatform productions (Ibrus, 2012). One of the case studies was a transmedia story titled "Superhero Blog" produced by German micro-studio, Fiction Zwei Null. I learned how the ability to output content to multiple platforms was initially believed to be an empowering experience by the producers of the project. They welcomed the possibilities to acquire new roles (online-distribution), to disrupt the normal industry career-paths, to build an intimate relationship with audiences and how the hope of multiple revenue-streams enabled them to make long-term plans for their projects. As was articulated by interviewees, in film schools one does not learn about distribution, not even the producers. However, with the use of multiple Internet channels and platforms, for the first time the control over distribution is in the hands of content owners. By building the product and its brand independently of any channel or platform producers, and by utilising a variety of platforms for connecting with audiences, the content owners may eventually have somewhat more bargaining power in relation to any of the individual platform owners (see Feldmann, 2002, 2005; Yang, Kim, Nam, & Moon, 2004).

The findings at the late stage of the project execution, however, pointed to a conflicting sentiment among the producers; that it is a challenge to make a small transmedia project visible in the crowded Internet. They, therefore, implied that in the future they would attach their transmedia projects to established media brands, to ease the marketing effort as well as to increase the limits to budgets. This could be understood as potentially enforcing the path dependency of the oligopolistic market structure. This would mean that 'convergence' may still condition the endurance of the broadcaster-dependent-compacts suggested by Bennet et al., rather than the more fragmented structure potentially characteristic to the network era.

Regarding the evolution of new kinds of 'compacts', we asked with our next study (that focused on the clustering tendencies of the AV-industry in Tallinn, Estonia—see Tafel-Viia et al., 2012) whether further clustering, cooperation with ICT industries and related prospects of crossmedia related innovations would seem appealing. The majority of them voiced support for this scenario. That is, they saw an opportunity in industry convergence, but only very few of them had previous experience of working on such 'inter-disciplinary' projects or with partners from the ICT sector. The context is that the public funding for Estonian film is relatively small, only a handful of feature films are produced each year and, therefore, the studios are all very small and struggling to make the ends meet. Therefore, developing new kinds of distribution and business models was perceived as a positive opportunity, not surprising in the country with a lively start-up scene and highly regarded IT-infrastructures and online services. However, when these same film entrepreneurs, who mostly were also filmmakers themselves, were asked as to what are the 'true goals' of their companies it was always to make another film, or better films, etc. That is, what became apparent in Tallinn, as well as in the interviews with their colleagues in neighbouring Nordic countries was the very strong identity of film professionals. The main goal is to make good films, which is supported by the belief that their training provides them the skills of telling stories well—better than any 'prosumer' could potentially do.

One instance of this was the definition of 'participatory storytelling', as suggested by the above mentioned 'Superhero Blog' crew. Their superhero-themed sitcom unfolded as a series of video-blog posts by a fictional protagonist Robin W. Schrader (a superhero apprentice) on social networks Facebook and Twitter (Ibrus, 2012). According to them, however, their story was meant to entertain and they, as film professionals, knew best how to do it well. Hence, Facebook friends or Twitter followers were enabled to react to the events, but not to have dialogic relationships with the main character so that to potentially influence the narrative or the course of the story development. This phenomenon that, despite the industry discourses that celebrate user empowerment, the realities are often such that users' actual possibilities for meaningful contribution are strictly limited has been identified before in much bigger productions. Perryman (2008) has demonstrated how the fan cultures, their online activities and remix practices that evolved around *Doctor Who* property have been strictly disciplined by the BBC, especially after *Doctor Who* was turned into BBC's transmedia showcase. Bolin (2010) has demonstrated the same with two Swedish major TV/transmedia productions, *The Truth about Marika* and *Labyrint*. He argues that despite the users being celebrated as 'prosumers' their commitment is in reality limited to being mere contributors of raw material that may subsequently be tooled into final commodities by the professionals. Evidence shows, he argues, that viewers' options in terms of creativity are restricted, and the gatekeeping function of broadcasting institutions has been revitalised in context of their emergent multiplatform strategies. It is, therefore, interesting to learn that such attitudes did not hold true only for major media enterprises, but tended to prevail also among the AV industry's start-ups that had only recently started to experiment with transmedia storytelling. Take, for instance, this quote from the producers of 'Dream Equilibrium', another studied transmedia fiction project from Denmark:

> We are willing to engage the audience and allow them to affect the story, but on our conditions. ... Also, we do not have a lot of experience in letting the audience to co-create or affect the story and hence it is something we want to build in with ease so that we can control the process.

What we could perhaps learn from all this is that the strong professional identity of film professionals and their confidence that their methods and craft could also work on the most nascent of media platforms effected certain path dependency of the 'film subsystem' also in the condition of 'convergence culture'. What is more, also their uncertainty about letting go and loosening their control of the story, experimenting with more flexible forms of interactivity and user involvement, could be interpreted as having the same effect. This is significant taken that the 'convergence condition' normally associated with the understanding that the 'new rules of new media' would also presuppose the change of game for the older media, genres and professions. However, here we saw resistance to these new rules and recognized identity-conditioned and auto-communicatively confirmed path dependency for the old practices. That is, convergence was neither absolute nor unidirectional, instead the film subsystem's auto-communicative functioning enabled its re-creation in the new situation and so it agreed to converge only on its own terms. It was perhaps unsurprising that similar to the Estonian filmmakers, young German producers of the Superhero Blog we interviewed, still saw as the core of their whole project the eventual feature film to be produced, despite their experiments with transmedia storytelling. This is where their passion lay and getting the film distributed by established distributers and shown in cinemas and on television was still their main objective.

Relating to this, our study also identified certain frustrations among filmmakers who had ventured to work on their first crossmedia projects. A pair of Swedish filmmakers who had worked for a year on their transmedia documentary titled Ghost Rockets,[2] for instance, expressed aggravation deriving from their lack of experience of transmedia that made the whole process much more time-consuming and challenging:

> I don't think that we had any understanding what we were getting ourselves into. I feel that for the last six months or the last year we have spent way too much time on transmedia in the sense that we have no time to work on the film. . . . We discuss things too much. Next time we should be like: yes, there should be a Facebook ad and yes it should come at this point in time and it should look like this and let's try these five different strategies. ('Ghost Rockets' producer)

At the time of interview, the Ghost Rockets team had just hired two recognised experts to take care of the transmedia extensions of their project and they were fairly satisfied with this development. The sentiment was emerging that 'transmedia producer' should be a profession on its own that could be potentially incorporated into a film project. The role of this professional would be to master a set of standardised solutions on how to execute specific crossmedia strategies or transmedia narratives, consequently rendering the whole process more 'professional' and, therefore, more straightforward for others. The articulated need for such a professional could be interpreted to refer to the expected gradual codification of the respective field of skills and knowledge. That is, a separate field. The creation of the 'transmedia producer' credit by the Producers Guild of America indicates that the new practice and professional identity was added to the existing set-up of professions, but in this way, also kept distinct from other more traditional roles and cultural institutions. That is, the technical convergence has eventually conditioned the emergence of new roles, in effect divergence, pluralisation of roles. What we

[2] See: http://www.ghostrockets.se/

should also recognise is the auto-communicative codification of this divergence—there is already a penumbra of textbooks, written by the first exemplars and envoys of this new professional breed as well as training programmes run by the same people that all in their own ways define and teach this new profession to motivated individuals around much of the world. Including Estonia, where the local film school (Baltic Film and Media School) has launched study programmes on Crossmedia Production. Graduates of the programme, their work and projects are in turn seen to have a relaxing effect on the cautious resistance to multiplatform strategies on behalf of the older segments of the local AV-industry, described earlier. That is, we could recognise another instance of the convergence paradox; the 'convergence culture' has conditioned the divergence of a new profession, the one that is supposed to 'keep the convergence process together', to connect the autonomous media, arts and associated professional identities. Convergence has conditioned divergence in order to 'ease' its effects.

6 Conclusions

At the beginning of this chapter I discussed the multiple dimensions of convergence processes. Convergence is a meta-term that refers to an 'ensemble of dynamics' that are all, to an extent, interdependent; they co-evolve and do this mostly rather organically. What this chapter also aimed to demonstrate is that convergence may be viewed as a limited metaphor since the dynamics that it denotes are not linear and may be often even conflicting. That is, convergence of network infrastructures may condition, for instance, converged services, but they may also condition the emergence of entirely new media forms, new industries, practices and professions—various dynamics that could be understood as processes of divergence. For instance, as two or more older media phenomena (industries, services, genres) converge to be something qualitatively different and this new structure/form is either well received or is persistent in its auto-communicative self-establishment there is a potential for an additional divergent trajectory within the broader process of media evolution. In relation this, the chapter proposed that 'convergence' should in effect be interpreted as normally referring to paradoxical trinity of 'convergence', 'emergence' and 'divergence'—processes that are to an extent the 'same', or at least mutually conditioning; one cannot exist without the other when it comes to the evolutionary dynamics of culture.

The chapter also proposed that the currently buzzing emergence of crossmedia or transmedia phenomena is indicative of the above-mentioned dynamics. These new ways of content development, media consumption and related evolving industry organisation are conditioned by the parallel movement towards convergence and divergence, following the logic of culture's centrifugal and centripetal development described by Lotman. Crossmedia and transmedia phenomena are enabled by the convergence in networks, services and markets, to an extent also in industries, and they might even condition certain connectivity or similarity between access terminals or media forms, but at the same time they rely heavily on distinctions between the devices and forms. That is, crossmedia strategies are mostly built on utilising the differences in affordances, use cases, 'model readers' (Eco, 1979), usage patterns and audience expectations—the platforms and their different use cases are designed to be complementary by most crossmedia strategies. Although the exploitations of such distinctions allow for innovative artistic expressions, these are also motivated by straightforward need to multiply revenue streams and develop business models that accommodate the ongoing multiplication of platforms and related audience fragmentation. Such principal need to keep the experiences or offerings distinct contributes to the relative autonomy of different media, supported by the auto-communicative functioning of the industry fractions that 'produce' the particular media together with their particular distinctions.

What is more, again deriving from the logic of culture's parallel and mutually conditioned centripetal and centrifugal development, the contemporary convergence processes tend to condition the emergence (or divergence) of new terminal types (smartphones, tablet computers, various forms of videogame consoles, etc.). As demonstrated in this chapter, these developments may also be accompanied by divergence in practices that results in multiplication of professions. What we also recognised, based on Lotman's conceptualisations, was how convergence may be resisted by recursive autonomy re-creation by acts of auto-communication. For instance, our interviewed Estonian filmmakers, although being curious about the opportunities brought about by technical convergence and potential industry convergence, also emphasised a need to be cautious not to lose too much of their own identity and integrity of their craft and art. There are no signs of fields such as filmmaking being ready to sacrifice much of their distinguished (and popular) cultural institution to the altar of convergence. Convergence is only tolerated as long as it enables them to extend their practice and to amplify its presence in the cultural space. Therefore, in relation to the conceptual framework laid out in this chapter, media evolution, the dynamics of convergence and divergence therein, are conditioned by both the dialogic as well as auto-communicative processes in culture. While convergence is facilitated by dialogic processes between existing media subsystems, divergence of these new 'convergent systems' is enabled by their ability for auto-communication, for their being able to define, distinguish and confirm themselves. Yet, auto-communication is also needed for creating stability, for securing the relative autonomies of all existing systems.

However, the chapter also addressed the degrees of freedom that subsystems may have for autonomy creation in the general condition of 'convergence culture'. If convergence could be understood as a homogenising process conducted by the globally dominant industry fractions/enterprises, the ability of independent content producers to secure their autonomy may be understood as crucial from the perspective of sustaining culture's inherent pluralities. Our reported studies demonstrate that the endeavours to develop 'independent crossmedia strategies in the saturated Internet marketplace have been failing, at least in Europe's affluent North. This tends to motivate the micro-studios, instead, to further their cooperation with large brands and companies dominating the oligopolistic media markets and to accept the standards for crossmedia strategies set by the large content houses, as well as by international providers of infrastructural services. These latter tendencies may, however, inspire the questions if securing further divergence processes (i.e. plurality) in culture may justify public interventions and support for independent crossmedia producers.

References

Alexandrov, V. (2000). Biology, semiosis, and cultural difference in lotman's semiosphere. *Comparative Literature, 52*(4), 339–362.

Baudrillard, J. (1983). *Simulations*. New York: Semiotext(e).

Belsunces Gonçalves, A. (2012). Fringe: Playful transmedia. In I. Ibrus & C. A. Scolari (Eds.), *Crossmedia innovations: Texts, markets, institutions* (pp. 93–110). Frankfurt: Peter Lang.

Bennet, J., Strange, N., Kerr, P., & Medrado, A. (2012). *Multiplatforming public service broadcasting: The economic and cultural role of UK Digital and TV Independents* (p. 60). London: Royal Holloway, University of Sussex, London Metropolitan University.

Bolin, G. (2010). Digitization, multiplatform texts, and audience reception. *Popular Communication, 8*, 72–83.

Britton, J. N. H. (2007). Path dependence and cluster adaptation: A case study of Toronto's new media industry. *International Journal of Entrepreneurship and Innovation Management, 7*(2-5), 272–297.

Broms, H., & Gahmberg, H. (1983). Communication to self in organizations and cultures. *Administrative Science Quarterly, 28*(3), 482–495.

Chan-Olmsted, S. M., & Chang, B.-H. (2003). Diversification strategy of global media conglomerates: Examining its patterns and determinants. *Journal of Media Economics, 16* (4), 213–233.

Chan-Olmsted, S. M., & Chang, B.-H. (2006). Mobile wireless strategy of media firms: Examining the wireless diversification patterns of leading global media conglomerates. In J. Groebel, E. M. Noam, & V. Feldmann (Eds.), *Mobile media: Content and services for wireless communications* (pp. 165–183). Mahwah, NJ: Lawrence Erlbaum Associates.

Christensen, L. T. (1997). Marketing as auto-communication. *Consumption, Markets and Culture, 1*(3), 197–302.

Congdon, T., Graham, A., Green, D., & Robinson, B. (1995). *The cross media revolution: Ownership and control*. London: John Libbey.

David, P. A. (2000). Path dependence, its critics and the quest for 'historical economics'. In P. Garrouste & S. Ioannides (Eds.), *Evolution and path dependence in economic ideas: Past and present* (pp. 15–39). Cheltenham: Edward Elgar.

Doyle, G. (2010). From television to multi-platform: Less from more or more for less? *Convergence, 16*(4), 431–449.

Eco, U. (1979). *The role of the reader: Explorations in the semiotics of texts*. Bloomington, IL: Indiana University Press.

Erdal, I. J. (2009). Cross-media (re)production cultures. *Convergence: The International Journal of Research into New Media Technologies, 15*(2), 215–231.

Evans, E. (2011). *Transmedia television: Audiences, new media, and daily life*. New York: Routledge.

Fagerjord, A., & Storsul, T. (2007). Questioning convergence. In T. Storsul & D. Stuedahl (Eds.), *Ambivalence towards convergence. Digitalization and media change* (pp. 19–31). Götenborg: Nordicom.

Feldmann, V. (2002). Competitive strategy for media companies in the mobile internet. *Schmalenbach Business Review, 54*(4), 351–371.

Feldmann, V. (2005). *Leveraging mobile media: Cross-media strategy and innovation policy for mobile media communication*. Heidelberg: Physica-Verlag.
Filmby Aarhus, Alexandra Institute, & High Tech Accelerator Innovation Center University of Lodz Foundation. (2011). *Report on business models, value chains and business development services in the audiovisual/creative industries: Cases of the Łódź and Małopolska provinces and West Denmark* (95 p.). Aarhus, Łódz.
Garrouste, P., & Ioannides, S. (2000). Evolution and path dependence in economic ideas: Past and present. In P. Garrouste & S. Ioannides (Eds.), *Evolution and path dependence in economic ideas: Past and present* (pp. 1–14). Cheltenham: Edward Elgar.
Goldhaber, M. H. (1997). The attention economy and the net. *First Monday, 2*(4).
Goldhaber, M. H. (2006). The value of openness in an attention economy. *First Monday, 11*(6–5).
Hartley, J. (1996). *Popular reality: Journalism, modernity, popular culture*. London: Arnold.
Ibrus, I. (2008). *Collective memory and media innovation: Interdependencies* (Vol. 14, EDS Innovation Research Programme. Discussion Papers. Media, Connectivity, Literacy & Ethics). London: EDS-London School of Economics and Political Science. URL: http://www.lse.ac.uk/collections/EDSInnovationResearchProgramme/pdf/EDSdp014/edsdp014.pdf
Ibrus, I. (2010). *Evolutonary dynamics of new media forms: The case of the open mobile web*. Unpublished PhD thesis. London School of Economics and Political Science. URL: http://etheses.lse.ac.uk/53/
Ibrus, I. (2012). The AV industry's microcompanies encounter multiplatform production. In I. Ibrus & C. A. Scolari (Eds.), *Crossmedia innovations: Texts, markets, institutions*. Hamburg: Peter Lang.
Ibrus, I. (2013a). Evolutionary dynamics of media convergence: Early mobile web and its standardisation at W3C. *Telematics and Informatics, 30*(2), 66–73.
Ibrus, I. (2013b). Evolutionary dynamics of the mobile web. In J. Hartley, J. Burgess, & A. Bruns (Eds.), *A companion to new media dynamics* (pp. 277–289). Malden, MA and Oxford: Wiley-Blackwell.
Jenkins, H. (2006). *Convergence culture: Where old and new media collide*. New York: New York University Press.
Kotov, K. (2002). Semiosphere: A chemistry of being. *Sign System Studies, 30*(1), 327–336.
Krippendorff, K. (1995). Redesigning design: An invitation to a responsible future. In P. Tahkokallio & S. Vihma (Eds.), *Design—Pleasure or responsibility* (pp. 138–162). Helsinki: Helsinki University of Art and Design.
Krippendorff, K. (2008). *Reconciling radical constructivism with social organisations as networks of conversations and of stakeholders*. Paper presented at the American Society for Cybernetics 2008 Conference, Urbana, IL, 11-15.05.2008
Lotman, Y. (1990). *Universe of the mind: A semiotic theory of culture*. Bloomington, IL and Indianapolis, IN: Indiana University Press.
Luhmann, N. (1995). *Social systems*. Stanford, CA: Stanford University Press.
Mandelker, A. (1994). Semiotizing the sphere: Organicist theory in Lotman, Bakhtin, and Vernadsky. *PMLA, 109*(3), 385–396.
Morsing, M. (2006). Corporate social responsibility as strategic auto-communication: On the role of external stakeholders for member identification. *Business Ethics: A European Review, 15* (2), 171–182.
Nöth, W. (2006). Yuri Lotman on metaphors and cultures as self-referential semiospheres. *Semiotica, 161*(1/4), 249–263.
Perryman, N. (2008). Doctor who and the convergence of media: A case study in 'Transmedia Storytelling'. *Convergence: The International Journal of Research into New Media Technologies, 14*(1), 21–39.
Potts, J., Cunningham, S., Hartley, J., & Ormerod, P. (2008). Social network markets: A new definition of the creative industries. *Journal of Cultural Economy, 32*, 167–185.
Potts, J., & Keane, M. (2011). Creative clusters and innovation. In J. Potts (Ed.), *Creative industries and economic evolution* (pp. 152–161). Chelthenham: Edward Elgar.

Schönle, A., & Shine, J. (2006). Introduction. In A. Schönle (Ed.), *Lotman and cultural studies: Encounters and extensions* (pp. 3–35). Madison, WI: The University of Wisconsin Press.

Steedman, M. D. (2006). State power, hegemony, and memory: Lotman and Gramsci. In A. Schönle (Ed.), *Lotman and cultural studies: Encounters and extensions* (pp. 136–158). Madison, WI: The University of Wisconsin Press.

Suárez Candel, R. (2012). Adapting public service broadcasting to the multiplatform scenario—Challenges, opportunities & risks. In *Working Papers of the Hans Bredow Institute* (Vol. 25, 83 p.). Hamburg: Hans Bredow Institute for Media Research in Hamburg.

Tafel-Viia, K., Lassur, S., Ibrus, I., Tafel, A., & Terk, E. (2012). *Audiovisuaalvaldkonna klasterdumise perspektiivid* (p. 97). Tallinn: Tallinna Ülikooli Eesti Tuleviku-uuringute Instituut.

Terranova, T. (2012). Attention, economy and the brain. *Culture Machine, 13*, 1–19.

Tether, B. S. (1998). Small and large firms: Sources of unequal innovations? *Research Policy, 27* (7), 829–841.

Tether, B. S., Smith, J., & Thwaites, A. T. (1997). Smaller enterprises and innovation in the UK: The SPRU innovations database revisited. *Research Policy, 26*(1), 19–32.

Torop, P. (1999). Cultural semiotics and culture. *Sign System Studies, 27*, 9–23.

Yang, D.-H., Kim, S., Nam, C., & Moon, J.-S. (2004). Fixed and mobile service convergence and reconfiguration of telecommunications value chains. *IEEE Wireless Communications, 1*(5), 42–47.

Digital Doppelgängers: Converging Technologies and Techniques in 3D World Modeling, Video Game Design and Urban Design

Shaun Foster and Jacob Brostoff

1 Introduction

Real-time 3D computer graphics gaming technology has been used to create entire worlds. Some of these worlds have cities of people simultaneously interacting with each other across the real world. As these gamers interact online they generate large amounts of data. At the same time there has been enormous growth in data gathered from and representing the real world. This data has been analyzed and used by urban design and planning practitioners ("planners") in developing better urban living environments. Tools such as Geographic Information Systems (GIS) and urban form modeling software can help structure and mine this data for more sophisticated decision-making and more meaningful public involvement in planning processes.

While the primary goals of video game designers and urban planners differ, a convergence has begun in some areas between these fields. The nature of technology has historically been a driver of convergence. Human society has reached a point where new technologies are constantly being developed. These new tools allow for new or deeper areas of a field to be explored. As new areas of a field are explored there emerges the need for new technologies. Sometimes the complexities and technologies being used in one field find application in other fields. In the cases of video game design and urban planning, one discipline seeks to create a compelling entertainment experience, while the other seeks to influence the future through

S. Foster (✉)
Rochester Institute of Technology, Rochester, NY, USA
e-mail: scffaa@rit.edu

J. Brostoff
MURP, Portland, OR, USA
e-mail: j.brostoff@gmail.com

A. Lugmayr, C. Dal Zotto (eds.), *Media Convergence Handbook - Vol. 2*, Media Business and Innovation, DOI 10.1007/978-3-642-54487-3_10

the shaping of urban form. Both share the goal of constructing a cohesive, credible and legible visual narrative about an alternate reality. Until now, both fields have had their own methods of visualizing their alternate worlds. New spatial visualization techniques and the growing complexity of data in both fields create the need for new tools which can help us further understand, model and play with alternate worlds in the service of divergent goals.

Because of these converging technologies, there are new opportunities for cross-disciplinary knowledge and collaboration between practitioners in the two fields.

This chapter examines the nature of technology as a driving force for convergence. It then gives a brief overview of the goals, economic underpinnings and history of each field. Finally, we look at specific convergences between 3D graphics, video game technologies and urban planning tools as well as some synergies between the two fields and some analysis of where additional emerging technologies may take the two fields.

2 The Nature of Technology and Convergence

From rockets to microchips, giving examples of technology is easy. Defining technology is more challenging. For the purpose of this discussion we will define technology as: a continuously evolving process of harnessing and amplifying sources of power which were originally part of the natural world in the form of tools or processes which serve to increase human abilities to fulfill our needs and desires. Technology has evolved in tandem with human culture. The development of human culture is directly linked to the creation and development of new technologies. The evolutionary nature of technology reflects cycles similar to biological development; many new technologies are tested and the most useful combinations move ahead.

2.1 Technology Cycles

A technology cycle is the process by which new tools and techniques are created, reducing the time it takes to accomplish a task. This cycle revolves around tool creators, technical tool users, and designers (Fig. 1).

The toolmakers develop the equivalent of hardware, software and workflow processes. As new hardware improvements are made new tools are created to take advantage of the hardware. Once the tools reach widespread use, technical tool users figure out how to automate and optimize their use. Designers test creative ways to use the tools, looking for new ways to find synergy through combination and integration of technologies from other fields and disciplines. This convergence starts a new technology cycle.

Fig. 1 Technology cycle

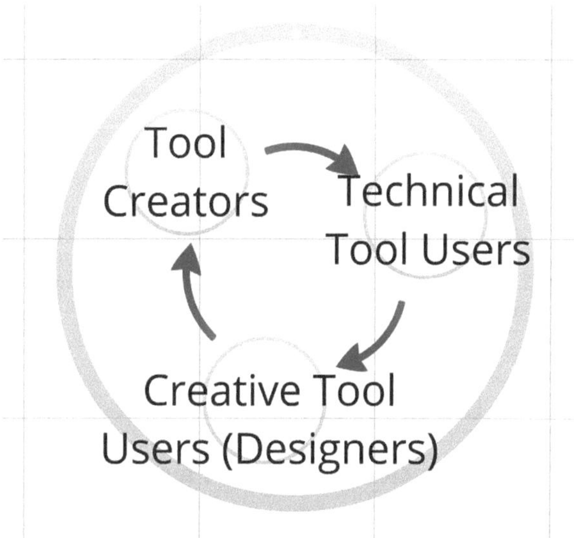

When a new technology cycle starts many new tools emerge, all trying to find an optimal integration and application. Most are incremental tool improvements. They result in small increases in productivity. Some changes in technology completely revolutionize a process. These "paradigm shifting" (Kuhn—*The Structure of Scientific Revolutions*) technologies result in such dramatic changes in productivity or cost which that they sometimes completely change or disrupt a field. There has been emerging research which points to a logarithmic (see illustration below) increase in the number of paradigm shifting and disruptive technologies (Kurtzweil "The Singularity" and Christensen, Bower "Catching The Wave"), increasing the speed of adoption of new technologies (MIT Article on Smart phone adoption) (Fig. 2).

The nature of technology is both convergent and divergent. In evolutionary biology some traits of mammals have arisen independently from one gene-pool or from different geographic areas. The same is true of technology and has been noted by both historians and sociologists. Multiple technologies of the same type have been invented in independent contexts. Some of the most famous examples: Newton and Leibniz with calculus, Scheele, Priestley and Lavosier discovering oxygen and Darwin and Wallace and the theory of evolution. As tools and technologies become increasingly complex there may be long periods where technologies develop independently of each other. In the recent past many new technologies have allowed huge amounts of new data to be collected. Other technologies have allowed new visualizations of space. These technologies are now being combined and used by the fields of urban planning and 3D computer graphics and games design.

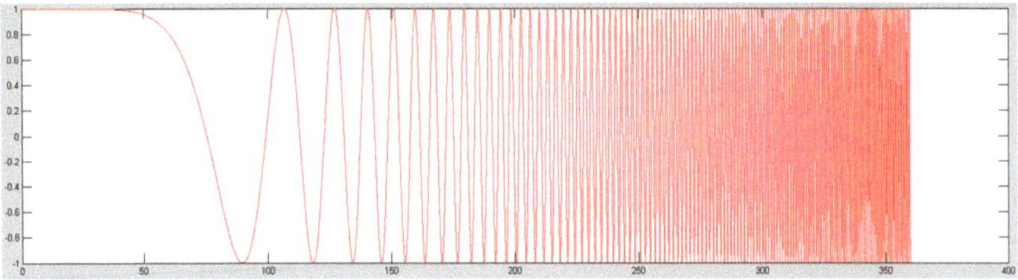

Fig. 2 Accelerating pace of technology is similar to a sin wave with logarithmically decreasing wavelengths

3 Urban Planning

3.1 Basic Goals of Urban Planning

As with all forms of planning ahead, the primary goal of urban planning is to create a future that is different in some way from the present. Especially in the North American context, the history of urban planning has revolved around interventions in the built and natural environments that seek to alter present conditions in order to create a future that is different from today. Arguably, this is true for all planning practice, including even such endeavors as historic preservation, which can be seen as an effort to change current development dynamics in order to "preserve" aspects of the built environment that might otherwise disappear.

3.2 Economic and Social Goals

In North America, the policies, practice and tools of urban planning and its affiliated professions, including urban design, architecture, landscape architecture and engineering, have generally been framed as a way to "improve" upon present conditions in order to create a "better" future. In general, because urban planning in North America currently exists primarily as a set of policy tools designed to shape private development in order to achieve a desired future condition for the urban environment, e.g. zoning codes and associated regulations, it is a tool of political power. Therefore, the ideas of a "better" future are typically shaped by the powerful, who exert control over the built and natural environments in order to achieve specific future conditions. The idealized future conditions are described using the language of amelioration. In some cases, "better" means "better for all stakeholders". In other cases, urban planning is intended to improve things for specific groups of people, e.g. developers or neighborhood residents.

3.3 Historical and Current Tools

Historically, planning practice has been intertwined with architecture, landscape architecture and engineering. Perhaps the most familiar example of a planning tool is the zoning code, common to most local governments in the United States. Many zoning codes regulate "development" and "land use". These are the aesthetics of and changes to a site. Development is distinct from land uses, which are the ways in which the site is used. An example of development is a shopping mall. The site is developed with a mall. Inside the mall, there are commercial land uses (stores). To develop a hypothetical shopping mall, the land on which the mall is planned would have to be zoned (designated by a government) for commercial development. Thus, a typical zoning code regulates the aesthetics of development as well as the intensity and mix of uses allowed in a given area.

The regulation of development and uses in the United States can be traced back to the New York City Zoning Code of 1916 and the 1926 Village of Euclid, Ohio v. Ambler Realty Co. United States Supreme Court case. Before the advent of zoning, development in the United States was essentially unregulated. After, incompatible land uses were segregated and zoning regulated the aesthetics of development. In the United States, zoning was created as part of a broader public health movement in the late nineteenth and early twentieth centuries that also included the widespread adoption of building codes, and the planning and development of many local parks systems.

Planning requires some basic knowledge about the world. This knowledge includes information about urban form, as well as geography. Thus, cartography is and has been an extremely important technology for planners. In the past, mapping was a laborious and slow process that involved recording information about the world onto paper maps using cartographic projections. More recently, with the advent of Geographic Information Systems (GIS), digital mapping has become feasible and is now widely used by planners. GIS involves combining a database of information about the world (for example, tax lot information and zoning boundaries) with digital mapping data. This newer technology allows planners to query the database and get useful information about things in the world, combined with geographic information about the location of those things.

Another technological development related to GIS is digital urban form modeling tools, such as ESRI's CityEngine. These tools allow planners to model current or proposed urban form (the shape of the built environment). In the past, modeling was also a laborious, time- and labor-intensive process, often involving the creation of miniature, hand-built models or complex renderings. With tools like CityEngine, planners can easily see how changes to one urban design parameter affect the modeled urban form. This in turn allows planners to answer complex "what-if" questions, and to show the public and clients how various changes to existing or proposed conditions would result in different urban forms over time. This kind of quick "window into the future" is unprecedented in the history of planning. Similar to the implications of the development of the spreadsheet for the financial sector, these tools are revolutionizing how quickly and easily we can visualize alternate urban design futures.

3.4 Recent Example of Effective Urban Planning

Generally, planning efforts can be considered effective when they produce the outcome envisioned by the powerful stakeholders of the original process, without other undesirable outcomes or negative unintended consequences, measured over the planning horizon. A recognized example of an urban planning success story is the regional light rail system in Portland, Oregon USA ("MAX"). Key to the strategy behind MAX is a strong connection between transportation investment decisions and land use policy decisions at a regional level. Many other regions in the USA do not coordinate land use and transportation decision-making. By using transportation investments to intentionally support pedestrian-scale development in desired locations, the Portland region was able to achieve public policy goals of growing compactly, accommodating population growth, strengthening downtown, reducing greenhouse gas emissions and ensuring mobility for its residents. (*Source*: http://trimet.org/pdfs/publications/beyonddreams.pdf).

4 Video Games

4.1 The Rise of Video Games

Video games have existed since the early 1970s. Playing with pixels has evolved from rudimentary black and white graphics and nominally interactive experiences to fully realized, highly complex, three-dimensional multi-sensory experiences. Several aspects of video games have remained constant throughout their history. There have been strong connections between the latest technologies, graphics, audio, and interactivity. In order to play the newest games there are often additional hardware and software requirements.

Rapid improvements within the software used to develop video games have also transformed the creation process. In order to increase the speed of development, the industry has created "game engines". They include standardized tools for building new levels, animations and narrative. Many games now include relatively easy access to the game engine and world-building toolkits, which have encouraged players and community members to create custom levels and spaces.

While the individual goals of video games differ, the general goal of videogame play is to have an enjoyable, compelling and immersive experience. Games have done this in numerous ways. Early video games focused on simple hand-eye coordination. Growing levels of complexity introduced involved narratives and entire worlds to explore as well as multiple genres of games. Beyond puzzles and hand-eye coordination, many new games give players a strong sense of "agency", meaning that their actions have significance on the world they are playing in. These actions, good or bad, affect the world.

Over the last 15 years video games have begun to move beyond single- or two-player experiences. Large online communities have developed, connecting players around the world. Game designers continue to face a combination of technical and aesthetic challenges in their work.

4.2 Challenges Faced by Video Games

Video game developers face many challenges and design questions during the development process: What hardware to develop on? How much testing and time for refinement will be allowed before a game is released for sale? Two of the most subjective aspects of video game design are art direction and aesthetics. Initially, the goals of game aesthetic development revolved around technical accomplishments such as more realistic graphics, higher fidelity, and improved workflows for production teams. Finding the right balance between realism and abstraction and the correct level of interactive complexity have been key factors for game design.

A specific challenge for designers has been how to populate the new virtual worlds with characters, and whole crowds, that need to move, look and react in a realistically "human" way. The idea is that as game designers get better at creating something approaching a real human, players will be much more able to relate to the non-player-characters and thereby have a more fun game experience. However, when virtual characters get close to "human" but don't actually achieve fully human traits, it can actually cause a negative player reaction. Approaching a realistic representation of human behavior but not fully achieving it has been called the "uncanny valley". The current state of technology allows a fairly close reproduction of human physiology and also a fairly close approximation of human movement through motion capture. However neither of these technologies are 100 % exact representations of what it is to be human.

A recent student published in the Journal of Social Cognitive and Affective Neuroscience showed that graphics, which fall into the "uncanny valley", trigger an increase in brain activity in areas of the brain associated with human interaction and empathy. These human like figures are next identified as non-humans, often causing an unsettled feeling or an inverse empathy reaction for viewers and players. One solution has been to step back from the highly realistic and pursue artistically stylized characters and environments. These have often proved to be extremely enjoyable without the need for hyperrealism. However, research also continues on the development of the virtual human which can completely fool other humans as well as developing accurate ways of modeling reactions of whole groups of humans in virtual spaces.

In many ways accurately modeling human behavior graphically with 3D avatars and with groups of game characters parallels the test for successful artificial intelligence (AI) known as "The Turing Test". The Turing Test is a series of questions used to determine "human like" behavior from a computer. Games often also implement AI the combination of interactive avatars with artificial intelligence that can react to players in complex and realistic ways is one of the biggest challenges in game development, as well as an area to look forward to further convergence in technologies in the future.

In a basic way, the gap between the representations of reality depicted in video games and the human ability to discern "close-but-not-quite" verisimilitude is a locus of convergence between video game design and planning. To a lesser degree than video game designers, planners have to depict their alternate futures in a way that is sufficiently recognizable to humans as real, or at least potentially real. The goal of both disciplines is convergent: create a compelling narrative that immerses the reader in an alternate reality.

4.3 Economic and Social Implications

The economic impact of video games is large. The industry generates billions of dollars in revenue in sales of both software and hardware. In 2011 consumers spent over USD 24B. Video games sales have far surpassed other entertainment industries such as film. They have also spawned multiple additional markets. Previously there were two markets: the console market and the PC market. More recently, the market for mobile device applications ("apps") has developed. These new platforms expand the interactivity of gaming systems. The Nintendo Wii Fit Plus and the Microsoft Kinect are key examples of this. The Wii Fit Plus is an interactive system specifically geared toward exercise applications and the Microsoft Kinect is a computer vision based system, which allows touch-free interactivity. The expanded interactivity of new gaming systems have broadened previous markets and opened up new ones. Previously a male dominated pastime, now over 47 % of all game players are women. People of all ages now regularly play video games in a social context.

Social interactivity has become a core part of the gaming experience. The key to this has been the development of large online communities. Potentially increasing the move to online social spaces has been declining social spaces in real life. In the past people would socialize by going to dances or religious services to meet and interact with others. Now games are giving crowds the ability to enter and customize their own virtual online spaces.

4.4 Tools and Processes: Historical to Present

The tools used to create games have become increasingly sophisticated. Initially one individual would program every pixel and every sound effect for a video game. This has changed dramatically. Now teams of specialists work to develop multiple facets of each game. Game designers, 3D modelers, layout artists, user interface artists, sound technicians, computer programmers and other specialists work together to develop a final product with a production cycle sometimes lasting more than 2 years.

The tools the specialists use continue to increase in sophistication. This allows each person working on a project to do more but also requires constant retraining in the new tools and technologies as they emerge. Quite a few of these tools, originating in other disciplines, are creating a convergence and overlapping of technologies. They also imply ever-broadening fields of study necessary to contribute to an increased sophistication within games and to meet the rising expectations of players.

4.5 Game Genres and Environments

As the video game field has matured and become more technologically complex, multiple genres have evolved, each allowing for different types of game play challenges and interactivity. Currently a fundamental set of six genres is usually recognized: action, action-adventure, adventure, role-playing, simulation, and strategy. Each of these can be further subdivided into multiple sub-genres. While there may be additional genres and sub-genres in the future, a core component of nearly all-current video games are focus on experiencing and navigating different environments. These environments cover an increasingly large scope of different types of worlds. An almost uniform component to designing virtual worlds has been the construction of multiple and varying types of space for the players to interact in.

4.6 The Rules and Nature of Space in Games and 3D Worlds

The use of space, and how players interact within 3D worlds has a multitude of variations in games. From the way buildings are constructed or represented to rules of physics, nearly all variables are explored and played with in video games. This flexibility allows for new ways to move through and explore the spaces within games. New explorations of space are a function of advances in gaming technology. Early game spaces were constrained to a single "screen shot" constraining the player to the borders of a TV screen. As games became more advanced they moved beyond single screens to multiple "screen shots", or scrolling graphics, which could follow a moving character giving the illusion of traveling across space in a linear way. These "side scrollers" eventually included ways to simulate motion parallax, creating the illusion of traveling across a plane within a 3D space. As technology advanced new, larger and more complex spaces needed to be designed. Often greater amounts of research were done into different fields to help understand what makes different spaces feel more or less real. Some spaces have been designed that are rule-based. Other worlds are constantly created by the computer as needed, built by algorithms, and populated by modular geometry. Other spaces are always predefined. How the player traverses these new spaces also varies greatly.

In the game Portal the player is encouraged to transform their thinking of how to move through space. In the real world walls are barriers to entry. Portal gives the user a device that allows them to fire two shots. Instead of being destructive, these shots open up two doors which instantly teleport the player across any spaces. A door in one wall and then another fired far across a chasm allows the player to bypass dangers or difficult terrain. This thought process inverts the classic notion of walls as barriers and transforms them into portals.

Another function of space is to hold objects. In the real world there are a huge variety of objects that serve many purposes. In most games spaces and objects usually have a much more narrowly focused reason for existence, which often connects, directly to the story or purpose of the game.

The game Little Big Planet deals with objects, space and a redefines a sense of scale: In the game you are a very small doll-like creature. The camera's aesthetic, featuring a narrow depth-of-field, and the objects around you let you know you are small in a word of big objects.

Many games now offer the ability of the players to build their own customizable levels. What had previously been complicated coding tasks have now been given graphical user interfaces and strong toolsets toward allowing players to become their own 3D level builders. While the tool expands the number of users who can participate in game design, they tend to constrain the designer by their limitations. Thus, the tools have a strong impact on the spaces that are created. One level builder, Unreal 2004–2007, required the builder to carve out space (subtractive creation), which often gave 3D level a very cave-like feel. While it was possible to carve out large "skyboxes" then add back geometry, the "medium often drove the message". The growing rules and tools of constructing virtual worlds have many parallels with the accepted rules of planning which also constrain or allow growth of environments..

As the complexity of games grows, graphical fidelity increases, and there are ever more physical simulation of natural phenomena. Games have not yet reached the level of visual complexity and interactivity of reality. One work-around solution of not being able to have anywhere near the kind of complexity present in real life, has been to program multiple levels of detail which allow the user to "zoom in" selectively seeing more details. The field of game design continues to work to increase the visual complexity of virtual worlds. However, one different approach was taken by the game Minecraft. Its reductive world allows for higher amounts of customizability by the players but offers a much lower degree of resolution. This has created a much greater feeling of agency and control over environments by the players. It also offers a greater number of creative iterations. Players can try various approaches easily and get instant feedback. This move towards realism with greater ease of simulation mixed with external datasets is where the fields of 3D graphics and game design are headed. In order for this to happen, key connections need to be made between the way the real world is designed and the way people and traffic flows. These information sets is one area where there is convergence between games and urban planning.

5 Convergence

Similar technologies that are used for visualizing space, collecting and utilizing data, as well as similar rules being used to design in both the fields of 3D game design and urban planning. We will discuss and contrast some of these parallels in terms of how these have been used in different ways. We will conclude by some speculation as to where the fields are going and what further convergence is possible as well as convergence with additional fields.

5.1 Spatial Visualization

Urban planning has previously focused on visualizations that have been mostly graphics poor but represented significant data accumulation. The opposite was true in the realms of video game construction. Levels were constructed with the goal being that they specifically serve the visuals, narrative and gameplay. In the area of space, greater visual realism has entered both realms. Obviously there are great variations in the types of spaces within games. However, growing amounts of attention within video game worlds are given to understanding the underlying rules of the way spaces affect their players. For urban planning and architecture being able model and visualize how new structures will fit into the environment have increased effectiveness of decision making. With 3D Graphics game technology, the range of accuracy varies greatly. There has been some use of GIS data to create accurate historical locations (cite ESRI source) however there are also many instances where urban visualizations are created by rule based systems (see illustration below). ICQTown is a procedural script written by Ingo Clemens. It uses many rules to randomly build an urban environment. The environment may look realistic when seen from an aerial view (see the second illustration below). However, the cityscape does not include roads, or many of the elements that would give it further understanding as a planned, real and constructed environment (Figs. 3 and 4).

Many 3D game worlds are intensely urban environments, with a touch of the unreal. In Bioshock, for instance, a dystopian, anti-Randian tale unfolds in an intricately imagined undersea city whose buildings are in the Modernist style, with Art Deco-inspired façades and interior spaces. Uncharted features scenes and puzzles that take place in Middle Eastern cityscapes. And many military-inspired first-person shooter (FPS) games feature gritty urban combat scenes. The game Mirrors Edge challenges the player to explore and navigate the rooftops of an urban environment in the obstacle style course of someone who practices Parkour, the French style of navigating an environment negotiating the obstacles in-between.

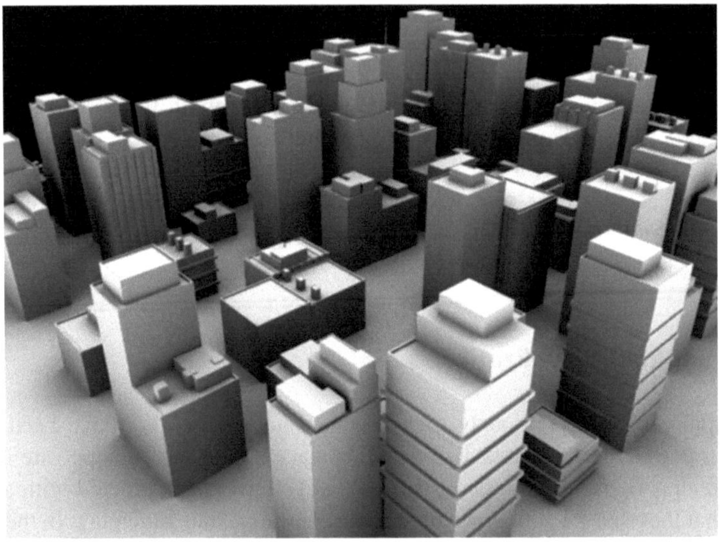

Fig. 3 *Side* view of procedurally generated environment (Foster)

Fig. 4 *Top down* view of procedurally generated environment (Foster)

5.2 Collection and Utility of Data

There has been increasingly greater attention to the way space works for people and vehicles by how the space affects the flow and direction of players through that space. Human pedestrian flow and vehicular traffic have been modeled for many years and used in planning applications. In recent years video game engines have also been used to incorporate some of this data. Bergman and Associates used virtual design and traffic modeling with the Unity 3D game engine in their development of transportation and civil designs of systems. Examples of some of their work can be viewed at: http://www.baviz.com/. Game systems use "flocking behaviors" to model complex movement of birds, insects, etc. through space. Urban planners have also used these same tools. Game systems have also begun to collect vast amounts of data for use in analysis and design. Player movement visualizers (see image below for path visualization) as well as "Heatmaps" which show levels of time and interaction between players on a map are being used to optimize design strategy for the construction of space within game levels.

The New Urbanism movement in the United States advocates for walkable urban environments, in contrast to the widespread post-WWII practice of auto-oriented, suburban development. This kind of environment is often consistent with what is offered in 3D video games, where walking through intensely urban environments is is common. Video games allow the player to experience environments first hand, often from multiple points of view. Video games set in immersive environments often replicate the ideals of New Urbanism, with highly walkable, intricately detailed environments that engage the pedestrian as she passes by.

Visualizing cities and real-life urban spaces are beginning to be done in greater detail inside of various mapping programs. Google maps allow multiple views and overlays: traffic, weather, terrain, satellite, and "street views". Google Maps 3D on android devices allows further interaction with environments in a 3D view. One convergence where gaming has intruded on the realm of cartography is the app for Google Earth called "Monster Milk Truck" which allows the player to "drive" across google maps. Another driving convergence is the intelliDrive program that is being developed by the US Department of Transportation. It uses the UDK game engine to deliver "true-life community to test road conditions in suburban communities and calculate the effects of bad weather".

5.3 Rules and Simulation

As previously discussed, 3D video game world designers create immersive, coherent and legible experiences for players. The games follow rules connected to metaphors from reality, techniques from film and video production, and creative license to construct alternate realities that are both clearly unreal and easily read by human cognition systems. While in many games, plot development is scripted and part of a fairly linear narrative there are also games which feature a simulation component. Games like Sim City and Sid Meier's Civilization allow the user to build and manage complex rule based systems. The cities grow, shrink, prosper or perish based on a mix of decisions made by the player as well as a combination of scripted and random events. In many ways urban planning has begun to do this in ways that are convergent to games. There have been urban simulations which test how changing roads will affect traffic flow, how putting in new buildings or industry will noise, light and pollution. (Fraunhaufer Iao) and also (Frasca—Rethinking Agency and Immersion: videogames as a means of consciousness-raising) Tools and rules from urban planning have also worked to help create visualizations for entertainment purposes: "They created a basic model of a glacier and then designers created a program that added elements like roads, intersections and ravines until they had a complex, yet organic-looking, cityscape." (Wikipedia article: Inception, Visual Effects section, second paragraph.)

Looking towards the future new tools and technology will continue parallel convergence between video game design and urban planning. GPS, augmented reality are already being used for both games (eg Geocaching) and in Google's New game which combines Urban Exploration with Augmented reality "Ingress".

6 Conclusions

As technology cycles continue creating new technologies, there will be further convergence where formerly discrete fields are connected to new technologies. Video games connection to urban planning for fun and simulation will only increase as new data is gathered both about how humans explore virtual worlds as well as how cities have grown and evolved using GIS data. A similar connection of datasets and tools for simulation, which are already happening to test the results of traffic patterns, noise, economics, etc using optimized graphics and the field of Urban Planning will also utilize game simulation technology. In the foreseeable future increasing levels of data and detail will be added to the games and simulations within both fields from increasingly large sensor networks, to the integration of Building Information Modeling (BIM) to both game environments and Urban Planning. It may only be a matter of time until real world urban planning and game simulation converge to the point where both are subsets of some larger field of study.

Additional References and Links

Arrington, G. B., Jr. (1996). *Beyond the field of dreams: Light rail and growth management in Portland*. Portland: Tri-met. http://trimet.org/pdfs/publications/beyonddreams.pdf.

Arthur, W. B. (2009). *The nature of technology: What it is and how it evolves*. New York: Free Press.

Aschwanden, G., Halatsch, J., & Schmitt, G. (2008). Crowd simulation for urban planning. In *Architecture in computro [26th eCAADe Conference Proceedings/ISBN 978-0-9541183-7-2], Antwerpen (Belgium) 17–20 September 2008* (pp. 493–500). http://www.academia.edu/283856/CROWD_SIMULATION_FOR_URBAN_PLANNING

Badler, W., & Yang, (2010). Vehicle Traffic Simulation. http://www.joeweinhoffer.com/Weinhoffer_Yang_Final_Paper.pdf

Degusta, M. (2012, May 9). Are smart phones spreading faster than any technology in human history? *Technology Review,* Published by MIT. http://www.technologyreview.com/business/40321/

Frasca, G. (2001). Rethinking agency and immersion: Video games as a means of consciousness–raising. *Digital Creativity, 12*(3), 167–174. http://www.siggraph.org/artdesign/gallery/S01/essays/0378.pdf.

Kuhn, T. (1996). *The structure of scientific revolutions* (3rd ed.). Chicago, IL: University of Chicago Press.

Surowiecki, J. (2005). *The wisdom of crowds doubleday*. New York: Anchor Books.

Yaukey P. (2005). Patterns of avian population density, habitat use, and flocking behavior in urban and rural habitats during winter. http://onlinelibrary.wiley.com/doi/10.1111/j.0033-0124.1996.00070.x/abstract

"Monster Milk Truck"

Bergman and Associates: virtual design and traffic modeling

DOT Traffic Patterns for Urban Planning: http://www.navmart.com/navteq_traffic_patterns.php

Google Maps 3D: http://www.google.com/mobile/maps/3d/

http://en.wikipedia.org/wiki/Inception, Visual Effects section, second paragraph.

intelliDrive program

Rethinking Agency and Immersion: videogames as a means of consciousness-raising

Unity 3D and urban planning/architecture: http://archvirtual.com/2010/10/21/realtime-architectural-visualization-with-unity3d/

Unreal Heatmaps

Part III

Media User Interfaces, Experiences and Perspectives

User Perspectives on Media Convergence: Results from Quantitative and Qualitative Studies

Alexandre Fleury

1 What We Talk About When We Talk About Convergence

Convergence is an ambiguous and hard-to-define concept because it can refer to many different ideas and processes. "It is used differently, both with regard to what is converging (networks, terminals, social practices, etc), and with what happens when something converges (merging, new complexities, etc)" (Storsul & Stuedahl, 2007, p. 13). In academic and industrial publications, authors have used the term to describe any phenomena related to the development of digital media. Most however acknowledge that convergence occurs across media organizations (institutional convergence), across consumers (cultural convergence), and across platforms (technological convergence), in a carefully regulated environment (regulatory convergence).

Studying the blurring of mass and interpersonal communication, Lüders argues that converging forms of communication rely on three processes: interaction, participation, and social integration (Lüders, 2007). Firstly, interaction offers symmetrical communication channels, which typically contrasts with the asymmetrical communication process of mass media (to which new media is opposed). The bi-directional dialogue resulting from interactions should however not be over acclaimed or promoted as superior to mass media's monologue, which is expected to keep a significant role in future media consumption. Secondly, audience participation in new media differs from that of mass media as it is unformatted and unedited. Despite the issues raised by the integration of unedited material into their in-house created content, user generated content has become vital to media corporations who encourage audiences to voice their opinion and share stories and multimedia creations. Finally, social integration refers to the ritual functions of mediated communication, which Lüders considers reflecting a desire to situate oneself in relation to others. Overall, if "the opposite ends

A. Fleury (✉)
Aalborg University – Section for Multimedia and Signal Processing, Niels Jernes Vej 12, 9220, Aalborg, Denmark
e-mail: alexandre.m.fleury@gmail.com

A. Lugmayr, C. Dal Zotto (eds.), *Media Convergence Handbook - Vol. 2*, Media Business and Innovation, DOI 10.1007/978-3-642-54487-3_11

of the interpersonal and mass-mediated communication axis are still easily distinguishable" (Lüders, 2007, p. 195)—placing phone calls versus reading newspapers, an increasing grey area exists within which the boundaries between mass and interpersonal communication have become somehow irrelevant.

According to Jenkins, convergence is more than the tendency of various types of media to be available for consumption on a single device thanks to technological evolution. Instead, convergence occurs on five levels: technological, economic, social, cultural, and global (Jenkins, 2001). He argues against the early vision of convergence, describing the phenomenon as the advent of an "übercomputer", a singular device managing all accesses to media: "No single medium is going to win the battle for our ears and eyeballs. And when will we get all of our media funnelled to us through one box? Never." (Jenkins, 2001, p. 93). More than a technological process, convergence is mostly a cultural shift as audiences reach out for new information and build bridges between dispersed content.

While acknowledging "how central mobiles have become to the process of media convergence" (Jenkins, 2006, p. 5), the above definition clearly emphasizes the cultural aspect of media convergence over the accumulation of functions into today's mobile devices. Moreover, Convergence culture argues that media consumption has become a collective process, what Pierre Lévy called "collective intelligence".[1] To exemplify the outcome of this process, Jenkins explores the world of The Matrix as transmedia storytelling. The franchise indeed spread across a multitude of media including the three films, an animation series, two video games, and numerous web sites and online forums. The idea behind its success was to offer audiences multiple asynchronous entry points to the story, while disseminating subplots and hidden messages throughout the intrigue. This triggered the audience interest and desire to search for more about the story, which resulted in countless discussions and generation of fan made material that participated in further enhancing an already extremely rich universe.

Numerous authors are in agreement with Jenkins's perspective on convergence. According to Pavlik & McIntosh, convergence is "the coming together of computing, telecommunications, and media in a digital environment" (Pavlik & McIntosh, 2010, p. 8) and happens at three overlapping levels: technological, economic, and cultural. Firstly, "technological convergence refers to specific types of media, such as print, audio, and video, all converging into a digital media form" (Pavlik & McIntosh, 2010, p. 8). The resulting digital media transforms not only the nature of the original media but also the way it is generated and distributed. Considering a newspaper as an example, today's journalists often produce content using various audiovisual technologies, and readers increasingly access the content on portable media devices (such as the Amazon Kindle and tablets), which extend the basic functionality of simply reading an article to sharing it through social networks and searching the Internet for further information. Secondly, "economic convergence refers to the

[1] Pierre Lévy on Collective Intelligence Literacy, interview by Howard Rheingold: http://www.youtube.com/watch?v=8kCV4EEy2IE (31st January, 2013).

merging of Internet or telecommunication companies with traditional media companies" (Pavlik & McIntosh, 2010, p. 9). The outcome of this convergence is new corporations that control not only the content but also the distribution of media. A potential censorship issue however arises from such organizations. If in appearance the audience has access to a broader selection of content (visible for instance in the number of television channels available now compared to the early 1990s), choice might in fact be dictated by conflicts of interest inside media giants and distribution strategies focusing on the most profitable types of content. Finally, "one aspect of cultural convergence comes from the process of globalization of media content [, but] we can also look at cultural convergence from the perspective of how we consume, create, and distribute media content" (Pavlik & McIntosh, 2010, pp. 10–11). Concerning media globalization, this process helps content owners reach a broader audience. The latter perspective mostly concerns the fading border between mass and interpersonal communications. With an increasing ability to produce their own content and distribute it through online social platforms, audiences now play an important role in producing multimedia content that complements that of traditional media companies. Finally, by using the same distribution channels as their audience, media giants can increasingly adapt their communication strategies to better deliver their content to individuals.

1.1 Convergence or Divergence?

The diversification and blurring of delivery platforms put forward by Jenkins' description of transmedia storytelling can be argued to illustrate divergence or fragmentation rather than convergence. A first argument for a more diverging depiction of new media is the apparition of ever more technological apparatuses, putting consumers in front of a virtually infinite range of possibilities when acquiring new media equipment. In turn this diversity in supporting technology generates an equally vast range of options concerning how to access and create media. At the same time, one might argue that those devices integrate a limited list of functions, the packaging being the only differential element of the object. Nevertheless, taking the case of mobile phones, however similar in functions, different devices are perceived differently by consumers. This difference in perception is socially constructed (Campbell & Russo, 2003) and heavily influenced by the physical design of the device (Chuang, Chang, & Hsu, 2001, Suteja & Tedjohartoko, 2011).

From a content production perspective, the accessibility of tools and the professionalism claimed by audiences generating content gives birth to another converging yet diverging dimension of the media landscape, between the highly cinematic look of television productions, and the intimacy of online personal productions. Taking the example of online drama, Creeber argues for "the return of the intimate screen", which blends traditional and modern media broadcasting practices (Creeber, 2011). The spreading of participatory culture and the breaking of barriers between consumers and producers create great changes in the media production industry, a process thoroughly investigated by Caldwell in Production culture (Caldwell, 2008). Media producers need to acknowledge and leverage the power of online media alternatives to traditional distribution channels such as YouTube or Last.fm in their content distribution strategy. Considering YouTube, more than a platform for broadcasting home-made videos generated by non-professionals, it acts as a social filter, putting forward the material that people find interesting, disregarding its level of professionalism. In fact, despite the large number of home-made videos available, the most viewed are actually the professionally edited ones (Kruitbosch & Nack, 2008). At the same time, YouTube and television are getting more similar in terms of flow, liveness, broadcasting, form, and content (Vermeer, 2011). It is thus crucial to recognize YouTube's role in bringing professional content owners and producers together with end users, an increasing number of whom are becoming "produsers" (Bruns, 2007). As a final comment on YouTube, some consider the platform as the future evolutionary step of television (see for instance Uricchio (2009)), a vision that makes sense in the light of the increasing presence of YouTube applications on today's TV sets.

In the introduction to Convergence and Fragmentation: Media Technology and the Information Society, Ludes focuses on the European Information Knowledge Society and identifies both converging and diverging trends allowing "unity in diversity" (Ludes, 2008). According to him, achieving European media unity requires better understanding of culture-specific media and more transcultural initiatives to offer access to a common pool of knowledge. Elaborating on this idea, Heller (2008) reviews the trends in the global and European ICT development. Heller contrasts converging trends in ICT development (increased availability of similar devices and the possibility to access the same content on various media) to the risk of creating a greater digital divide as well as increased public fragmentation. As a result of these observations, she asks: "Do users/citizens/individuals become more dominated and passive by the entertaining functions of the new devices or do they become more active and partaking in growing democratic participatory discussions?" (Heller, 2008, p. 30) The many recent accounts of new media use as a mean of organizing citizens and expressing public opinion tend to support the latter development.

Looking specifically at the Scandinavian media landscape, Lund notes the common attributes shared by the Danish, Swedish and Norwegian media ecosystems. For instance the development of traditional mass media (press, radio and television) adopting a democratic-corporate model that differs greatly from that of Southern European states and to an even greater extend to the United States'. In this model, "the mass media are assigned a political role as proxy citizens, representing a diversity of voices and connecting the masses with competing elites" (Lund, 2007, p. 123). Mass media is thus supposed to avoid concentration of power and engage citizens in the democratic process. It also encourages news consumption, especially concerning public affairs and international issues (Curran, Iyengar, Brink Lund, & Salovaara-Moring, 2009). This perspective reminds of Castells' concept of mass self-communication, according to which the horizontal network of communication supported by online media brings about the possibility to participate in political discussions: "The media are not the holders of power, but they constitute by and large the space where power is decided." (Castells, 2007, p. 242) In practical terms however, the voices expressed on this network have good chances to remain unheard or ignored.

Finally, in order to appreciate the central role of mobile phones in convergence cultures, one may look at Japan and the "mobile-izing" process (Miyata, Boase, Wellman, & Ikeda, 2004, 2005, Miyata, Wellman, & Boase, 2005) started in the late 1990s. Not specific to Japan however, mobile users who use multiple media "tend to have stronger and more frequently contacted ties" (Miyata et al., 2004, p. 159). At the time Miyata et al.'s study was conducted (late 2002), mobile phones in Japan were primarily used to reinforce social ties within close physical proximity, rather than to maintain weak social connections, to develop one's network further, to reach for information, or to participate in online discussions. This supports the assumption that modern societies are embracing 'networked individualism' in which the individual is the unit of connectivity, rather than 'neighbourhood and kinship solidarities' (in which a group such as a household is the unit). The evolution of information societies from 'little boxes' to 'glocalized' networks and eventually to 'networked individualism' as observed in Wellman (2002) and Wellman et al. (2003) is partly due to the shift between place-to-place to person-to-person connections, illustrated by the shift between fixed to mobile telephony. It has been for instance argued that ICT "has the potential for transforming societies into networked communities or community networks" (Hofkirchner, 2009, p. 71). As a result of this process, Wellman wonders if individuals take responsibility for only strong existing relationships and not for weaker ties; and whether new social needs emerge from the reshaping of digital lives. The mobile phone is thus a socio-technical device reflecting the cultural context within which it is used (Ito, 2005).

The relative confusion in the above summary of the various cultural dimensions of convergence does not come as a surprise. In fact, Jenkins warned us already in 2006: "For the foreseeable future, convergence will be a kind of kludge—a jerry-rigged relationship among different media technologies—rather than a fully integrated system" (Jenkins, 2006, p 17). Keeping in mind that "the tools available to a culture matter, but what that culture chooses to do with those tools matters more" (Jenkins, 2009, p. 8), and that today's digital cultures rely on participation, remediation and bricolage (Deuze, 2006), this chapter therefore addresses how end users make sense of this bricolage and how the perceive media technology (focusing on television and mobile technology), as well as how they adapt their content consumption practices to the new multi-platform environment that is now available to them. In particular, we will explore how end users perceive and integrate second screens, which we understand as any device (smartphone, tablet, laptop) allowing TV consumers to interact with TV content displayed on a primary screen (typically a home television set). The three case studies reported later in this chapter have been conducted in Denmark and Japan, two highly connected societies where broadcasters and end users are faced with rather similar issues.

1.2 Status on Second Screens

Television-related technologies have evolved vastly lately, and so have evolved TV consumption habits (Cesar, Knoche, & Bulterman, 2010). On the one hand TV is changing its form, with the consumer moving from passive reception of one way broadcasts to being a part of an interactive media experience. The TV audience is starting to get used to having a much larger degree of control over the TV content. On the other hand, TV remains a social experience, yet the changes occurring at the interactivity level are intertwined with the social aspect of TV (Cesar, Bulterman, & Jansen, 2008). This evolution of television has also been established in Tsekleves, Whitham, Kondo, & Hill (2009, 2011): television is the dominant entertainment medium in the home for it allows relaxation and it is familiar and easy to use. In addition it is a shared resource and is associated with social interactions. A first observed converging media practice appeared between television and Internet. According to Hess et al., "TV and Internet are interwoven in their usage, especially for: (1) searching for information, (2) staying in contact with friends or colleagues and (3) the selection of devices for media consumption"(Hess, Ley, Ogonowski, Wan, & Wulf, 2011, p. 15).

At the same time, smart phones and tablets are making their way into the living room. As a result more TV viewers engage in media multitasking activities such as browsing the web while watching TV: In Denmark, 48 % of people above 15 years old report browsing the web on a smartphone of tablet at least weekly while watching TV, and 45 % of them focus on their Internet activity when doing so (Larsen & Svenningsen, 2011). Similarly in the US, Nielsen reported in 2011 that 40 % of tablet and smartphone owners use them while watching TV.[2] Therefore and according to Tsekleves, et al., second screens are positively welcomed for sharing photos, TV shows, other videos, and films. At the same time, TVs themselves would be used to share TV shows, photos, films, and other videos. Today broadcasters are striving to support this evolution and provide cross-platform solutions to deliver content to their audience, as it is the case with BBC's iPlayer in the UK (Thompson, 2010) and ARD and ZDF's Mediathek portals in Germany (Berthold, Schmidt, & Kirchknopf, 2010). Communication between content providers and end viewers increasingly becomes two-way instead of one way. From a research perspective, it is therefore interesting to investigate how to successfully combine television and mobile technologies in a cross media, or second screen environment.

Second screens have been on the agenda of interactive TV researchers since the mid-1990s, focusing on various aspects of the integration of the two devices. A number of prototypes have been designed and tested with potential end users, ranging from multimedia content manipulation for real estate agents (Robertson, Wharton, Ashworth, & Franzke, 1996) to content augmentation for language learning (Fallahkhair, Pemberton, & Masthoff, 2004). However prior to pushing such innovative products to the market, one needs to investigate if viewers actually want to have the opportunity to interact while watching a TV show, and if this provides added value to the TV experience. Such an experiment was recently conducted with eleven households, and reported in Basapur et al. (2011). In this

---

[2] 40 % of tablet and smartphone owners use them while watching TV: http://blog.nielsen.com/nielsenwire/online_mobile/40-of-tablet-and-smartphone-owners-use-them-while-watching-tv/ (31st January, 2013).

study the families were provided with a second screen prototype with which they were to interact while watching various TV shows for a period of three weeks. The enhancement of TV experience was found to be due to two factors: (1) the possibility of accessing extra relevant information immediately and after the show, and (2) the broadening of the experience to outside the TV room and to an extended social circle. Synchronization and relevance of content, variety in information sources, filtering of user generated content, and personalization of information were other characteristics found necessary to ensure the success of such service.

Recently, the extensive work by Cesar et al. on the concept of secondary screen illustrates the diversity of possibilities offered by multi-device media environments by exploring new interaction paradigms (Cesar, Chorianopoulos, & Jensen, 2008). According to the authors, the possibility to bring media content along on a mobile device when leaving the home TV set has been mainly investigated through a technological perspective, disregarding user studies. Cesar et al. introduce a taxonomy describing user behaviours in multi-devices media environment. The taxonomy includes content control (deciding what and how to consume TV content), content authoring (actively manipulating the TV content), and content sharing (socially communicating with others).

2 Case Study 1: Online Survey on Media Practices

This first user study investigates how the convergence between television and mobile technologies is experienced in highly connected societies. It relies on an online survey conducted in the spring of 2012, aiming at discovering everyday practices with televisions and mobile phones in Denmark and Japan. The survey explored mobile phone and television usage separately as well as jointly, in configurations of mobile television and of a second screen as defined previously.

The survey was intentionally short (17 topical questions in addition to three demographics-related ones) and presented in three languages (Danish, Japanese, and English). It ran for 7 weeks between March and April 2012 and was answered by 116 Danish and 102 Japanese respondents. Gender parity was respected in both regions, however the age ranges differed slightly in the sense that Danish pupils under 20 years old participated in the survey, whereas this population is not present in the Japanese panel.

2.1 The Use of Mobile Phones

Survey participants were first asked to describe the type of mobile phone(s) they carry around with them. The vast majority of Danes and Japanese only use a personal device, while few use both a personal and a professional one. This result is consistent across country and gender, but significantly[3] varies with age in Denmark: the 25 year old and younger Danes only use personal devices, contrasting with the other age groups. When it comes to everyday use of mobile phones, Japanese surf Internet, check their emails or calendar and use their mobile phone's camera significantly more often than Danes do. Not surprisingly, both populations reported surfing Internet and checking emails more often than they download content for or take pictures with their phone. Both populations also reported downloading content such as ringtones, games and apps rarely and in similar proportions, especially among respondents above 50 years old. Gender appears to have no incidence on the frequency at which Danes and Japanese perform these activities, except when it comes to downloading content, which seems to be a significantly more masculine activity in Denmark. Additionally, the following differences across age groups appear in the frequency at which each activity is performed.

Internet Surfing

- Danes aged 35 and under access Internet on their mobile phone more often than Danes above 50
- Japanese under 26 use mobile Internet more often than any other age group

Checking Email/Calendar

- Danes between 26 and 35 check their emails on their mobile phone more often than Danes under 26
- Japanese under 26 do so more often than Japanese above 50

Downloading Content

- Danes under 26 download content for their mobile phone more often than any other age group
- In Japan, they do so more often than people aged 26–50

Taking Pictures

- Danes above 50 use their mobile phone's camera less often than any other age group
- Japanese between 26 and 35 use their mobile camera less often than Japanese under 26 and between 36 and 50 years old

Important buying factors are also found comparable in both countries: functions, design and price are the top three criterion influencing device acquisition. On the contrary, experience continuity (similarity to previously owned phones) and social

[3] All differences expressed related to age, gender, and country are statistically significant at a level $p < 0.05$.

conformance (similarity to devices owned by colleagues and friends) are of negligible importance.

These buying criterion are coherent with what respondents reported liking and disliking about their mobile phone, even though differences exist between the two countries. Indeed the vast majority of Danes like mostly the phone's functions (calling, accessing Internet) and applications in the case of smartphones, whereas Japanese are mostly positive about the design of the phone, slightly more than its functions. The importance of mobile phone functions in the perception of mobile phones had been previously established for Japanese and North Europeans (Swedes) in Baron (2009). In both countries issues related to performance and usability are the main sources of dissatisfaction with mobile phones. The phone specifications are also a source of concern for Danes.

Finally, Danish and Japanese respondents expect to react slightly differently in case they could not use their mobile phone any longer. Among the six basic emotions (anger, sadness, surprise, joy, disgust and fear) identified by Ekman, Friesen, and Ellsworth (1972, 1992), and validated cross-culturally (Ekman & Friesen, 1971, Ekman, 1989, Elfenbein & Ambady, 2002, Sauter, Eisner, Ekman, & Scott, 2010), both Danes and Japanese chose sadness as the emotion they would most likely experience if they were suddenly deprived of mobile phone. However, while the sadness experienced in Denmark would be tinted with anger, Japanese sadness would be mixed with disgust. The varying responses in Denmark and Japan concerning disgust might be partially explained by how this feeling is interpreted in the two cultures. The Japanese translation of disgust includes the notion of getting tired of something, of losing interest and getting bored, while the Danish translation reflects loathing and aversion. Surprise, fear (and anger in the case of Japan) follow closely these feelings. Furthermore, participants mentioned a large amount of additional feelings they anticipate to experience if they were to no longer have access to their mobile phone. Here again the two populations vary slightly in their answers. Danes cited in majority irritation and a mixture of powerlessness, frustration and loss, while Japanese referred to a combination of anxiety, impatience and trouble. It could be argued that the consensus reached by participants on those emotions cited spontaneously makes a compelling argument for their accuracy at describing how one might feel without mobile phone, in complement to the basic emotions.

2.2 TV/Video Consumption Habits

Survey respondents from both regions reported similar amounts of daily video consumption. The terms video is used here in a broad sense and includes live and recorded TV, videos from online services or stored in a personal archive. The survey shows that 52 % of Danes and 48 % of Japanese spend on average 2-4 hours daily watching such types of video content, while 43 % and 48 % do so for less than one hour. Only a small fraction of respondents (5 % and 4 %) reported dedicating more than four hours per day consuming videos. No gender effect appeared in the answers, and only the Danish population displayed significant age differences: Danes below 35 reported watching significantly more video content than their older peers.

The type of device used to watch this video content was also investigated. Participants should indicate the frequency at which they consume videos on a home TV set, a computer, and a mobile device. Overall results show again very similar trends in Denmark and Japan, with home TV sets being the favoured devices for video consumption, shortly followed by computers, mobile devices being less often used. The amount of use for each device is in line with current global trends, as reported by Nielsen Media Research.[4] However a closer look at these results brings up interesting findings. Danes above 50 use the home TV set significantly less often than between 26 and 50, and Japanese women use the TV significantly more than Japanese men. Males in both countries use PCs significantly more often than females, and Danes below 35 use them significantly more than their elders. Finally, Danish men watch video on mobile devices significantly more than Danish women, and Danes between 26 and 35 years old do so significantly more often than Danes above 50 years of age.

With regard to the type of content watched on television, news is clearly the most popular genre watched in both Denmark and Japan, with 70 % and 79 % of respondents mentioning it, respectively. Then, while movies are following closely in Denmark (69 %), entertainment and reality shows appears to be the second most popular genre in Japan (67 %). Interestingly, series and sitcoms are mostly popular in Denmark, and while documentaries are often watched in both regions, talkshows are seldom mentioned.

Similar to with mobile phones, respondents were asked to imagine that they had no longer access to their television. Ekman, et al.'s six basic emotions were again used as a basis for the analysis, which yielded slightly different results than for mobile phones. Danes expect to experience a mixture of sadness (22 %), anger (19 %) and surprise (18 %), whereas in Japan, sadness (30 %) would be tinted with disgust (16 %) and surprise (13 %). Similarly than with mobile phones, disgust might be understood slightly differently in Japanese (the translation here being close in meaning to irritation) than in Danish (loathing, aversion). Additionally, an especially large number of respondents (34 % of Danes and 48 % of Japanese) found that none of the six basic emotions suggested represented well their expected feeling in this situation. A potential explanation is actually the lack of reaction that would experience the respondents with no longer TV access, as suggested by the additional emotions freely cited by participants. Indeed, indifference was explicitly mentioned by 8 % of Danes and 11 % of Japanese. In addition, irritation was cited by 11 % of Danes. Finally, respondents described what they like and dislike about their television. Answers collected in Denmark differed visibly from those collected in Japan. The two predominant sources of positive feedback in Denmark are

[4] Nielsen, Global Report: Multi-Screen Media Usage: http://blog.nielsen.com/nielsenwire/global/global-report-multi-screen-media-usage/ (31st January, 2013).

the content (cited by 28 %) and the purpose of watching TV (23 %), which includes catching up with news and being entertained. In Japan however, the TV's performances are the main source of satisfaction with television (30 %), followed by content (18 %) and purpose (14 %). The two populations also differ in the features they dislike about TV: it is considered a time waster by 22 % of Danes, and the content available is a source of dissatisfaction for 20 % of them. Japanese complain mostly about the content available (32 %) and the specifications of their TV set (26 %). As it could be expected, television content holds thus an ambiguous yet critical role in the TV experience, as it is the source of both positive and negative feelings. Tensions were also visible in comments in which TV was considered a time waster: Often TV is appreciated for its relaxing or informative purpose but watchers tend to get caught up and keep on watching even though they lose interest in the programme, which leads to frustration.

2.3 Mobile TV Use

The survey briefly investigated mobile TV usage, asking respondents to report their interest in watching television on their mobile phone in various situations. In both countries, commuting time as well as waiting situations (for public transport, at an airport, or at a café) are the most likely situations for watching mobile TV. At home before sleeping is additionally considered a favourable situation for mobile TV consumption. These results are consistent with previous research (see for instance Repo, Hyvönen, Pantzar, and Timonen (2004), O'Hara, Mitchell, and Vorbau (2007) and Oksman, Ollikainen, Noppari, Herrero, and Tammela (2008)). Danish respondents expressed more interest in watching TV on mobile phones while commuting, during short breaks at work, and while waiting for public transports than Japanese. Somehow surprisingly, the younger population expressed as limited an interest in watching mobile TV in these situations as their elder did. The only age effects noticed are the following: Japanese aged 36–50 are less interested in mobile TV during work breaks than Japanese aged 25 and younger, and above 50; and Danes aged 26–35 are more interested in mobile TV while waiting for public transports than Danes younger than 25 years old.

As seen previously, interest in mobile television remains quite low, at least considering the situations investigated in the previous question. Respondents were then asked to name additional situations in which they are interested in watching TV on a mobile device. Danes and Japanese have different interest in mobile TV: travelling and waiting situations are prioritized in Denmark by respectively 24 % and 21 % of those who cited other situations of interest. This further confirms the results to the previous question. It should also be noted that 26 % of Danes explicitly mentioned not being interested in watching television on mobile phones. Comparatively in Japan, mobile television is appealing in case of emergency situations and to be informed in case of a disaster occurring. Additionally, mobile TV is seen as a potential companion during specific activities, especially at home, such as taking a bath.

2.4 Second Screen Viewing Practices

The final part of the survey addressed the use of a mobile device in front of TV, as a second screen. First, survey participants were asked to indicate whether they use other electronic devices (such as mobile phones, tablets, fixed or mobile computers) while watching television. This appears to be a common practice among both Danes and Japanese, as about half of all respondents reported doing so often, and more than one in three sometimes. Danish respondents aged 50 and above engage in such media multitasking less often than members of the other age groups. Once again, the findings are comparable across region and gender.

The survey then investigated the activities one can engage in while the TV is on. First, respondents were to indicate how often they discuss content with co-viewers, check their emails, surf Internet, and chat with remote friends. Overall, checking emails and surfing Internet are the most popular activities in both regions, with more than three respondents in four reporting doing so at least sometimes. However, Danes chat with remote peers more often than Japanese do. Gender differences also appear in Japan: women discuss content with co-viewers and check emails more often than Japanese men. Moreover, age differences are found in Denmark: Danes above 50 years old surf Internet less often than any other age group, and when it comes to chatting online, young Danes under 25 do so more often than Danes above 35.

When asked to cite other activities they usually engage in while watching TV (including non-media related tasks such as cleaning or cooking), 30 % and 47 % of Danish and Japanese respondents mentioned eating, followed by housework (22 % and 19 %). Communicating (via mobile phones), and playing games (also on mobile devices) are other frequent activities in Denmark, as they were mentioned by 20 % and 19 % of Danish respondents. In Japan however, cooking (16 %) and working/studying (16 %) follow in the list of activities conducted while watching TV.

Additionally, respondents were prompted for their interest in a selection of second screen activities, by expressing how likely they would perform these activities if their television was connected to a mobile phone or tablet computer. Overall, checking other channels and accessing more information about the current TV programme are the two second screen activities Danes and Japanese alike most likely envision engaging in while watching TV. Regional differences appear only concerning interest in checking other channels with the secondary device. Danes reported being more likely to do so than Japanese. Gender is a source of differences regarding interest in participating in a poll related to news, and to send comments to a TV show reviewing consumer products: in both cases, Danish males are more likely to engage in those activities than Danish females. Finally, Danish age groups differ in how likely they would send comments to a consumer product show (Danes between 26 and 35 are more likely to do so than Danes below 25 and above 50 years old), access more information about the content of the show being watched (Danes between 26 and 35 are more likely to do so than all other age groups), and chat with friends about the programme being watched (Danes 25 years old and younger are more likely to do so than Danes above 35).

Finally, a few participants mentioned other activities they would be interested in for second screen services. These include reaching more information about the current programme, which further emphasizes the finding from the previous question, as well as communicating with peers via social network. Specific functions appeared also relevant to participants, such as using the second screen as a remote control.

2.5 Culture-Specific Considerations

The answers collected through this survey on everyday media use in Denmark and Japan reveal similarities and differences with regard to how the two regions perceive and use mobile phones and televisions, and apprehend the convergence of the two technologies. In terms of current technology use, it is not surprising that the Japanese access Internet, check emails and use the phone's camera significantly more often than the Danes do. The similarity and low level of mobile video consumption in the two countries is however surprising, considered the longer history of popular access to video content on Japanese handsets. Similarly, mobile television generated only a limited interest in both regions, and Danes actually expressed more interest in what could be considered typical mobile TV settings (such as commuting, or waiting for public transports), in which Japanese actually have the possibility to watch TV on mobile phones (and have been able to do so for about six years). Potential explanations include (1) the novelty effect on the Danish consumers who are starting to experience TV on the go, (2) the lack of valuable content on the Japanese mobile TV market, limiting the use and interest in the service as it is now, and (3) the lack of infrastructure for mobile TV on the commuting network within metropolitan Tokyo, the main area Japanese respondents are from.

As deduced from the answers concerning second screen television viewing, both the Danes and the Japanese are very active while watching television. They engage in both media and non-media related activities, which confirms results from earlier studies about multitasking in general (Brasel & Gips, 2011, Tokan, 2011) and on the Danish population in particular (Thunø, 2012). For instance both Brasel & Gips and Thunø found that people involved in computer based activities while watching TV attend more to the computer than to the TV. Another interesting finding regarding media multitasking is the extent to which Japanese report working while watching TV compared to Danes. This can be explained through the lens of early anthropological work by Hall on polychronicity (preference for conducting multiple activities simultaneously) versus monochronicity (preference for finishing an activity before starting another one), showing that Japan is an exception in the developed countries in that it is inherently more polychron than monochron (Hall, 1984) (although this statement has been contested since, for example in Lindquist, Knieling, and Kaufman-Scarborough (2001)). Regarding specific use of second screen activities however, only few of the suggested activities were considered at least likely to be performed by respondents. Checking what happens on other channels as well as accessing more information about current programmes have received the most favourable likeability scores, perhaps due to the ease to grasp and comprehend those activities without actual experience with them, compared to the other suggested hypothetical activities. Moreover, the high interest in news content on television reported by respondents from both countries and the interest in accessing news on mobile phones in those regions discussed in Westlund (2010) demonstrate the importance of this content genre for those two regions representative of ICT societies.

3 Case Study 2: Workshops on Second Screen Applications

A series of workshops has been designed for investigating not only acceptability of second screens from a conceptual point of view, but also two more specific issues inherent to interactive television services: Prompting strategies, and separation or combination of content and interactive functions. On the one hand, prompting strategies address the issue of engaging viewers in using the interactive service relevant to the TV show currently watched. For instance, assuming a quiz game show offers its audience the possibility to play along live participants, how to grasp viewers' attention toward the second screen where the interaction takes place? On the second hand, content/interaction separation refers to the distribution (merging or splitting) of the live TV stream and interactive elements between the primary and secondary devices. In order to explore those issues, we invited 23 Danish media consumers between 18 and 60 years old to try prototypes and engage in card-sorting like activities, in four groups of 5–6 people. The prototypes illustrate interaction schemes associated with four genres of TV shows: (1) a news show (TV Avisen, a national news bulletin), (2) a consumer show (So Ein Ding, which reviews new technology products), (3) a talk show (Aftenshowet, in which a couple of hosts discuss news related topics with guests), and (4) a quiz show (Hvem Vil Være Millionær?, Danish version of Who Wants to Be a Millionaire?). Table 1 presents the types of interaction associated with each of these shows.

The tasks performed by workshop participants with the prototypes consisted in engaging with the interactive activities described in Table 1 while an extract of the corresponding TV show was played on the wall-mounted TV screen. In addition to

Table 1 The four TV shows and associated interaction schemes

TV show	Interaction type
TV Avisen	Retrieve more information about the news items presented. The extra information retrieved is displayed on the second screen
So Ein Ding	Participate in poll (like/dislike) about a product being reviewed. Results from the poll are displayed on the primary screen, as an overlay to the live video feed
Aftenshowet	Submit comments related to the programme. The comments are displayed in a ticker rolling at the bottom of the live video feed on the primary screen
Hvem Vil Være Millionær?	Answer questions simultaneously with live participants. Answers (correct/incorrect) are displayed on the secondary screen

the four basic scenarios, the Aftenshowet and Hvem Vil Være Millionær? prototypes have been modified to address specifically the issues of prompting strategies and content/interaction separation. Participants experienced different prompting strategies through the Aftenshowet prototype, while Hvem Vil Være Millionære? illustrated two variations of the game where content and interactivity are either separated or combined on the secondary screen.

Prompting Strategies The first prompting method investigated occurs on the primary screen by way of a ticker message encouraging viewers to send comments via the app running on the secondary device. The second method occurred on the secondary screen, via a message popping up while the user is using the app. In both cases the task scenario required participants to be using the app when the prompting occurred. This helped not only keeping the test conditions similar, but also illustrating an issue inherent to prompting on secondary devices: how to draw the attention of viewers toward the secondary device when it is not already in use.

Content/Interaction Separation To illustrate how interactivity and content can remain separated, the interactive content (that is the questions to the quiz show and the clickable options to cast answers) are displayed on the secondary device while the live TV show plays on the TV screen. To illustrate a case where those two are combined, both the interactive content and the live video feed are displayed on the secondary device, in sync with the TV video feed.

The prototypes developed (depicted in Fig. 1) were tested by two populations. In the first two workshops (W1 and W2), 5 males and 6 females between 38 and 58 years old were recruited from the "DR panel", which consists of approximately 7000 voluntary Danish citizens above 15 years of age. Comparatively, 6 male and 6 female students from Aalborg University in their early 20s were equally distributed among the last two workshops (W3 and W4). All 23 participants are active media consumers and demonstrated a medium to high level of IT literacy.

The four workshops generated 813 statements from the 23 participants trying out the prototypes and engaging in discussions following a semi-structured interview approach. The statements were systematically classified according to five categories: (1) overall user experience (9 % of all comments), (2) interaction and content (42 %), (3) prompting strategies (9 %), (4) user-generated content (19 %), and (5) usability and practical issues (18 %).

3.1 Overall User Experience

Participants showed interest in the interactive features they experienced during the workshop and expect such innovations to be part of the future television experience, under the condition that they are easy to access and use. Some older participants however felt that they were not the main target audience for such features. Critics were emitted toward the possibility of disrupting the TV experience, by removing

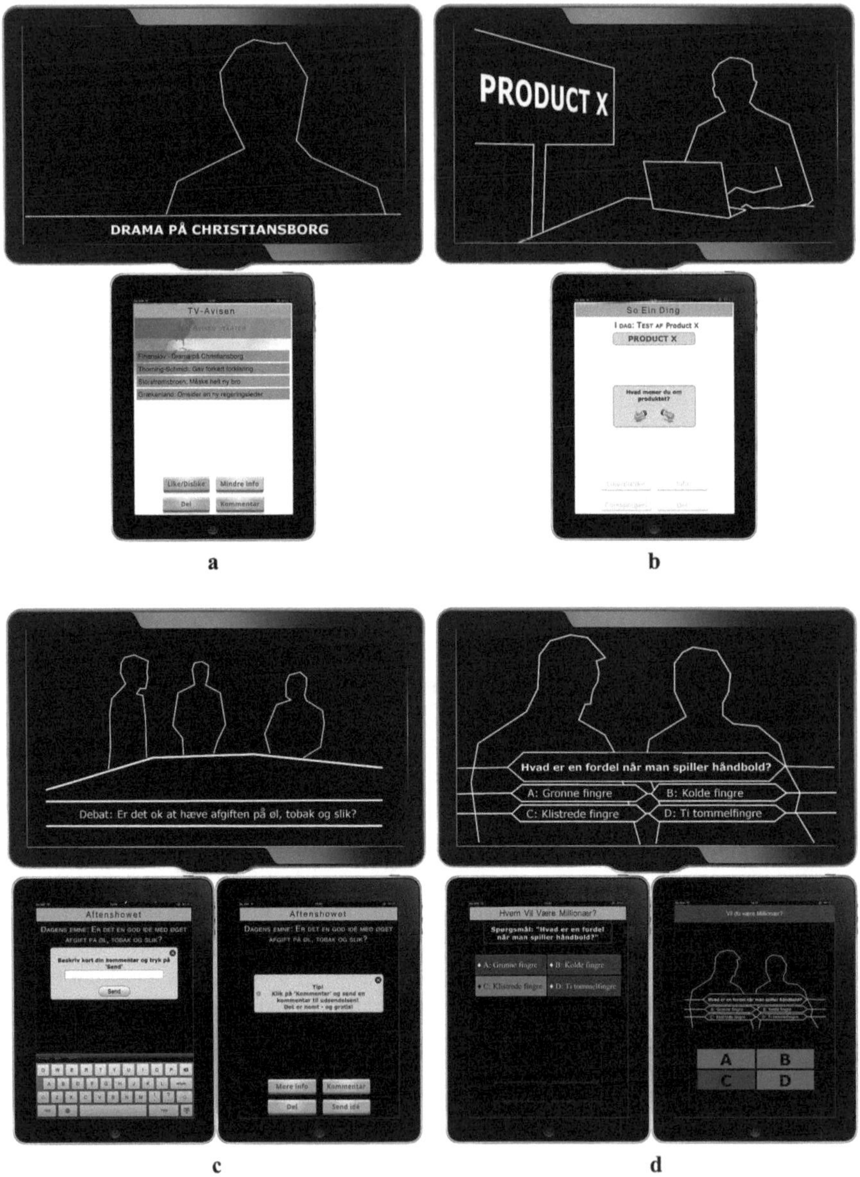

Fig. 1 Prototypes of interactive second screen services designed for the workshop: (**a**) TV Avisen, (**b**) So Ein Ding, (**c**) Aftenshowet, (**d**) Hvem Vil Være Millionær? (Images altered for copyrighting)

its relaxing property or shifting the viewers' focus away from the broadcast content. Participation should be left up to the audience, who should thus be able to switch on and off the interactive services as desired. General interaction between audience and broadcasters was also discussed: some programmes (such as talk shows) are suitable for submitting user-generated content, while others (such as news programmes) are seen as "receive-only" programmes.

3.2 Interaction and Content

Participants from all workshops stressed that certain interaction types are only relevant to certain programmes, for instance voting for quiz shows. In general entertaining programmes are better suited for interaction than informative ones. The possibilities to access more information about the current programme and to bookmark a programme for later revision or as a reminder to further explore are clearly seen as useful features. Uploading user-generated content and sending comments/questions to the programme editors also generated interest. This requires (1) the user generated content to match quality requirements, and (2) a dialogue with the editors is established, instead of using the feature as a mere feedback channel. Discussions about having the TV show running on the secondary device generated much debate. Participants clearly agreed that the TV content belongs on the primary screen and interactive content and controls on the secondary device. Furthermore having both content and control on the secondary device may render the primary screen irrelevant, which corresponds to watching mobile TV.

Nevertheless, watching TV shows on the secondary device is considered convenient under certain circumstances, such as when leaving the room where the primary screen resides without missing out on the TV experience or when the primary screen is occupied by other viewers watching a conflicting program. The participants wanted to be able to control if the show should be running on the secondary device, in sync with the content broadcast on the TV screen.

3.3 Prompting Strategies

Participant agreed that they need direct solicitation for using the interactive services available on the second screen, as this is not yet a common practice. At first, almost all participants agreed that prompting should not happen on the primary screen. However, participants also wondered how one would be made aware of the opportunity to interact with a TV show. Discreet prompting solutions such as a soft sound, an icon in the corner of the TV screen, slightly animating the channel logo, or specific timings (before programmes or during breaks) were suggested.

Moreover, participants generally agreed that the second screen setup does not render the primary screen superfluous. This ambiguity is clearly due to the fact that TV consists of both audio and video, and in many cases the audio is quite sufficient for viewers to continue following a TV show even when engaged in other activities on a second screen. This is expected to be especially true with low engagement TV shows such as entertainment or sport, which are particularly suited for second screen services.

3.4 User-Generated Content

This topic was the most discussed one in the third workshop, and generated diverging opinions in all groups. All however agreed on the need to make access to user-generated content optional. UGC should also be clearly indicated as such, moderated (even though who should do it is unclear), possibly edited (to keep the broadcast professional looking), and dissociated from personal identifiers (although contributors should be held responsible for their submission). As previously discussed, content submitted should be acknowledged by programme editors to encourage further participation. Risks related to spam, hidden advertising, and censorship were also recurrently mentioned. These issues are perceived less critical if the content is available on the secondary device, which leaves us to believe that people consider content displayed on primary (TV) screens as more authoritative, and requiring less critical thinking.

3.5 Usability and Practical Issues

All groups confirmed that interactive features should be optional or filtered, and that it should be easy to access and use. The content on the primary screen should be kept minimal and uncluttered. The presence of a ticker for displaying viewer comments was much debated in the second workshop, but no clear agreement on whether its presence on the primary screen is useful or disturbing was reached. Keeping broadcast content and interactive features separated appears sensible when sitting in front of the primary screen; however it should also be possible to access them both on the secondary device when leaving the primary screen. Especially with quiz shows, using only the secondary screen made sense to most participants. Moreover, they suggested binding all interactive functions into a single application, regardless of the channel, programme, or type of interaction to be accessed, and to offer a Twitter-inspired feed of extra information available running on the secondary screen as design proposals. Additional statements about long lasting battery time, big screen size, fast system response, and general ease of use were coherent with previous findings concerning second screens (Basapur et al., 2011), mobile TV (Knoche & Sasse, 2007), and general usability principles (Nielsen, 1994).

3.6 Social Context

In workshops 3 and 4 (involving younger participants), the social context took an important part of the discussions. It is expected to drive viewers' desire to engage with interactive features. According to the participants, primary interaction (happing face-to-face between co-viewers) would take precedence over mediated interaction, except when all co-viewers can engage in for instance a quiz show.

3.7 Findings Summary

Overall the interactive features were well received, although some programmes are considered more suitable than others for supporting interaction. Audiences are not yet used to interact with TV shows through secondary screens and thus require to be prompted when interactive features are available. Such prompting should be very discreetly advertised on the primary screen, in order to only redirect potentially interested users toward the secondary device. This discussion puts forward an ambiguity inherent to the second screen paradigm: How to involve viewers in a secondary activity that takes away their attention from the primary screen while keeping their focus on the broadcast program? Similarly, study participants demonstrated little interest in mixing live TV content and interactive functionalities on the secondary screen, unless in specific situations. The primary screen can be used to display collaborative contributions such as poll results, but not personal opinions. The TV receiver is thus mostly dedicated to content playback, while value adding interactive services belong to the second screen.

On a side note, the workshops revealed discrepancies concerning how individuals perceive TV audiences: some believe that television consumers are critical and capable of making sense of what is reliable and criticize what is not; while others expressed their concern about audiences potentially being either flooded with debilitating content or otherwise manipulated through inappropriate censorship. These issues have been widely discussed by mass media scholars although we believe they require further attention within transmedia contexts. Indeed the increasing amount of information available via second screen services makes critical interpretation of content more difficult, especially when sources for content increase in number and types.

4 Case Study 3: Drawings on Personal Relationships with Media

This last case study explores personal relationships between end users and media technology, once again focusing on television and mobile technologies. The following reports the results from four drawing exercises conducted in Denmark and Japan, for which participants (mostly young and highly IT literate) were provided drawing material and were instructed to:

1. Draw the layout of their home, placing media devices where they are mostly used, and indicating those used simultaneously
2. Illustrate an impressive memory involving television
3. Depict themselves and the mobile devices they usually carry around with them
4. Illustrate an impressive memory involving mobile phone

Although drawings are rare in human-computer interaction studies, they have been extensively used in children psychology and more generally in medical studies as well as in education, and they are a common communication tool for designers. The benefits and challenges of the method have been more extensively discussed in (Fleury, 2012) and can be briefly summarized as follows:

- Drawing helps create a relaxed and comfortable atmosphere in which test participants are willing to express personal matters
- The absence of boundaries in drawings further encourages participants to reveal personal aspects of their life
- Responses are influenced by the experimental setup
- Drawings should be used in triangulation with other research methods

4.1 Media Use at Home

First and similarly to the study in Brush and Inkpen (2007), participants were asked to draw the layout of their home, and to place media devices where they are normally used. The media equipment depicted in the drawings (see examples in Fig. 2) collected is quite consistent across countries. Televisions are present in the large majority of the drawings, followed by computers (fixed and mobile), mobile phones and radios. Despite analytical caution preventing a simple count of the devices drawn, it appears quite clearly from the drawings collected that televisions and computers are the centre of Danish and Japanese in-home media life. They are sometimes connected to each other, or used simultaneously. In Japan more than in Denmark, computers are classified either as fixed (referred to as "PC" or "desktop") or as mobile ("laptop" or "notebook").

Even though all participants reported owning at least one mobile phone, only slightly less than half of them included mobile phones in their drawings. Japanese represented fixed phones more often than Danes, which can arguably be attributed to the age difference between the two populations of test participants. Younger respondents are expected to use mobile phones more often than fixed ones, even though a fixed phone might actually be present at home.

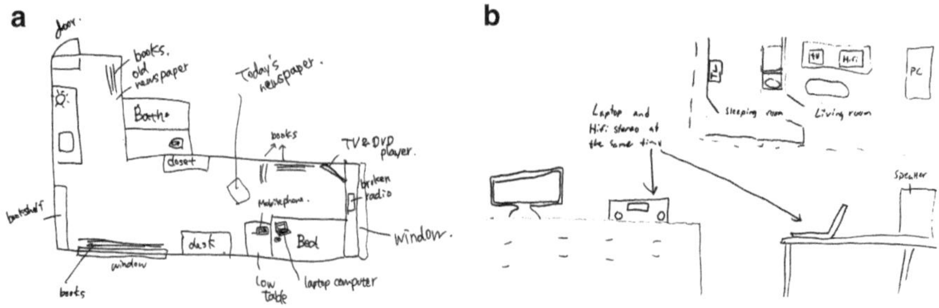

Fig. 2 Illustrations of home depictions with various media equipment: (**a**) Home depiction (Female, 28—Japan), (**b**) *Laptop and Hifi stereo at the same time* (Male, 22—Denmark)

4.2 Personal Memories About Television

The stories depicting memories related to television collected in Denmark and Japan (illustrated in Fig. 3) can be classified according to the following three categories. Most of the drawings created in Japan belong to the content category, while Danish drawing are more equally spread across the three categories.

Device Drawings in this category depict events related to the device itself, either in terms of ownership, acquisition, or use. The Danish submissions in this category cover all three subcategories, while the Japanese ones are all related to situations in which TV is used.

Content Content related contributions are associated either with a specific genre (most often tragic news) or the action of producing material, participating in a show, or watching a specific show.

Social Submissions in this category focus on the social environment of the memory illustrated in the drawing. Friends, family members and strangers are equally represented in both countries.

Another possible categorization of the drawings submitted consists in considering only who is depicted, notwithstanding the topic illustrated. By doing so small

Fig. 3 Illustrations of TV related memories associated with the device itself: (**a**) *When I was 8~11 years old. I like TV very much. I sat nearby TV, changed channels using my toe skilfully* (Female, 31—Japan), (**b**) *When we took our new TV home at the back of the cycle* (Female, 22—Denmark)

differences between Danish and Japanese drawings emerge. While the majority of Japanese illustrations depict the drawer alone, and to a lesser extent with family members, Danish drawings represent the drawer alone or among strangers in equal proportions.

Studying the mood expressed in the drawings on a negative-neutral-positive scale reveals other differences between the two countries. Danish drawings are mostly emotionally neutral while the Japanese ones are equally distributed among the three emotions.

4.3 Self-Depictions Carrying Mobile Devices

When asked to sketch themselves with the mobile devices they usually carry around, Danes and Japanese produced quite different drawings. These differences concern the number and types of devices represented in the drawings, and the place around the body where mobile phones are depicted. The analysis of the types of devices included in the drawings informs how participants understand the concept of "mobile device", the definition of which being left to the participant.

Firstly, the number of devices included in the drawings is age- and gender-independent, however Japanese tend to include more devices in their drawing (2.5 devices per drawing in average) than Danes (1.8). This means either that Japanese actually carry more mobile equipment in their everyday lives or that the mental association mobile device equals mobile phone is stronger in Denmark, which hinders the consideration of other types of apparatuses as being mobile devices.

Secondly, all drawings in both countries include at least one mobile phone. This might not be surprising given the ubiquity of such device in Japan and Denmark, and considering that all participants reported owning at least one mobile phone. Still, this reinforces the established importance of mobile phones in everyday mobile lives in digital societies. Laptops were represented equally by one third of the participants from both countries. Equipment used to listen to music such as iPods were depicted in one out of four Danish drawing, while the same proportion of Japanese though about other IT equipment (including iPads and external drives). Finally, wrist watches were illustrated by one in five Japanese participants together with other non-IT equipment such as transportation cards and pens (one participant even mentioned his wedding ring), while none of the Danish participants mentioned any of such objects. See examples of submissions in Fig. 4.

4.4 Personal Memories About Mobile Phone

Similarly to television related drawings, depictions of mobile phone related memories can be categorized according to either the topic described in the picture, the social surrounding depicted, or the emotional charge of the illustration (see examples in Fig. 5). For what concerns the topics, Danes and Japanese reported in

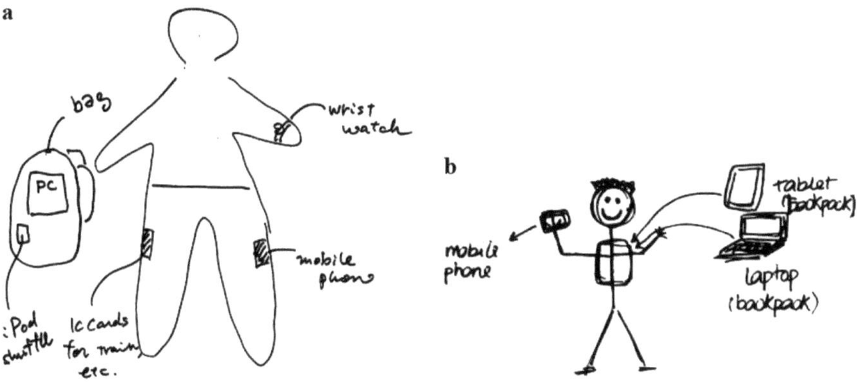

Fig. 4 Illustrations of self-depictions with mobile devices usually carried around: (**a**) Self-depiction including a *PC* and *iPod shuffle* in a *bag*; *IC cards for train, etc.* and a *mobile phone* in pants pockets; and a *wrist watch* (Male, 47—Japan), (**b**) Self-depiction including a *mobile phone* in the hand; and a *tablet* and a *laptop* in a *backpack* (Male, 22—Denmark)

Fig. 5 Illustrations of memories of mobile phone use: (**a**) *Google everything every time everywhere* (Male, 31—Japan), (**b**) *Capture memorable moments* (Male, 22—Denmark)

similar proportions stories in which mobile phones have been used or misused. On the one hand, "using" includes specific situations such as taking pictures, or following directions from the embedded navigation system. On the other hand, "misusing" situations reflect circumstances in which the device has been lost, broken, or subject to inappropriate handling (such as dropping the phone in water).

Common to the Danish and Japanese drawings is the social settings the memories are set up in. In both cases drawers depicted themselves alone, and only few mentioned either friends, family members, or strangers. Situations involving colleagues were sketched only by Japanese participants, while some Danes represented nobody in their drawing.

With regard to the emotional load carried by the drawings, the Japanese clearly focused on negative feelings, twice as much represented than positive ones, and four times more frequent than neutral ones. In Denmark however, the analysis is difficult due to the high level of ambiguity in the feelings depicted. It is not possible to conclude the emotional stance of almost half the drawings. As for the remaining half, they tend to be mostly positive oriented, slightly over neutral.

5 Near Future Trends for Converging Media

Considering the findings from the three previous case studies, the following identifies possible development of media consumption in the near future. On a global scale, economies are converging, which is to say that the growth rate of smaller economies is higher than that of the larger ones. Technology is responsible for the increasing rate at which this happens. The big players of the media economy such as Google and Facebook achieved internationally recognition, similar to Coca-Cola or Nike a few decades ago. For end users, those brands become technology ambassadors and remodel our relationship with media. People "google" things (including themselves), and exchange links to Facebook profiles when meeting for the first time. The most popular online social networks aggregate communities and offer tools for them to develop and for their members to communicate.

It is thus crucial to study these emerging practices as they reshape the way people relate to and access media. From the outcome of the studies reported in this chapter, we have identified not only general trends concerning future media technology, content, and human-computer interaction, but also specific usability aspects of upcoming services such as mobile TV and second screen features.

Overall, the trends identified in Miller (2012) concerning the near future mobile TV landscape have been generally confirmed. However, we believe that these trends in fact reflect a more general shift in media consumption paradigms than the mobile TV market alone. Following the report's nomenclature, one can expect the five following directions to drive the future of the converging media market.

Changing Consumer Attitudes Compared to a few years ago, people have included new media-related habits to their everyday life. By studying the media consumption patterns of today's early technology adopters, this chapter provides an informed insight of the soon to become standard media life in ICT societies. Nowadays' one expects more flexibility in where, when, and how to access digital content. Furthermore, audiences expect more interactivity with content providers, and a growing mass of end-users turn into produsers, taking part in the national (and global) debate by contributing with user-generated content.

Ubiquitous WiFi Internet is becoming a commodity, and "free" access to the web is now the norm at home, at the office, in most hotels and in an increasing number of public places. Connectivity on-the-go is however still an issue according to the participants in our studies, for whom network availability, quality of service, and price are major concerns. The rapid development of mobile broadband technology is however expected to alleviate these issues and rapidly fully support mobile lifestyles.

Entrance of Tablets to the Market Tablets are expected to play a major role in front of televisions, and broadcasters see the platform as a game changer in how audiences interact with TV content. In fact, interactive services are being tested and receive franc success by technology-minded audiences.

Internet TV and IPTV Soon the switchover from analogue to digital television will be completed on a global scale. All-digital TV content means easier distribution over a larger number of channels: TVs, PCs, and mobile terminals are now equally capable of delivering TV content to end users. Relying on IP technology to transmit content also means that a return channel becomes available, supporting the interactivity called for by audiences and content providers. Both parties indeed have a growing interest in developing a bi-directional communication channel.

The Trend for Multiscreen The concept of multiscreen can be understood in two related ways. Firstly, from the perspective of the ubiquitous media paradigm, according to which content can be consumed regardless of the platform it is accessed from. Secondly, from the perspective of the second screen paradigm, according to which multiple devices are used at the same time, for performing either dependent or independent tasks.

One way of summing up ongoing converging trends is to consider these various aspects as the manifestation of Jenkins's vision. The old and new media have collided, and as a result new media genres have emerged. Interactive television is back on the agenda, mobiles are more than ever a critical player in media distribution, and tablets transform couch potatoes into active participants to UGC-based interactive shows. However it can be argued that not everything converges, and that convergence may not be the optimal model describing new media. Complementing Miller's predictions, a quick review of converging and diverging trends is discussed in the following.

5.1 Increased Mobility

Recent evidence tends to show that technological convergence currently revolves around mobility. When considering mobile phones, the recent release of smartphones sporting quad core processors and the likely following up of this trend in the coming months by other manufacturers allows new generations of mobile devices to act as personal ubiquitous computers capable of supporting users in their context-independent, cross-platform, cloud-based multimedia life. This confirms earlier anticipations of the new role mobile phones would play in the ubiquitous media landscape. As mobile phones fulfil more and more functions, upcoming generations of devices can be expected to increase their role in people's everyday lives by granting access to an increased number of default services (such as banking or personal identification). Previously restricted to high-end models, phones' secondary functions such as high-speed Internet access and high resolution cameras become mainstream due to technology improvement and reduced costs, and are now equally important buying criteria as the communication features, taken for granted.

A number of converging media-related practices are also observable in ICT societies, for which mobile devices play an essential role. Media consumers now expect to be able to access content on any device, no matter which context they find themselves in. Until recently mobile phones were about "communicating with anyone at any time and from anywhere", today's smartphones add the possibility to "access anything at any time and from anywhere, and to share it with anyone". As one of the consequences of this "liquid" life, work and leisure activities are increasingly distributed across time and place. This explains why privacy issues are so important to tackle, as this is a big concern for end users. People want to use this distributed media infrastructure, which allow them to access all their "digital stuff" and share their online experiences with friends. However they also want to stay in control of who has access to what.

When it comes to watching television on-the-go, mobile TV is indeed being developed and can be expected to be soon a common feature. The most successful delivery method (broadcast, multicast, or streamed over IP) is yet to be determined, even if IP-based technologies seem to take the most attention at the moment. This technological advance however did not happen in the form of the revolution that was expected circa 2005. On the one hand, broadcasters see mobile devices as another content distribution channel, while on the other hand, end viewers expect to be able to watch media on their personal smartphones. Being able to watch live TV on a mobile phone is thus only the logical result of these tendencies reflecting maturing technology and evolving consumption practices.

By enabling ubiquitous access to one's digital world, smartphones essentially act as the converging point for work- and leisure-related activities while on the move. The tendency of these activities to overlap has been extensively discussed in recent literature, focusing on a variety of topics such as redefining the "elastic construction" of work and life (Cohen, Duberley, & Musson, 2009), the implication of homeworking (Wapshott & Mallett, 2012), or workers' attitude toward work according to their place of work (Redman, Snape, & Ashurst, 2009). In addition to this, work and leisure are increasingly distributed across location and time, partly due to and certainly supported by mobile technologies. As a result, work and play can also be said to converge, in a way that let users manage the threshold between the two.

5.2 Personalization

An aspect of new media that can be said to diverge is personalization. People want the devices they own not only to fulfil the functions they are expected to enable, but also to make a statement about themselves either through a specific design, the secondary functions they support, or by allowing customization. A typical example is again the mobile phone, to which can be added physical accessories, customized ringtones, backgrounds, and applications. When looking at the design of recent mobile phones and especially smartphones, it has recently converged under the influence of major actors such as Apple. In response to the iPhone's worldwide success, other manufacturers have rapidly aligned their device design on a similar template. As Colton's saying goes, imitation is the sincerest form of flattery, although in the mobile world such practice is more likely to end up in patent infringement battles. Perhaps partly to escape this, and also to satisfy customers' call for diversity, the recent apparition of more "exotic" smartphone designs supports the claim that mobile phones are integrated into people's life to the extent that they act as an extension of the self and should therefore closely match public representation of individuals. For a long time less visible in other devices, this need for personalization also apply to today's TVs and perhaps more so to computers, which are now selected not only for their performances, but also their look-and-feel.

Content is also reflective of one's personality. Similarly to the diversification of niche TV stations focusing on specific genres or topics, online communities nowadays exist for virtually any interest. As an illustration, in June 2012 Wikipedia listed 199 social networking sites dedicated to an incredibly diverse range of topics. By choosing to join and contribute to a number of these online communities, people reinforce their individuality while contributing to the digital debate on subjects that matter to them. This attitude of picking personally relevant content from a vast choice of sources available is occurring online as much as in front of TV, where on-demand viewing and time-shifting are becoming more common practices. These practices however remain marginal, and flow TV is still by far the most common way of watching television today. Nevertheless, additional degrees of personalization integrating these practices and relying for instance on the paradigm of cloud computing are to be expected in the years to come.

5.3 Less Flow, More Interaction

In front of TV, audiences are now introduced to more engaging services, and some TV shows rely more on more on their input. Examples of such interactive shows are still rare both in Japan and northern Europe and are still mostly trials, from which broadcasters not only gain experience with technology, but also test the reaction of viewers. According to these preliminary experiments, entertainment, news, and sports seem to be strong candidates for such new formats, perhaps leading to the definition of new TV genres. It is not to say that traditional, passive television watching is about to disappear, but in some certain cases, more interactivity is desirable and beneficial for both content providers and consumers.

In that regard, second screens are playing an eminent role in establishing a two-way communication between production teams and audiences. Interactive features accessible on tablets and smartphones offer access to further information about a topic discussed, or the possibility to send comments to contestants during a song contest, or to cast a vote for a poll. The information, traditionally travelling from the TV set directly to the viewer, is thus rechannelled via this second screen.

For broadcasters, the balance between added value and experience disruption needs to be carefully considered in order to keep the audience focused on the content while deliberately shifting its attention away from the TV set. Early trials however successfully showed that guiding viewers through multiple devices during a live show is not only possible but also seems to result in a more dedicated audience.

While people welcome such new programme formats, television viewing remains a social activity among co-viewers, and the social interactions that take place in front of TV screens will not suffer from the introduction of second screens. On the contrary, these are expected to increase interactions also among viewers, for instance in cases of a family jointly participating in a quiz show at the same time as live contestants, and/or competing against remote peers.

5.4 Cultural Perspective

Cultural convergence is also happening, as visible in the new media environment in Japan and Denmark. Broadcasting corporations in both countries are undergoing structural changes and tackle the cultural gap between traditional media oriented mindsets focusing on newspapers, radio and TV, and new media oriented ones who try to push their corporation to fully embrace the paradigm shift toward the web and mobile. This contradiction in broadcasting corporations' management reflects the generational gap also observable in consumers, where younger generations (especially in Japan) tend to access Internet-based content primarily on their mobile devices, while older generation primarily rely on televisions, radios, and newspapers.

Technology adoption converges toward smartphones, currently more rapidly in Japan, although it currently stands far behind Denmark in terms of penetration rate. One potential factor explaining this difference is the definition of smartphones. Indeed the notion of smartphone in Japan is debatable, given the technological advances of feature phones, far more advanced than their western equivalent. For what concerns mobile Internet access however, Japan is indeed more advanced in terms of number of users, and other regions are catching up.

Even though the two countries differ concerning their use and personal relationship with TV and mobile technology, similar trends are also observable. At home, Danes and Japanese alike place TVs and PCs at the center of their in-home media life. However it appears that mobile devices are used frequently also at home; although people tend not to consciously report it. This indicates that when reflecting on home media use, one might omit or lower the importance of mobile devices, which are not inherently dedicated to home settings. Confirming previous findings from South Korea, mobile TV is popular at home. Furthermore, a better integration of mobile devices with home media equipment is expected and desired. Two examples illustrating this, second screen applications generate much interest among early adopters and improved mechanisms for transferring content across devices are considered useful.

6 Concluding Remarks

In the debate of media convergence versus divergence, our studies illustrate the current ambivalence in the literature. On the one hand, the general tendency is for platforms, content, and practices to converge toward popular designs, genres, and habits. On the other hand, the increasing growth of personalized services, devices, and the many ways in which audiences use those supports divergence. Thus, if convergence goes in pair with divergence, the research questions to be tackled should maybe less concerned with which tendency is taking over the new media ecosystem, but instead how to understand and support end users (or depending on the interest, content or technology providers) in making sense and benefiting from the evolution media is currently going through.

Considering the viewpoint of end users, the studies reported in this chapter show that the combination of television and mobile technology benefits from both the very personal character of mobile phones and from the shared condition of TVs, and of the different roles attributed to each (e.g. communication versus information). On the one hand, some media consumption is becoming increasingly tight to individual preferences and activities, while on the other hand, media consumers are increasingly bound to content providers and fellow consumers by way of community platforms and direct communication channels.

At the same time, those studies participated in discovering potential opportunities for new or improved media services. In particular, second screens should rapidly become companions to the television experience. Additionally, social communities (for their bounding and sharing aspects) and games (for their motivation and reward aspects) are expected to be valuable sources of inspiration when it comes to rethinking the relationship between media content and end users, and developing new genres of content.

The remaining challenges that need to be to tackled are mostly of a usability nature. A typical example of such challenge is the issue of content synchronization across devices, which is highly relevant not only for end users but also for broadcasters as future second screen services will heavily rely on a good timing of the content delivered.

Acknowledgments The author would like to thank the CAMMP project and the Danish National Advanced Technology Foundation for supporting the work presented in this chapter, as well as all the participants who took part in the user studies reported.

References

Baron, N. S. (2009). Three words about mobile phones: Findings from Sweden, the US, Italy, Japan, and Korea. The good, the bad, and the challenging: The user and the future of information and communication technologies. *COST Action, 298*, 273–282.

Berthold, A., Schmidt, J., & Kirchknopf, R. (2010). The ARD and ZDF Mediathek portals. EBU Technical Review (2010 Q1). Retrieved January 12, 2012, from http://bit. ly/f5HsdV

Basapur, S., et al. (2011). *Field trial of a dual device user experience for iTV* (pp. 127–136). New York: ACM.
Brasel, S. A., & Gips, J. (2011). Media multitasking behavior: Concurrent television and computer usage. *Cyberpsychology, Behavior and Social Networking, 14*(9), 527–534.
Bruns, A. (2007). Produsage. In *Proceedings of the 6th ACM SIGCHI Conference on Creativity & Cognition (C&C '07)* (pp. 99–106). New York: ACM.
Brush, A., & Inkpen, K. (2007). Yours, mine and ours? Sharing and use of technology in domestic environments. In J. Krumm, G. Abowd, A. Seneviratne, & T. Strang (Eds.), *UbiComp 2007: Ubiquitous computing* (pp. 109–126). Berlin: Springer.
Caldwell, J. T. (2008). *Production culture: Industrial reflexivity and critical practice in film and television*. Durham, NC: Duke University Press.
Campbell, S. W., & Russo, T. C. (2003). The social construction of mobile telephony: An application of the social influence model to perceptions and uses of mobile phones within personal communication networks. *Communication Monographs, 70*(4), 317–334.
Castells, M. (2007). Communication. *Power and Counter-power in the Network Society. International Journal of Communication, 1*, 238–266.
Cesar, P., Bulterman, D. C., & Jansen, A. J. (2008). Usages of the secondary screen in an interactive television environment: Control, enrich, share, and transfer television content. In *Changing television environments* (pp. 168–177). Berlin: Springer.
Cesar, P., Chorianopoulos, K., & Jensen, J. F. (2008). Social television and user interaction. *Computers in Entertainment, 6*(1), 4:1–4:10.
Cesar, P., Knoche, H., & Bulterman, D. C. A. (2010). From one to many boxes: Mobile devices as primary and secondary screens. In A. Marcus, R. Sala, & A. C. Roibás (Eds.), *Mobile TV: Customizing content and experience* (pp. 327–348). London: Springer.
Chuang, M. C., Chang, C. C., & Hsu, S. H. (2001). Perceptual factors underlying user preferences toward product form of mobile phones. *International Journal of Industrial Ergonomics, 27*(4), 247–258.
Cohen, L., Duberley, J., & Musson, G. (2009). Work-Life Balance? *Journal of Management Inquiry, 18*(3), 229–241.
Creeber, G. (2011). I's not TV, it's online drama: The return of the intimate screen. *International Journal of Cultural Studies, 14*(6), 591–606.
Curran, J., Iyengar, S., Brink Lund, A., & Salovaara-Moring, I. (2009). Media system, public knowledge and democracy. *European Journal of Communication, 24*(1), 5–26.
Deuze, M. (2006). Participation, remediation, bricolage: Considering principal components of a digital culture. *The Information Society, 22*(2), 63–75.
Ekman, P. (1989). The argument and evidence about universals in facial expressions of emotion. *Handbook of Social Psychophysiology, 58*, 342–353.
Ekman, P. (1992). Are there basic emotions? *Psychological Review, 99*(3), 550–553.
Ekman, P., & Friesen, W. (1971). Constants across cultures in the face and emotion. *Journal of Personality and Social Psychology, 17*(2), 124–129.
Ekman, P., Friesen, W. V., & Ellsworth, P. (1972). *Emotion in the human face: Guidelines for research and an integration of findings*. New York: Pergamon Press.
Elfenbein, H., & Ambady, N. (2002). On the universality and cultural specificity of emotion recognition: A meta-analysis. *Psychological Bulletin, 128*(2), 203–235.
Fallahkhair, S., Pemberton, L,. & Masthoff, J. (2004). A dual device scenario for informal language learning: interactive television meets the mobile phone, pp. 16–20
Fleury, A. (2012). Drawing and acting as user experience research tools. In *Proceedings of the 10th Asia Pacific Conference on Computer Human Interaction (APCHI '12)* (pp. 269–278). New York: ACM.
Hall, E. (1984). *The dance of life: The other dimension of time*. New York: Anchor Books.
Heller, M. (2008). Global and European information society. In P. Ludes (Ed.), *Convergence and fragmentation: Media technology and the information society* (pp. 29–46). Bristol: Intellect.

Hess, J., Ley, B., Ogonowski, C., Wan, L., & Wulf, V. (2011). Jumping between devices and services: towards an integrated concept for social tv. In *Proceddings of the 9th International Interactive Conference on Interactive Television* (pp. 11–20). New York: ACM.

Hofkirchner, W. (2009). "Community"—Where to from Here? From "networked individualism" towards "community networks". *Journal of Sociocybernetics, 7*(2), 62–71.

Ito, M. (2005). Mobile phones, Japanese youth, and the re-placement of social contact. In *Mobile communications* (pp. 131–148). London: Springer.

Jenkins, H. (2001). Convergence? I diverge. *Technology Review, 104*, 93.

Jenkins, H., 2006. *Convergence culture: Where old and new media collide* (Rev éd.). New York: NYU Press.

Jenkins, H., et al. (2009). *Confronting the challenges of participatory culture: Media education for the 21st century*. Cambridge, MA: MIT Press.

Knoche, H., & Sasse, M. A. (2007). Getting the big picture on small screens: Quality of experience in Mobile TV. In G. Lekakos, K. Chorianopoulos, & G. Doukidis (Eds.), *Interactive digital television: Technologies and applications* (pp. 242–260). Hershey, PA : IGI Publishing.

Kruitbosch, G., & Nack, F. (2008). *Broadcast yourself on YouTube: Really?* (pp. 7–10). New York: ACM.

Larsen, A. L., & Svenningsen, U. H. (2011). Many Danes surf the internet while watching TV. In: *The development of media 2011*. DR Medieforskning, pp. 26–31.

Lindquist, J. D., Knieling, J., & Kaufman-Scarborough, C. (2001). Polychronicity and consumer behavior outcomes among Japanese and US students: A study of response to culture in a US university setting. In *Proceedings of the tenth biennial world marketing congress* (Vol. 10, No. 4.4, p. 4).

Lüders, M. (2007). Converging forms of communication? In T. Storsul & D. Stuedahl (Eds.), *Ambivalence towards convergence: Digitalization and media change* (pp. 179–198). Göteborg: Nordicom.

Ludes, P. (Ed.). (2008). *Convergence and fragmentation: Media technology and the information society*. Bristol: Intellect.

Lund, A. B. (2007). Media markets in Scandinavia: Political economy aspects of convergence and divergence. *Nordicom Review, 28*, 121–134.

Miller, C. (2012). *Mobile TV—What's on?* Juniper Research.

Miyata, K., Boase, J., Wellman, B., & Ikeda, K. (2004). The mobile-izing Japanese: Connecting to the internet by PC and webphone in Yamanashi. In M. Ito, M. Matsuda, & D. Okabe (Eds.), *Personal, portable, pedestrian—Mobile phones in Japanese life* (pp. 143–164). Cambridge, MA: MIT Press.

Miyata, K., Wellman, B., & Boase, J. (2005). The wired and wireless Japanese: Webphones, PCs and social networks. In R. S. Ling & P. E. Pedersen (Eds.), *Mobile communications: Re-negotiation of the social sphere* (pp. 427–449). London: Springer.

Nielsen, J. (1994). *Usability engineering*. San Francisco: Morgan Kaufmann.

O'Hara, K., Mitchell, A. S., & Vorbau, A. (2007). *Consuming video on mobile devices* (pp. 857–866). New York: ACM.

Oksman, V., Ollikainen, V., Noppari, E., Herrero, C., & Tammela, A. (2008). 'Podracing': Experimenting with mobile TV content consumption and delivery methods. *Multimedia Systems, 14*(2), 105–114.

Pavlik, J. V., & McIntosh, S. (2010). *Converging media: A new introduction to mass communication* (2nd ed.). New York: Oxford University Press.

Redman, T., Snape, E., & Ashurst, C. (2009). Location, location, location: Does place of work really matter? *British Journal of Management, 20*, S171–S181.

Repo, P., Hyvönen, K., Pantzar, M., & Timonen, P. (2004). *Users inventing ways to enjoy new mobile services—The case of watching mobile videos* (p. 40096.3). Washington, DC: IEEE Computer Society.

Robertson, S., Wharton, C., Ashworth, C., & Franzke, M. (1996). *Dual device user interface design: PDAs and interactive television* (pp. 79–86). New York: ACM.

Sauter, D. A., Eisner, F., Ekman, P., & Scott, S. K. (2010). Cross-cultural recognition of basic emotions through nonverbal emotional vocalizations. *Proceedings of the National Academy of Sciences, January, 107*(6), 2408–2412.

Storsul, T., & Stuedahl, D. (Eds.). (2007). *Ambivalence towards convergence: Digitalization and media change*. Nordicom: Göteborg.

Suteja, J., & Tedjohartoko, S. (2011). The influence of mobile phone's forms in the user perception. *World Academy of Science, Engineering and Technology, 75*, 473–476.

Thompson, M. (2010). Putting quality first-The BBC and public space. *BBC*. Retrieved January 12, 2011, from http://bbc. in/wTwWj6

Thunø, L. (2012). Medieudviklingen 2011. DR Medieforsknings. Retrieved February 13, 2012, from http://bit.ly/20GaCFW

Tokan, F. (2011). *Media as multitasking: An exploratory study on capturing audiences' media multitasking and multiple media use behaviours*. Helsinki, Finland: Aalto University School of Economics.

Tsekleves, E., Whitham, R., Kondo, K., & Hill, A. (2009). *Bringing the television experience to other media in the home: An ethnographic study* (pp. 201–210). New York: ACM.

Tsekleves, E., Whitham, R., Kondo, K., & Hill, A. (2011). Investigating media use and the television user experience in the home. *Entertainment Computing, 2*(3), 151–161.

Uricchio, W. (2009). The future of a medium once known as television. *The YouTube Reader* (pp. 24–39).

Vermeer, F. (2011). *What's on the Tube today?—On the ontology of television and YouTube*. Netherlands: Utrecht University, Faculty of Humanities.

Wapshott, R., & Mallett, O. (2012). The spatial implications of homeworking: A Lefebvrian approach to the rewards and challenges of home-based work. *Organization, 19*(1), 63–79.

Wellman, B. (2002). Little boxes, glocalization, and networked individualism. Digital cities II: Computational and sociological approaches, pp. 337–343.

Wellman, B., et al. (2003). The social affordances of the Internet for networked individualism. *Journal of Computer-Mediated Communication, 8*(3).

Westlund, O. (2010). New(s) functions for the mobile: A cross-cultural study. *New Media & Society, 12*(1), 91–108.

Quality of Experience in Convergent Communication Ecosystems

Peter Reichl

1 Introduction

Research in telecommunications is currently undergoing a significant paradigm change. While for several decades, it has been strongly dominated by engineering-driven challenges and solutions (and deservedly so, especially as long as there were still fundamental technological problems which had to be solved, eventually allowing the broad introduction of mobile networks and the subsequent convergence of fixed-line and mobile telephony with Internet technology), over the last couple of years a counterbalancing evolution has started to gradually integrate also socio-economic and user-centric perspectives into a broad holistic research approach. Already more than a decade ago, this development has been identified as a "triangle of forces" between Network Economics, Usability and Technology, altogether leading to the notion of the so-called "NUT Trilemma" (Reichl, Hausheer, & Stiller, 2003). More recently, this emerging interdisciplinary research field at the cross-section of communication technology, microeconomics and social sciences has been termed "Communication Ecosystems" (Kilkki, 2012), in order to further underline the dynamic and lively character of the corresponding system of actors and their technological environment plus the mutual interactions between them, which eventually lead to the generation of innovative products and services, while at the same time requiring a deep understanding of human needs and how to serve them through appropriate technological and economical progress (Kilkki, 2012).

P. Reichl (✉)
Department of Computer Science, University of Vienna, Währinger Strasse 29, 1090 Wien, Vienna, Austria

Department RSM, Université Européenne de Bretagne/Télécom Bretagne Rennes, 2 rue de la Châtaigneraie, 35576 Cesson-Sevigné, France
e-mail: peter.reichl@univie.ac.at; peter.reichl@telecom-bretagne.eu

A. Lugmayr, C. Dal Zotto (eds.), *Media Convergence Handbook - Vol. 2*, Media Business and Innovation, DOI 10.1007/978-3-642-54487-3_12

In this chapter, we argue that, in fact, the "ecosystems" metaphor may be understood in an even broader sense, i.e., as general paradigm for the entire structure of today's telecommunication landscape. To this end, let us briefly remind that the concept of an ecosystem originates from biology, where it describes a community of organisms together with their environment, viewed as a system of interaction and interdependent relationships. As we all know, typical examples for such ecosystems include coral reefs, rainforests, deserts, savannas, lakes or garden ponds. Note that the main structural characteristic of all these ecosystems is based on strict hierarchies, along which interaction takes place in two dimensions: in a horizontal sense, mainly as collaboration or competition between organisms, and in a vertical sense, eventually forming the so-called "trophic pyramid", which is used to describe various types of producers and consumers along a food chain (with "eating" and "being eaten" as main types of interaction).

It is now interesting to observe that similar hierarchical structures are very well-known and characteristic also for the world of communication networks and services. A particularly important example is the OSI reference model (ITU-T Rec. X.200, 1994), which describes the functions of a communication systems in the abstract form of seven layers (physical layer, data link layer, network layer, transport layer, session layer, presentation layer, application layer), where each layer and the corresponding protocols are served from the layer(s) below and serve the layer(s) above. In terms of business entities, a comparable hierarchical structure comprises infrastructure providers, network service providers, application service providers and content providers (a.k.a. OTT—Over The Top providers), which altogether bridge the gap between physical network infrastructure and customer premises equipment.

Hence, going a step beyond (Kilkki, 2012), we characterize a communication ecosystem as a hierarchically structured dynamic system of technological infrastructure and its user communities (both businesses and end customers) plus the interactions within and between the communities as well as with the infrastructure. In this sense, we understand the notion of communication ecosystems as structural paradigm rather than a illustrative metaphor for describing certain innovation processes in the telco business. As a consequence, we argue that research in this area requires a truly interdisciplinary approach, with a close methodological integration of disciplines as different as computer science, electrical engineering, sociology, Human-Computer Interaction (HCI), microeconomics, psychology and communication sciences, and should aim at the provision of quantitative models and their empirical validation way beyond the investigation of innovation processes.

This chapter aims at illustrating the corresponding research field, with a strong emphasis on the recent paradigm change from Quality of Service (QoS) to Quality of Experience (QoE), which is perceived as a key example for interdisciplinary research on communication ecosystems. Hence, our purpose and contribution is threefold: firstly, we present a comprehensive introduction into the current state and recent advances in QoE research in general; secondly, we provide a discussion of specific convergence-divergence aspects of QoE; and thirdly, we elaborate on two specific dimensions of actual QoE research, i.e., the quest for fundamental underlying laws of quality as well as the economics of QoE, with a specific focus on communication ecosystem aspects.

Therefore, Sect. 2 describes the mentioned paradigm change in more detail, including a discussion of related literature, models, metrics and methodologies. Section 3 presents a survey on some recent results on fundamental laws for QoE, while Sect. 4 describes the consequences of this transition if it comes to the question of how to charge for QoE. Section 5 concludes with a brief summary and outlook. Note that, due to the wealth of relevant material, our account deliberately stays on a rather general and abstract level, and the interested reader is kindly invited to delve into the numerous references provided for further illustration.

2 From Quality of Service (QoS) to Quality of Experience (QoE): An "Anti-Copernican Revolution"

Providing an attractive service quality has always been amongst the core goals of modern communication technology. It is however interesting to observe that the underlying basic quality concepts have significantly changed over time. The present section therefore discusses Quality of Service (QoS) and Quality of Experience (QoE) as main steps of this evolution, and surveys related models, metrics and methodologies.

2.1 A Brief History of Service Quality

For decades, service quality has been captured mainly by the notion of Quality of Service (QoS), which originally has included a strong subjective component. This becomes immediately evident for instance from (ITU-T Rec. E.800, 1994), where QoS has been defined as *collective effect of service performance which determines the degree of satisfaction of a user of the service* (ITU-T Rec. E.800, 1994). Very soon, however, this clearly user-centric position in practice has been abandoned in favor of a QoS concept which focuses rather on corresponding technical parameters than on immanent quality itself. For example, the definition provided by (ITU-T Rec. X.200, 1994) reads instead *Quality of Service (QoS) is the collective name given to a set of parameters associated with data transmission among service-access-points*, and subsequently mentions expected transmission delay, probability of corruption, probability of loss or duplication probability of wrong delivery, expected throughput, probability of out of sequence delivery, connection establishment delay, connection establishment failure probability, connection release delay, connection release failure probability and connection resilience (plus some more) as potential QoS-related parameters.

This strictly technocratic view on service quality has only recently been reversed by a countermovement, which could easily be called an "Anti-Copernican Revolution" (Reichl, 2009), as it puts the user back into the center of the technological universe. In this sense, it seems pretty much comparable to Nicolaus Copernicus who neglected the human-centricity of the universe in favor of its solar-centricity (Copernicus, 1543), or Immanuel Kant who used a similar term to describe his fundamental epistemological turn (Kant, 1787). As a consequence, in the meantime the research community has started to use the notion of "Quality of Experience" (QoE) rather than QoS for an appropriate description of service quality.

For a standardized definition of QoE, the reader is often referred to (ITU-T Rec. P.10, 2008), where QoE is defined as *overall acceptability of an application or service, as perceived subjectively by the end user*, which *includes the complete end-to-end system effects* and *may be influenced by user expectations and context.* However, this definition has been strongly debated, for instance concerning the notion of "acceptability", which basically is a binary term (something is either acceptable or not), while many researchers would prefer measuring QoE on a more fine-grained (discrete or even continuous) scale. Therefore, quite a few additional definitions have been proposed in the meantime, see e.g., Reichl, Tuffin, and Schatz (2011) for a brief discussion. Most recently, the European COST Action QualiNet has been working on a radically new and much broader definition, which has been initiated during two related Dagstuhl Seminars (Fiedler, Kilkki, & Reichl, 2009; Fiedler, Möller, & Reichl, 2012) and reads as follows: *Quality of Experience is the degree of delight or annoyance of the user of an application or service. It results from the fulfillment of his or her expectations with respect to the utility and/or enjoyment of the application or service in the light of the user's personality and current state* (Le Callet, Möller, & Perkis, 2012).

2.2 Models—Metrics—Methodologies

Along these (still ongoing) discussions, also the question of how to model QoE appropriately has not yet received a unanimous answer. While, for instance, the model presented in Kilkki (2008) mainly focuses on the different user roles, and thus distinguishes the general notion of QoE from related subconcepts like Quality-of-User-Experience, Quality-of-Customer-Experience and Quality-of-Group-Member-Experience, the chain model introduced in Reichl (2007) rather emphasizes the end-to-end aspect, and distinguishes QoS in the core and network accesses from the concept of Quality-of-Design, which addresses the interface between end device and the user. From yet another perspective, the application-oriented framework developed in the project ACE[1] (see Reichl et al., 2011 for an overview) links Key Performance Indicators (KPIs) from three different layers (network, application, user) towards an application/service-specific QoE prediction

[1] Advancing the Customer Experience (ACE) is the title of a series of COMET projects which are performed at the Telecommunications Research Center Vienna (FTW) and investigate the link between technical network parameters and the customer's Quality of Experience (QoE). For further details, please refer to the project web site at http://ace.ftw.at.

function. Even more comprehensively, Fiedler et al. (2012) presents a model for the quality formation process and distinguishes a perception and a reference path: while the reference path relates to former quality experiences as well as the temporal and contextual nature of quality, the perception path is strictly based on a concrete physical signal as trigger for subjective quality perception which subsequently is processed into the perceived nature of the signal under the constraints of the reference path.

It is important that all these models have one thing in common: the strict focus on a subjective, i.e., user-centric, perspective. In this sense, this understanding of QoE is strictly different from a wide-spread alternative usage (not to speak of "misuse") of this acronym, which can be observed within large parts of the research community, who—describing it in a slightly polemic way—reduce the mentioned paradigm change towards a mere replacement of one buzzword (QoS) by another one (QoE), without any subsequent adoption of e.g., metrics and methodologies (Reichl, 2009). In clear contrast to these trends, we firmly argue for a strict user-centric QoE approach, which consequently has to be based on comprehensive user trials as main instrument for the quantitative determination of perceived service quality, at least as long as no appropriate alternative evaluation methods of comparable expressiveness have been developed.

This clear position is also reflected if it comes to metrics and methodologies for measuring QoE. Traditionally, the most important metric for evaluating perceived service quality is the *Mean Opinion Score* (MOS), a so-called absolute category rating (ACR) scale distinguishing five quality levels (5 = excellent, 4 = good, 3 = fair, 2 = poor, 1 = bad), which trial subjects are asked to employ for expressing their quality evaluation according to a standardized trial protocol (ITU-T Rec. P.800, 1996). While the MOS approach tries to achieve a certain degree of objectivity by averaging test results over a sufficient number of individual user evaluations, it has been alternately criticized for fundamental flaws (Watson & Sasse, 1996) or for being too simplistic (Hossfeld, Schatz, & Egger, 2011). However, no adequate alternative has been broadly accepted in practice yet, and MOS grading still represents the most widespread metric used during subjective quality trials.

The disadvantages of quality tests involving actual users are obvious and range from high cost for appropriate lab infrastructures, recruitment and motivation of trial subjects, over large time requirements for test performance to difficulties in the acquisition of significant results due to a certain data fuzziness. Some of these problems may be addressed by replacing the direct user involvement with "educated guesses" about the probable evaluation results. Approaches in this direction differ mainly in terms of complexity and accuracy, and range from standardized methods for specific use cases like, e.g., Perceptual Evaluation of Speech Quality (PESQ) (ITU-T Rec. P.862, 2001) or Perceptual Objective Listening Quality Assessment (POLQA) (ITU-T Rec. P.863, 2011), to proposals for learning tools, for instance based on specific random neural networks, like, e.g., the Pseudo-Subjective Quality Assessment (PSQA) tool (Rubino, Tirilly, & Varela, 2006). In this way, it is possible to avoid most disadvantages of direct user trials while providing at least a reliable approximation for the probable user judgment, which, by the way, for a broad range of scenarios turns out to be astonishingly accurate (Rubino et al., 2006).

Another recent approach for gathering massive amounts of QoE data without relying on expensive lab infrastructures uses the concept of *crowdsourcing* as a very efficient, quick and flexible way for conducting trials with large user numbers. To this end, the evaluation task is split into small pieces which are then outsourced to a huge crowd of anonymous users via an open call through the Internet. Despite of several problematic aspects, for instance concerning the reliability of the data, which is severely endangered due to the anonymity of the trial subjects, crowdsourcing has been successfully applied e.g., for determining the QoE of Youtube videos (Hossfeld et al., 2011), and exhibits significant potential for further applications, especially with respect to online services which provide by definition a rather realistic context for this novel evaluation methodology.

2.3 The Convergence-Divergence Dilemma of QoE

Summarizing what has been said so far, the transition from QoS to QoE is considered a veritable paradigm change which is putting the user back into the center of quality evaluation, and at the same time requires a strictly interdisciplinary approach for evaluating subjective quality perception—in this sense, we can even interpret this paradigm change as the hour of birth for the ecosystem paradigm discussed in the introductory section. While trials with real users will always be subject of budget constraints (both in terms of time and money), semi-automatic evaluation tools like PESQ or PSQA provide an interesting alternative. Another issue which is still not solved adequately concerns the standard metric to be applied, where MOS seems to be still the "scale of choice", despite the mentioned disadvantages.

We may easily interpret the ongoing fuzziness around some key QoE concepts as a straightforward indication that this is a field in its infancy, which has still some way to go until reaching a mature state. However, on the other hand, this instability could also point to an inherent characteristic of an otherwise already pretty much consolidated research topic, which in this case could well be described within the "convergent divergence" framework that is pivotal for the entire present book. Hence, we will conclude this section with a brief discussion of some related aspects.

To start with a trivial remark, QoE research is clearly a very interdisciplinary endeavor (at least if we consider it seriously, i.e., beyond a mere buzzword replacement as sketched above). The broad scope of involved research disciplines, ranging from communication technology over economics to social sciences, requires profound and long-standing efforts in finding a common language and a mutual understanding of rather divergent concepts and methodologies. On the other hand, this divergence constitutes the source for the predominant part of intellectual appeal and scientific excitement of this field, as interdisciplinarity in this case does not remain on the superficial level of mixing a few different approaches, hoping to gain additional insight on a small and well-defined research topic. Instead, the experience of the past years shows that a critical mass of researchers from very different origins has already joined forces to establish lively common exchange and discussion platforms [like the Dagstuhl[2] seminars (Fiedler et al., 2009, 2012) or the European COST Action QualiNet[3]], which subsequently have already started to produce significant joint results, like, for instance Le Callet et al. (2012).

Beyond this particular aspect of strong interdisciplinarity, the dilemma between convergence and divergence also forms an intrinsic part of the concept of QoE as such. Remember that, earlier in this chapter, we have complained that the evolution of QoS research has been shaped by a strong emphasis on QoS parameters. As far as network engineering is concerned, this simplifies of course the direct integration of QoS mechanisms into the network protocols. On the other hand, we have to acknowledge that there is not much of a common denominator between parameters like packet loss rate, delay or jitter. Instead, different services and applications relate to different subsets of QoS parameters: interactive voice and video applications focus on delay, streaming video requires low jitter, file download depends on low packet loss rates, etc. QoE, on the contrary, starts from the fundamental idea that it is the *quality in the eye of the beholder*[4] which eventually plays the decisive role in determining whether a service or an application is accepted by the end customer. Intrinsically, this type of quality evaluation cannot be restricted to one or a few technical network parameters, but rather depends on the totality of them via some implicit tradeoff, which is performed by the individual user.

Hence, we may conclude that the transition from QoS to QoE implies convergence on the level of parameters, which in a QoS framework are treated orthogonally, while in a QoE framework they contribute altogether to forming a unified quality concept. At the same time, QoS parameters are much more well-defined and objectively measurable than any QoE metric, thus an objective measurement of QoE is more or less impossible a priori, and any subjective results produce additional divergence on the user level.

Attempts to overcome this fundamental dilemma are still far from being satisfactory. On the one hand, aggregating a sufficient number of user evaluations to an average value (like with MOS) reduces the impact of individual errors and deviations during the quality evaluation process, but at the same time reduces the expressive value of the collected data (Hossfeld, Schatz & Egger, 2011). On the other hand, progress with *instrumental QoE metrics*, which aim at increasing objectivity through automatic or semi-automatic evaluation processes, is currently stagnating, and moreover their applicability is usually restricted to rather narrow use cases. However, at least significant advances have recently been achieved with

[2] See http://www.dagstuhl.de for further details.

[3] See http://www.qualinet.eu/ for further details.

[4] slightly varying the famous bon mot ascribed to David Hume that "beauty is in the eye of the beholder"(Essays: moral, political, literary, London 1742)

respect to discovering and quantifying analytical relationships and laws for QoE, as will be discussed in the next section.

3 Towards Fundamental Laws for Quality of Experience

The question of general laws for the quantitative determination of individual measurement values can be asked in many and very different circumstances, and accordingly there are more than a few ways to give a proper answer. Measuring the quantitative value of a physical parameter or fundamental physical constant is the core business of any natural scientist, while fixing the monetary value of a resource for the end user forms one of the key challenges of micro-economics and leads—if performed correctly—to pricing schemes which allow the market to optimize social welfare according to Adam Smith's principle of the *invisible hand* (Smith, 1776). The same question posed in psychology leads directly to the mechanisms of human perception, which around 1860 has prompted the German physicist and philosopher Gustav Theodor Fechner to establish the area of *psychophysics* as one of the first fields of genuinely psychological research, based on the earlier work of his colleague Ernst Heinrich Weber (1834). Values may also be considered from a moral perspective; this business is mainly addressed by philosophy and theology, while practical consequences for human society serve as main trigger for sociological discussion and/or political action.

It is certainly out of scope for this chapter, if not impossible at all, to integrate all these divergent directions into a joint framework. Instead, we will limit our discussion to those aspects which are relevant for communication ecosystems as defined in the introduction, and thus focus on the case of the end user and her quality perception.

3.1 Stimulus–Response Models

Stimulus–response models are fundamental for dealing with user experience, and therefore provide a natural starting point also for our discussion of experimental work on QoE. Essentially, the concept of stimulus and response originates from behavioral psychology and describes the human organism, especially the brain, as a black box, which reacts to external stimuli in a measurable and predictable way. Without engaging in the controversial discussion whether such a reductionist view of the human mind as subject to strict causality is appropriate or not, we note nevertheless that the basic concept can be easily transferred to the field of communication ecosystem, especially for investigating the relationship between QoS parameters and QoE metrics.

To better understand the usefulness of this concept, consider a typical situation in QoE user trials. Suppose the experiment aims at determining the QoE of a specific Internet service, like, for instance, interactive Voice over IP (VoIP). To this end, a sufficient number of test users (usually between 10 and 40) is invited to spend some time in a dedicated lab environment, where a couple of test cases have been prepared. These test cases usually include certain tasks to be fulfilled by the test users (listen to speech quality, leading structured conversations on predefined topics, etc.), while, during the runtime of the trial, the parameterization of the test environment (in our case, for instance, codecs, bitrates, transmission capacities, packet loss rates, delay, jitter, etc.) changes according to a detailed specification fixed before the start of the trial. During the trial, the test user's reactions and experiences are observed, using, for instance, hardware for direct feedback, physiological measurements, questionnaires etc. Finally, as many as possible resulting data are collected, and significant correlations are determined.

While in such a setting the user hardly ever has a direct perception of changes in the parameterization of the network or application, this setting nevertheless allows us interpreting these parameters as underlying stimuli which trigger user response (perception) in a *mediated* way, i.e., through the actual interface of the application (screen, mobile device etc.). Hence, based on this general approach, we now discuss two interesting cases of resulting laws for QoE.

3.2 IQX Hypothesis

As a first example for an analytical law of QoE, we briefly introduce the so-called *IQX hypothesis* (exponential interdependency of Quality of Experience and Quality of Service) due to Hossfeld and Fiedler (Fiedler, Hossfeld, & Tran-Gia, 2010; Hossfeld, Hock, Tran-Gia, Tutschku, & Fiedler, 2008). In their model of QoE for VoIP applications, a *QoS impairment factor* (e.g., packet loss rate) takes over the role of the stimulus and triggers a QoE response (in form of a MOS score). As essential result, it turns out that this response basically follows a negative exponential function, as documented in Hossfeld et al. (2008) and Fiedler et al. (2010).

Let us consider this model in a bit more detail. Basically, the authors aim at deriving a quantitative relationship between QoS and QoE in the form of a function $x(q)$ for QoE, which depends on a single QoS parameter q serving as "impairment factor", i.e., as a trigger reducing the user's perceived quality. Note that packet loss rate provides a typical example for such an impairment factor. Moreover, it is assumed that small deviations of the impairment factor q lead to small changes of the QoE function $x(q)$. More specifically Fiedler et al. (2010) assumes that the size of these changes depends on the corresponding current QoE level. In other words: if the perceived service quality is high, small changes of the impairment factor may cause relatively large changes of QoE, while in a situation where the QoE is already low, a change of the impairment factor has relatively low impact.

Following this rationale leads to a function $x(q)$ whose shape is monotonically decreasing and convex. With the additional assumption that the QoE sensitivity is *directly proportional* to the actually attained level of QoE, this function may be described with a simple differential equation as follows:

$$\frac{dx}{dq} = -\beta \cdot (x - 1) \tag{1}$$

Here, the lowest possible QoE is assumed to be 1 (following the convention of the MOS scale), hence the derivative of $x(q)$ is proportional to $(x–1)$, while the minus sign originates from the fact that $x(q)$ is a monotonically decreasing function.

Solving Eq. (1) leads to the negative exponential function

$$x(q) = \alpha \cdot \exp(-\beta q) + 1 \tag{2}$$

In order to validate their model, the authors of Hossfeld et al. (2008) present results from an experiment with iLBC-coded speech samples, which are sent over a network emulator and thus subject to various packet loss rates. The resulting degraded speech samples, together with their original versions, are used as input to the standardized PESQ algorithm (ITU-T Rec. P.862, 2001), which calculates the corresponding MOS values. Finally, the resulting (packet loss rate, MOS) pairs are subject to a negative exponential regression (Hossfeld et al., 2008), resulting in a function which precisely corresponds to Eq. (2):

$$x(q) = 3.010 \cdot \exp(-4.473q) + 1.065 \tag{3}$$

For space reasons, the corresponding results cannot be reproduced here, hence the reader is kindly invited to refer to Hossfeld et al. (2008) and Fiedler et al. (2010) for more details and graphical illustration. Finally note that, while this result provides a clear justification for the underlying assumptions, we refer to Reichl et al. (2011) for a critical account of merits and problems of the IQX hypothesis.

3.3 Weber–Fechner Law

Summarizing the basic idea of the IQX hypothesis, it supposes that the functional dependency of QoE on QoS impairment is *monotonically decreasing*, while the *QoE sensitivity* is *directly proportional to the actual QoE level*. Altogether, these two assumptions are sufficient to derive a negative exponential function as representation of QoE.

Similarly, we now consider the case where QoE is a *monotonically increasing* function of QoS, which is the case e.g., for bandwidth as QoS parameter (=stimulus) (Reichl, Egger, Schatz, & D'Alconzo, 2010). In fact, this case seems to be much more natural and consistent with the general mechanisms of human perception which is built upon increasing responses to increasing stimuli throughout. Note in addition that, as a very particular feature, the overwhelming majority of human senses (including skin pressure, vision, hearing and smelling) seem to be based on the principle of *just noticeable differences* of perception, which in fact are *proportional* to the actual size of the stimulus. In other words: for a given stimulus S and the corresponding perception P, a stimulus of size $S+\Delta S$ with $\Delta S \propto S$ is required to achieve a significantly different perception $P + \Delta P$. Altogether, this leads to a fundamental law of human perception, which has been formulated by the founders of psychophysics as mentioned already earlier, and which hence is known as *Weber–Fechner Law*:

$$\Delta P = k \cdot \frac{\Delta S}{S} \quad \Rightarrow \quad P = k \cdot \log \frac{S}{S_0} \tag{4}$$

with k as factor of proportionality between ΔS and S, and a constant of integration S_0 which may be interpreted as the limiting value of a perceptible stimulus change, if we assume that a lack of stimulus leads to vanishing perception.

Coming back to our original context, let us replace S and P by the QoS stimulus q and the QoE metric x, respectively, and furthermore rely on our convention that the lowest QoE level is represented by a numerical value of 1. Then, following (Reichl et al., 2010), the QoE function x is described by the following differential equation:

$$\frac{d(x-1)}{dq} = \frac{k}{q} \quad \Rightarrow \quad x(q) = k \cdot \log \frac{q}{q_0} + 1 \text{ for } q \geq q_0 \tag{5}$$

Among the many examples complying with such a logarithmic relationship between QoS stimulus and perceptional QoE response (Reichl et al., 2010), we would like to refer in more detail to two important cases: perceptual speech quality under VoIP for varying transmission rates, and the ACE user trials on file download for mobile broadband.

In his PhD thesis (Varela, 2005), Varela presents MOS evaluation results for the speech quality of VoIP under varying bit rates. Using the Speex codec with FEC level of 2, and for bitrates ranging from 2.4 to 24.8 kpbs, the perceived speech quality has been evaluated using the PSQA tool already mentioned before (Rubino et al., 2006). The results of this study under logarithmic scaling are depicted in Reichl et al. (2011) and demonstrate a logarithmic behavior of the resulting QoE for a broad range of bitrates and loss rates.

Logarithmic laws of Weber–Fechner type have also been substantiated during actual user trials focusing on QoE for mobile broadband scenarios as performed within the ACE project (cf. footnote 1) and reported in Reichl et al. (2011) as well. A clear logarithmic relationship exists for instance for the task of file download, which is demonstrated for two common download scenarios, i.e., mp3 downloads (2.5 MB) and download of larger zip files (10 MB). Note that, in contrast to the IQX hypothesis as well as Varela's work, in the ACE study real user ratings have been determined, see Reichl et al. (2011) for more details. As an additional result, it turns out that also the size of the file is influencing the QoE, probably via user expectations for the probably download time. Altogether, file download QoE

x may be described as a function of normalized file size f and normalized download bandwidth q as follows (Reichl et al., 2011):

$$x(q, f) = 0.775 \cdot \ln q + 2.068 \cdot \sqrt{f} \tag{6}$$

Note that this approximation of trial results by logarithmic regression results in a root-mean-square deviation (RMSD) of only 0.063, hence a quite close fitting (Reichl et al., 2011).

4 Charging for QoE

While in the previous sections we have discussed QoE mainly from the perspective of cognitive science and psychology, we should not neglect the fact that the question of the value of resources is also a deeply economical one. For instance, any standard textbook on microeconomics defines the notion of a *utility function* for describing the value of a certain resource (like, for instance, bandwidth) or service (like, for instance, VoIP or mobile broadband) to the end customer (Kreps, 1990). Utility functions are usually assumed to be monotonically increasing, but may have different shapes, depending on the scenario: for the example of network traffic, Shenker (1995) has introduced a basic distinction between elastic traffic (which is tolerant towards packet delays and packet losses and has some flexibility in the case of network congestion) and non-elastic traffic (which is for instance subject to real-time constraints or non-trivial bandwidth constraints).

While non-elastic traffic can be described with non-concave (e.g., sigmoid) or non-continuous (e.g., step) functions (Hande, Zhang, & Chiang, 2007), elastic traffic is usually linked to a smooth and concave function, like, for instance, the logarithm. Indeed, mainly for reasons of mathematical tractability, logarithmic functions are often used as standard example for illustrating the concept of utility functions in general, and may lead to rather illustrative results. As a particularly interesting example, let us point to the work of Kelly, Maulloo, and Tan (1998), who have demonstrated how TCP congestion control may be interpreted as an optimization problem for underlying logarithmic utility functions, which after all guarantee a specific fairness property (the so-called "proportional fairness") for the users competing for bandwidth.

Finally, the concept of utility functions can also be useful for performing user trials. There are several different ways to quantitatively determine utility functions—for instance by determining the user's willingness to pay or the expected resale revenue, while decision theory provides more or less sophisticated approaches to elicit user valuations, see, for instance, the Direct Rating or the Midvalue Splitting Technique approaches (Keeney & Raiffa, 1976). Another interesting concept has been followed during the EU FP5 project M3I (Hands, 2002), where test subjects have received a small amount of cash money which they could decide to spend (or not) during a user trial for enhancing the quality of videos they were asked to watch. For a detailed illustration and discussion of the results we refer to Hands (2002).

4.1 QoS-based Charging

It has been repeatedly argued that QoS mechanisms and corresponding charging schemes are intimately linked (Jain, 2006), not least because providing differentiated QoS classes is useless without price differentiation. As a consequence, a set of more or less sophisticated QoS-based charging mechanisms has been proposed in the literature. As it is out of scope for this chapter to provide a detailed survey, we refer to two comprehensive overview papers (Stiller, Reichl, & Leinen, 2001; Tuffin, 2003) and references therein.

Instead, we would like to discuss a simple scenario where network capacity acts as QoS parameter. Hence, suppose a network with limited bandwidth resources and a set of users competing for them, according to their utility functions which we assume to be identical and concave. For this scenario, there exists the following characteristic feedback loop, which is sketched in Fig. 1 left: The user demands are satisfied by offering them QoS, i.e., bandwidth. The more bandwidth a user receives, the higher she is charged. Finally, depending on the user budget, her demand increases or decreases as a function of the charge, thus closing the loop.

Putting it more formally, we follow (Reichl, Maillé, Zwickl, & Sackl, 2013) and define the *demand function* d, the *QoS function* q and the *price function* p with

$$q = q(d), \quad p = p(q), \quad d = d(p) \tag{7}$$

Note that the QoS function basically describes congestion effects and hence can be supposed to be concave and decreasing, while we further assume a standard convex demand function which is monotonically decreasing, together with a tariff that is increasing linearly. Then, as illustrated in Reichl et al. (2013), the system possesses two trivial fixed points:

- an unstable (repelling) one where high demand leads to universal congestion (i.e., zero QoS) which is for free (i.e., vanishing price), and
- a stable (attractive) one where the demand is zero, and both QoS and prices at their maximum.

Note that the system, if started from an arbitrary constellation (except for the unstable fixed point), converges directly to this attractive fixed point, which serves as a Nash equilibrium, i.e., no user has an incentive to deviate from this solution.

4.2 Fixpoint Model of Charging for QoE

Based on the results of the last subsection, we will now discuss the same scenario, however from a QoE point of view. This introduces a major change in the feedback

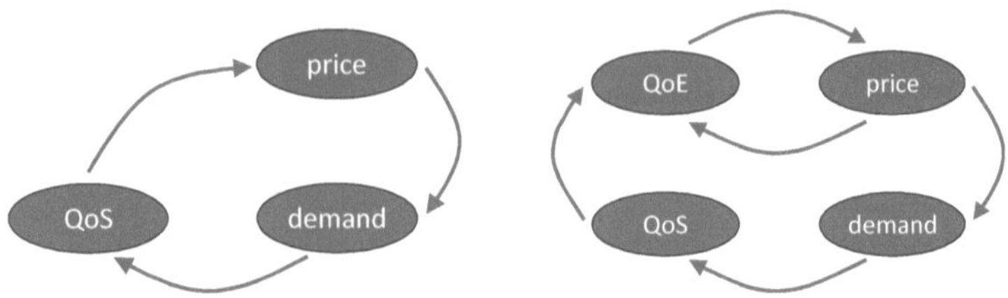

Fig. 1 Fixed point models for charging based on QoS (*left*) and QoE (*right*)

loop, see Fig. 1 right: the subjective perception of the offered quality does not longer depend on the offered QoS only, but at the same time also on the tariff charged from the user. Here, the impact is rather straightforward (Reichl, 2010): the higher the tariff, the higher the user expectations, and, as a direct consequence, the lower the perceived QoE resulting from the offered QoS.

Hence, our model for QoE-based charging consists, as previously, of a QoS function q, a price function p and a demand function d, with the slight change that the price function p does no longer depend on the QoS parameter q, but on a new parameter x which we call *QoE function*:

$$q = q(d), \quad p = p(x), \quad d = d(p) \tag{8}$$

Consequently, we have to define this additional (two-dimensional) QoE function x as follows (Reichl et al., 2013):

$$x = x(q, p) \tag{9}$$

For further simplification, we assume that the two dimensions of x are separable into a product between the *quality function* x_1 and the *expectation function* x_2:

$$x(q, p) = x_1(q) \cdot x_2(p) \tag{10}$$

where x_1 represents the mapping from QoS to QoE, and x_2 the negative impact of high charges on QoE, due to increased user expectations.

Altogether, the system described in Eqs. (7)–(9) determines again a feedback loop for QoE-based charging which is more complex than its analogy in the context of QoS-based charges. Like with (6), also this second system possesses two fixed points [for a detailed illustration and the corresponding mathematical analysis we refer to Reichl et al. (2013)]:

- a fixed point where very high demand leads to complete congestion (i.e., no QoS), which on the other hand is for free (zero price) and corresponds to total lack of QoE. As with QoS-based charging, this trivial fixed point is unstable and causes the system to diverge.

- a second fixed point which is stable (attractive) but not trivial, i.e., results from a convergence process which resembles the movement of a dampened pendulum, and essentially results from resolving the individual trade-off between the QoE raised at the user's side from receiving a certain positive QoS versus the simultaneous mental impact of the price charged for this QoS level. If this price is considered too high, the resulting QoE moves down, whereas for low prices the QoE goes up etc., until reaching an equilibrium between these two forces (which then again represents a Nash equilibrium).

4.3 Trial-Based Validation

In order to validate the results from the previous subsections, we have conducted two different user trials in the framework of the EU FP7 project ETICS[5] (Le Sauze et al., 2010). The trials have taken place at FTW's i:lab premises,[6] in order to investigate the users' purchasing behavior for different quality levels of real-time video streams.

The first trial was focusing on UDP streams, which have been impaired by random packet loss with logarithmic spacing (Sackl, Egger, Zwickl, & Reichl, 2012). Summarized briefly, during this trial a total of 43 test users have been asked to view three streaming videos. For each of these videos, they could decide amongst four different quality classes (i.e., different loss rates). At the beginning of the test, the test users were handed out an amount of 10 € in cash, which they could either spend for improving the video quality (the lowest quality was for free, the best quality was charged 1.50 € per movie) or take home afterwards. Interestingly, around 20 % of the test users spent as much money as possible on quality, while only 10 % of them decided not to spend anything, and consequently suffered from the worst quality level throughout the entire duration of the movies. The remaining 70 % of the user population has been distributed rather evenly over the entire tariff spectrum. For further details about the results of this trial we refer to Sackl, Egger et al. (2012).

Based on the lessons learned from this first trial, we have specified a second user trial in a more sophisticated way, aiming at results on a much more fine grained scale. In contrast to the first trial, we have used HD video quality instead of SD and chose video bit rate instead of loss rate as independent parameter. Moreover, the number of quality classes (and hence also tariff steps) has been increased drastically from 4 to 20. Each user was assigned one out of three pricing plans for every movie, where pricing plan A included charges in the range 0...2 €, B in the range 0...3 €, and C in the range 0...4 €. Finally, each user started each movie from the lowest

[5] Economics and Technologies for Inter-Carrier Services (ETICS) is a European Community's Seventh Framework Program running 2010–2013 under grant agreement no. 248567. Further information is available at www.ict-etics.eu.

[6] http://www.ftw.at/portfolio/i-lab?set_language=en

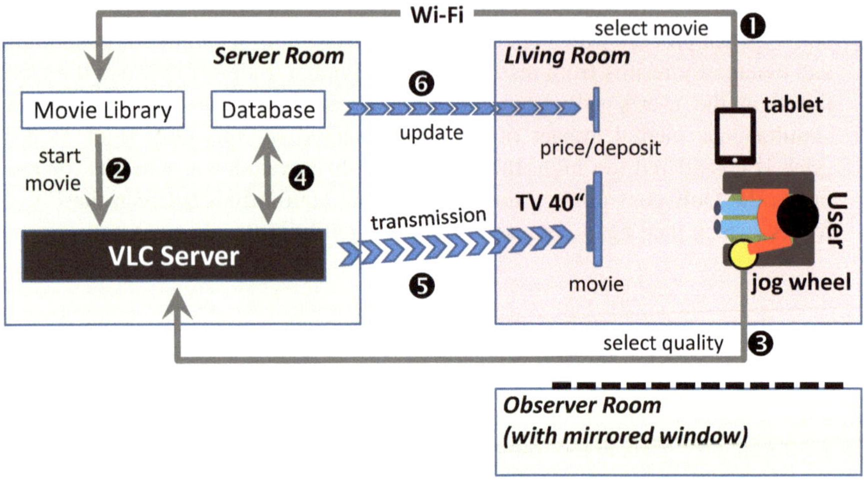

Fig. 2 User trial setup (reproduced from Reichl et al., 2013)

quality level and, during a complimentary test phase of 5 min, could experiment with all quality classes using a jog wheel (no charges during this period). As an important feature, quality changes triggered by moving the jog wheel happened almost in real-time (i.e., within max. 1 s of delay).

The technical trial setup is depicted in Fig. 2. As an interesting result, it turned out that most users (with the exception of less than 20 %) displayed a rather consistent convergence behavior over all three movies and could be assigned to one of three convergence classes of almost equal size (Reichl et al., 2013): (1) users who approached step by step, but relatively fast, their preferred quality class and remained there more or less until the end of the decision phase, (2) users who explored quickly the entire range of available quality and continued their oscillation with decreasing amplitude until convergence, and (3) users who used most of the trial period to jump around in the entire quality space with relatively late convergence towards their final choice. For further results of this trial, the reader is referred to Reichl et al. (2013) and Zwickl, Sackl, and Reichl (2013).

5 Summary and Conclusions

While convergence aspects in telecommunications are usually discussed either on a technology and/or services and/or market level,[7] one of the main goals of the present chapter was to introduce the notion of "communication ecosystems" as a

[7] For a comprehensive example including all three dimensions, we would like to refer for instance to the evolution and deployment of the IP Multimedia Subsystem (Camarillo & García-Martín, 2006) aiming at the convergence of mobile and fixed networks and services with the Internet (Reichl, Bessler, Fabini et al., 2006).

much broader and truly multidisciplinary framework for addressing convergence issues, and at the same time to use the recently emerging interest in Quality of Experience both in industry and academia to highlight a concrete research topic as example for the resulting holistic approach. To this end, we have started with a description of the paradigm change from QoS to QoE, in terms of the (still ongoing) discussion over proper definitions as well as with respect to QoE models, metrics and measurement methodologies. Based on that, we surveyed recent advances concerning the underlying fundamental laws for QoE, including the IQX hypothesis and perception laws of Weber–Fechner type. Finally, the spectrum of ecosystem dimensions has been completed by integrating the microeconomic perspective and discussing open issues with respect to charging for QoE, which still is somehow neglected by mainstream QoE research. Here, as a main outcome, we have presented related fixed point problems and surveyed the results of corresponding user trials for exploring user purchasing behavior for service quality.

After almost a decade of intense QoE research, related material and literature has grown rather abundantly, hence the present chapter does by no means claim any completeness in covering the entire topic. Instead, we have aimed at providing both a general starting point together with a few selected concrete directions of current research, in order to facilitate the reader's own exploration of this interdisciplinary field. Once more, especially as far as practical examples and further applications are concerned, we would like to refer explicitly to further related work as listed in the bibliography, which has been compiled in order to complement the sometimes superficial description of concepts and results with sufficient guidance for deeper understanding and insight.

Among the open issues already touched earlier in the chapter, the question of how to reconcile the different approaches of mathematical mappings between QoS and QoE deserves particular attention, especially if it comes to going beyond a purely statistical formulation of this interrelationship, i.e., an *explanation and understanding* rather than a mere *description* of what is going on. We have already seen in our discussion of the IQX hypothesis and the Weber–Fechner law that, depending on the type of underlying stimulus, human quality perception seems to follow differential equations which are different and rather similar at the same time, respectively relating the change of stimulus to the current situation and deriving relatively simple functional forms of the mapping between the corresponding QoS parameter and the QoE metric. Despite some recent attempts in providing a unified formulation of these different laws, for instance in (Khorsandroo, Md Noor, & Khorsandroo, 2013), a fully satisfying answer to this problem could not be presented yet. Maybe this also requires further strengthening the links to the psychology of human perception, as has already been sketched for the case of time perception in Egger, Reichl, Hossfeld, and Schatz (2012).

The same is valid for future research on economic implications of the transition from QoS to QoE. Here, it is especially the notion of user context which needs further refinement, where charging, as sketched above in Fig. 1, represents only one out of several relevant dimensions (which additionally are strongly interrelated to each other). In this context, cognitive dissonance phenomena provide an especially interesting topic (Sackl, Zwickl, Egger, & Reichl, 2012), which directly links human decision-making to subjective quality perception.

Finally, while QoE has been introduced as a key example for the new interdisciplinary research field established by the mentioned "Anti-Copernican Revolution", current and future research is targeting also further dimensions of communication ecosystems in general, like, for instance, the question of service innovation models and corresponding value networks, both viewed from a user-centric perspective (Kilkki, 2012). All in all, there is still significant progress to be expected from the endeavor of putting the end user into the center of interest, overcoming the corresponding problems of convergent divergence on the consumer level and delving deeper into this highly innovative and fruitful holistic research area, where communication technology meets microeconomics and user-centered research.

Acknowledgments This paper is mainly based on the inaugural lecture of the RBUCE WEST International Research Chair NICE (Network-based Information and Communication Ecosystems) at Université Européenne de Bretagne, which has been given by the author on January 23, 2013, at Télécom Bretagne Rennes, France. RBUCE WEST is jointly financed by Institut Mines-Télécom Paris and the EU FP7 Marie Curie program. Additional support by the Telecommunications Research Center Vienna (FTW) in the framework of the EU FP7 project ETICS and the CELTIC project QuEEN as well as COST Action IC 1003 (QualiNet) is gratefully acknowledged. The author would like to thank his colleagues Sebastian Möller, Kalevi Kilkki, Markus Fiedler, Bruno Tuffin, Patrick Maillé, Gerardo Rubino, Martin Varela, Andreas Sackl, Patrick Zwickl, Sebastian Egger, Raimund Schatz and Alexander Raake for many valuable discussions.

References

Camarillo, G., & García-Martín, M. (2006). *The 3G IP multimedia subsystem (IMS): Merging the internet and the cellular worlds* (2nd ed.). Chichester: Wiley.

Copernicus, N. (1543). *De Revolutionibus Orbium Coelestium*. Nuremberg: Johann Petreius.

Egger, S., Reichl, P., Hossfeld, T., & Schatz, R. (2012, June). "Time is bandwidth"? narrowing the gap between subjective time perception and quality of experience. In *Proceeding of IEEE International Conference on Communications (ICC'12)*, Ottawa, Canada.

Fiedler, M., Hossfeld, T., & Tran-Gia, P. (2010, June). A generic quantitative relationship between quality of experience and quality of service. In *IEEE Network Special Issue on Improving QoE for Network Services* (Vol. 24). NJ, USA: IEEE Press Piscataway.

Fiedler, M., Kilkki, K., & Reichl, P. (Eds.). (2009). From quality of service to quality of experience. In *Dagstuhl Seminar Proceedings 09192*, Schloss Dagstuhl, Germany.

Fiedler, M., Möller, S., & Reichl, P. (Eds.) (2012). Quality of experience: From user perception to instrumental metrics. In *Dagstuhl Seminar Proceedings 12181*. Doi: 10.4230/DagRep. 2.5.1

Hande, P., Zhang, S., & Chiang, M. (2007). Distributed rate allocation for inelastic flows. *IEEE/ACM Transactions on Networking, 15*(6), 1240–1253.

Hands, D. (Ed.). (2002). *M3I user experiment results*. Deliverable 15/2, FP5 Project M3I, IST1999-11429.

Hossfeld, T., Hock, D., Tran-Gia, P., Tutschku, K., & Fiedler, M. (2008, May). Testing the IQX hypothesis for exponential interdependency between QoS and QoE of voice codecs iLBC and G.711. In *Proceeding of 18th ITC Specialist Seminar*, Karlskrona, Sweden.

Hossfeld, T., Schatz, R., Seufert, M., Hirth, M., Zinner, T., & Tran-Gia, P. (2011, December). Quantification of YouTube QoE via crowdsourcing. In *Proceeding of IEEE International Workshop on Multimedia Quality of Experience: Modeling, Evaluation, and Directions*, Dana Point, CA.

Hossfeld, T., Schatz, R., & Egger S. (2011, September). *SOS: The MOS is not Enough! 3rd International Workshop on Quality of Multimedia Experience (QoMEX'11)*, Mechelen, Belgium.

ITU-T Rec. E.800. (1994). *Definitions of terms related to quality of service*. Geneva: International Telecommunications Union.

ITU-T Rec. P.10. (2008). *Vocabulary for performance and quality of service, Amendment 2: New definitions for inclusion in Recommendation ITU-T P.10/G.100*. Geneva: International Telecommunications Union.

ITU-T Rec. P.800. (1996, August). *Methods for subjective determination of transmission quality*. Geneva: International Telecommunications Union.

ITU-T Rec. P.862. (2001). *Perceptual evaluation of speech quality (PESQ), an objective method for end-to-end speech quality assessment of narrow-band telephone networks and speech codecs*. Geneva: International Telecommunications Union.

ITU-T Rec. P.863. (2011). *Perceptual objective listening quality assessment*. Geneva: International Telecommunications Union.

ITU-T Rec. X.200. (1994 E). ISO/IEC 7498–1 *Information technology—Open systems interconnection—Basic reference model: The basic model*. ITU-T Rec. X.200 (1994 E).

Jain, R. (2006, October). Internet 3.0: Ten problems with current internet architecture and solutions for the next generation. In *Proceeding of IEEE Military Communications Conference (MILCOM'06)*, New York, NY, USA.

Kant, I. (1787). *Kritik der Reinen Vernunft* (2nd ed.). Königsberg: Johann Friedrich Hartknoch.

Keeney, R., & Raiffa, H. (1976). *Decisions with multiple objectives*. New York: Wiley.

Kelly, F. P., Maulloo, A., & Tan, D. (1998). Rate control for communication networks: Shadow prices, proportional fairness and stability. *Journal of the Operational Research Society, 49*(3), 237–252.

Khorsandroo, S., Md Noor, R., & Khorsandroo, S. (2013). A generic quantitative relationship to assess interdependency of QoE and QoS. *KSII Transactions on Internet and Information Systems, 7*(2), 327–346.

Kilkki, K. (2008, Spring). Quality of experience in communications ecosystems. *Journal of Universal Computer Science*, special issue on socio-economic aspects of next generation Internet.

Kilkki, K. (2012). *An introduction to communication ecosystems*. CreateSpace: Helsinki. ISBN 978–1478253303.

Kreps, D. (1990). *A course in microeconomic theory*. Harlow: Pearson.

Le Callet, P., Möller, S., & Perkis, A. (Eds.). (2012, June). Qualinet white paper on definitions of quality of experience. In *European Network on Quality of Experience in Multimedia Systems and Services (COST Action IC 1003)*, Lausanne, Switzerland.

Le Sauze, N., Chiosi, A., Douville, R., Pouyllau, H., Lonsethagen, H., Fantini, P., Palas-ciano, C., Cimmino, A., Callejo Rodriguez, M. A., Dugeon, O., Kofman, D., Gadefait, X., Cuer, P., Ciulli, N., Carrozzo, G., Soppera, A., Briscoe, B., von Bornstädt, F., Andreou, M., Stamoulis, G., Courcoubetis, C., Reichl, P., Gojmerac, I., Rougier, J. L., Vaton, S., Barth, D., & Orda, A. (2010, June). ETICS—QoS-enabled interconnection for future internet services. In *Proceeding of Future Network & Mobile Summit 2010*, Florence, Italy.

Reichl, P. (2007, September). From 'Quality-of-Service' and 'Quality-of-Design' to 'Quality-of-Experience': A holistic view on future interactive telecommunication services. *Invited Paper, IEEE SoftCOM'07*, Split, Croatia.

Reichl, P. (2009, December). From QoS to QoE: Buzzword issue or anti-copernican revolution? Keynote abstract. In *Proceedings EuroNF Workshop on Traffic Management and Traffic Engineering for the Future Internet* (p. 23).

Reichl, P. (2010). From charging for quality-of-service to charging for quality-of-experience. *Annals of Telecommunications, 65*(3–4), 189–199.
Reichl, P., Bessler, S., Fabini, J., Pailer, R., Poropatich, A., Jordan, N., et al. (2006). Practical experiences with an IMS-aware location service enabler on top of an experimental open source IMS core implementation. *Journal of Mobile Multimedia (JMM), 2*(3), 189–224.
Reichl, P., Egger, S., Schatz, R., & D'Alconzo, A. (2010, May). The logarithmic nature of QoE and the role of the Weber–Fechner law in QoE assessment. In *Proceeding of IEEE International Conference on Communications (ICC'10)*, Cape Town, South Africa.
Reichl, P., Hausheer, D., & Stiller, B. (2003). The cumulus pricing model as an adaptive framework for feasible, efficient and user-friendly tariffing of internet services. *Journal of Computer Networks, 43*(1), 3–24.
Reichl, P., Maillé, P., Zwickl, P., & Sackl, A. (2013, August). A fixed-point model for QoE-based charging. In *Proceeding of ACM SIGCOMM 2013 Workshop on Future Human-Centric Multimedia Networking* (pp. 33–38), Hong Kong, China.
Reichl, P., Tuffin, B., & Schatz, R. (2011). Logarithmic laws in service quality perception: Where microeconomics meets psychophysics and quality of experience. *Telecommunication systems*. Heidelberg: Springer (to appear 2014, electronically published 18 June 2011).
Rubino, G., Tirilly, P., & Varela, M. (2006). Evaluating users' satisfaction in packet networks using random neural networks. *ICANN, 1*, 303–312.
Sackl, A., Egger, S., Zwickl, P., & Reichl, P. (2012, July). QoE alchemy: Turning quality into money. Experiences with a refined methodology for the evaluation of willingness-to-pay. *4th International Workshop on Quality of Multimedia Experience (QoMEX'12)*, Yarra Valley, Australia.
Sackl, A., Zwickl, P., Egger, S., & Reichl, P. (2012). The role of cognitive dissonance for QoE evaluation of multimedia services. In *Proceeding on Quality of Experience for Multimedia Communications Workshop (QoEMC), IEEE Globecom 2012*, Anaheim, CA.
Shenker, S. (1995). Fundamental design issues for the future internet. *IEEE Journal of Selected Areas in Communications (JSAC), 13*(7), 1176–1188.
Smith, A. (1776). *An inquiry into the nature and causes of the wealth of nations*. London: Methuen.
Stiller, B., Reichl, P., & Leinen, S. (2001). Pricing and cost recovery for internet services: Practical review, classification and application of relevant models. *Netnomics, 3*(1), 149–171.
Tuffin, B. (2003). Charging the internet without bandwidth reservation: An overview and bibliography of mathematical approaches. *Journal of Information Science and Engineering, 19*, 775–786.
Varela, M. (2005, November). *Pseudo-subjective quality assessment of multimedia streams and its applications in control*. Ph.D. Dissertation, Université de Rennes 1.
Watson, A., & Sasse, A. (1996). Evaluating audio and video quality in low-cost multimedia conferencing systems. *Interacting with Computers, 8*(3), 255–275.
Weber, E. H. (1834). *De pulsu, resorptione, auditu et tactu. Annotationes anatomicae et physiologicae*. Leipzig: Koehler.
Zwickl, P., Sackl, A., & Reichl, P. (2013, December). Market entrance, user interaction and willingness-to-pay: Exploring fundamentals of QoE-based charging for VoD services. In *Proceeding of IEEE GLOBECOM'13*, Atlanta, GA, USA.

Interaction Design for Convergence Medias and Devices: A Multisensory Challenge

Tatiana Aires Tavares and Damian Schofield

1 Introduction

Information and Communication Technologies (ICT) have developed in unexpected ways in recent decades. Changes in the physical environments, hardware, communication and transmission modes have allowed for the development of techniques for transmitting information with very high performance, maximizing speed and reducing data loss. These technologies have enabled high speed networks and the development of applications based on multimedia objects, in particular, interactive real-time applications (Balbinot et al. 2000).

These applications have manifested in our society in recent years in the form of video streamers and recorders, videotext, telephone-based voice systems, on-line services, information kiosks, 'intelligent' household appliances and multimedia systems (Jensen, 1998).

New media is "invading" the Internet everyday. Images, videos, texts, animations are used to share any kind of human feeling (happiness, sadness, passion, hunger, pride, etc.). Also, the Internet is not open to only computers anymore. Internet access is a feature of modern cellphones, domestic equipment, cars, and many other devices. Mobility and connectivity are now requirements for many daily activities.

This is changing the way that society deals with networked content. It is transforming user's needs, actions and reactions. Figure 1 illustrates the new reality for billions of users nowadays: there are many things to read and share, many ways to distribute information and many people to notice and react to it.

T.A. Tavares (✉)
Federal University of Pelotas, Brazil
e-mail: tatianaires@gmail.com

D. Schofield
State University of New York, Oswego, New York, USA
e-mail: schofield@cs.oswego.edu

A. Lugmayr, C. Dal Zotto (eds.), *Media Convergence Handbook - Vol. 2*, Media Business and Innovation, DOI 10.1007/978-3-642-54487-3_13

Fig. 1 The evolution of Information on the Internet (or "Reliable Unreliability!"), a particular view of Neal Yamamoto (2013)

Digital convergence/divergence has many different perspectives and definitions. One of them takes in account the "*shapeshifting*". For Rachel Hinman (2011):

> *Just like the Wonder Twins transforming into "the form of" a convenient animal/water configuration that will save the day, convergence is what enables experiences to shapeshift between different devices and environments.*

So, thinking about convergence means that designers have to allow user experiences to move fluidly through multiple content and devices. Hinman (2011) also pointed out a proposal for convergence levels which defines convergence as a number of procedures: activity convergence (what users do), media convergence (what users perceive) and technology convergence (what users experience).

Instead of being an isolated experience, the technology convergence enables the fluidity capable to allow the user to move across multiple devices. Combining media and technology convergences in this manner, we have the tools which make possible user perception and experience. In Fig. 2 is shown how this convergence levels could be applied in practice. Netflix is a tool for media convergence that promotes entertainment activities to users, such as movies and additional information. Netflix offers to users, by different technologies convergence, media content. Users can use mobile devices, TV sets or cellphones to connect to Netflix services. The TV still being the social hub for entertainment, but another resources are easily connected to improve interaction and preserve the individual participation.

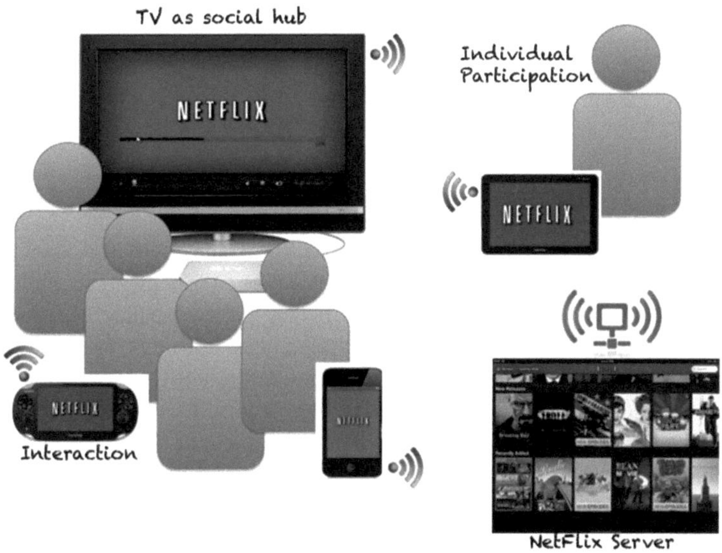

Fig. 2 An example of covergence levels in practice: using NetFlix (MailOnline, 2012)

For Lund (2011) this "toolbox" can be seen as a massively networked ecosystem that gives us increasing power to connect with others, to accelerate the growth of intelligence and to shape the world with our ideas.

Social media, the Internet of things, "quantified self", robots, gestural user interfaces, homebrew or Do It Yourself (DIY) technology are examples of elements that make part of this convergence toolbox according with Lund perspective (Lund, 2011).

In the digital age, social media is a good way to think about and understand human beings. According to Mayfield (2008) "sharing ideas, cooperating and collaborating to create art, thinking and commerce, debate and discourse, finding people" are the main benefits from social media. These benefits make virtual things seem like natural things. That is why it spreads so quickly.

Social media has traditionally been a term used to describe user-generated content that can be shared with others online. It can include blogs, wikis, social networks, and a variety of other platform types and applications. Over recent years, there has been a huge growth of social media networks and a corresponding increase in the number of taxonomies to classify them (Kaplan & Haenlein, 2010).

For Kaplan and Haenlein (2010) social media can be grouped into: blogs, social networks, social multimedia sites, wikis, discussion forums, social bookmarking websites, location based services. These tools are designed to eliminate friction and make communication more accessible for anyone using any device anytime. As devices mediating communication become smaller and less intrusive, the closer communication approaches the potential ideal of telepathy (Lund, 2011).

The Internet of Things is a simple name to define a network of complex and sophisticated systems, usually systems that refer objects with sensors attached. McKinsey et al. cited in (Atzori, Iera, & Morabito, 2010) described it as:

> *When objects can both sense the environment and communicate, they become tools for understanding complexity and responding to it swiftly. What's revolutionary in all this is that these physical information systems are now beginning to be deployed, and some of them even work largely without human intervention.*

The Internet of Things involves both objects and connectivity. In other words, it adds new tangibly interconnected elements to the online environment. So, Internet tools can control domestic equipment using information gathered online or from sensors (perhaps using weather forecasts, user food preferences or location information).

This term "The Quantified Self" has emerged in the USA to define a collaboration of users and tool makers who share an interest in self knowledge through self-tracking. This movement believes that self-improvement of technological tools can be used successfully to persistently monitor and record all facets of human life.

A recent publication of BBC Future (Weintraub, 2013) highlights that today it is easier to track everything, from diet to mood to sleep quality, then it was in the past. Features like GPS, accelerometers, cameras, microphones and gyroscopes can record human activity, location and other vital statistics. The data from this constant monitoring can reveal the patterns and habits of individuals and also of the community. this constantly available information can be used to identify opportunities for health improvement or to prevent natural disasters. For Amy Robinson (Weintraub, 2013): "The insights that we could learn from having all this quantified self data available are almost unfathomable".

In a recent ambitious project, the University of Zurich's Artificial Intelligence Lab plans to create a robot in 9 months, it is expected to be "born" in March 2013. This robot, called "Roboy" brings up many daily activities inside and it aims to interact with people. As noted in Mail Online (2012): "Roboy is the robotic boy set to help humans with everyday tasks".

This kind of effort supports the hypothesis that one of the primary goals for robot development is communication. However, for humans the most natural communication media still being humans, it happens using human senses. For Kanda and Ishiguro (2013) the human brain does not react emotionally to artificial objects such as computers or smartphones in the same way as it reacts to an image of a human face. For this reason, there are many projects investigating and building solutions for humanoids.

Thus, in general robots are getting "smarter" and more functional, they are learning how to deal with in our environment, with more sophisticated monitoring systems and sensor. The robots are being prepare to human interaction and communication as social elements of our convenience.

Human communication always used gestures, movements and expressions as oral language support. Certain gestures are so commonly used around the world that are understood throughout different cultures and times, such as a wave or thumbs up. Natural Interaction is a way to apply this concept to user interfaces in computer systems

Making a brief retrospect we can identify the evolution of these devices through the command-based languages, through the graphical user interfaces (GUIs), and finally the direct manipulation with the advent of the use of the pointers (mouse). From the rise of touch screens, the use of cameras in the analysis of user actions and the creation of devices that allow us to use technology more easily. For example, the Kinect that enables us to think about the development of more sophisticated and natural user interfaces. The Natural Interaction (NI) studies ways that humans can interact through humans five senses, be that with gestures, voice commands, corporal expressions or human body parts detection and identification (Frati, 2011; Hewett et al., 2009; Rauterberg et al., 1996; Smith & Waterman, 1981; The Engineer, 2012).

The Do-It-Yourself (DIY) culture has been continuously articulated since mid-1920s. Nowadays, DIY is an evidence across many disciplines as Health, Publishing, Production, Projects. DIY is improved by Internet tools for write, edit, publish and distribute content. The current DIY perspective uses these new sharing mechanisms for enables communities, creativity efforts and social capital. For Kuznetsov and Paulo (2010) this accessibility and decentralization has enabled large communities to form around the transfer of DIY information, attracting individuals who are curious, passionate and/or heavily involved in DIY work. Lund (2011) added that the DIY culture associated with convergence trends makes possible turning thoughts into things at a easier and faster way.

For all the elements briefly described before, one common feature is observed: the challenge goes further technological issues, it is more related with how we think about the use of technology.

2 Multisensory Interaction Design

Recently, the area of "Interaction Design" (Preece, Rogers, & Sharp, 2005) became an active concept in the design of interactive systems. Increasingly system design takes into account not only technical aspects, but especially the appropriation of these technologies in the daily users activities. Another way to analyze the interaction design is to approach it as a space for communication between people and computer systems.

The interaction design integrates knowledge from different fields of study (cognitive engineering, ethnography, communication, psyoology, among others). Another important point is to provide a new understanding of user-centered approach in terms of productivity, efficiency and usability for empathy, fun, beauty, loyalty and users involvement.

For De Paula (2003) the interaction design concept is closed related with users' needs, the design of interactive technologies has also to encompasses user's feelings and thoughts.

The technology is now increasingly available, miniaturized, transparent and ubiquitous. And the challenge for designers is to make information and functionality more easily understood by users. A practical scenario for this situation is the cell phone. Cell phones are available for a large portion of the population with different backgrounds, age or social class. The models of the cell phones are very diverse in colors and shapes, however, over time the miniaturization process is evident. Increasingly, mobile devices with mobile internet are used by the population, thus check emails, send messages to social networks or post a video. These activities are often facilitated by technology. Finally, the cell phone is present everywhere, providing information at any time and in different formats.

In this simple example, it is observed that:

- Increasing functionality contrasts directly with the reduction of the device sizes (thumbnail). The interaction with the interface elements getting smaller is a challenge to our physical capacity (the size of our hands and finger dexterity).
- Increased functionality overlaps artifacts and their uses. For example, cell phones are commonly found in football matches to replace traditional radio batteries. The use of watches has been declining due to recent habit of consultation the time by phone.
- The ubiquitous connectivity adds a new dimension to our daily activities. The information is everywhere, anytime on any device. It generates a cognitive overload and contrasts with the growing demand from its use.

Therefore, computing becomes ubiquitous, which leverages the use of our shared environments that are also enriched with new possibilities of communication and interaction. This reality brings to the field of Human-Computer Interaction a series of challenges and new opportunities.

In literature this discussion for multiplicity and overload scenario has many ramifications. In this study, we will initiate the concept of distributed cognition.

The term "distributed cognition" is explored in the literature (Hutchins, 1995; Norman, 1993; Salomon, 1996) since 1993. In 2000, James Hollan, Edwin Hutchins, and David Kirsh discussed that distributed cognition provide a theoretical basis for effective understanding of Human-Computer Interaction. Besides, distributed cognition is a fertile arena for discussing design and evaluation of digital artifacts (Hollan, Hutchins, & Kirsh, 2000).

According to Hollan et al (2000), in contrast to traditional theories, distributed cognition extends the range of cognitive possibilities because instead of considering only the individual, it covers the interactions between people, systems and devices into the environment. It is important to understand that distributed cognition is closed to the "whole" perspective, instead of a particular type of cognition. One of the aspects evaluated is the embodied cognition. From the point of view of distributed cognition, the organization of the human mind in development and operation modes is an emergent property of interactions between internal and external resources. In this perspective, the human body and the virtual environments are central rather than peripheral roles in the interaction. An example of this concept in practices is the daily multiplicity in terms of digital information, which increasingly requires different methods of acquisition and display of information (Hollan et al., 2000).

One approach to working with different modalities of interaction is the development of multimodal interfaces (MMUI: MultiModal User Interface) (Sun, Chen, Shi, & Chung, 2006). Multimodal interfaces are centered on the user and allow the user to interact with a computer using their own natural styles of communication, such as speaking, writing, touch, gestures and looking. A multimodal interface is easier to use because the interface is more natural and intuitive. Moreover, this type of interface has a potential to make complex technology more accessible to a wider range of users. However, such interfaces development also is complex and requires the integration of different technologies for effective use of different senses simultaneously.

To Chang and Ishii (2006) the solution for the digital information overload is the use of other modes of interaction—the sensory interfaces. The sensory interfaces are digital enlargements of existing physical objects by adding sensory mappings. Designers of these interfaces are less concerned with the design of new physical shapes to manipulate digital information, but rather they are concerned with expanding the expressive power of family well knowned artifacts. The interface design is focused on sensory real world, rather than suggesting new mappings for the virtual world objects. As a consequence, the designer must address the physical and aesthetic limitations of the existing devices. Finally, the sensory interfaces design process depends on three aspects: understanding the senses, understanding the physical objects semantic, and understanding the ritual use of the object (Chang & Ishii, 2006).

A more practical view of sensory interfaces is presented by Keith V, Nesbitt and Ian Hoskens in 2008. In this work, a multi-sensory interface could better users information retrieval. One particular issue is that the sensory interaction where the sensitive (user) combines information and insight in an unique direction. Sensory interaction is related to sensory perception. Our perceptions produce our experiences and this is consistent with our senses. When perception and realization do not work together the interaction is comprometida. Moreover, enrich interaction with estímulos in perception can facilitate and accelerate the understanding and implementation of user tasks. An example is an application that uses the main screen and an additional screen for complementary interaction. This multisensory interaction can serve to increase user confidence and reduce perceived workload, since redundant information can collaborate in knowledge retention (Nesbitt & Hoskens, 2008).

Another constant presence in mobile devices today is the touchscreen or Touch Interaction. Usually these devices were treated as binary input devices. Nowadays, works in the detection and processing of additional dimensions of touch, for the purposes of enriching interaction on touchscreens, including pressure, orientation, posture, hands movements and configuration added new possibilities of interaction in this mode. Chris Harrison and Scott Hudson (2012) discuss the many facets of Touch Interaction. According to the authors, this new style of interaction allows a variety of interaction techniques and could be used to better the user experience required in these devices today.

Multisensory Interaction Design (MID) should consider the user interface design as a intersection of three main aspects: pluralistic, adaptability and cognitive ability, as shown in Fig. 3. Apply a pluralistic design means to consider the "multi" inherent to the digital convergence world. Multimedia, multiuser, multidevice and also multiuse are examples of things that should be considered. Even when the application is not designed for a "multi" perspective, it can be used like that. We have many site designed for personal computer screens and today they

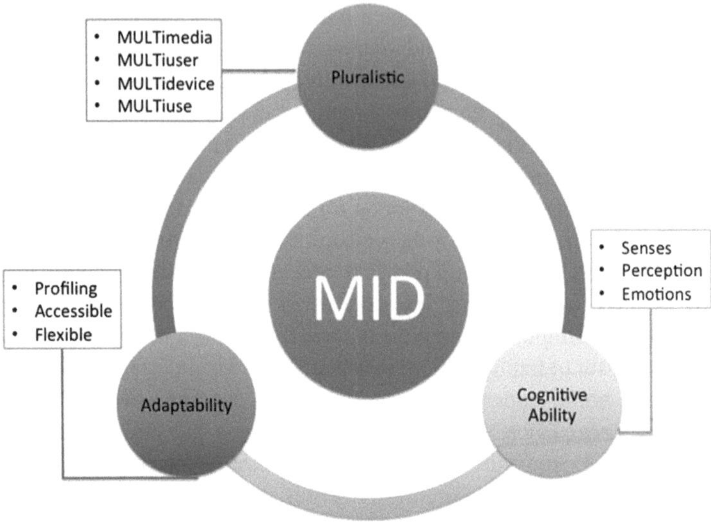

Fig. 3 Multisensory interaction design

are accessed by smartphones. Adaptability or flexible interfaces brings up a fluid design. In practice, it means that an user interface is capable to be transformed by itself. For design, it should consider profiling (or user profile), accessible interfaces (accessibility issues) and flexible features. We have many complex software techniques to develop this kind of feature, but all of them should begin in the interaction design. And finally, but not less important, the users and their cognitive ability. Cognitive ability considers how all this stuff is absorbed by users. For do that, we have to think about users' senses, perception capabilities and also emotions. For example, a user with some perception limitation will need some special feature of the user interface. This kind of assessment is treated in accessibility issues, for sure. But, accessibility does not consider users' senses. An user with some impaired has other senses more sensitives. So, to think about cognitive ability is a way to approach all the user experience abilities.

The multisensory design interaction considers sensory interfaces, devices, media content and mainly users' senses. For any kind of media content there one or more users senses used to interpret it. Deal with multiple users senses efficiently is a golden goal for designers' expertise. In the other hand, users have many possibilities of interaction and perception which gave them a full multimedia experience. For example, Fig. 4 presents the recent technology promoted by Google called Google Glasses (Google, 2013). This case explores a multisensory and fully connected user. This "new" user is always connected with Internet and has the property of receiving and sending information. The media publishing is something almost transparent to users. Information about time or a historical building can enrich user reality. GPS functionalities associated with other resources can make a no borders world immediately, changing language or time zones.

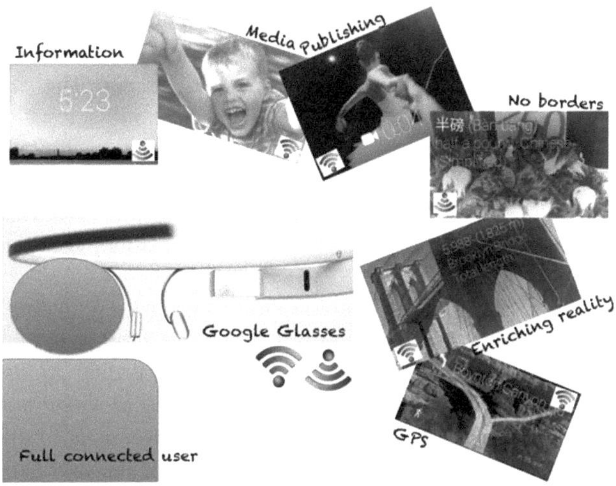

Fig. 4 Multisensory interaction design in practice: Google glass project (Google, 2013)

3 Multimedia User Experience

According to Tim Morris (2000), a multimedia system can be considered a computer system designed to play multimedia content whether it is audio, image, video or graphic simply text. From the perspective of users, multimedia means a combination of two or more continuous media being played in a time interval—usually audio and video. Integrating all these media on a computer, allows us to use the computing power to represent the information interactively. According to Jensen (1998), interactivity is:

> *a measure of the potential ability of a media to allow the user to exercise influence over the content or form of the mediated communication.*

Today, the digital convergence is suitable with multiple capabilities (mobility, hypertext, 3D, natural interaction) in order to provide an enriched interaction experience to users. This fact puts the multimedia experience beyond a passive interaction where users receive multimedia objects.

For Liu (1999) a multimedia experience, inherent of its nature, will include different types of media content. For a seamless experience, each element's timing should be coordinated with the other element's timing. A synchronization function is also required to provide the delivery of the multimedia experience.

Universal Multimedia Experience (UME) concept brings up the notion that users should have an experience anytime and anywhere. Thus, the user is the focus and network is purely a vehicle of the content (Pereira & Burnett, 2003). Hudgeons and Lindley (2010) added that an interactive multimedia experience is compromised with the audience interaction and its specification includes a experience segment having a plurality of multimedia elements and their attributes.

So, a multimedia user experience is closed related to users' senses. A multimedia experience can use two or more senses, individual or at the same time. Typically, multimedia experience involves differents types of media. The medias can be presented individual or synchronized. The users can interact with the medias, with other users and also coordinate medias exhibition (multiple views). This view enables us to think about a surgery transmission as a multimedia experience in telemedicine. The goal pointed out here is how we can recover this multimedia experience making use of multiple medias, synchronization, coordination (user's views) and interaction.

For Nalin Sharda (2003) the technology used for creating, coding, storing and transmitting multimedia content has an important role to play in any multimedia experience. A multimedia document is classified according to their temporal characteristics as dynamic media (or continuous), such as video and audio, and static media (or discrete), such as images and text (Hudgeons & Lindley, 2010). A hypermedia document is a multimedia document where the relationship between the components, meaning its logical structure and presentation, is set based on the hypertext paradigm, with the reservation that in the case of hypermedia documents, nodes contain information represented in different media (Sharda, 2003).

The incorporation of multiple views allow the combination of several static and dynamic media, their interactions and its presentation in different ways. In fact, create user interfaces to deal with a multisensory interaction makes the activity of design more complex. The designer has to understand human sensation and perception systems. The design still answering to users needs but now considering how the content is absorbed by senses of users. In the end, the multimedia user experience becomes a pluralistic experience applied for any field of knowledge according with the user cognitive ability. By the way, user experiences always enrich the design process. It is not different in multisensory interfaces as discussed in the following subsections.

3.1 Telematic Dancing

In 1966, the show "9 Evenings: Theatre and Engineering" organized by EAT (English, Experiments in Art and Technology) showed a series of performance art presentations that united artists and engineers, where they were exploited technological resources to the theater (Augusto, 2004). The technological poetics in the 50 decade were marked by early experiences with synthetic art made by Abraham Palatinik. In the 60 decade, the emergence of electroacoustic music, the initiative of Jorge Antunes and the entry of the computer in art by Waldemar Cordeiro marked the convergence between digital and artistic worlds in Brazil (Machado, 2005).

The show "*Versus*" (see Fig. 5) consisted of a distributed multimedia experience, where artists located in three distinct Brazilian cities (João Pessoa, Salvador and Brasilia) interacted through a high definition video session. Dancers located in Salvador and Brasilia interacted in real time via video streams. The created

Fig. 5 Versus show scenes

Fig. 6 "E-Pormundos Afeto" show scenes

atmosphere turned possible the feeling of being in the same physical place. In parallel, musicians located in a third place (João Pessoa) generated audio from laptops and transmitted simultaneously to Salvador and Brasilia (RNP, 2005). In this sense, the show "*(In) Toque*" connected two dancers in Rio de Janeiro, a dancer in Salvador, a DJ in Sao Paulo and a robot in Natal. The use of robots as a part of the show activated other level for user experience. For the audience, it was possible to observe the entanglement between synthetic and human bodies. For the artists, it was possible to design movements for non-humans participants (robots) could interact with the human ones (dancers). For the designers, it was possible to evaluate how users (audience and dancers) perceive and interact with robots (Murilo, 2008).

The show "*E-Pormundos Afeto*" can be seen as an evolution of "*Versus*" and "*(In) Toque*" because it was managed by a distributed multimedia system called Arthron (Silva et al., 2011). "*E-Pormundos Afeto*" discussed questions about the changes in our behavior, in our understanding of near and far, present and past, together and separately. Arthron was used to allow real time switching of high-definition video. The show integrated distributed participants in two Brazilian cities (Fortaleza and Natal) also in Spain (Barcelona). In this multimedia experience 3D models, natural interaction and video streams worked together with dancers as we can see in Fig. 6 (UFBA, 2015).

3.2 eHealth

Video-based applications are increasingly popular. A interesting area is Telemedicine or eHealth applications, such as clinical sessions, second medical opinion, interactive training or surgery transmission. This scenario is featured by handle multiple video streams. Also, other objects, as clinical images, animations, video-based exams can be used to enrich the multimedia experience (Coury, Messina, Filho, & Simões, 2010; Silva et al., 2011).

Nowadays the digital content is largely used in clinical exams. For example, exams like x-rays, ultrasound and laparoscopy are generated in digital format. Also, these exams (images or videos) are used by surgeons during a medical procedure. Multimedia objects make part of medical scenarios and are used to increase or make possible some clinical procedures. So, a live surgery transmission requires also a support for integrating and reproducing another multimedia objects as medical images or 3D models. A telemedicine system for live surgeries should consider this plurality which makes surgery a real multimedia experience.

Live surgery transmissions are useful in many medical fields but some surgeons believe it has potential value for educational benefit (Gandsas, McIntire, Palli, & Park, 2002). Besides, the connectivity support has expanded opportunities for the provision of a flexible, convenient and interactive form of continuing medical education (Curran & Fleet, 2005).

The idea of recording and playing multimedia experiences on Arthron was motivated as a way to enrich the Medicine students experience in their surgery classes. The transmission of surgeries in real time, using the Arthron at University Hospital Lauro Wanderley has been a successful practice. In Fig. 7 we can observe students during a class using the Arthron. In Fig. 7b (a) slides illustrate the traditional mode of display contents and in (b) the real time visualization. In addition, students can ask questions to the surgeon who is conducting the operation in the surgery room. Also, it is possible to switch the video streams during the

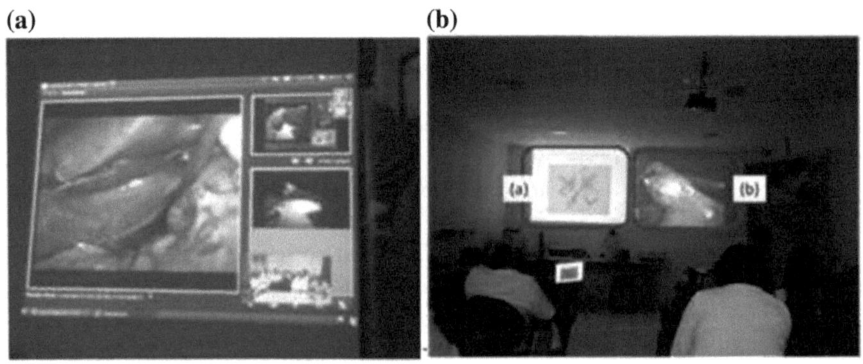

Fig. 7 Multimedia Experience for Telemedicine Scenario. In (**a**) we have the captured nodes (video streams) managed remotely by Arthron. In (**b**) we have the telemedicine room where students and the professor could interact in real-time with the surgery room

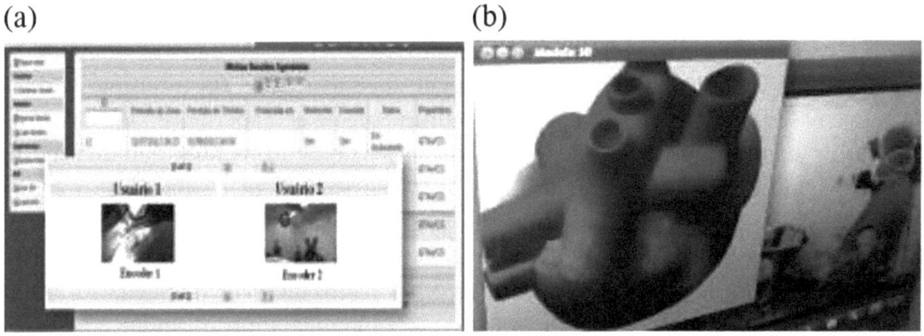

Fig. 8 Arthron User Interfaces. In (**a**) The interface for video manipulation and in (**b**) the 3D models used in a surgery transmission with Arthron

surgery transmission. The Arthron main feature is to offer the user a simple interface for handling different sources/streams of media simultaneously. Therefore the user can remotely add, remove, configure the presentation format and schedule the exhibition in time and space of media streams as shown in Fig. 7a.

An innovation of Arthron is to provide the possibility to manipulate 3D objects, especially human anatomical structures, while viewing other streams, such as video. The addition of these 3D models is especially useful as a didactic resource focused to distance training and learning. Through this feature the surgeon can show students in an integrated mode live video, 3D models that demonstrate the normal organs function, tissues or structures of the human body. In Fig. 8b we present the integration of 3D models to Arthron tool.

4 Discussion

The interaction design of digital convergence resources is a new challenge for designers and users. Designers have to think about new solutions made of pluralistic medias and devices. Users have to adapt their multiple senses required to perceive an embedded digital world to day-by-day life. The technology convergence enables the design of multisensory user interfaces capable to move across multiple devices and medias.

Multisensory Interaction Design (MID) is one approach to address the plurality inherent to digital convergence domain. This approach considers the user interface design through plurality, adaptability and cognitive ability aspects. A pluralistic design means to consider multimedia, multiuser, multidevice and also multiuse features. Adaptability is closed related with profiling (or user profile), accessible interfaces (accessibility issues) and flexible features. The cognitive ability involves the users capability of perception or users' senses. It means that user experience abilities should be observed and represented in the interaction design.

The user experience in digital convergence is also a multimedia user experience. A multimedia experience can use two or more senses, individual or at the same time. Typically, multimedia experience involves differents types of media. The medias can be presented individual or synchronized. The users can interact with the medias, with other users and also coordinate medias exhibition (multiple views). We discussed about two scenarios of multimedia user experience: the arts and technology and the eHealth one. These scenarios show to us how rich and multisensory a user experience can be. We used the same tool, Arthron, to explore different user's senses in the both cases. The first one (Arts and Technology) has shown how to integrate high quality video and audio streams with human movements (dancers) and robots. On the other hand, the eHealth scenario brings up 3D objects and natural interaction as a complementation of audio and video streams.

Therefore, the digital convergence is an ongoing phenomenon that changes our reality in terms of media, devices, applications. In the other side, users are changing, too. The expectations, the needs, the profiles, the use of perception. It is all in a transformation process stimulated by the digital convergence issues. However, user's senses still the same but working with them is the emergent challenge. Multisensory interaction design can be a primary effort to understand the digital convergence design and its integration with everyday applications.

Acknowledgments The current work has been made possible thanks to the financial support provided by CAPES and CNPq. Thank RNP (National Network for Education and Research) for funding the workgroups cited on this work, specially, GTMDA (Workgroup of Digital Media and Arts) and GTAVCS (Workgroup of Video Collaboration in Health). I would like to thank the "Brazilian Science without Borders" program and the State University of New York—Oswego for welcome me. Finally, thanks to all LAVID (Digital Video Apps Lab) "family" for the work indispensable partnership.

References

Atzori, L., Iera, A., & Morabito, G. (2010). The internet of things: A survey. *Computer Network, 54*(15, October 2010), 2787–2805. doi:10.1016/j.comnet.2010.05.010.

Augusto, S. (2004, Fevereiro 06). Um pouco da história da fusão entre Arte e Tecnologia. ARTE E TECNOLOGIA. [Online] *Jornal Express*. http://www.jornalexpress.com.br/noticias/detalhes.php?id_jornal=10046&id_noticia=2

Balbinot, R., Silveira, J. G., & Nunes, P. L. (2000). Desenvolvimento De Aplicaçõdes Multimídia Para Redes De Alta Velocidade. In *Workshop da Rede Nacional de Pesquisa*, Belo Horizonte: Federal University of Minas Gerais, May 23–24, 2000. Retrieved from http://www.rnp.br/wrnp2/2000/posters/multimidiapara%20redesdealta.pdf

Chang, A., & Ishii, H. (2006). Sensorial interfaces. In *Proceedings of the 6th Conference on Designing Interactive Systems (DIS '06)* (pp. 50–59). New York: ACM. doi: 10.1145/1142405.1142415.

Coury, W., Messina, L. A., Filho, J. L. R., & Simões, N. (2010). Implementing RUTE's usability the Brazilian Telemedicine University Network. In *6th World Congress on Services*.

Curran, V. R., & Fleet, L. (2005). A review of evaluation outcomes of web-based continuing medical education. *Medical Education, 39*, 561–567.

De Paula, R. (2003). A new era in human computer interaction: The challenges of technology as a social proxy. In *Proceedings of the Latin American Conference on Human-Computer Interaction (CLIHC '03)* (pp. 219–222). New York: ACM.

Frati, V. (2011). Using Kinect for hand tracking and rendering in wearable haptics. In *World Haptics Conference (WHC) at IEEE.*

Gandsas, A., McIntire, K., Palli, G., & Park, A. (2002). Live streaming video for medical education: A laboratory model. *Journal of Laparoendoscopic and Advanced Surgical Techniques, 12*(5), 377–382.

Google. (2013). Google glass project. Retrieved March 05, 2013, from http://www.google.com/glass/start/

Harrison, C., & Hudson, S. (2012). Using shear as a supplemental two-dimensional input channel for rich touchscreen interaction. In *Proceedings of the 2012 ACM Annual Conference on Human Factors in Computing Systems (CHI '12)* (pp. 3149–3152). New York: ACM. doi: 10.1145/2207676.2208730.

Hewett, T. T., Baecker, R., Card, S., Carey, T., Gasen, J., & Mantei, M., et al. (2009). Curricula for human-computer interaction. In *The Association for Computing Machinery, Special Interest Group on Computer Human Interaction.* Retrieved September 12, 2012, from http://old.sigchi.org/cdg/cdg2.html#2_1

Hinman, R. (2011, September 11). What does "Convergence" mean to you? In *Rosenfeld Media—The Mobile Frontier.* Retrieved March 03, 2013, from http://rosenfeldmedia.com/books/mobile-design/blog/what_does_convergence_mean_to/

Hollan, J., Hutchins, E., & Kirsh, D. (2000). Distributed cognition: Toward a new foundation for human-computer interaction research. *ACM Transactions on Computer-Human Interaction, 7* (2), 174–196. doi:10.1145/353485.353487. http://doi.acm.org/10.1145/353485.353487.

Hudgeons, B. L., & Lindley, H. (2010). Method and system for facilitating interactive multimedia experiences. *U.S. Patent No. US7650623B2.*

Hutchins, E. (1995). *Cognition in the wild.* Cambridge, MA: MIT Press.

Jensen, J. F. (1998). Interactivity: Tracing a new concept in media and communication studies. *Nordicom Review, 19*, 185–204.

Kanda, T., & Ishiguro, H. (2013). *Human robots interaction in social robotics.* Boca Raton, FL: Taylor and Francis Group.

Kaplan, A. M., & Haenlein, M. (2010). Users of the world, unite! The challenges and opportunities of Social Media. *Business Horizons, 53*(1), 59–68.

Kuznetsov, S., & Paulos, E. (2010). Rise of the expert amateur: DIY projects, communities, and cultures. In *Proceedings of the 6th Nordic Conference on Human-Computer Interaction: Extending Boundaries (NordiCHI '10)* (pp. 295–304). New York: ACM. doi: 10.1145/1868914.1868950.

Liu, J. C. (1999). System and method for online multimedia access. *U.S. Patent No. US5953005A.*

Lund, P. (2011). *Massively networked how the convergence of social media and technology is changing your life* (1st ed.). San Francisco, CA: PLI Media.

Machado, A. (2005). TECNOLOGIA E ARTE CONTEMPORÂNEA: COMO POLITIZAR O DEBATE. *Revista de Estudios Sociales, 22*, 71–79.

MailOnline. (2012). Roboy, the robotic 'boy' set to help humans with everyday tasks (and scientists hope to build him in just nine months). Retrieved March 03, 2013, from http://www.dailymail.co.uk/sciencetech/article-2253652/Roboy-robot-boy-set-born-months-help-humans-everyday-tasks.html#ixzz2MVGARdrE

Mayfield, A. (2008). What is social media? [e-book] iCrossing. Updated 01.08.08.

Morris, T. (2000). *Multimedia systems: Delivering, generating, and interacting with multimedia.* New York: Springer.

Murilo, J., Jr. (2008). (In)TOQue—dança telemática. Join Vimeo. [Online] Vimeo, LCC. http://www.vimeo.com/1137759

Nesbitt, K. V., & Hoskens, I. (2008). Multi-sensory game interface improves player satisfaction but not performance. In *Proceedings of the Ninth Conference on Australasian User Interface—Volume 76 (AUIC '08)* (Vol. 76, pp. 13–18). Darlinghurst, Australia: Australian Computer Society.

Norman, D. (1993). *Things that make us smart: Defending human attributes in the age of the machine*. Reading, MA: Addison-Wesley.

Pereira, F., & Burnett, I. (2003, March). Universal multimedia experiences for tomorrow. *IEEE Signal Processing Magazine, 20*(2), 63–73. Copyright IEEE 2003.

Preece, J., Rogers, Y., & Sharp, H. (2005). *Design de interação: além da interação homem-computador* (V. Possamai, Tradução de). Porto Alegre: Bookman.

Rauterberg, M., Mauch, T., & Stebler, R. (1996). How to improve the quality of human performance with natural user interfaces as a case study for augmented reality. *Advances in Occupational Ergonomics and Safety, I*, 150–153.

RNP. (2005). RNP versus. *Notícias da RNP*. [Online] http://www.rnp.br/noticias/2005/not-051121-fotos.html

Salomon, G. (Ed.). (1996). *Distributed cognitions: Psychological and educational considerations*. Cambridge: Cambridge University Press.

Sharda, N. (2003). *Multimedia: Fundamentals and design principles. Computer Science and Multimedia School of Computer Science and Mathematics–Notes* (pp. 111–126). Melbourne, VIC: Victoria University.

Silva, J. C. F., Ferreira, A., Vieira, E., Passos, M., Melo, E. A., & Tavares, T. A., et al. (2011). A multi-stream tool to support transmission in surgery applied to telemedicine. In *International Workshop on Health and Social Care Information Systems and Technologies—HCIST2011, 2011, Algarve—Portugal. Proceedings of HCIST2011*.

Smith, T. F., & Waterman, M. S. (1981). Identification of common molecular subsequences. *Journal of Molecular Biology, 147*, 195–197.

Sun, Y., Chen, F., Shi, Y. (. D.)., & Chung, V. (2006). A novel method for multi-sensory data fusion in multimodal human computer interaction. In J. Kjeldskov & J. Paay (Eds.), *Proceedings of the 18th Australia Conference on Computer-Human Interaction: Design: Activities, Artefacts and Environments (OZCHI '06)* (pp. 401–404). New York: ACM. doi:10.1145/1228175.1228257.

The Engineer. (2012). Researchers use Kinect gesture control in stroke rehab system. Retrieved September 12, 2012, from http://www.theengineer.co.uk/medical-and-healthcare/news/researchers-use-kinect-gesture-control-in-stroke-rehab-system/1012902.article

UFBA. (2015). *e-PORMUNDOS AFETO*. Grupo de Pesquisa Poéticas Tecnologicas. [Online] http://www.poeticatecnologica.ufba.br

Weintraub, K. (2013, January 03). Quantified self: The tech-based route to a better life? *BBC Future*. Retrieved March 03, 2013, from http://www.bbc.com/future/story/20130102-self-track-route-to-a-better-life

Yamamoto, N. (2013) *The evolution of information on the internet (or "reliable unreliability!")*. Retrieved March 05, 2013, from http://www.discovernikkei.org/en/journal/2013/2/2/my-name-is-neal/

Multimodal Interface for Effective Man Machine Interaction

N.S. Sreekanth, Nobby Varghese, C.H. Pradeepkumar, Pal Vaishali, R. Ganga Prasad, N. Pal Supriya, and N.K. Narayanan

1 Introduction

Computers are used extensively in day-to-day life; hence more focus is expected to make human computer interaction as natural as possible (Sreekanth, Supriya, Thomas, Hassan, & Narayanan, 2009). It is still a dream to interact with your electronic gadgets like how you interact with your friends. Providing intelligence to machines is being a research area for past few decades. There have been tremendous achievements in this area. The significant progress in the areas of automatic speech recognition, natural language processing and computer vision, facilitate the man-machine interaction environment more intelligent. In the past few decades there have been lots of initiatives for improving human computer interaction. As more powerful and complex computer systems emerged, efforts to make computer user interfaces more simple and natural become important. The effort behind all these works has been to make the interaction between computer and human as natural as the way human beings communicate with each other (Thomas, Hassan, Sreekanth, & Supriya, 2008).

Bringing the research outcomes to practical applications requires massive effort. If we review the progress in the individual threads of machine intelligence like speech recognition, language processing and computer vision, quite a good amount of performance is guaranteed. Human beings are accustomed to convey ideas through various modalities. The five modalities namely speech, hearing, vision, taste, smell and touch are involved in human–human interaction. If you consider

N.S. Sreekanth (✉) • N. Varghese • C.H. Pradeepkumar • P. Vaishali • R. Ganga Prasad • N.P. Supriya
Center for Development of Advanced Computing (C-DAC), #68 Electronics City, Bangalore 560100, Karnataka, India
e-mail: nssreekanth@gmail.com

N.K. Narayanan
Department of Information Technology, Kannur University, Kannur, Kerala, India

A. Lugmayr, C. Dal Zotto (eds.), *Media Convergence Handbook - Vol. 2*, Media Business and Innovation, DOI 10.1007/978-3-642-54487-3_14

human as a machine, it has two output mechanisms and five input mechanisms to send and receive various forms of communication signals. Speech and Gestures are two output mechanisms and hearing, vision, olfaction, taste and haptic (touch) are the five input receptors.

As we discussed when human beings communicate with each other we use various modalities like speech, gestures, text, and images in various combinations. Human cognitive systems are capable of recognizing the combination of various modalities and they can synchronize and understand it. Building user interfaces by mimicking the human way of communication, lead to thinking about multimodal interface. Multimodal interaction is a type of Human Computer Interaction, which combines multiple modalities or different modes of communication like speech, gestures, text and various other combinations. The most common multimodal interface combines a visual modality (e.g. a display, keyboard, and mouse) with a voice modality (speech recognition for input, speech synthesis and recorded audio for output). These devices have grown to be familiar but tend to restrict the information and command flow between the user and the computer system. However other modalities, such as haptic and olfactory can also be combined with the previous ones this limitation has become even more apparent with the emergence of novel display technologies such as virtual reality and wearable computers. Thus, in recent years, there has been a tremendous interest in introducing new modalities into HCI that will potentially resolve this interaction bottleneck.

Multimodal systems are sometimes designed based on one main modality, with the other modalities simply added on top. As handling several modalities together may result in cognitive overload and reduced usability, especially in the demanding usage situations that arise in mobile use. Providing the logical synchronization between the various signals such as speech, haptic, gesture, olfaction seems to be really challenging and this is where human cognition is still a black box to Multimodal researchers.

In this chapter we discuss about the convergence of various modalities to make human machine communication efficient and easier and the best available practices for designing a user friendly and effective multimodal interface.

2 Literature Review

Multimodal interfaces emerged approximately three decades ago within the field of human/computer interaction with Richard Bolt's "Put-That-There" application. First multimodal systems sought ways to go beyond the standard interaction mode at this time, which were graphical interfaces with keyboards and mice. Bolt's "Put-that-there" processed spoken commands linked to a pointing gesture using an armrest-mounted touchpad to move and change shapes displayed on a screen in front of the user (Bolt, 1980; Dumas, Lalanne, & Oviat, 2009).

Another interesting study, which has been done at Pennsylvania State University. "A Real-Time Framework for Natural Multimodal Interaction with Large Screen displays" in which they discussed about a framework, which uses speech and gesture to create a natural interface (Krahnstoever, Kettebekov, Yeasin, &

Sharma, 2002). The system is designed to accommodate the use natural gestures and speech commands of an experienced as well as an inexperienced user to increase the usability of the system in domains where user training is not feasible. Another important aspect is the use of a large screen display to provide appropriate feedback to the user. Large screen displays are a natural choice for many applications, especially interaction with spatial/geocentric data, immersive virtual reality environments and collaborative systems that allow interaction with multiple users simultaneously.

"Gaze-X: Adaptive Affective Multimodal Interface for Single-User Office Scenarios". This paper describes an intelligent system that they developed to support affective multimodal human–computer interaction (AMM-HCI) where the user's actions and emotions are modeled and then used to adapt the HCI and support the user in his or her activity (Maat & Pantic, 2007). The proposed system, which they named Gaze-X, is based on sensing and interpretation of the human part of the computer's context, known as W5+ (who, where, what, when, why, how). It integrates a number of natural human communicative modalities including speech, eye gaze direction, face and facial expression. To attain a system that can be educated, that can improve its knowledge and decision making through experience. To support concepts of concurrency, persistency, and mobility, Gaze-X has been built as an agent-based system where different agents are responsible for different parts of the processing. A usability study conducted in an office scenario with a number of users indicates that Gaze-X is perceived as effective, easy to use, useful, and affectively qualitative.

UI on the Fly is a system that dynamically presents coordinated multimodal content through natural language and a small-screen graphical user interface (Reitter, Panttaja, & Cummins, 2004). It adapts to the user's preferences and situation. Multimodal Functional Unification Grammar (MUG) is a unification-based formalism that uses rules to generate content that is coordinated across several communication modes. Faithful variants are scored with a heuristic function.

Another interesting work in the category of assistive technology in Kanagawa Rehabilitation Center, Japan "Multi-modal Interface with Voice and Head Tracking for Multiple Home Appliances" addresses a multi-modal interface that allows use of voice and gesture commands for controlling distributed home appliances used by people with disabilities (Ito, 2001). The main objective of this study is combined with nonverbal and verbal interface for intuitive and efficient control that uses hands-free operation. The pointing gesture by facing as nonverbal interface represents selecting one of the home appliances. The voice commands as verbal interface represent button operation of the remote controller such as the power on/off, the channel select and the volume up/down. The prototype system can provide a hands-free remote controller for people with quadriplegia who do not have to send verbal commands for selecting home appliances.

Researchers at AT&T labs are addressing this challenge by developing technologies to support truly multimodal interaction. Various products and prototypes from this lab brought a new dimension in the area of multimodal interaction. The prototypes include MATCH (Multimodal Access to City Help), Multimodal IPTV and Multimodal presentation dash board. Building these systems involves significant advances in the areas of multimodal integration, understanding, multimodal dialog management, and multimodal generation of sentences. These multimodal interface technologies have been applied to a broad range of different application areas, including local search, corporate directory access and messaging, medical informatics, accessing and controlling presentations, and searching and browsing for Internet Protocol television (IPTV) content such as movies-on-demand (http://www2.research.att.com/~johnston/).

3 Multimodal Interface- Methodology and Approach

Multimodal interface provides a very natural way for humans to perform tasks on a machine, using direct manipulation and speech interaction methods similar to those used daily in human-to-human communication. However, despite the availability of high accuracy speech recognizers and the available haptic and gesture-based devices such as gaze trackers, touch screens, and gesture trackers, very few applications take advantage of these technologies. One reason for this may be that the cost in time of implementing a multimodal interface is prohibitive (Flippo, Kerbs, & Marsic, 2003). Multimodal interaction can have many benefits compared to unimodal interaction. It may bring more bandwidth to the communication and provide alternative modalities for the same tasks, for example in the case of disabled users it provides speech based and haptic alternatives for graphical elements. Unfortunately, multimodal systems are sometimes designed based on one main modality, with the other modalities simply added on top. As handling several modalities together, may result in cognitive overload and reduced usability, especially in demanding usage situations that arise in mobile use (Turunen, Hakulinen, Kainulainen, Melto, & Hurtig, 2007).

The speech and visual modes are the most commonly used communication methods in information dissemination and perception process of human–human interaction. Addition of new modalities not only increases the bandwidth of communication, but also resolves the ambiguity in the primarily communicated message. The resolution of ambiguity in one mode of signal can be complemented by the other mode of signal. The best examples are using visual information to understand ambiguous speech (lip tracking for improving the accuracy of speech recognition).

Multimodal systems represent a new class of user-machine interfaces, different from standard WIMP interfaces. WIMP—"Windows Menu, Icon and pointing device"—is a style of Human Computer Interaction. The primary benefit of this style of system is to improve the HCI by enabling better usability for non-technical people, both novice and power users. The Multimodal system differs from WIMP by emphasizing the use of richer, natural ways of communication. Hence, the objectives of multimodal interfaces are to support and accommodate user's perceptual and communicative capabilities; and also to integrate computational skills of computers in the real world, by offering more natural ways of interaction to humans (Dumas, Lalanne, & Oviatt, 2009). The evolution of speech technologies and computer vision (gesture) technologies provides the way to implement naturalness in man–machine interaction. The component of a typical multimodal system is given in Fig. 1.

A typical Multimodal message has candidate elements from various modalities, which is defined as the dimensionality of multimodal signal. Consider speech, gesture, olfaction, taste, haptic and input via conventional input devices (keyboard, mouse) as various input modalities and candidate elements of these modality sets are

Speech {any spoken meaning full units}
Gestures {certain visual patterns generated by the human/ external object}
Haptic {touch input} e.g. Touch at a coordinate location 300,250
Olfaction {any smell}
Taste {any taste}
Conventional Input {input from Keyboard, mouse, joystick etc.}

Comprehension of a multimodal signal will be an appropriate synchronization of various elements drawn from the above said modality set. Dimensionality of a multimodal signal is defined as the number of the participating modality set. The following example gives detailed explanation of multimodal signal.

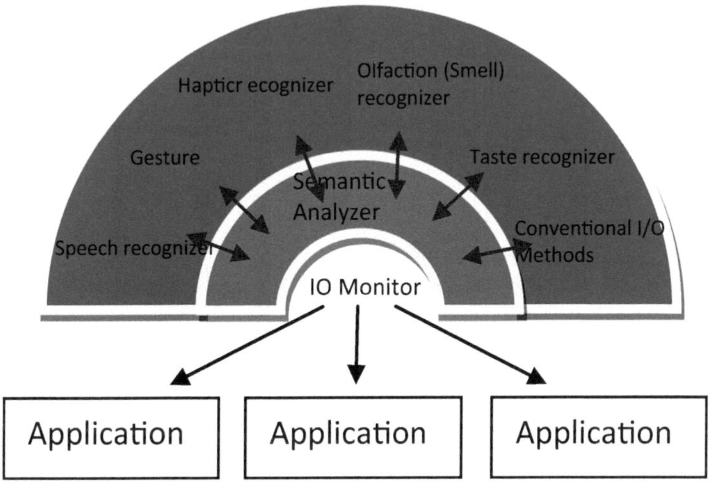

Fig. 1 Component diagram of multimodal interface system

3.1 A Sample Use Case: Multimodal Interface for Interacting with Desktop

A sample multimodal use case is simulated for interacting with the desktop. The user can communicate with the system through speech and hand gestures. Simple file operations like copy, delete etc. are considered as use cases here. The user issues a command to the system through speech "*Copy this file to that folder*" and gestures for the source and destination corresponds to the deictic "*this*" and "*that*" in the utterance. Here the user points twice, where the first pointing gesture is for the source and second one is for the destination. The generated Multimodal message signal has candidate elements of speech and gestures. The sequencing diagram for the above discussed scenario is shown in Fig. 2. The number of candidate modality sets in the communicated multimodal signal defines the input dimensionality of multimodal signal.

In the above example there are two input modalities involved so the dimension of input modality is two. The horizontal axis represents the time and the vertical axis represents the various input modalities involved. Here the speech signals started at time T_0 and ended at T_1. In between the speech event the gesture events *e1* and *e2* also took place. Pointing the source file event e1 happened in the interval T_a and T_b and the gesture which corresponds to pointing the destination folder happened in the interval T_c and T_d.

Here both gestures are identical in structure i.e. first one may be Pointer (200,175) the icon location of the source file to be copied and Pointer(100,230) the location of the destination folder where the file has to be placed. But the chronological ordering of these two gestures is mandatory as they specify the source and destination folders sequentially.

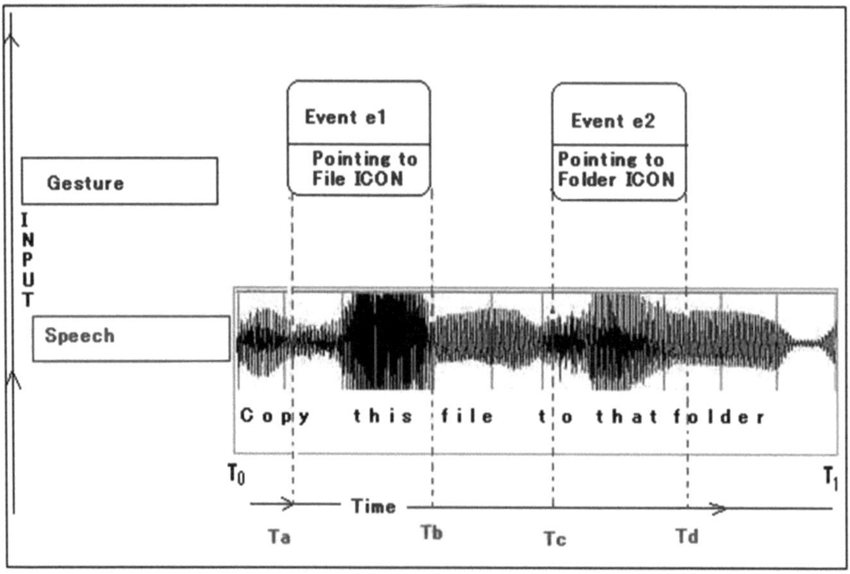

Fig. 2 Sequence diagram with occurrence of speech and gesture

4 Components of Multimodal Interface

Multimodal interface is composed of input modules, output modules and interaction manager (W3C, 2003). The input module is responsible for collecting input from the user and forwarding it to the interaction manager for processing. The processed input signal will be given to the user via output module. Speech, gestures (human body gestures, pen, handwritten gesture), olfaction, taste, haptic and input from conventional devices are the components of input module. The most popular output modalities are in audio and video formats. Interaction manager plays an important role in synchronization of various signals from independent sources. Interaction manager has two major components, which analyzes the semantics of the communicated message and manages the input output functionality. Semantic analyzer will analyze the meaning of the communicated message and I/O monitor will manage the input output functionality for interacting with the application program.

4.1 Speech

Speech is the most prominent mode of communication in human–human interaction. As we have discussed earlier, human prefers speech based interaction with the machine too because of ease of use. Automatic speech recognition is being a challenging research problem for past few decades. However the domain based speech recognition systems are available now with a reasonably good performance. High accuracy speaker independent speech recognition with emotion identification is still a research problem. There are many matured speech recognition systems/frameworks available where you can plug the desired language and acoustic models. CMU's Sphnix-4 is one of the widely used open source system for speech recognition. User can create his own acoustic and language models and plug in to the sphinx speech engine (http://cmusphinx.sourceforge.net/sphinx4/). There are many commercially available systems like Dragon naturally speaking, IBM's ViaVoice etc.

Considering the usability point, speech recognition system can be classified into two types, namely, small vocabulary/large users and large vocabulary/limited users. The small vocabulary program is perfect for automated answering on the telephone. It can identify different accents and variations in speech patterns. Sensibly, it is restricted to basic menu and generic responses. In larger vocabulary program, the system can identify more words with greater accuracy but it can identify fewer users (http://www.accuconference.com/resources/speech-recognition.aspx).

Automatic Speech Recognition (ASR) is a technology that allows a computer to identify the words that a person speaks into a microphone or telephone. The analog signal received through microphone will be digitized and sent to the pre-processing module. The system filters out unwanted noises, categorizes the frequency levels and normalizes the sound. Since people speak at different speeds, this should be aligned to the speed stored in the computer. It is really difficult to identify exact phoneme boundary in the spoken unit (word). Hence comparing the phonemes in the context to other phonemes is really a computationally intensive task which involved complex mathematical and statistical methods.

In the initial stages of speech recognition studies, speech recognition was a mere signal processing problem where language attributes were not considered. Later statistical and mathematical methods were incorporated which considerably improved the recognition accuracy. Statistical models like Hidden Markov models and neural networks are extensively used to solve various non-linear pattern recognition problems. In this chapter we have considered a small vocabulary speech recognition system with semantic analyzer for explaining the multimodal concept.

4.2 Gesture

Gesture is one of the ancient modes of communication before the evolution of spoken languages. The idea of interacting with machines via gestures is also a four decade old problem. Gesture recognition enables humans to communicate with the machine (HMI) and interact naturally without any mechanical devices. Gestures have long been considered an interaction technique that can potentially deliver more natural, creative and intuitive methods for communicating with computers (Sreekanth, Gopinath, Supriya, & Narayanan, 2011). The gestures are commonly used in human-human interaction and plays a major role in communication when the participants are unable to speak, or the situation does not allow the participant to speak etc. The gestures also play as an offset-input to the other mode of communication, for example gesture and speech are co-expressive and they form a part of rich human conversational features (Quek, 2003; Quek et al., 2002). If we look at the evolution of gesture based interaction initially there were glove based devices, but they lacked the naturalness factor as they had introduced an additional hardware constraints on the user. The models employed for gesture processing are either 3D models or image based processing. The former lacks the computational efficiency and the simplicity compared to other. In the image based processing method there are several techniques based on color, contour and correlation for identifying gestures (Chen, 2008).

Gestures are mainly classified into two based on the origin. Gestures are generated with the help of external objects like pen, coloured objects or with parts of the human body. Generally gesture recognition process means ability to track and identify the movements of user's body parts, which plays a major role in gesture based communication. As far as human user is concerned the hands, face, lips and heads are the major organs that take part in gesture based interaction. As we discussed earlier the device or object based gesture lacks the naturalness, focus is shifted more towards the computer vision based gesture processing. In gesture recognition technology, a camera reads the movements of the human body and communicates the data to a computer that uses the gestures as input to control devices or applications. Complex image processing techniques are used for extracting the meaning of the communicated message.

Gestures can be classified into three based on their functionality as

1. Symbolic gestures which convey a single meaning like American Sign Language gestures
2. Deictic gestures are pointing gestures which are mainly used in HCI
3. Iconic gestures which convey information about the size, shape or orientation of the object. Iconic gestures cannot be understood without accompanying speech.

Gesture recognition finds application in the emerging gaming scenarios, as it highly enhances the entertainment experience as well as simplifies the human–computer interaction. SixthSense is a wearable gestural interface device by Pranav Mistry, a PhD candidate in the Fluid Interfaces Group at the MIT Media Lab. SixthSense augments the physical world around us with digital information and lets us use natural hand gestures to interact with that information (http://www.pranavmistry.com/projects/sixthsense/).

Pen based gestures are another form of input mechanism for computers. Computer interaction through the drawing of symbols with a pointing device like a pen is taken as an input pen gesture. It provides an alternative to the direct manipulation or point and click method of interacting with a computer, allowing gestures or strokes of the pen to be translated into direct commands. The current technology is advanced in such a way that the user can even write the commands directly to the console.

4.3 Haptic

The word "haptics" refers to the capability to sense a natural or synthetic mechanical environment through touch. Haptic technology or haptics refers to the technology that connects the user to a computerized system by the application of sense of touch such as force, vibration or motion. Haptic information is a combination of tactile information as well as kinesthetic information. Kinesthesia is the ability to perceive one's body position, movement and weight. Haptic interfaces generate mechanical signals that stimulate human kinesthetic and touch channels and thus enable the human–machine communication through touch in response to user movements. The applications of haptic devices are mainly in mobiles, games, medicine, robotics etc. Haptics provides improved usability, enhanced realism and restoration of mechanical feel (http://www.cim.mcgill.ca/~haptic/pub/VH-ET-AL-SR-04.pdf).

4.4 Artificial Nose and Tongue (e-Nose, and e-Tongue)

Electronics nose and electronics tongue functionally imitate human nose and tongue for detecting the smell (odour) and taste respectively. The sensing system in the both the devices can be an array of several different sensing elements (e.g., chemical sensors), where each element measures a different property of the sensed chemical, or it can be a single sensing device (e.g., spectrometer) that produces an array of measurements for each chemical, or it can be a combination. Each chemical vapour/substance presented to the sensor array produces a signature or pattern characteristic of the vapour or substance. By presenting many different chemicals to the sensor array, a database of signatures is built up. Like any pattern classification system database of labeled signatures for various vapours or gases or substance should be maintained for training (Keller, Kangas, Liden, Hashem, & Kouzes, 1995).

4.5 Semantic Analyzer

As we have discussed, multimodal signal will be a combination of signals from independent sources which should be logically synchronized to achieve the goal of communication. Human cognition is capable of analyzing the signals from various sensors (eye, ear, skin, tongue, and nose) as well as synchronizing them logically. Functionally, the semantic analyzer module will mimic the human cognition system. Signals from individual sensors are recognized and it will to be sent to the semantic analyzer for understanding the message. Multimodal grammar has to be defined to understand and parse the multimodal signal. Before analyzing the meaning of the message, the signals from various sources has to be combined and it has to be represented in the system understandable format. Generation of a multimodal signal is discussed in following section. The recognized signals from various sensors are converted and represented in XML format. The recognized words from the speech are embedded in the tag < s>,</s>. The tag < w1>,<w2 > contains the recognized words in the communication. Similarly for gesture, the recognized gesture will be converted and embedded in between the tags < G> <w1 > . . ...</ w1>, <w2 > . . ..</w2 > . . .. .. </G>, where w1, w2 are the recognized words in the gesture vocabulary. The temporal information will be encoded with each event for proper synchronization. Semantic analyzer will convert the communicated messages from user vocabulary space to system vocabulary space (Sreekanth, Supriya, Girish, Arunjith, & Narayanan, 2008).

4.6 Input Output Monitor (I/O Monitor)

The multimodal sentence generated by semantic analyzer will be given to I/O monitor for issuing necessary signals to perform desired operations on the application program. Error or ambiguity in the input signal will be communicated to user via the feedback module. For example, if the user issued a command via speech *"copy this file"*, but the gesture corresponds to the deictic *"this"* is not given, i.e., gesture is missing, and then it should be notified to the user. The semantic analyzer will seek signals from the input modules to substitute the word "this" in speech. If it is not found in the stipulated time, the same will be notified to the user. The user can give the corresponding gesture if possible during this notification period. If nothing is received during a given interval, the system will go back to idle or safe mode so that it can listen to new input. If an invalid input pattern is recognized, the system will not respond to it.

4.7 Output Modality

While interacting with the machine, the interaction cycle will get complete only if the user gets a valid output, in the desirable format. The preferable output formats are audio visual signals. The current display technology along with speech synthesizer can provide the output in user desirable format. Advancements in the 3D display technology actually added to the effective ways of information display. 3-D displays are really effective for product design, complex scientific simulations, DNA/ chemical structure analysis, aircraft design and gaming. Coupling of speech synthesizer with 3D display will virtually create a real-time world for a better user experience. Discussion of output modalities is presently not considered in the scope of this chapter.

5 Implementation of Multimodal Interface

There are plenty of interface mechanisms available for HCI and the effort to couple the various modalities for enriching the interaction mechanism with computers lead to the development of Multimodal Frame work. It has been proven that adding more modalities always improves the quality of interaction and also helps to resolve the ambiguity in communications. By mimicking the way of human–human interaction via speech, researchers are more interested in incorporating Automatic Speech Recognition system, for natural and easy way of man–machine interaction. For example adding gesture recognition improves the quality of speech recognition also. The ambiguity in decoding an input speech signal can be resolved by accepting the gestures so that the more accurate word or sentence can be picked up from the vocabulary list. The lips reading or lips modeling is one among the several other gesture offset methods to improve the recognition accuracy of a speech recognition system (Cetingul, Erzin, Yemez, & Tekalp, 2006). In addition to that, hand and head gestures also improve the performance of interaction with computers by coupling with the Automatic Speech Recognition system. Simply adding or overloading the modalities one up on the other will not improve the quality of interaction. More systematic and intelligent models are required to couple various modalities and identify the semantics of the communicated message.

The typical implementation scenario for interacting with desktop systems for normal operations like simple file operations (open, close, copy, delete, move etc.) and other operations like search for a key word, zooming, copy a selected image location in the image, selection etc are discussed here. The present case study is based on a system which accepts input through speech, hand gesture and pen gesture.

5.1 Multimodal Signal Representation

As we have discussed earlier a typical multimodal signal have candidate elements from various participating modalities e.g. from speech, gesture, olfaction etc. Physically all these signals are independent in nature but semantically they are coupled. To process this signal from independent sources, this has to be synchronized properly. Moreover in real time scenario, these individual signals have a high temporal relation. The combined multimodal signal can be represented as an XML file or a markup language (W3C, 2009). Consider a scenario for deleting a file from the system, user can choose various ways since various modalities are incorporated. Suppose a user says "*Delete this file*" followed by a gesture pointing to the required icon using hand or finger. The string generated for processing will have the candidate members from speech vocabulary as well as gesture vocabulary. In this example the words "*delete*", "*this*" and "*file*" are the members of the set of speech vocabulary set S which are recognized by a speech recognition system. Similarly the gesture recognition system will return a string with location reference and temporal information to a semantic analyzer. An example of a gesture vocabulary data base is given in Fig. 3. Depending on the recognized gesture, corresponding strings will be generated as per the database.

The generated multimodal message for "*delete this file*" is given below (Fig. 4).

	Pointer (x,y)		Select Icon at the Present location of the courser
	Zoom Out The Selected area	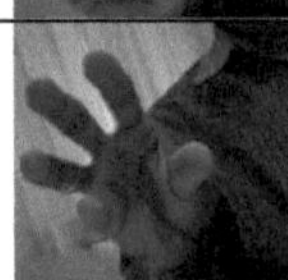	Zoom the selected area
	Page Down (Hand Expected to move from top to bottom	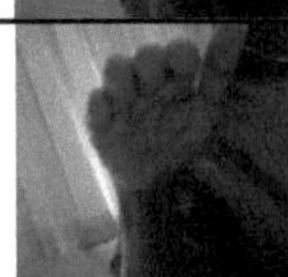	Page up(Hand Expected to move from bottom to top
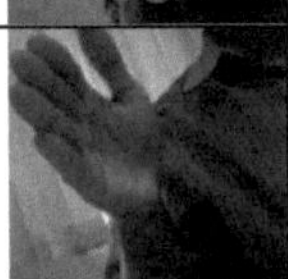	Previous / Move right Hand moves left to right		Next / Move left Hand moves right to left
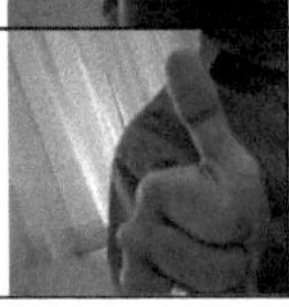	OK	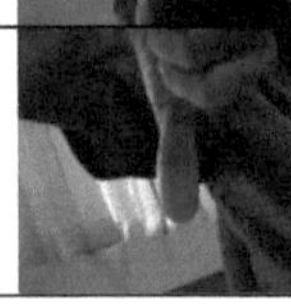	Cancel

Fig. 3 Issuing commands through gesture

```
<BEG_OF_MES>
    <S>
        <w1>Delete</w1>
        <w2>this</w2>
        <w3>file</w3>
    </S>
    <G>
        <w1>pointer (200,175)</w1>
    </G>
<END_OF_MES>
```

Fig. 4 Multimodal Message

5.2 Message Parsing and Understanding

Once multimodal message is generated, it will be sent to the semantic analyzer for understanding the meaning of the communicated message. Consider the message "Delete this file" followed by a gesture to point the required icon through hand or finger, the operational keyword "delete" is the significant part of this multimodal message. This can be represented as DELETE arg1, [arg2, arg3,. . .] (here at least one argument is must and others are optional) where DELETE is the operational key word.

In the above message the word after "*delete*" is "*this*", which is an deictic whose resolution is done using a multimodal reference resolution parse tree. The three level multi modal reference resolution parse tree for the above string can be represented as follows. This parse tree has a root node with Multimodal message and has three or more immediate child nodes depending on the number of input modalities involved in a typical multimodal message as shown in Fig. 5.

Here the circled word "this" is an deictic and after applying the reference resolution the antecedent equivalent to the deictic "this" is "pointer (200,175)". In the above example the pen gesture input is not present so it is marked as ε (null).

The above discussed is a simple multimodal reference resolution parse tree. Consider a multimodal parse tree which is highly time dependent which means the structure of the tree depends on the time of occurrence of input event. The user issues a command to the system as speech "*Copy this file to that folder*" and gestures for the source and destination corresponds to the deictic "*this*" and "*that*" in the utterance. Here the pointing gesture is used twice where the first points to source and second to destination. The corresponding multimodal string is generated by the system considering the time of occurrence of the event and the corresponding time stamp is added. The sequencing diagram for the above discussed scenario is shown in Fig. 2. The chronological ordering of these two gestures is mandatory as they specify the source and destination folders sequentially. So generation of the multimodal string should be with respect to the temporal aspects of the event. The multimodal message with temporal information for the

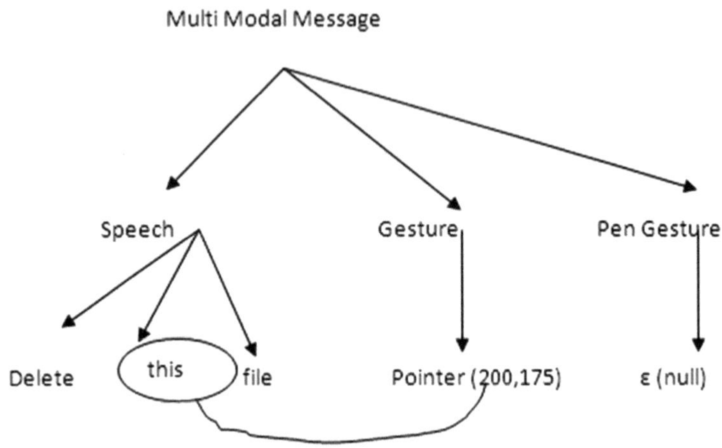

Fig. 5 3-level multimodal reference resolution parse tree

```
<BEG_OF_MES>
  <S, st= "T0", et= "Tn">
    <w1> copy </w1>
    <w2> this </w2>
    <w3> file </w3>
    <w2> to </w4>
    <w5> that </w5>
    <w6> folder</w6>
  </S>
  <G >
      <w1 st="Ta" et = "Tb"> Pointing (x,y)</w1>
      <w2 st="Tc" et = "Td"> Pointing (p,q)</w2>
  </G>
  <EOF_MES>
```

Fig. 6 Multimodal message with Temporal Information

above tasks can be represented as follows. In this example "st" and "et" are start time and end time respectively (Fig. 6).

From the above multimodal message the operational keyword, the time stamp and the attributes associated with various tags can be found by performing left to right parsing. The multimodal dependency parse tree for the above example is given in Fig. 7.

Consider a case in which pen-gesture is also involved.

"*Open this file and search for*"

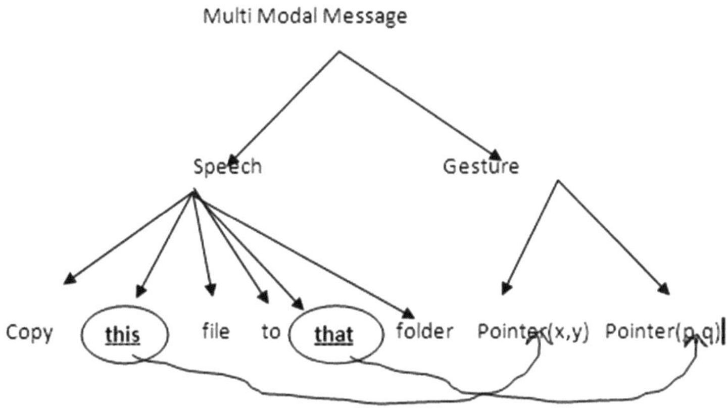

Fig. 7 Multi modal dependency parse tree

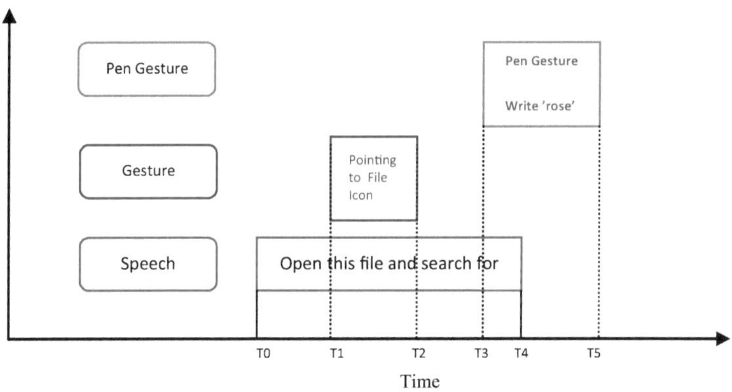

Fig. 8 Sequence diagram with the occurrence of speech gesture and pen gesture

In this case three modalities are involved, they are speech, finger gesture and pen gesture. The speech recognition system will return the string "*Open this file and search for*", and finger gesture will give the value of the location of "*this*" in the string. The pen gesture will recognize the stroke and will return the word "*rose*". The sequence diagram is shown in Fig. 8.

The inclusion of pen gesture in multimodal message is indicated by the tag <PG> </PG>. The multimodal message is represented in Fig. 9.

In this message the operational keyword is "*search*" which is defined as SEARCH (arg1, pattern1,[pattern2, pattern3,….]) where arg1 is the file to be searched and pattern1 is the pattern to be searched in the file specified in arg1. The operational keyword vocabulary for multimodal desktop interaction is listed in Fig. 10

Fig. 9 Multimodal message with many modalities

```
<BEG_OF_MES>
  <S, st= "T0", et= "T4">
    <w1> Open</w1>
    <w2> this </w2>
    <w3> file </w3>
    <w2> and </w4>
    <w5> search </w5>
    <w6> for</w6>
  </S>
  <G >
      <w1 st= "T1" et = "T2"> Pointing (x,y)</w1>
  </G>
  <PG>
      <w1 st= "T3" et = "T5">word(rose)</w1>
  </PG>
  <EOF_MES>
```

Fig. 10 Commands through speech

Operational Keywords	**Command**
Search	SEARCH (KeyWord , [arg1, [arg2,….]],)
Open	OPEN (arg1,[arg2,…})
Open with	OPEN_WITH (arg1,[arg2,…], argn)
Delete	DELETE (arg1, [arg2,….])
Delete All	DELETE_ALL
Copy	COPY (Src,[srcs..],dest]
Zoom	ZOOM(x1,y1,x2,y2) x1,y1,x2,y2 are location coordinates of a image

6 Multimodal Interaction in Consumer Electronics

6.1 Mobile and Hand-Held Devices

As a result of increasingly capable networks, devices, and speech recognition technology, the number of existing multimodal applications, especially mobile applications, is rapidly accelerating. Especially while using mobiles, user has very limited access to input space, and it is a cumbersome effort to use mobile keypad for text compilation. The developments in the haptic and speech based technologies have given a different dimension for interaction with the mobile systems. Speech offers one-handed and hands-free operation (W3C n. d.).

A related effort has recently been completed in the W3C by the HTML Speech Incubator Group (HTML Speech XG). The focus of the XG was developing proposals for accessing speech recognition and speech synthesis from HTML5 browsers, and Voice Search and Speech Command Interfaces are possible use cases for these technologies in the browser. However, the XG did not attempt to address modalities other than speech, such as handwriting, emotion, or the wide variety of present and future input modalities. Similarly, it didn't attempt to address non-browser contexts. In contrast, the Multimodal Architecture provides a generic framework for modality integration and control. Speech in the browser can be seen as a special case of modality integration covered by the MMI Architecture (W3C, 2014).

Dynamic gestures like waving and fist hitting gesture recognition are integrated by Microsoft Kinet, Sony PSP, etc. in their consoles. Dynamic hand-shape recognition is addressed in American Sign Language recognition in game development for deaf children (Brashear et al., 2006).

6.2 Home Appliances, e.g., TV, and Home Networks

There has been a tremendous effort from the players of consumer electronics industries for incorporating the multimodal interface to interact with the electronic gadget. Multimodal interfaces are expected to function as a remote control for home appliance and entertainment systems. The smart TV introduced by Samsung is a good example for this, where speech and gesture based interaction provides a hands free interaction. The fusions of various modalities for interacting with the system are the remarkable changes that we can observe in the forthcoming versions of electronic gadgets and home appliances. The gesture pendant (Starner, Auxier, Ashbrook, & Gandy, 2000) is a wearable device for the control of home automation systems via hand gestures. This solution has many advantages over traditional home automation interfaces as it can be used by those with loss of vision, motor skills, and mobility.

6.3 Enterprise Office Applications and Devices

Multimodal has benefits for desktops, wall mounted interactive displays, multi-function copiers and other office equipments which offer a richer user experience and the chance to use additional modalities like speech and pens to existing modalities like keyboards and mice. W3C's standardization work in this area should be of interest to companies developing client software and application authoring technologies, and who wish to ensure that the resulting standards live up to their needs.

Dialogue-Assisted Visual Environment for Geoinformation (DAVE_G) (Rauschert et al., 2002) that uses different interaction modalities, domain knowledge and task context for a dialog management that supports collaborative group work with GIS in emergency management situations. DAVE_G, a multimodal, multiuser geographical information system (GIS), has an interface that supports decision making based on geospatial data to be shown on a large-screen display. Interactions with Robot assistants (Rogalla, Ehrenmann, Zöllner, Becher, &

Dillmann, 2002) will be effective if they resemble natural human dialogue with gestures and speech.

6.4 Intelligent IT Ready Cars

With the emergence of dashboard integrated high resolution colour displays for navigation, communication and entertainment services, W3C's work on open standards for multimodal interaction should be of interest to companies working on developing the next generation intelligent car systems. Ford Model U Concept Vehicle (Pieraccini, Dayanidhi, Bloom, Dahan, & Phillips, 2003) was first shown at the 2003 North American International Auto Show in Detroit. The system, including a touch screen and a speech recognizer, is used for controlling several non critical automobile operations, such as climate, entertainment, navigation, and telephone. The prototype implements a natural language spoken dialog interface integrated with an intuitive graphical user interface, as opposed to the traditional, speech only, command-and-control interfaces deployed in some of the vehicles currently on the market. Hyundai has also come up with their concept car HCD-14 with integrated eye-tracking and 3-D hand-gesture recognition to satisfy driver commands.

6.5 Medical Applications

Mobile healthcare professionals and practitioners of telemedicine will benefit from multimodal standards for interactions with remote patients as well as for collaboration with distant colleagues. Wheelchairs, as mobility aids, have been enhanced through robotic/intelligent vehicles (Kuno, Murashima, Shimada, & Shirai, 2000) able to recognize hand-gesture commands. "Gestix",(Wachs et al., 2008) a vision-based hand gesture capture and recognition system that interprets in real-time the user's gestures for navigation and manipulation of images in an electronic medical record (EMR) database. Navigation and other gestures are translated into commands, based on their temporal trajectories, through video capture. A novel human–machine interface, called "FAce MOUSe"(Nishikawa et al., 2003), for controlling the position of a laparoscope was designed which allows nonintrusive, nonverbal, hands off and feet off laparoscope operations, which is more convenient for the surgeon.

7 Conclusion

The advances in information and communications technologies, computing and proliferation in use of internet have been one of the biggest contributors to the media convergence phenomenon. It helped to bring various modes of communication like audio, video, text based communication etc. under a single platform.

It introduced a different perspective of information sharing over conventional media like newspaper, radio and television. All those media were unidirectional, monotonous and had very limited scope for user interaction. The emergence of online version of newspaper, converged with social media like facebook, twitter, etc. provided a platform for people to interact and communicate their thoughts on a topic effectively. It is found that the convergence of various communication technologies have wider acceptance in the present society. The standard input output mechanisms for man machine interaction created a bottleneck in effective utilization of the full potentials of convergence of communication technologies and media. This can be overcome by the convergence of various modalities. The multimodal interaction provides varieties of input modalities like speech, gesture, haptic, etc. along with standard input output mechanisms.

Multimodal interaction framework provides a natural way to interact with the computers and electronic systems. The notion of converging different electronic and mobile devices accelerate the necessity of investing more time of researchers, to bring naturalness in human–machine interaction. By mimicking the human information perception and dissemination model we can design systems that are intelligent and effectively user friendly. From a human computer interaction point of view it is interesting to look at the various multimodal ways people interact with the environment and each other and to design systems that are sensitive to what the user wants without having been given explicit commands. The advancement of multimodal interaction paves the way to progress towards interfaces that are capable of human like perception.

References

Bolt, R. A. (1980). Put-that-there: Voice and gesture at the graphics interface. *Computer Graphics, 14*(3), 262–270.

Brashear, H., Henderson, V., Park, K., Hamilton, H., Lee, S., & Starner, T. (2006). American Sign Language recognition in game development for deaf children. In *Proceedings of ACM SIGACCESS Conference on Assistive Technologies (Portland, OR, Oct. 23–25)*. ACM Press, New York, 2006, pp. 79–86

Cetingul, H. E., Erzin, E., Yemez, Y., & Tekalp, A. M. (2006). Multimodal speaker/speech recognition using lip motion, lip texture and audio. *Signal Processing, 86*, 3549–3558. Science direct, Elsevier.

Chen, Q. (2008). PhD thesis, Canada.

Dumas, B., Lalanne, D., & Oviat, S. (2009). *Human machine interaction* (pp. 3–26). Berlin: Springer.

Dumas, B., Lalanne, D., & Oviatt, S. (2009). Multimodal interfaces: A survey of principles, models and frameworks. In D. Lalanne & J. Kohlas (Eds.), *Human machine interaction: Research results of the MMI Program* (LNCS, Vol. 5440, pp. 3–26). Berlin: Springer.

Flippo, F., Kerbs, A., & Marsic, I. (2003). A framework for rapid development of multimodal interface. In *Proceedings of ICMI'03, Vancouver, British Columbia, Canada. November 5–7, 2003*. http://www.caip.rutgers.edu/disciple/Publications/icmi2003.pdf

Ito, E. (2001). Multi-modal interface with voice and head tracking for multiple home appliances. In P*roceedings of INTERACT2001 8th IFIP TC.13 Conference on Human-Computer Interaction*, pp. 727–728.

Keller, P. E., Kangas, L. J., Liden, L. H., Hashem, S., & Kouzes, R. T. (1995). Electronic noses and their applications. In *IEEE Northcon/Technical Applications Conference (TAC'95) in Portland, OR, USA on 12 October 1995*.

Krahnstoever, N., Kettebekov, S., Yeasin, M., & Sharma, R. (2002). A real-time framework for natural multimodal interaction with large screen displays. In *International Conference on Multimodal Interfaces Proceedings of the 4th IEEE International Conference on Multimodal Interfaces-2002*, pp. 349–354.

Kuno, Y., Murashima, T., Shimada, N., & Shirai, Y. (2000). Intelligent wheelchair remotely controlled by interactive gestures. In *Proceedings of 15th International Conference on Pattern Recognition (Barcelona, Sept. 3–7, 2000)*, pp. 672–675.

Maat, L., & Pantic, M. (2007). Gaze-X: Adaptive affective multimodal interface for single-user office scenarios. In T. S. Huang, A. Nijholt, M. Pantic, & A. Pentland (Eds.), *Artificial intelligence for human computing* (Lecture Notes in Computer Science, Vol. 4451, pp. 251–271). Berlin: Springer.

Nishikawa, A., Hosoi, T., Koara, K., Negoro, D., Hikita, A., Asano, S., et al. (2003). FAce MOUSe: A novel human-machine interface for controlling the position of a laparoscope. *IEEE Transactions on Robotics and Automation, 19*(5), 825–841.

Pieraccini, R., Dayanidhi, K., Bloom, J., Dahan, J.-G., & Phillips, M. (2003). A Multimodal Conversational Interface for a Concept Vehicle. In *Eurospeech 2003*.

Quek, F. (2003). The catchment feature model for multimodal language analysis. In *Proceedings of the Ninth IEEE International Conference on Computer Vision (ICCV 2003) 2-Volume Set*.

Quek, F., McNeill, D., Bryll, R., Duncan, S., Ma, X. F., Kirbas, C., et al. (2002). Multimodal human discourse: Gesture and speech. *ACM Transactions on ComputerHuman Interaction (TOCHI), 9*(3), 171–193.

Rauschert, I., Agrawal, P., Sharma, R., Fuhrmann, S., Brewer, I., & MacEachren, A. M. (2002). Designing a human-centered, multimodal GIS interface to support emergency management. In *Proceedings of the 10th ACM International Symposium on Advances in Geographic Information Systems (McLean, VA, Nov. 8–9)*. ACM Press, New York, 2002, pp. 119–124.

Reitter, D., Panttaja, E. M., & Cummins, F. (2004). UI on the Fly: Generating a multimodal user interface. In *Human Language Technology Conference Proceedings of HLT-NAACL 2004*. pp. 45–48.

Rogalla, O., Ehrenmann, M., Zöllner, R., Becher, R., & Dillmann, R. (2002). Using gesture and speech control for commanding a robot assistant. In *Proceedings of the IEEE International Workshop on Robot and Human Interactive Communication (Berlin, Sept. 25–27, 2002)*, pp. 454–459.

Sreekanth, N. S., Gopinath, P., Supriya, N. P., & Narayanan, N. K. (2011). Gesture based desktop interaction. *International Journal of Machine Intelligence, 3*(4), 268–271. ISSN: 0975–2927, E-ISSN: 0975–9166.

Sreekanth, N. S., Supriya, N. P., Girish, K. G, Arunjith, A, & Narayanan, N. K. (2008). Performing operations on graph through multimodal interface: An agent based architecture. In *Proceedings of ICADIWT 2008. First International Conference on the Applications of Digital Information and Web Technologies*, pp. 74–77.

Sreekanth, N. S., Supriya, N. P., Thomas, M., Haassan, A., & Narayanan, N. K. (2009). Multimodal interface: Fusion of various modalities, multimodal interface fusion of various modalities. *International Journal of Information Studies, 1*(2), 131–137.

Starner, T., Auxier, J., Ashbrook, D., & Gandy, M. (2000). The gesture pendant: A self-illuminating, wearable, infrared computer-vision system for home-automation control and medical monitoring. In *Proceedings of the Fourth International Symposium on Wearable Computers (Atlanta, Oct. 2000)*, pp. 87–94.

Thomas, M., Hassan, A., Sreekanth N. S., Supriya, N. P. (2008). Multimodal interface to desktop. In *Proceedings of International Conference on Opensource Computing-2008, NMAMIT-Nitte Mangalor, Indiae*, 2008, pp. 26–30.

Turunen, M., Hakulinen, J., Kainulainen, A., Melto, A., & Hurtig, T. (2007) Design of a rich multimodal interface for mobile spoken route guidance. In *Proceedings of Interspeech 2007–Eurospeech*, pp. 2193–2196.

W3C. (2003). Multimodal interaction frame work. http://www.w3.org/TR/mmi-framework/

W3C. (2009). EMMA: Extensible multimodal annotation markup language http://www.w3.org/TR/emma/

W3C. (2014). Multimodal interaction working group charter. http://www.w3.org/2011/03/mmi-charter

W3C. (n. d.). Multimodal Access http://www.w3.org/standards/webofdevices/multimodal

Wachs, J., Stern, H., Edan, Y., Gillam, M., Feied, C., Smith, M., et al. (2008). A hand-gesture sterile tool for browsing MRI images in the OR. *Journal of the American Medical Informatics Association, 15*(3), 321–323.

Gearing Up the Knowledge Engineers: Experience Design Through Effective Human-Computer Interaction

Elspeth McKay

1 Introduction

There is an assumption that a good experience design will descend logically from a predetermined set of goals and objectives. In this scenario, both the user and the sourcing organization will have their own goals for an interaction. In some cases, the user may want to buy a book from a retailer. In others, the user may be trying to become better informed about dietary choices by seeking out an expert. In still others, the user may be attempting to better the lives of others by donating or contributing to a cause. In all cases, both the user and the sourcing organization have explicit goals. Therefore, an effectively designed experience will do its best to satisfy the goals of both parties (Henkle, 2009). Given the apparent ease in which we access everyday business processes online (Quintanilla & Wahl, 2014), it makes perfect sense to suppose that experience design also provides the necessary framework to enable successful media convergence.

As such, in a traditional business model, a key ingredient for experience design is to satisfy the needs of both parties involved in a business transaction (the sourcing organization/seller and their user/customers). In the modern world, advertising has increased awareness for such economic activity (Wharton, 2013). Put more simply this means to remain in business, the seller calculates a healthy cost profit margin for their service/product to attract customers, while satisfying their corporate goals and business objectives. A common sense approach on the part of the seller is to consider their net profit as a percentage of revenue. On the part of the customer it is all about how well the item purchased measures up to expectations. In the first instance, a business will continually monitor customer satisfaction with a view to increasing sales, while the customer will tend to browse the competition to ensure

E. McKay (✉)
School of Business Information Technology & Logistics, RMIT University, GPO Box 2476 V, Melbourne, VIC 3106, Australia
e-mail: Elspeth.mckay@rmit.edu.au

A. Lugmayr, C. Dal Zotto (eds.), *Media Convergence Handbook - Vol. 2*, Media Business and Innovation, DOI 10.1007/978-3-642-54487-3_15

they get the best deal. It is vital therefore to know how to manage this type of cognitive dissonance that is involved in the market place. This is where the notions of convergent divergence of digital media provide the necessary means to manage intersecting relationships.

To this end the concept of knowledge engineering was first noted as the integration of an expert's knowledge into the digital environment to solve complex problems. With the advent of convergent media as described within this book, knowledge engineering has evolved beyond the initial domain of computer science where the main activities involved: artificial intelligence (AI), high-level data bases, expert systems and decision support systems.

First we need to explain a little about how the digital media as it operates through the Internet has changed the entire face of business.

2 Background

Before the advent of computerized (online) business processes (now referred to as eCommerce), the traditional business model stood the test of time through careful attention to maintaining customer satisfaction and subsequent market share. Managing the relationship between the customer and the seller involved traditional nonverbal and verbal communication. However trouble emerges for experience design when we consider that eCommerce is fraught with complexity (Merali, 2004). In an eCommerce environment the aforementioned relationship shifts from the traditional one of supplier/customer to one that vacillates between both business parties when the service/product involves human-computer interaction (HCI). Not only are vagaries of human beings' interactions difficult to define (Vandyme, 2008); such interactions with computers sets out new dilemmas, as computer users have individual preferences for dealing with the screen-based information they receive (McKay, 2008). Within eCommerce, the goals and objectives of each party/stakeholder will differ significantly from the traditional business approach (McBride, 2012). The digital environment affords a plethora of interactive-media alternatives (Howcroft, Mitev, & Wilson, 2004; McKay, 2008). Except they do not blend easily to form a convergent dogma that confirms the original view of the principles of experience design outlined at the beginning of this introduction.

There is a dynamic relationship between the various business stakeholders that changes according to need. However, online experience design needs to be flexible enough to cater for these changing relationships. For instance, end-user payments in an online shopping environment are subliminal to the Web-mediated resource developer/supplier and as such are not paid to the digital artefact (media) producer. Digital artefacts by their very nature have a physical presence and therefore affect the process of how people do their jobs. Placing this type of materialistic view of digital media design taps into the very core of why eCommerce is such a complex environment. It is essential to instantiate the abstractness of such digital artefacts into physical examples to afford correct understanding (Leonardi, 2012). The real nightmare for the HCI/content-producer is therefore to manage the wide ranging and divergent interactive media. If the Web-developer's information system

(IS) design brief lacks the full description of the individual usability needs of the end-user (the paying customers), the organization's goals and objectives will not be met either. Nobody wins.

Experience design involving converging online media broadens the original concept of design to include a more dynamic profile. Take for instance managing the online warehouse catalogue updates once an order is placed by a customer. Imagine how many converging media tools are necessary to maintain a stable screen-based order-entry IS. In this case, instead of thinking of the end-user as the traditional customer making up front product payments, the eCommerce customers can entail vendor business entities buying the technical know-how of a Web-mediated resource/content developer to sell the vendor's merchandize online. No longer is it sufficient to design and implement an IS that is divorced from the vagaries of human interaction online.

The primary aim of this chapter is to understand how experience design, as seen through a social-contextual lens, serves to enhance knowledge engineering. It will give a critical view of convergence to discuss shortcomings/pitfalls. The objective is to ascertain whether or not the practice of convergent media leads to divergence. The overall perspective of this chapter is therefore designed to provide a stance on convergent divergence as it relates to the ways HCI affects IS design to reveal new thought on how to manage convergence by adopting better IS development strategies.

3 Knowledge Engineers Are Centre Stage for IS Development

Knowledge engineers are therefore centre stage for IS development to synthesize the experience of subject matter experts with the technical requirements of an IS. It cannot be overstated that the knowledge engineering role is important in crafting efficient and effective experience design that affords a balanced approach to IS development. This development cost will provide many experience design rewards. The resulting media functionality will be truly user-centered, or at least provide the user with this perception.

3.1 Unravelling the Expert Knowledge

If we were able to peek inside a Web-developer's media realm these days, sooner or later the need for building some kind of artificial intelligence tools (otherwise known as AI) would surface as an important part of their business model. Similarly, we would also find a preoccupation within the general AI/computer science community on how to capture certain aspects of the human endeavour for digital replication. The reason for this preoccupation on intelligent media development is because information communications technology (ICT) media tools that offer the perfect environment for capturing information for reuse as digital artefacts. As you can see in Fig. 1 that was produced at the Learning Agents Centre (Tecuci et al., 2004), knowledge engineers are central to unravelling the knowledge from other people. For the purposes of the discussion for this book on media convergent divergence, let us refer to these 'other people' as the subject matter experts or SME. Consequently the knowledge engineer needs to study how a SME reasons and solves problems. The resulting information gleaned from the SME is then encoded digitally by the knowledge engineer into a database, feeding into a problem solving application for use by others. Provided the problem solving logic engine has been designed with robust AI rules, the SME can examine the solutions generated by a type of intelligent agent (the information database and the problem solving application) to identify errors; passing them back to the knowledge engineer for correction.

The tricky part of this AI focused knowledge extraction scenario as depicted in Fig. 1, is twofold. Firstly, the person taking the role of the knowledge engineer needs to immerse themselves in each new genre to become deeply involved within the SME's problem solving expertise. Typical methods people use to articulate their knowledge in a form that may be useful for others' use includes: natural language, visual schemas and common sense. However, this leads to the second dilemma facing the knowledge engineer: experts may use very informal ways to express their knowledge, at times leaving out critical pieces of information considered too obvious to record. Moreover, according to Tecuci et al. (2004) this AI systems development process often causes a knowledge acquisition bottleneck. The transferring and transformation of information captured from the domain expert, passing through the knowledge engineer to the digital agent is often a long, painful and inefficient pathway.

However, it is necessary to leave the computer science domain to bring this discussion back to the argument for managing the effects of convergent media by more ordinary folk. It is more usual to find people lacking the ICT developer skills to cope with the rigour of the AI discipline. For instance, within the corporate sector, a SME is more often called upon to devise the digital business processing content as an appropriate media resource for the online corporate Web-site. As a

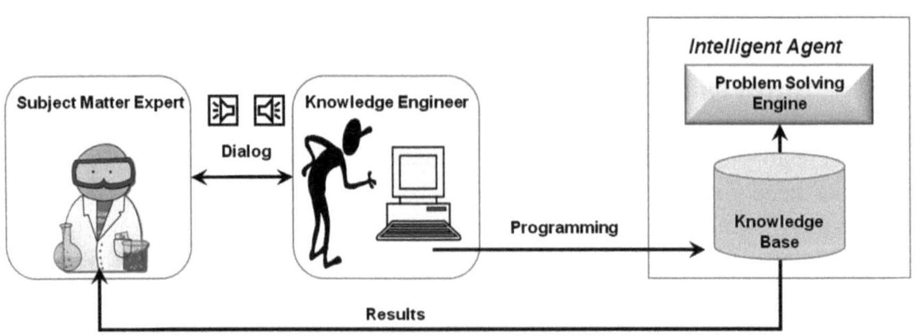

Fig. 1 Knowledge engineers taking centre stage for IS development (adapted from Tecuci, Boicu, Ayers, & Cammons, 2004)

consequence, their role shifts from the AI context, where their SME expertise was gathered to form an informational database, to a person who is directly responsible for the instructional design and development of electronic media for the so called end-users/customers. In this type of IS design setting, the SME provides the business processes to support the corporate Web-site. This means devising and implementing the experience design to achieve the goals and objectives of the business entity.

When people want to develop online material, in ideal conditions they should have access to instructional media developers (Caplan, 2004). Sadly this is not the perfect world we all desire and it is rare to find someone with enough educational technology savvy to create suitable electronic resources (Martin, McKay, Hawkins, & Murthy, 2007). In using the educational sector to explain how Caplan instead suggests that online course development can include para-academics who, as the metaphor suggests, are first on the educational technology scene. As such they liaise with the course owner or SME. This interactive relationship (knowledge engineering) can therefore keep track of the knowledge extraction (or content authoring process), removing distracting barriers to the learning context as they arise along the digital media developmental pathway. Retrieving information from a SME is not the end of the story either, especially these days when interaction with the Internet is a daily event. Para-academics can also look after many of those tricky little jobs that bog down ordinary folk, such as obtaining copyright permissions for images and content that the SME wants to include.

Thus far we have examined ways that a knowledge engineer acts as the intermediary between SME and their media development tools (problem solving engine and digital knowledge base). Next we look at strategies to manage the various convergent media involved in building digital artefacts.

4 Managing Media Convergence

Thus far we have shown how business firms in general can benefit from experience design which is sensitive to effective HCI. Rather than deal with the technical aspects of converging media, that would extend beyond the aims of this chapter, the preference is to detail instead the significance of the HCI environment. Subsequently, a knowledge engineer versed in the individuality of the interacting relationships between the various layers of the model can proceed to tool up the design brief accordingly. Figure 2 shows how the IS users' needs play directly into the design brief. The trick is to provide an experience design in which the user perceives that they have control over the functionality of the media. Every effort should therefore be expended to ensure that the project management decisions from the proof-of-concept to trial of the media are meticulous.

At first glance the whole idea of the user taking control of the IS media design process may seem far fetched. However, places of business where people with disabilities are having their special needs met include the theatres and sporting

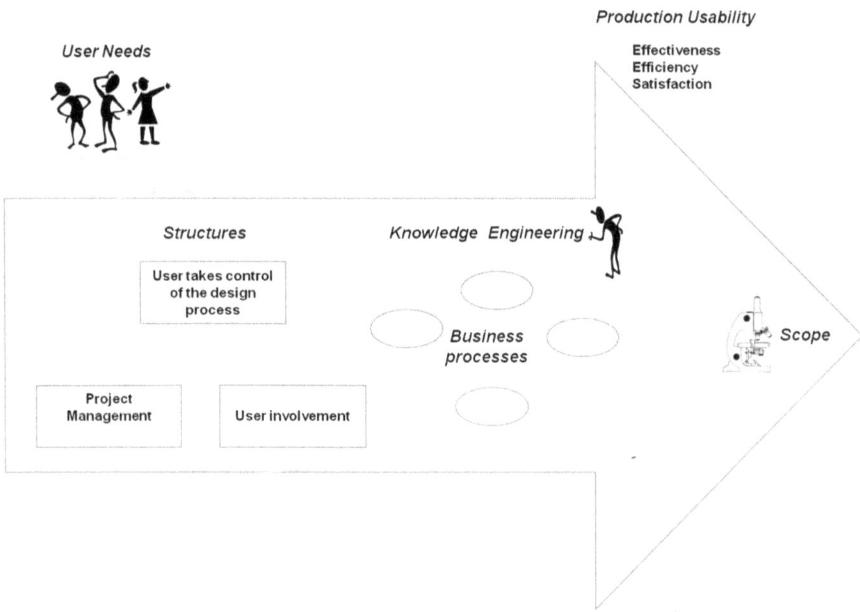

Fig. 2 Management of convergent media (adapted from McKay, 2008)

stadiums (Law, Jaeger, & McKay, 2009). According to Law et al. (2009) successful implementations of increased accessibility (media) products are the result of including the end-user in the initial design decisions (Law & McKay, 2007). Yet counting the end-user in the initial project management team must be a corporate mindset; while there appears to be plenty of inclusive rhetoric, the opposite may be the reality. Traditional project management teams operate in clearly defined processes that are highly controlled. Events are predictable and well understood; much patience is required when an end-user is put into this design scenario. However due to convergent media, business processes are more complex and interconnected than in the past. Whereas in projects where there is a significant software component, agile project management (APM) is better suited to involve the unpredictable user (Hass, 2007). In an APM team, instead of remaining outside the boundaries as shown in Fig. 2, all project stakeholders are actively engaged and have their say throughout the design and development processes. Each APM process ends with a review and lesson learned audit, thereby leading to a production usability sign off that ensures the end-users'/corporate stakeholders' needs reflect the design brief for effectiveness, efficiency and satisfaction.

Having set the context to argue for the importance of knowledge engineering and experience design to manage media convergence; it is time now to examine how the social shaping of media is driving a new approach to IS thinking (Howcroft et al., 2004).

5 Convergent Media Emerges Through a Social Contextual Lens

The convergence of our roles within society is noticeable through an applied social-contextual lens (Bradley, 2006). Since the mid 1980s there has been a shift from the way we would tend to isolate our work, professional practice and community citizenship. These roles are instead, more integrated with their lines of demarcation blurred. Figure 3 shows how these roles impact the quality of life.

Convergent media serves to support the merging of these role relationships. Although the Bradley model identified some of the key processes, the addition of mental health awareness in higher education has been added more recently by McKay and Martin (2010).

Be that as it may, can we assume there has been a parallel convergence of ICT tools/media? Bradley describes a tri-level view of continuous information technology (IT) development that mirrors the gradual computerization of our work content and work organization since the mid 1980s. These socio-technology levels include: mainframe computing where batch processing media commenced a new found reliance by the corporate sector on IT for business processing; screen-based online IS that widened access to more community-based activities; and micro-computing. Thus, computerization led a revolution in smaller computing machinery, yet such IT enhancements gave us the faster applications for the online access that we have become accustomed to.

It can be said that within each IT development level there has been a shifting towards convergent media that mirrors our changing roles as we carry out our private lives, community activities and professional practice. The problem is that

Fig. 3 Convergence of ICT—Humans—Society (adapted from Bradley, 2006)

we are captured by the market forces that continually bombard us with advertisements announcing the next technological gizmo. For the more techno savvy people, procurement of the latest ICT tools seems to have taken on cultural significance.

The fastest tools adopted are the ones that rely upon mobile technologies. This is one example of where convergence of media is not leading to divergent technologies. Consider how just one of these mobile devices can work as phone (frequency), send/receive eMails (data transmission), take digital pictures, provide interactive location maps, tell you the time around the world, provide flashlights and magnification. Now that's convergent media.

We can say that the blending of human's societal roles has given rise to a convergent media evolution. Figure 4 demonstrates how these roles integrate work, play and community. As such, they rely on the convergent evolution of ICT tools. Through this scenario we can witness socio-networking that is leading to important collaborative projects where historians, linguists, educators and gaming developers form collaborative Web-mediated teams united by ICT.

However convergent media presents a dilemma for people who wish to use off-line ICT tools in a narrow and limited manner. What happens when people want to carry out a role which does not require blended media activity? Perhaps they

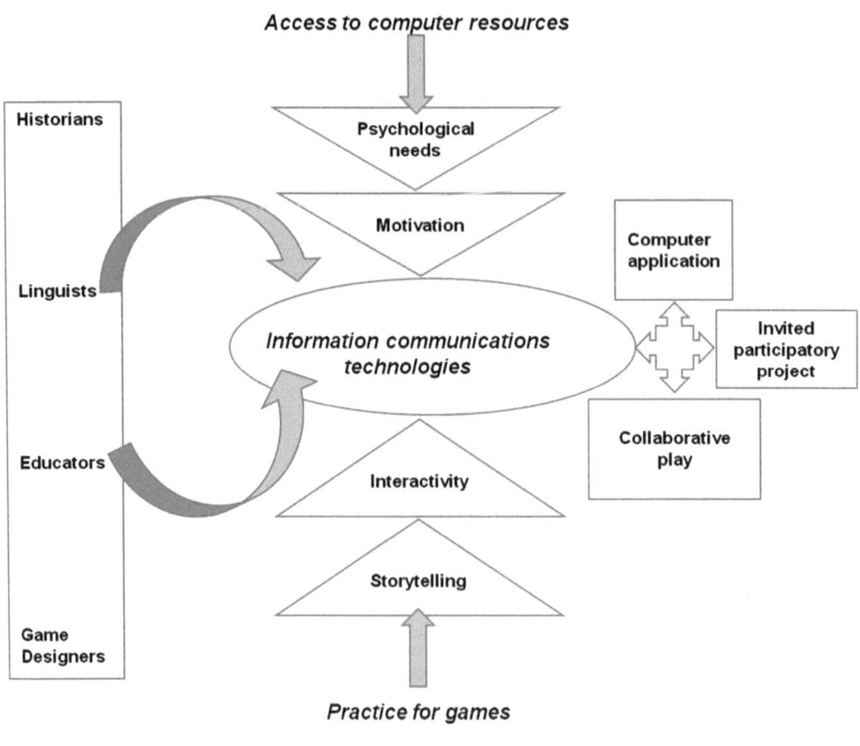

Fig. 4 Convergence of media

possess limited computer literacy and/or cognitive functioning. The problem is that truly convergent media requires sophisticated mobile radio data transmission. The solution for off-line performance is to maintain computing resources where localized applications do not conform to a convergent media framework.

Nevertheless, in order to better understand how to manage convergence in a media firm, we need to dig deeper to see whether the notions of convergent divergence media affects IS experience design?

6 Changeable Nature of the HCI Environment

Our adoption of ICT tools has led to the socio-enculturation of Web-mediated communications. Yet we cannot say that the changeable nature of the HCI environment is fully understood by knowledge engineers as they craft IS development (Mayhew, 2008) to enable efficient and effective media convergence. In an ideal world, Web-mediated IS should enhance dissemination and global accessibility. Furthermore, some of us believe the Internet should promote distributed people to work in a collaborative manner. This is where the notion of convergent media becomes apparent. However, in the past this type of distributed work group has required some form of physical collaborative IS, usually configured to provide two ways of working together. The first, through direct communication and collaboration, and the second, by indirect communication through shared artefacts and common workplaces (Allen, 2009). As such, collaborative IS can be created to run in different environments, adapted to cater for different groups of users, and made to be usable for various tasks. Moreover, small group instruction in a traditional face-to-face context can also be implemented with people over a distance using Web-mediated technology (E. McKay, Garner, & Okamoto, 2002). However, in drawing on an ICT tool-set as the vehicle for creating a successful collaborative climate, it should be noted that Web-mediated collaboration is very different from the collaboration involved in the more traditional face-to-face interaction (Anderson, 2008). For instance, face-to-face collaboration typically makes use of a rich cultural variety of verbal and non-verbal cues, such as facial expressions, glances, tones, and hand gestures (Metykova, 2013). By virtue of the enforced separation of people (Gannon & Pillai, 2013), Web-mediated collaboration will lack sufficient perceptual clues.

It has been shown by many of the educational technologists who are the front runners in dealing with online collaborative learning, that this lack of the usual communication framework is not only awkward, it is also quite stilted and can be extremely frustrating compared to face-to-face settings (Anderson, 2008). In Web-mediated collaborative events, there should be some capacity for visual and audio communications to enable this seemingly natural tendency (at least in some cultures (Gannon & Pillai, 2013) that humans tend to eye-ball their communications partner for a conceptual acknowledgement and/or agreement/disagreement!

However, management of convergent media presents the IS developer with the problem of maintaining a stable data link to keep the virtual communication running (Beeson Jr, Helmers, & Stein, 2011). It has been proposed that one advantage of face-to-face group communication is the tendency for a more effective interaction to occur than in computer-mediated groups, despite the fact that the former tends to deviate from the task at hand. These breaks in concentration provide humans with the opportunity to relax and attend to the emotional needs of the group that may only be noticed through visual cues (Wallace, 1999). Furthermore, people engaged in Web-mediated IS have broader discussions than the face-to-face groups (Benbunan-Fich, Hiltz, & Turoff, 2001). Face-to-face interaction can involve effective sequential problem solving activities. It is thus suggested that Web-mediated facilitation can become focused on solving general disagreements rather than on the questions at hand (Gackenbach, 2007). Therefore, it should be understood when Web-mediated collaboration is applied to the education/training sectors, the instructional context is quite dissimilar to the collaborative context of a more traditional instructional setting. Because of this, carefully planned Web-mediated instructional strategies need to be devised to ensure that the people participating benefit as much as from the more traditional type of physical collaborative setting; that is, where face-to-face communication takes place (Anderson, 2008).

In order to achieve the effective use of Web-mediated collaborative strategies, behavioural changes in the patterns of learning among group participants should be well understood (Wallace, 1999). Further research is needed to determine how people learn in a technology based collaborative setting, and more specifically how they accomplish common tasks while involved in Web-mediated IS. Much more work is needed on the effects of HCI and the interactivity of the learning process. The key to understanding collaborative learning lies in understanding the rich interaction between individual participants (Dillenbourg, 2005).

However, while Patricia Wallace (1999) was the first to uncover the dark side of people's behaviour on the Internet; where at times the nasty side of some online communications is left wanting, yet the number of positive interactions win out (Imaz & Benyon, 2007). Indeed to the extent that collaborative communities are springing up worldwide (Zaphiris, Ang, & Laghos, 2008). Our discussion now turns to how people are sharing their knowledge and experience online.

6.1 Collaborative Media Tools

When it comes to describing collaborative knowledge sharing on the Internet these days, some people have become entrenched in the terminology. For instance: see how easy it is to forget that we have all been using collaborative tools in our everyday lives for many years (Imaz & Benyon, 2007). Consider how accustomed we are to sharing our problems over the telephone with a friend. In the same sense, collaborative media tools provide a shared digital experiential platform—otherwise known as computer-supported cooperative work (CSCW). Here, people can reconstruct their thoughts to generate new meanings (Nguyen & Fussell, 2013). These days, Google.com shows how easy it is to find the techno-tools that enable people to come together using common off the shelf communication packages. Moreover, it helps things along if one has a Web-camera to transmit an image of one speaking, while tapping into the computer. Many mobile devices have a camera built in. As such, live shows that are broadcast online have become quite common amongst the techno-geeks as well as more common folk. These media tools are now so advanced that we can choose between real-time communication sessions (synchronous) where there is a two way flow of information, and the ones that are delayed (asynchronous) where the information is saved and may be received at another time as described on http://www.linfo.org/asynchronous.html. While it may be a new revelation to the IS/IT focused reader, at one point, educational technologists were flooding the literature with research papers that explain the differences between synchronous and asynchronous communication episodes (Anderson, 2008).

Therefore in keeping with the purpose of this chapter to explain how experience design affords the knowledge engineering of effective HCI, we now turn to look at how digital media communications may be sent differently.

6.2 Lifestyle Dichotomy Synchronous/Asychronous Flexibility

With the advent of more powerful media tools, there has been a softer approach taken towards experience design to provide lifestyle flexibility where individuals welcome the synchronous/asynchronous nature of their HCI interactions (Merali, 2004). This type of flexible media communications is typical of the evolving human-dimensions/social context of HCI (Lee, 2004).

6.3 Synchronous

Here is a simple diagram to depict a typical synchronous communication model. Notice the continuous stream of data passing from the sender to the receiver. This is a real-time conduit that transmits the communication transaction leaving a sender and travelling directly to the receiver. On the phone for instance, it may become congested if two people talk together (McKay, 2008) (Fig. 5).

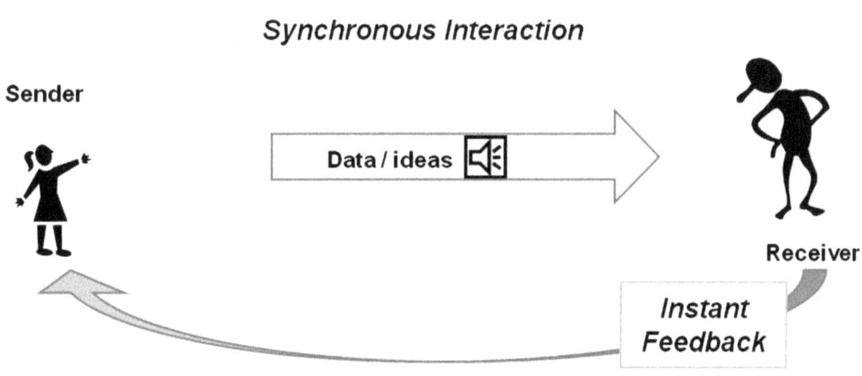

Fig. 5 Synchronous interaction (adapted from McKay, 2008)

With synchronous communication, the talker mostly waits for some sign that their speech has been received. Over the phone, non visual acknowledgement from the receiver of the message may take many forms of hearable communication techniques/cues; like uttering a '*Hmm*' or taking a breath. Over the Internet, synchronous feedback also comes in many different forms. When using a camera attached to one's computer, there are all the usual ways that we can indicate agreement or otherwise; like raising an eyebrow and twisting one's head, or looking downwards while one thinks. The beauty of our digital communications landscape means that we can also store these precious moments for later retrieval (McKay, 2008).

6.4 Asynchronous

When the receiver needs to wait for a response from another communication session, the flow of data/ideas becomes sporadic. This is known as asynchronous message transmission. The messages may leave the sender to be interrupted through a digital storage device, which means there may be a considerable passage of time before the receiver joins in. Typical asynchronous models can be found in threaded Internet discussion boards, where people take turns to respond in their own time. These delayed responses may also be sent to multiple people, who in turn respond at another time. eMailing is an example of this type of communication (McKay, 2008) (Fig. 6).

Thus far we have introduced the notion of experience design and knowledge engineering as an industrious partnership to enhance media convergence. A social-

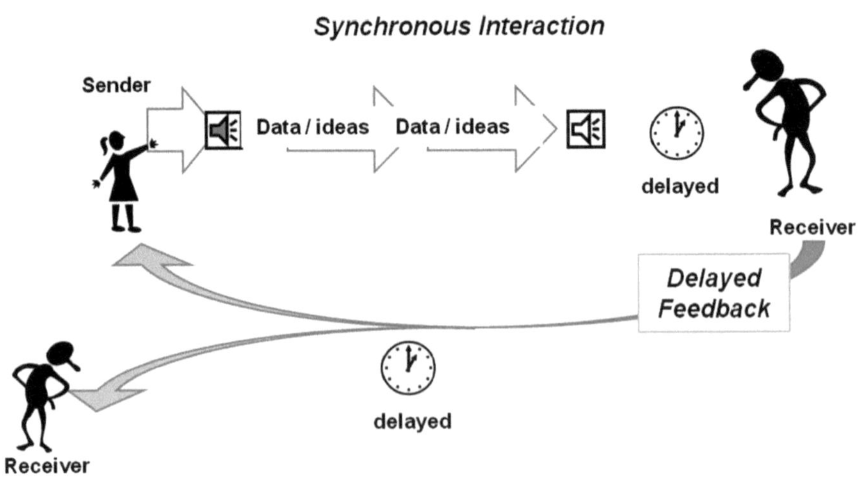

Fig. 6 Asynchronous interaction (adapted from McKay, 2008)

contextual lens was applied to demonstrate how far the acceptance of the human-dimension of HCI has travelled in recent times. However, while there is a general understanding of social-context (Bradley, 2006); we need to dig deeper into the modern media landscape.

7 Machine/Human Dichotomy of Effective HCI

We have seen in several places earlier in this chapter that effective HCI involves a dynamic dichotomous relationship between the machine-dimension and a human-dimension. Nevertheless the importance of balancing the HCI environment is more often lost along the IS-design pathway (McKay, 2008). Too often the resulting digital artefacts present divergent functionality, where the machine-dimension is separated completely from the human-dimension (Jones, Orlikowski, & Munir, 2004).

The human-dimension of HCI is complex with many environmental factors influencing how things turn out. The purpose of this following model is to suggest that there may be two layers of pressure that form the techno-humanoid landscape (McKay, 2008). Let us firstly take up the notion that the techno-humanoid landscape is the focal point for explaining the human-dimension of HCI. The more personal digital-awareness characteristics are shown here forming the closest layer, while the more distant external techno-humanoid contributors simmer away to complete the techno-scenario.

At any given moment the pressures from these layers can cause an interactive effect that ripples through to cause positive or negative outcomes for the comfort zone of this so called techno-landscape (McKay, 2008). We could of course extend the number of layers to include infrastructure elements like political and economic climate. For the moment however, this discussion will concentrate on the aspects of the techno-humanoid landscape that are closest to home.

It is fitting for the discussion on gearing up the knowledge engineers for experience design projects, that they are aware of the interactive effects of this model. It is proposed that the following descriptions form the basis of a media usability profiling tool.

7.1 External Techno-Humanoid Contributors

At the outer edge of this metaphorical universe there is a never ending supply of contributors that impact on our personal computing space. When they are perfectly meshed together they combine to form a powerful comfort zone that affords the user relative control over their computing environment. Mostly they will tick away in the background, unnoticed until something goes wrong (McKay, 2008).

7.2 Cultural Dogma and Professional Practice

The terms cultural dogma and professional practice are used here in the sense that a given culture (or professional practice) provides parameters for individual members to live and interact with others. By contrast, in the past we have thought about culture in a more closed environment bounded by traditional elements that dictate values and norms: religious belief and geographic location, for example. These days we also acknowledge that cultural groupings can conform to a number of interesting sub-groups. Still it would be fair to say there are a number of contributing factors that combine to form cultural specific parameters that may include:

- administrative systems—establish the rules of the social environment;
- economic systems—determine our survival and personal fulfillment;
- technology—ability to survive and prosper;
- physical assets—education systems to handle generational values; and
- creative pursuit—an endless list of activities that varies on circumstances.

7.3 Machine Sophistication

There can be no doubt that ever since the techno-generations commenced their march through the passage of time (when IS were definitely not convergent with their beefed up microprocessor brain power), nevertheless they did afford people access to everyday computing that became almost commonplace; including all the general whiz-bangery of the Internet. Strange as it may be to any of the techno-luddites left behind, as the computing power increased, the machinery and costs were downsizing. All this digi-freedom however came at a cost to the quality of our travels around the information highway. Unless one's machine had the ability to deal with the connectivity issues that involved hauling the huge bits per second of the broadcasting channels of the time, one could forget about using applications that involve image and sound.

7.4 Social Networking Availability

Various ICT tools have emerged to support our cyberspace wanderings. Prowling the Internet for opportunities to find and share information is a thing of the (recent) past. While it is very useful knowing how to push an Internet search engine around, it affords more of a solitary peek into the knowledge pool. It is no longer sufficient to rely upon these devices to do this type of information seeking hack-work. Instead these days, there is a whole new approach to sharing information that involves social networking sites. These sites are where ordinary people can and do let themselves go, revealing all sorts of personal information to anybody out there in cyberspace. Some of the more interesting and successful places that involve this

Table 1 Social networking Web-sites

Online social networking sites	Human-dimensions of HCI
http://www.myspace.com/ http://www.flickr.com/ http://www.youtube.com/ http://video.google.com.au/ http://www.livejournal.com/	Exploits the more traditional Web searching techniques that involve hyperlink navigation. At the same time relies upon information from implicit and explicit user feedback. Enables the leveraging costs of conventional Web searches, while ranking results that are relative to the interests of the social grouping

type of social networking include (provided your IS administrator supports these social networking sites): finding friends; sharing videos and photos; and writing blogs (Table 1).

Another way people go about their social networking online effectively is to simply pass URLs to interested people you know. This could be considered a divergent method of communication that is typical of a hidden mass of private knowledge navigation events, where the original author of the content found in a particular Web page has no control over their material spreading through cyber-space. One of the drawbacks of this type of social networking relates to the timeliness of the postings. There is no guarantee that URLs remain constant over time.

7.5 Accessible Information

The ability to access and use information through the various ICT tools has become a major daily activity for ordinary folk around the world. It has become an essential ingredient for the full participation of students in academic and other school-related activities. Sadly for most of the educational institutions, accessibility is commonly addressed as an afterthought and on an individual basis, often making it difficult, time-consuming, and costly to provide adequate access. A better approach is to consider the needs of all possible users in the planning phase, when considering the role that technologies will play in an educational entity's future. This type of approach is known as universal design, which should result in a technology-enhanced learning environment that benefits all users, including those with and without disabilities.

If an educational entity's technology environment has been established with a universal design process, all students and staff can participate in, and benefit equally, from the activities of that entity. Below are a few examples:

- **accessible web pages** allow people with disabilities to access information; share their work; communicate with peers, teachers, and mentors; and take advantage of distance learning options;
- **accessible instructional software** (on disks, CDs, or other media) allows students with disabilities to participate side by side with their peers in computer labs and classrooms as they complete assignments; collaborate with peers; create and view presentations, documents, and spreadsheets; and actively participate in simulations and all other academic activities;
- **accessible *multi-media***, including captions and audio descriptions, is accessible to all students, including those with visual or hearing impairments, and is more easily understood by people for whom the video is not in their first language or who learn best or otherwise benefit from multi-sensory input; and
- **accessible telecommunications and office equipment** make communication and educational administrative functions accessible to everyone, including those with mobility, visual, and hearing impairments.

7.6 Computing Activity Required

Before we endeavour to accomplish something, it is usual for us to seek out how we will know when we have achieved a successful outcome. This is the leading WYSIWYG principle of *What You See Is What You Get*. People deciding to undertake any sort of task on the Internet especially, will expect feedback along the way. For instance, in a Web-mediated environment there should be some type of correct performance indicator that typically has a form of knowledge of results or outcome feedback available. This is the most common type of feedback that lets someone know whether their performance is heading towards the desired goals (Stiller, Petzold, & Zinnbauer, 2011) Due to the speed at which we cognitively move around in cyberspace, these knowledge results should be crisp and to the point, carrying no additional information about the performed task other than its direct outcomes.

Consequently timely knowledge results are an inherent aspect of performing a recurrent constituent skill. Like when people are learning how to perform a word processing task. Performing particular procedures will typically yield information indicating when the particular goals have been reached. This is also very true within an eCommerce transaction, where the buyer needs to know that the correct item has been purchased and payment and delivery details are correct.

7.7 Ergonomics

Safety, comfort and reliability are central to making tasks easier to perform. Interest in human factors relating to physical comfort, or ergonomics, was activated through a combination of a number of otherwise disparate disciplines during World War II. These days, ergonomics as it refers to HCI is about ensuring safe workstation design and readability of screen-based information (Sharp, Rogers, & Preece, 2007), and how long individuals spend at a computer without taking adequate breaks. Similarly, much has been printed about the perils of long haul flights, where passengers sit for hours in cramped conditions, placing them at risk of developing deep vein thrombosis. Of particular interest in recent years is the likelihood of radiation levels emitted from visual display units causing harmful effects on unwary computer users. Other types of health threats from non-CRT technology include: '*repetitive stress injuries*' known as carpal tunnel syndrome.

7.8 Software Applications

In defining what a software application is, we must first mention that most often a computer program or group of programs are designed to make life easier for ordinary computer users (or end users). Software is divided into two general classes: systems software and applications software as shown in Fig. 7. Systems software consists of low-level programs that interact with the computer at a very basic level. This includes machine-related things like the computer's operating

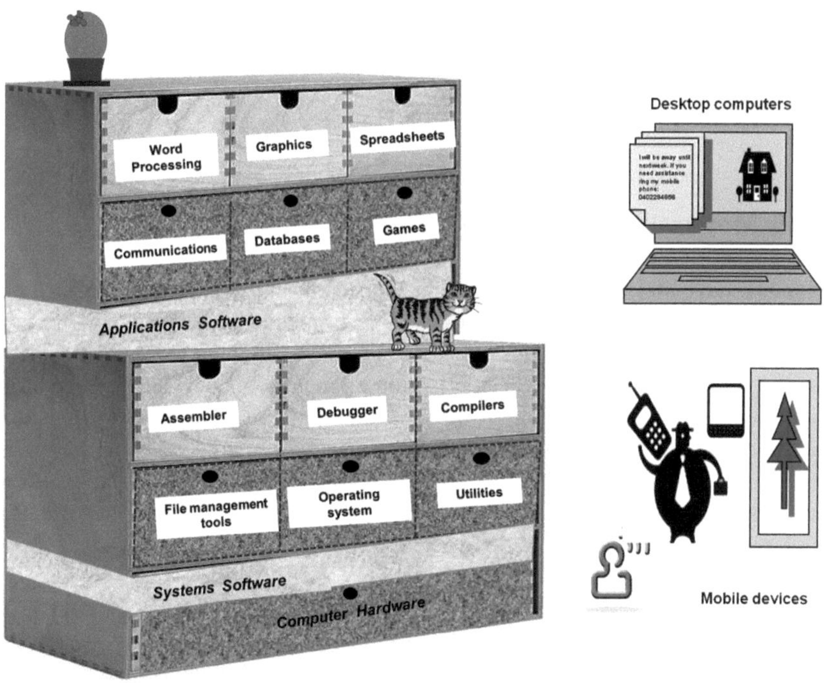

Fig. 7 Information system's software classification (adapted from McKay, 2008)

systems, compilers, and utilities for managing computer resources. Knowledge of how these programs function is unnecessary in most modern media tools. However, it is very useful for people to know how to reboot their mobile device when data corruption occurs. In contrast to the systems software, applications software (are also called end-user programs) they include: database programs, word processors, and spreadsheets. Figuratively speaking, applications software sit on top of systems software because they are unable to run without the operating system and system utilities.

We will leave this discussion on the outer edge of our metaphorical techno-universe where there are many hidden contributors that affect personal computing. In the main, unless the user has technical expertise, it is best, when asked by a software application, to simply accept the system defaults. This is one time when blissful ignorance should maintain your techno-status quo. Now moving to the inner circle of the techno-humanoid landscape, let's examine some of the more personal aspects of HCI (McKay, 2008).

These user-defined characteristics are possibly the most difficult to predict. Nevertheless the knowledge engineer should have an awareness of the types of issues that affect individuals' interactions with their media.

7.9 Personal Digital-Awareness Characteristics

Closer to the core of the techno-humanoid landscape is one's personal orientation or slice of the HCI action. For the purposes of explaining the more personalized contextual factors, one needs to note that it is usually easier for lay people to alter their HCI environment than to do anything else.

7.10 Motivation to Learn New Things

Humans have an innate drive to master their environment that remains a life-long trait, both at a personal and at a worldly level. Some people report huge satisfaction in completing a somewhat tricky task such as mastering how to download software from the Internet and install the application on their home pc. Competence motivation has a broadly based biological origin that explains human behaviour through our persistence of activities that constitute effective interaction with the environment (Al-Eraky, 2012). These include: exploration, activity and manipulation. In business circles the widely acclaimed Maslow's *"hierarchy of needs"* is still held in high regard today (Stahgnini, 2013). These relatively common sense steps to self-actualization include:

1. basic needs—play;
2. security needs—seniority plans, unions and severance payments;

Table 2 Herzberg's two-factor theory of motivation

Hygiene factors	Motivators
Company policy & administration	Achievement
Supervision, technical	Recognition
Salary	Work itself
Interpersonal relations, supervisor	Responsibility
Working conditions	Advancement

3. belonging needs—formal and informal work-groups;
4. esteem needs—titles, status symbols, promotions, banquets; and
5. self-actualization—leading to a person's motivation to transform perception of self to reality.

However it was Herzberg who extended Maslow's work to develop a specific content theory of work motivation. Herzberg identified satisfiers (as motivators) and dissatisfiers (hygiene factors). Taken together, these became known as Herzberg's two-factor theory of motivation (Maidani, 2012; Luthan, 1985, p. 201) (Table 2).

7.11 Novice/Experienced Dichotomy

Knowing how to learn new skills is something that we get better at as we grow older. For the most part, as we journey down the life-long learning path, it becomes easier to differentiate which instructional strategies are likely to suit us best. The trouble we run into, especially when using Web-mediated instructional environments, is likely to be whether there are any fast-tracking options for the learning tasks. It is well known that novice-learners require the full range of rules and information relating to learning something new, whereas an experienced person may only require a quick brush up. Novice learners will therefore respond best to measured amounts of guidance through progressively more complex instructional/learning content with strategic opportunities for interactive practice examples along the way (McKay, 2008). Alternatively, a person possessing a more experienced grasp of the particular task will want to experiment first, only dipping into the rules and basic information when they get into trouble. Unfortunately there are many Web-mediated instructional strategies that do not cater for this dichotomy. When the instructional systems cannot adapt to this important requirement, they run the risk of de-motivating both groups of learners. The result will produce confusion for novices when the primary rules and examples are not explicit enough. The same will cause utter boredom and frustration for the experienced when forced into involvement with the complete instructional strategy.

7.12 Information Processing Skills

While there are many well known theories that explain how our information processing per se (or thinking skills) work (McKay, 2008), researchers today remain keen to extend this earlier work as it would apply to Web-mediated learning. Although there are many high quality research investigations into various aspects of a user's psychological and physiological differences, they have not produced results that can be generalized to a wider Web-mediated learning community.

7.13 Physical Digital-Awareness

Perhaps an effective way to clarify the notion of a physical digital awareness as it pertains to the human-dimension of HCI, is to firstly rule out what it is not. Often digital awareness is published in the popular press (including the Internet) as being associated with new business practices that involve interactive ICT tools. Instead the definition should be one of a personalized awareness for the holistic nature of where a person cognitively fits within cyberspace, if at all. Accordingly this type of digi-cognitive presence means different things to different people. Furthermore, the notion of physical digital-awareness is predominantly a fluid environment that is constantly changing as the individual alters their position. As an example, consider the context of an individual's digital-awareness within wireless technologies.

7.14 Cognitive Capability

In keeping with the notion that people have an inherent learning capability that lasts for a life-time, we are able to extend this vision to qualify capability in terms of intellectual skills. To do this it is useful to return to Robert Gagne's five major categories of learned capabilities (see Table 3). They are known as: (1): Intellectual skills, (2): Verbal information, (3): Cognitive strategies, (4): Motor skills, and (5): Attitudes (Gagne, 1985; Kruse, 2012). At some point, while engaged in any learning event facilitated by some sort of ICT tool, it is likely that an individual will eventually need to operate using purely symbolic representations of physical objects. Therefore it can be said that acquiring such an ability requires transfer of symbolic operations within one's head (using imagination to operate a piece of machinery), or by writing (or typing at a keyboard about using the machine) and/or drawing. How a person uses the machine is very different from the actual physical actions that are required, where the learner is physically involved with an object. These capabilities are what Gagne means by intellectual skill.

Effective experience design should take account of this intellectual skill categorizing model. It is possible to translate the anticipated user reaction to digital artefacts drawing on this model.

Table 3 Robert Gagne's intellectual skills

Capability (learning outcome)	Examples of possible performance	Implications for human-dimension of HCI
Intellectual skill	Demonstrating symbol use for	In order to bring forward sufficient examples/non-examples; best practice in instructional systems design requires the need to consolidate a repository of reusable objects
Discrimination	Distinguishing printed m's and n's	
Concrete concept	Identifying the spatial relation 'underneath'; identifying a 'side' of an object	
Defined concept rule	Classifying a family, using a definition	
Rule	Demonstrating the agreement in number of subject and verb sentences	
Higher-order rule	Generating a rule for predicting the size of an image, given the distance of a light source and the curvature of a lens	
Cognitive strategy	Using an efficient method for recalling names, like originating a solution for the problem of conserving gasoline	The instructional architecture should cater for adaptable interactivity that provides timely and sufficient feedback to the user—responding to input from the keyboard, voice activation or touch screen tools
Verbal information	Stating the provisions of a nation's political agenda	
Motor skill	Printing the letter R; or skating a figure-8 on ice	
Attitude	Choosing to listen to classical music	

8 Conclusions

This chapter set out to identify convergent interactive media as seen through a social-contextual lens and in particular, concentrating on experience design. In so doing it gave a critical view of convergence of the IS development process showing where there are potential problems if left unnoticed by knowledge engineers. As to whether or not the practice of convergent media leads to divergence; it would seem this will be a never ending debate based upon the effectiveness of the media to converge in the first place. However, the perspective taken here was to discuss this likelihood as it relates to vagaries of HCI.

By adopting an agile project management (APM) approach, to information systems (IS) design, the knowledge engineer stands a better chance of capturing the tacit intellectual capital, which many corporate stakeholders are in fear of losing (DeLong, 2004). In our knowledge-based economy these days, while this hitherto lost information is more easily captured on mobile media devices; it is the finer detail of the imparted expert's knowledge that is more important for the experience design professional to retain.

After all, one of the challenges facing business sector's management of the media convergent divergence is to satisfy the goals and objectives of both the end-user and the sourcing organization. It would appear that the advertising industry have these softer skills off pat. For instance: in the way they link their profession (advertising) and culture (which is often expressed as global metaphors) as described by Gannon & Pillai (2013); by articulating them as two separate yet fully connected business processes. As such they rely upon language, not just as talking heads but a communications mechanism to provide them with the ability to talk about and share experiences (Wharton, 2013).

Management of media convergence by adopting better IS development strategies seems like a no brainer. Yet the Human-Computer Handbook (Sears & Jacko, 2008), is brimming full with excellent research studies that are collectively calling for the social shaping of media to drive this emerging shift in IS thinking (Howcroft et al., 2004).

Nevertheless, we saved the best until the last; in order to better understand how to better manage convergence in a corporate environment. We went deeper into the digital landscape to see whether the notions of convergent divergence media affects IS experience design. What we found was an HCI model (Fig. 8) which can be

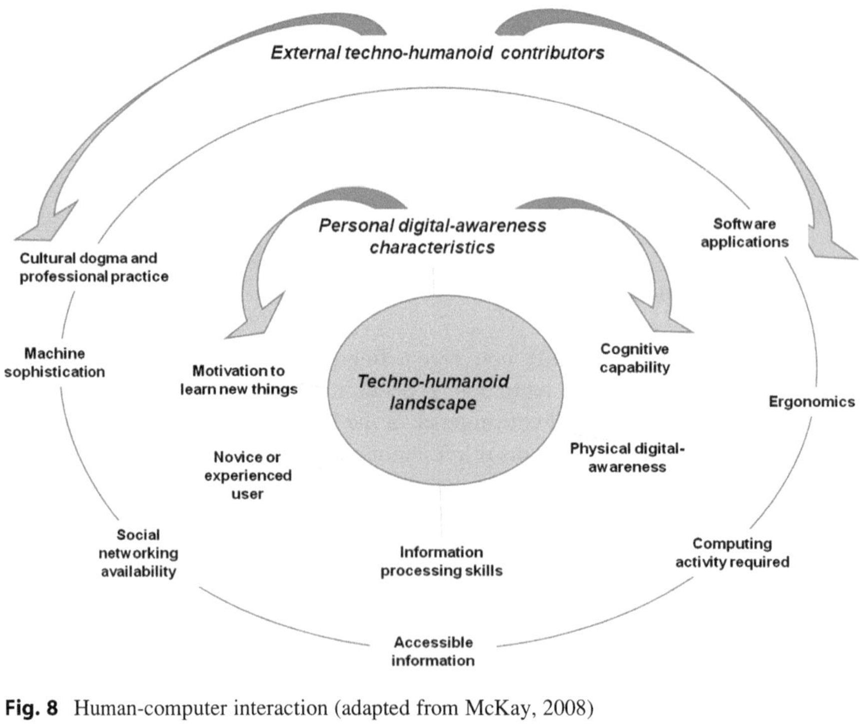

Fig. 8 Human-computer interaction (adapted from McKay, 2008)

interpreted as a digital profiling tool that knowledge engineers may wish to follow when trying to capture the all important tacit knowledge before it completely disappears.

References

Al-Eraky, M. M. (2012). AM last page: Robert Gagné's nine events of instruction, revisited. *Academic Medicine, 87*(5), 677.

Allen, C. (2009). *Creating shared language and shared artifacts*. Life with Alacrity. [Electronic version], from http://www.lifewithalacrity.com/2009/09/creating-shared-language-and-shared-artiifacts.html

Anderson, T. (2008). *Theory and practice of online learning: Issues in distance education*. From http://www.aupress.ca/index.php/books/120146

Beeson, R. B. Jr., Helmers, P. E., & Stein, M. V. (2011). *Maintaining stable virtual circuit data connection with spare protocol handler*. [Electronic version], Patent 5289179, from http://www.patentstorm.us/patents/5289179/description.html

Benbunan-Fich, R., Hiltz, S. R., & Turoff, M. (2001, January). *A comparative content analysis of face-to-face vs. ALM-mediated teamwork*. Paper presented at the 34th Hawaii International Conference on Systems Sciences (HICSS), Hawaii.

Bradley, G. (2006). *Social and community informatics: Humans on the net*. New York: Routledge.

Caplan, D. (2004). The development of online courses. In T. Anderson & F. Elloumi (Eds.), *Theory and practice of online learning*. Canada: Athabasca University. ISBN 0-919737-59-5.

DeLong, D. W. (2004). *Lost knowledge: Confronting the threat of an aging workforce*. New York: Oxford.

Dillenbourg, P. (2005). Designing biases that augment socio-cognitive interactions. In R. Bromme, F. W. Hesse, & H. Spada (Eds.), *Barriers and biases in computer-mediated knowledge communication* (pp. 243–264). New York: Springer.

Gackenbach, J. (2007). *Psychology and the internet: Intrapersonal* (2nd ed.). Amsterdam: Elsevier.

Gagne, R. M. (1985). *The conditions of learning: And the theory of instruction* (4th ed.). New York: Holt/Rinehart/Winston.

Gannon, M. J., & Pillai, R. (2013). *Understanding global cultures: Metaphoric journeys through 31 nations, clusters of nations, contents and diversity* (5th ed.). London: Sage.

Hass, K. (2007). *The blending of traditional and agile project management*. [Electronic version], from http://cs.anu.edu.au/courses/COMP3120/local_docs/readings/the-blending-of-traditional-and-agile-project-management.pdf

Henkle, D. (2009). *Experience design Part 1: definition*. [Electronic version], from http://devinhenkel.com/what-the/experience-design-part-1-definition/

Howcroft, D., Mitev, N., & Wilson, A. (2004). What we may learn from the social shaping of technology approach. In J. Mingers & L. Willcocks (Eds.), *Social theory and philosophy for information systems* (pp. 329–371). West Sussex: Wiley.

Imaz, M., & Benyon, D. (2007). *Designing with blends: Conceptual foundations of human-computer interaction and software engineering*. London: MIT Press.

Jones, M., Orlikowski, W., & Munir, K. (2004). Structuration theory and information systems: A critical reappraisal. In J. Mingers & L. Willcocks (Eds.), *Social theory and philosophy for information systems* (pp. 297–328). West Sussex: Wiley.

Kruse, K. (2012). *Gagnes nine events of instruction*. [Electronic version], from http://lrc.binus.ac.id/downloads/TE/Gagne.pdf

Law, C., Jaeger, P., & McKay, E. (2009). User Centered Design in Universal Design Resources. *Universal Access in the Information Society, International Journal, Constantine Stephanidis*

(Editor in Chief), ISSN: 1615-5289 ISSN: 1615-5297, Springer Berlin/Heidelberg, 9(4), 327–335.

Law, C., & McKay, E. (2007). *Taking account of the needs of software developers/programmers in universal access evaluations.* Paper presented at the 4th International Conference on Universal Access in Human-Computer Interaction, held at the 12th International Conference on Human-Computer Interaction (AC, UAHCI, HIMI, OCSC, VR, U&I, EPCE): Part 1, HCII 2007 LNCS 4554, Beijing, PR China, pp. 420–427.

Lee, A. S. (2004). Thinking about social theory and philosophy for information systems. In J. Mingers & L. Willcocks (Eds.), *Social theory and philosophy for information systems* (pp. 1–26). West Sussex: Wiley.

Leonardi, P. M. (2012). Digital materiality? *Firstmond Journal.* [Electronic version], *15* (6–7). From http://firstmonday.org/ojs/index.php/fm/article/view/3036

Luthan, F. (1985). *Organizational behavior*. New York: McGraw-Hill.

Maidani, E. A. (2012). *Motivation theory.* [Web page] from http://www.tutor2u.net/business/people/motivation_theory_herzberg.asp

Martin, J., McKay, E., Hawkins, L., & Murthy, V. K. (2007). Design-personae: Matching students' learning profiles in web-based education. In E. McKay (Ed.), *Enhancing learning through human-computer interaction* (p. 266). London: Idea Group Reference.

Mayhew, D. J. (2008). Requirements specifications within the usability engineering lifecycle. In A. Sears & J. A. Jacko (Eds.), *The human-computer interaction handbook: Fundamentals, evolving technologies and emerging applications* (2nd ed., Vol. 1, pp. 917–926). New York: Taylor & Francis Group.

McBride, S. (2012). *Companies like safemart grow market share by focusing on customer satisfaction.* MarketWire press release [Electronic version], from http://finance.yahoo.com/news/companies-safemart-grow-market-share-221858392.html

McKay, E. (2008). *The human-dimensions of human-computer interaction: Balancing the HCI equation* (1st ed., Vol. 3). Amsterdam: Ios Press.

McKay, E., Garner, B. J., & Okamoto, T. (2002, December 4–6). *Understanding the ontological requirements for collaborative web-based experiential learning.* Paper presented at the International Conference on Computers in Education 2002, Auckland, NZ

McKay, E., & Martin, J. (2010). Mental health and wellbeing: Converging HCI with human informatics in higher education. *Issues in Informing Science and Information Technology, 7*(3), 339–351.

Merali, Y. (2004). Complexitiy and information systems. In J. Mingers & L. Willcocks (Eds.), *Social theory and philosophy for information systems* (pp. 407–446). Chichester: Wiley.

Metykova, M. (2013). Media and advertising—The interests of citizens and consumers. In C. Wharton (Ed.), *Advertising as culture* (pp. 191–204). Bristol: Intellect.

Nguyen, D. T., & Fussell, S. R. (2013). Effect of message content on communication processes in intercultural and same-culture instant messaging conversations. In *CSCW 2013*. San Antonio, pp. 19–32.

Quintanilla, K. M., & Wahl, S. T. (2014). Technology in the workplace. In *business and professional communication: Keys for workplace excellence* (2nd ed., pp. 181–201). London: Sage.

Sears, A., & Jacko, J. A. (Eds.). (2008). *The human-computer interaction handbook: Fundamentals, evolving technologies and emerging applications* (2nd ed., Vol. 1). New York: Taylor & Francis Group.

Sharp, H., Rogers, Y., & Preece, J. (2007). *Interaction design: Beyond human-computer interaction* (3rd ed.). New York: Wiley.

Stahgnini, J. (2013). *Maslow's hierarchy of needs: Transcendence and belongingness, Part 1.* [Electronic version], from http://jeremiahstanghini.com/2013/07/16/maslows-hierarchy-of-needs-transcendence-and-belongingness-part-1/

Stiller, K. D., Petzold, K., & Zinnbauer, P. (2011). Presentation time concerning system-paced multimedia instructions and the superiority of learner pacing. *Australasian Journal of Educational Technology, 27*(4), 693–708.

Tecuci, G., Boicu, M., Ayers, C., & Cammons, D. (2004). Cognitive assistants for analysts. In *Proceedings of the 4th Annual Analysis & Productions's Analysis Conference: The future of Analysis*, NSA.

Vandyme, G. (2008). *The vagaries of human interaction*. Deviantart.com. [Electronic version], from http://greymanvandyme.deviantart.com/journal/The-vagaries-of-human-interaction-235043099

Wallace, P. (1999). *The psychology of the internet*. Cambridge: Cambridge University Press.

Wharton, C. (2013). Introduction—Advertising and culture. In C. Wharton (Ed.), *Advertising as culture* (pp. 3–11). Bristol: Intellect.

Zaphiris, P., Ang, C.-S., & Laghos, A. (2008). Online communities. In A. Sears & J. A. Jacko (Eds.), *The human-computer interaction handbook: Fundamentals, evolving technologies and emerging applications* (2nd ed., Vol. 1, pp. 603–619). New York: Taylor & Francis Group.

Transmedia Perspectives

Sabiha Ghellal, Annika Wiklund-Engblom, Ann Morrison, and Damjan Obal

1 Introduction

Media convergence and transmedia are closely related by theory, methods and application. Media convergence (1979) was introduced by Nicholas Negroponte during a presentation to raise funds for the MIT Media Research Laboratory.[1] Negroponte predicted that the digitalisation of content would lead to major industrial changes, forcing broadcasting and motion picture, computer and print or publishing industries to move closer towards one common content distribution model. He further outlined this theory in his book 'Being Digital' in 1990. Roger Fidler (1997) also researched the issue of digitalization in his book Mediamorphosis, where he anticipated how the digitalization of content could influence value chains and how convergence may impact upon more traditional media producers and service providers.

[1] MIT Media LAB (Media Research Laboratory) Massachusetts Institute of Technology. http://web.mit.edu/. Accessed 16 Dec 2012.

S. Ghellal (✉)
Department of Mobile Media, Stuttgart Media University, Stuttgart, Germany
e-mail: ghellal@hdm-stuttgart.de

A. Wiklund-Engblom, PhD
MediaCity, Åbo Akademi University, Vasa, Finland
e-mail: annika.wiklund@abo.fi

A. Morrison
Department of Architecture, Design and Media Technology, Aalborg University, Aalborg, Denmark
e-mail: morrison@create.aau.dk

D. Obal
Faculty of Electrical Engineering and Computer Science, University of Maribor, Maribor, Slovenia
e-mail: damjan@edgartells.me

A. Lugmayr, C. Dal Zotto (eds.), *Media Convergence Handbook - Vol. 2*, Media Business and Innovation, DOI 10.1007/978-3-642-54487-3_16

The concept 'transmedia', as coined by Marsha Kinder in 1993, describes a reproduction process of TV or movie characters in video games and how this reproduction might intensify and accelerate in commercial '*transmedia supersystems*'. In 2004, Henry Jenkins further defined 'transmedia storytelling' as *"the consumption of different content that is part of the same storyworld on several devices"* (Jenkins, 2006). He conversely discussed how consumer behaviour might cause a less predictable user dynamic by highlighting that convergence occurs via the social interaction amongst the users. Jenkins further defined convergence as *"...the flow of content across multiple media platforms, the cooperation between multiple media industries, and the migratory behaviour of media audiences who will go almost anywhere in search of the kinds of entertainment experiences they want. ... enables the same content to flow through many different channels and assume many different forms at the point of reception."* (Jenkins, 2006).

However, almost a decade after Henry Jenkins introduced his vision, storytellers and producers working with transmedia storytelling still struggle to live up to its full potential. Meanwhile, other interpretations of the term transmedia add confusion to this already complex emerging subject. Henry Jenkins himself has recently distinguished his vision from current transmedia branding 'buzz' by reminding us that transmedia content always expands rather than merely adopts existing stories (Jenkins, 2011).

While we have seen a number of independent transmedia storytelling productions and grassroot projects over the years we still have not seen a commercial transmedia storytelling project that is as successful as blockbuster movies. One possible reason might be that a commercial project that adheres strictly to Jenkins' vision would require very large corporate resources. The challenges within logistics, planning, scripting, marketing or distribution describes complex interdisciplinary issues.

In addition to the challenge of pinning down a new production model as it evolves comes the challenge of bringing together domains that have been firmly established as being separate but which now have to work together. Enabling processes, personnel and language to match up between game producers, film producers and book publishers involves having to change mind-sets, organizational structures as well as building new infrastructure for long established multimillion-dollar industries.

Transmedia content production requires scriptwriters, interaction designers and content producers to collaborate in order to come up with a complex storyworld that allows for interactive and non-interactive entertainment. This necessitates A-Shape skills (Leonard & Sensiper, 1998). The concept of A-shaped skills is a metaphor used in job recruitment to describe the abilities of a person. A-shaped individuals embody technological fusion; within the context of transmedia storytelling it describes individuals that merge creative writing and interaction design skills. The vision of 'growing up digital' (Tapscott, 2008) suggests that such talents might emerge over time. However, such talents or multi-talented teams do not fit within current production models. Current content financing and production models are limited to pre-produced content creation that might be distributed using crossmedia channels but would not allow for emerging storyworlds. Consequently, transmedia storytellers struggle to find funding—and those few that do, struggle with the challenge of combining traditional and new media production methods which in turn requires addressing a complex issue forcing major changes to long established multimillion dollar industries. Bearing in mind that transmedia storytelling describes only one possible entertainment format it would be unrealistic to claim that those industries would have to change in order to cater for transmedia only. What necessitate a change are user behaviours such as for example, video on demand (Cheng, Stein, Jin, & Zhang, 2008) trends that have a major impact on broadcasting value chains similar to the changes that the digitalization of music brought to that industry.

For the purpose of designing a conceptual transmedia methodology we conducted 20 international interviews including early adopters, scientific researchers, broadcasters, production experts as well as traditional storytellers. Each interview lasted 2 h and was focused on terminology in use, personal experiences as well as tools and techniques applied. During an international interdisciplinary 1-day workshop we consolidated the findings from these interviews and discussed perspectives of transmedia storytelling. We found that the wide array and differing perspectives that transmedia can entail often causes confusion from scientific, investigative and economic perspectives.

In this chapter we introduce several perspectives of transmedia content consumption, discuss its challenges, influences, current trends and illustrate an early version of our conceptual transmedia methodology.

1.1 The Problem of Complexity

Transmedia storytelling does not only require interdisciplinary creative teams to collaborate closely and form collective intelligence; it is also based on complex cross-domain know-how that forms complicated and diverse frame-works that require a multitude of tools and techniques.

One example of such complexity can be found in the area of cognition (Sternberg & Sternberg, 2011). Cognitive mental processes might include, attention, memory, producing and understanding language, learning, reasoning, problem solving and/or decision making, to name a few. A central problem that needs to be addressed is that multitalented teams, including scriptwriters, game designer, graphic designers, or user experience designers, use different mental models. Their understanding of cognition is related to their genre's perspective of how people perceive their creative work.

In the area of script writing, where story is key, cognition is often referred to as 'narratology' (Prince, 1994) utilizing frameworks such as 'heroes journey' (Vogler, 1992). The'heroes journey' claims that most stories can be boiled down to a series of narrative structures and character archetypes, which can be used to construct stories.

From an Interaction Design and User Interface perspective (Bolter & Gromala, 2005), cognition is discussed in terms of how well we understand and are able to manage an interface. Therefore concepts such as, 'cognitive load' (Paas, Renkel, & Sweller, 2004), researching the overwhelming overload of multimedia data and 'cognitive friction' (Cooper, 2004)—focusing on the resistance encountered by human intellect when it engages with a complex system—are relevant within this field. Graphic designers work with a different perspective of cognition in applying rules of Gestalt psychology (Humphrey, 1924), in order to determine how we perceive visual designs. Experience design, on the other hand focuses on cognition in relation to our emotions and the motivational affordances offered by a system in how well it fulfils our motivational needs (Pritchard & Ashwood, 2008).

1.1.1 Orchestration

Orchestration describes another complexity issue within the production of transmedia storytelling. It describes the issue of arrangement, management and illustration of complex story structures. During a set of international interviews, transmedia storytellers reported difficulties in keeping an overview of their own complex transmedia structures. Transmedia stories feature not only the common linear story-paths but also offer cross-paths to side stories and additional multimedia content that have to be orchestrated into holistic user experiences (Hassenzahl & Tractinsky, 2006). Keeping track of all the story and design elements within production teams is a time consuming and complex task. The main orchestration issues within transmedia storytelling are related to the multitude of media spaces and novel temporal aspects that have to be aligned to form rich transmedia experiences.

1.1.2 Media Disruption as 'Divergent Convergence'

Media disruption or discontinuity refers to a situation where a person seeking (or consuming) information is forced to change their mode of information processing or searching in order to comply with the media at hand. In divergence projects media disruption may play a major role and in order to deal with media disruption it has often been suggested to create seamless designs. In the context of transmedia this would refer to a situation where a person consuming information is disrupted and therefore no longer feels immersed in the story world. Media disruption should not necessarily be understood as an issue that arises when users switch media types since an interruption of an experience might be intentional within transmedia storytelling and might not necessarily result in breaking the illusion or 'flow' (Czikszentmihalyi & Nakamura, 2002). Switching media types is often introduced by designing for motivational affordance. For example, in the pervasive gaming project (Montola, Stenros, & Waern, 2009) 'Epidemic Menace' (Lindt, Ohlenburg, Pankoke-Babatz, & Ghellal, 2007) each media type featured a different role as part of a role-playing game. Here the user changed the mode of information processing if and when they felt the need to switch roles. Media disruption in the field of transmedia design could be defined as a phenomenon that needs to be integrated into the design of the storyworld in order to create one fictional world that provides a coherent experience in itself.

1.1.3 Storyworld or Fictional Universe

Fictional universe describes a self-consistent fictional setting that does not contain storylines or plots. Within a fictional universe all elements that form the universe such as theme, conflict, existence or genre are of equal importance (Jones, 2006). A cohesive fictional universe will consider all themes, but the trigger for an idea may of course vary in nature. A fictional universe is typically created prior to story writing and provides the frame for any type of story to occur. What distinguishes a fictional universe from simpler storytelling is the level of detail and internal consistency. Within linear storytelling frameworks, the story follows strictly along a line of engagement. In the case of TV series, such as 'The Sopranos',[2] it may also follow several storylines simultaneously, but each story line in itself is also based on a strict linear form of engagement. A fictional universe, on the other hand, has an established continuity and internal logic that must be adhered to throughout the work and even across separate works. Rather than adding interactive elements as an afterthought at the end of a production cycle, fictional universe based productions would allow considering user engagement from the outset of a production.

1.1.4 Convergence Trends as Opportunities for Transmedia Design

Above we suggested that convergence, divergence and transmedia describe separate intertwined theories. While convergence and divergence are used to describe both, a media distribution model and more general media consumption phenomena, transmedia describes one specific format of content consumption that inherits participatory elements. Television has drastically changed over the past 15 years due to the growth of digital modes of caption and transmission. Television content is no longer only delivered temporarily via networks, standardized broadcasting or cable/satellite solution providers. According to market researchers such as the 'Nielsen Company',[3] a growing number of households consume streamed rather than broadcasted content. It further suggests that many households no longer own TV-sets, but use their computer and high-resolution monitors as their main home entertainment device, consuming content as and when they want it.

Video on demand solutions impact current value chains, by excluding broadcasters and including network based content providers, such as for example, Netflix.[4] This trend may be relevant for transmedia storytelling by eradicating stipulated broadcasting times and allowing for evolving storyworlds. TV series content providers, such as for example HBO, have already adapted to the changing consumer behaviour. They now offer their content as video on demand solutions as well, and focus on more conceptualised content enabling multi-episode or even multi-season consumption. Here, the temporal aspect of content creation and

---

[2] The Sopranos HBO 1999–2007 http://www.hbo.com/the-sopranos#/. Accessed 16 Dec 2012.

[3] The Nielsen Company. http://www.nielsen.com. Accessed 15 Aug 2013.

[4] Netflix Inc. US Patent Nos. 6,584,450; 7,024,381; 7,631,323; 7,403,910; and 7,617,127. Accessed 15 Aug 2013.

consumption changes and provide new opportunities for stories to evolve across media types.

1.1.5 Divergence Trends as Opportunities for Transmedia Design

Due to the divergence of content, designers and business decision makers now face the task of figuring out how to deliver novel rich experiences across multiple screen sizes and devices, such as desktops, tablets, and smartphones. Responsive Web Design or RWD (Zhang & Urchurtu, 2011) offers automated device detection and optimized content displays by adopting CSS3 media queries.[5] Automated device detection in combination with environment and location detection could enable immersive responsive content selection. Responsive transmedia user experience designs could enable users to consume content, when, how and where they want.

2 Literature State of the Art

Various interpretations of transmedia cause confusion and scepticism in the commercial and scientific world. Scepticism is very often related to a misunderstanding of the role of transmedia. In order to explore why and how these interpretations arose we have collected various perspectives aligned them to the understanding of convergence or divergence. Please note that since a list of all current and recent projects, as well as all interpretations of transmedia storytelling would be too extensive to describe here, we focused on the most common interpretations only and grouped them into commercial, current transmedia example and the learning perspectives.

2.1 Commercial Perspectives

Commercial examples of transmedia storytelling differ from grassroot projects or other smaller case studies. Mostly produced for mass consumption commercial transmedia projects face interdisciplinary issues we outlined previously.

2.1.1 Divergence: Alternate Reality Games (ARG) and TV

One of the first transmedia storytelling examples can be found in alternate reality games (Bjork & Holopainen, 2005). The rise of the Internet usage in 1990s enabled new forms of interactive content consumption and storytelling. Early examples of alternate reality games such as 'Dreadnot' published by Sfgate.com in 1996,[6] introduce transmedia experiences that would entail fictional characters that moved in the real, and fictional world, leaving traces on websites, fake phone calls or clues in source code. Today we see a rise of TV series that are linked to

[5] CSS3 Media Queries. http://www.w3.org/TR/css3-mediaqueries/. Accessed 15 Aug 2013.

[6] 'Dreadnot' (1996) SFGate Productions www.SFGate.com. Accessed 13 Dec 2012.

alternate reality games. Mostly focused around crime and conspiracy type genres, real time alternate reality game extensions are linked to air times of TV episodes such as e.g. the truth about Marika 'Sanningen om Marika'.[7]

2.1.2 Pre-Divergence: 360° Pre-Release Marketing Solution

For pre-release marketing solutions, cross-media content is produced and distributed over the course of time involving several media types as part of e.g. a guerrilla advertising campaign. Here the cross-media content is diverted in order to maximise profit. However, even though the movie might not be part of an evolving storyworld the experience prior to the movie might well be perceived as a transmedia experience. For instance, prior to the premiere of 'The Dark Knight' Entertainment[8] launched an alternate reality game to promote the movie. This pre-release marketing campaign contained all elements of a rich transmedia project: a compelling story, fan participation, games and clues, the use of digital technologies as well as live and digital events.

2.1.3 Post Divergence: Merchandising and Remediation

Cross-media merchandising solutions are also frequently described as 'transmedia campaigns'. Henry Jenkins referred to this development as 'the transmedia buzz' (Jenkins, 2011). Content is reproduced to create additional profit from blockbuster movies or popular TV series. Here, the same storyline is reproduced and additional functional or unseen footage might be added as an extension but seldom actually expands a story. Remediation, on the other hand, describes a process where not only elements of a movie but the entire storyline are repurposed for a game or vice versa. Examples of such an approach include 'Tomb Raider' or 'Star Wars' where the story and the character are remediated from TV into a game or vice versa as outlined by the theory of transmedia supersystems (Kinder, 1991).

2.1.4 Divergence: Multi-Screeners, Social TV

Other convergence trends are more related to changing consumer behaviour, such as e.g. multi-screeners (IAB AB Europe, 2010). Describing, simultaneous consumption of content on different devices, mostly using one device for broadcasting and the other device for participation, in associated communities. This trend seems to utilize the growing involvement of users in social networks. Often used for Social TV (Abreu & Almeida, 2009), this genre enables multi-screeners to experience live broadcasts as part of a holistic experience across different media types, which by definition could be classified as transmedia experiences. Here the user does not only consume broadcasted data but can actually actively manipulate the broadcasted content via interaction in social networks or Internet based services. However, although Social TV formats highlight the participatory nature of transmedia content consumption, some transmedia evangelists criticize the quality of such experiences.

---

[7] http://truthaboutmarika.wordpress.com/. Accessed 13 Dec 2012.

[8] 42 Entertainment. http://42entertainment.com/. Accessed 13 Dec 2012.

2.1.5 Convergence: Multipurpose Devices and Transmedia Novels

In 2006 the pervasive gaming (Montola et al., 2009) prototype 'The Epidemic Menace'(Lindt et al., 2007), part of the EU funded research project IPerG,[9] combined a mobile, stationary, and augmented reality game into one storyworld that was introduced and interrupted by movie elements. The main purpose of the movie elements was to set the stage of the game world, ensure immersion into a fictional world, and afford switching game modes. At the time six different player devices and a multitude of backend devices were needed in order to enable the pervasive user experience. Soon after, mobile multimedia devices such as the iPhone entered the global market and we could witness a new convergence era. The implications for the game "the epidemic menace" would have been massive since all game elements (including the augmented reality game mode) could have been implemented on one single device.

This example illustrated how multipurpose devices (Murray, 2012) enable transmedia experiences by combining all functionalities associated with network computers; game consoles and conventionally delivered, episodic television, on one single platform. The shift from single purpose hardware to multipurpose hardware could not only enable immersive storyworlds on one single device but also allow the broadcasting, computing-, and publications industry to move even closer towards one common service platform.

One currently example that illustrates this perspective is the digitalization of books. The introduction of 'ePub' (Castro, 2010) technology enables transmedia experiences on eBooks by allowing for audio-visual experiences intertwined with more traditional reading experiences. Resent examples include 'Operation Ajax'[10] and 'Mirror World'[11] by Cornelia Funke. Both interactive books enable rich transmedia experiences on one single device.

2.2 Transmedia Cases

In addition to the highly commercial presentation of transmedia storytelling perspectives we outlined above, there are an increasing number of smaller more grassroot oriented transmedia productions. In order to portray their perspective of transmedia storytelling we asked two transmedia producers to provide us with a sneak preview of their latest projects. First, Christy Dena, a transmedia producer and expert, outlines her latest work 'AUTHENTIC IN ALL CAPS'.[12] Then the transmedia storyteller and producer Christian Fonnesbech, who has already launched several transmedia projects, such as 'TDC', 'The Galatheo Mystery' or 'The Climate Mystery', describes his latest work 'Cloud Chamber'.[13]

[9] IPerG, Integrated Project on Pervasive Gaming. http://iperg.sics.se/. Accessed 6 Jan 2013.

[10] Cognito Comics (2011) Operation Ajax. Accessed 6 August 2013.

[11] Funke, C & Mirade LLC (2013) ‚Mirror World'. Accessed 6 August 2013.

[12] Authentic in all Caps (2013) http://www.christydena.com. Accessed 6 Jan 2013.

[13] Investigate North (2012) 'Cloud Chamber'. Accessed 6 Jan 2013.

2.2.1 AUTHENTIC IN ALL CAPS

'"AUTHENTIC IN ALL CAPS" is a web audio adventure created independently. Independent in this context means it is not funded by a client (brand, publisher, broadcaster, or agency). The ideas are not constrained by the needs of an external client. The project is created in iOS and will be distributed through iTunes. The design of the project form, the web audio adventure, is influenced by my work on large-scale global transmedia projects and my PhD research. The fragmented nature of transmedia projects is an obstacle to many players/audience members. While on the one hand there is an undeniable tendency of humans to engage in multiple "media" or "modes" or "touch-points", engagement with the same storyworld (especially in the same session) is not as common. So what we see a lot of creators do is take away the fragmentation, the divergence, and "converge" all the elements. If they're all available in the same time and place, then it is more approachable. But this takes away the divergent side of transmedia. I wanted to figure out a way you could have both convergence and divergence operating at the same time. One thing I had discovered from my PhD research was that convergence happens at the intangible level. Divergence at the tangible. Usually transmedia projects are joined conceptually, with a continuing storyworld that the player/audience member takes with them/imagines. The combining force is the storyworld abstracted in the mind as triggered by the "text". What I realised is that there is a "media" equivalent of the intangible. Audio. Audio can be the glue that combines fragmented elements in a manner that their distinctness is maintained. And so I created the idea of the web audio adventure. I draw on the design principles of audio tours of museums and apply them to the web. AUTHENTIC IN ALL CAPS is a playful story delivered through a whole lot of fictional websites that I create (divergence), navigated by the guided audio, a radio drama layer (convergence)'.

2.2.2 Cloud Chamber

'Cloud Chamber is an online mystery inspired by space and electronic music. Players collaborate on a single website to uncover the story of a young scientist who has risked her sanity and betrayed her father in order to save humanity from itself. In Cloud Chamber, electronic music is the key to opening the secrets of the universe. Cloud Chamber orchestrates 'found footage' across a variety of media. Players navigate and collaborate and interact in order to access media (film, video, text, diagrams, photographs etc.)—they are detectives discovering what actually happened. The experience can be described as 'a single site Altered Reality Game.' We think of it as fiction for the Facebook generation—part social network, part filmed mystery and part game. The tone mixes psychological thriller elements with supernatural horror and real space science to draw players into a dark mystery. As they explore the story, the mysteries of the human psyche and the heartless voids and immensities of outer space become one.' (Fig. 1)

2.2.3 The Learning Perspective

When technological affordances change, it also changes how we interact with the world around us, as well as the way we learn and acquire knowledge. According to

Fig. 1 Cloud chamber Investigate North (2012)

this theory, learning stems from the relation between the collective and the individual (Säljö & Linderoth, 2002), similarly to the notion of collective intelligence.

There is no doubt that transmedia solutions will allow for new ways of acting and interacting. Therefore, we find it of interest to further elaborate on a perspective of transmedia learning, as we see that transmedia storytelling can add to our understanding of the learning process in the context of educational technology and how it is affected by media convergence and divergence. Jenkins states in his definition of transmedia that "each franchise entry needs to be self-contained enough to enable autonomous consumption". This echoes the notion of learning objects, which is often described as the smallest unit, which in itself can be seen as an entity, representing building blocks of content.

We argue that humans are by default transmedia storybuilders based on our multichannel sensory system and multimodal brain. Our perceptual and sensory systems are the source of our conscious experience (Fauconnier & Turner, 2002). Sight, hearing, touch, smell, and taste can metaphorically be seen as the tools of our bodies, which aid us in experiencing various perceptions from a variety of perspectives. Our cognitive and emotional brain translates these perceptions into experiences and learning is the process where knowledge is created through the transformation of experience (Kolb, 1984). Media tools function as extensions of our bodies and, hence, transmedia affordances can be described as extensions of our bodies and senses to assist this transformation of experiences into knowledge through learning.

In educational technology contexts, multimodal media platforms are often used for delivering information, as well as allowing for networking, collaboration, participation, and performance (Luckin et al., 2012). Multimodal stimulations allow for different perspectives and consequently optimize opportunities for learning (Sankey, Birch, & Gardiner, 2010), and thus, allow for 'additive comprehension' of a subject (Jenkins, 2006). Transmedia learning involves utilizing a variety of media tools that complement each other as a blended and dynamic content method to facilitate learning (Jenkins, 2006; Teske & Horstman, 2012). The intention is to put the learner through a constructive act, not merely transmit content.

The transmedia storyworld allows us to learn about the content through various transmedia affordances, i.e. options of interconnected content and possibilities to engage and interact with the content. One such example of a transmedia learning experience is provided by 'Inanimate Alice',[14] originally created for entertainment, but later adopted in education for teaching both global citizenship and digital literacy skills. However, in this transmedia storytelling, the storyworld is predetermined. Another angle is when the story in itself is a creation of the learner, enabled by powerful multipurpose devices. An example of this is 'Talking Tools' (Johansson & Porko-Hudd, 2013), a smartphone application, supporting multimodal transmedia storybuilding through documentation and communication. The aim is to make the learning process in itself transparent and taking the form of a learning story flowing between converging media options.

From a participator perspective, transmedia seams determine how you are able to make choices of actions, which path you can take, which bricks of content you can build your own story with. Ultimately, how you can add your own voice of co-creation in the transmedia conversation. From the participator's view this becomes a transmedia storybuilding process, as you as a participator may create your own version of how to experience the story depending on choices of content and choices of interaction. In these ways, you become a co-producer of your own experience.

3 Methodology and Approach

The goal of our conceptual transmedia storytelling methodology is first of all to enable storytellers to create better transmedia storyworlds—no matter which domain they draw upon, and no matter where they are in the complicated process of fusing media channels and/or media types. Secondly, our aim is to provide a visual tool to handle the involved complexity and to make narrative choices by displaying the storyworld and trajectories (Benford, Giannachi, Koleva, & Rodden, 2009) using infographics that are easy to read and manipulate. Thirdly, we want to help establishing a common language, as well as a structural understanding of the complexity of transmedia designs.

Figure 2 illustrates a possible conceptual transmedia methodology, and displays how more traditional storytelling frameworks such as the fictional universe could be merged with user experience design aspects of a transmedia experience. Our

[14] http://www.inanimatealice.com/index.html. Accessed 6 Jan 2013.

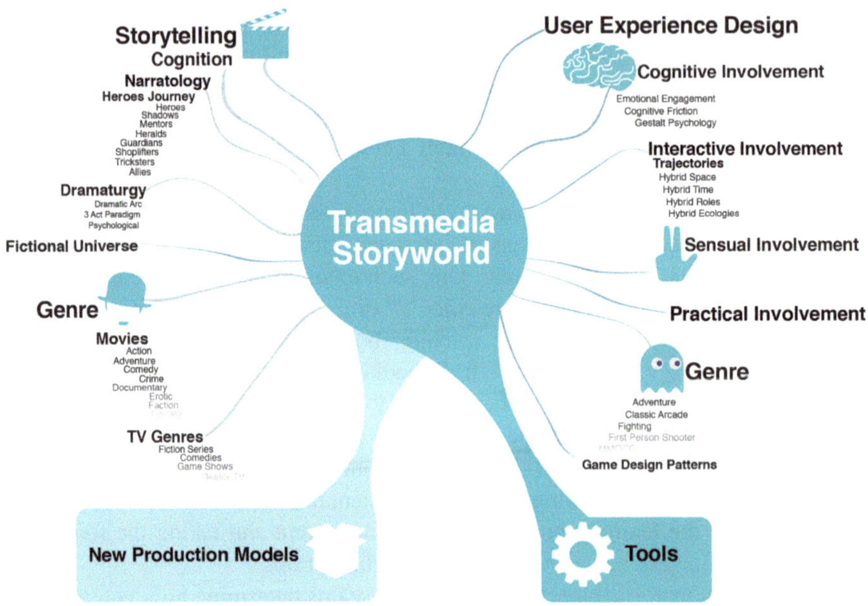

Fig. 2 Conceptual transmedia methodology

goal is to combine multi-disciplinary perspectives into one common conceptual framework. By placing storytelling frameworks and user experience design frameworks on opposite sides of a transmedia storyworld, we wish to emphasise that storytellers will have to design for user experience if they wish to produce a transmedia story.

3.1 Storytelling Frameworks

Many transmedia storytellers stress the importance of building a transmedia world that is based on a fictional universe. Fictional universe describe self-consistent fictional settings that are typically created without specific storylines and provide the frame for any type of story to occur. In contrast, more traditional storytelling frameworks such as 'heroes journey' (Vogler, 1992), 'three act paradigm' (Fields, 1979) or 'dramatic arc' (Freytag, 1836), which focus on telling one story that follows strictly along one storyline. Our evolving methodological approach, Fig. 2, readily allows for all storytelling frameworks. However, it stresses the necessity to consider user experience design theories such as cognitive, interactive and sensual involvement during creative writing in order to create a holistic and cohesive experience.

3.2 User Experience Design

For the purpose of drafting a conceptual methodology we started by focusing on the dimensions of a user experience encountered by the consumption of various media content. The goal of the transmedia storyteller could be to target specific user needs (Hassenzahl, 2010) in order to facilitate a great experience and user satisfaction. Experience can be regarded as 'meaning-making' (Dourish, 2004). According to Vyas and van der Veer (2006) there are at least four important areas to consider in the design of technology and digital content, in order to facilitate a positive user experience: the experience in interaction as a dynamic process, experience as interpretation of how we actively construct meaning, the experience as what the designers offer and what the users bring to it, and finally, the four inseparable dimensions of experience, which are practical involvement, cognitive involvement, emotional involvement, and sensual involvement. We use these four dimensions of experiences as headlines and overarching structure for the HCI aspects of transmedia storytelling and assigned below more tangible design patterns or frameworks such as 'game design patterns' (Bjork & Holopainen, 2005) or 'interactional trajectories' (Benford et al., 2009). Please note that this evolving methodology only illustrates a high-level approach as this point in time. Applying game design pattern or interactional trajectories will require a more abstract form of storyworld building. In order to apply game design patterns or interactional trajectories it will be necessary to translate those into the various contexts of a specific fictional universe.

3.3 Interactional Trajectories

Benford et al. introduced the concept of 'interactional trajectories' (Benford et al., 2009). It describes a trajectory as a journey through a user experience that passes through the following hybrid structures as outlined below (Benford et al., 2009, p. 71):

> "Multiple physical and virtual spaces may be adjacent, connected and overlaid to create a **hybrid space** that provides the stage for the experience.
>
> **Hybrid time** combines story time; plot time, schedule time, interaction time and perceived time to shape the overall timing of events.
>
> **Hybrid roles** define how different individuals engage, including the public roles of participant and spectator (audience and bystander) and the professional roles of actor, operator and orchestrator.
>
> **Hybrid ecologies** assemble different interfaces in an environment to enable interaction and collaboration"

While the theory of 'meaning-making' (Dourish, 2004) describes an overall approach of creating a user experience, the interactional trajectories framework could help creating transmedia user experiences on a more tangible level, such as defining orchestration tools that could enable transmedia storytellers to explore the overlaps of temporal and special aspects. In turn, this would enable transmedia teams to work collaboratively and assist in understanding both the interdisciplinary challenges and the consequences. Tools such as user journey simulations, an information visualization interface displaying narrative choices and media intersections based on user journeys, could enable designers to target user experience attributes pragmatically, by focusing on 'do-goals' and hedonistically by focusing on 'be-goals' (Hassenzahl, 2010). Do-goals describe utility-based functions such as casting a vote or searching for content while 'be-goals' focus on individual experience such as the subjective feeling of identification, being immersed, or feeling a sense of urgency.

4 Conclusion

In this chapter, we propose several transmedia perspectives, including commercial, non-commercial and learning perspectives, to expand and clarify the current terminology in use in the transmedia field and introduce a high-level conceptual transmedia storytelling methodology.

With regards to terminology, we argue that the term transmedia may provide a limiting view of future convergence centered content consumption. On the other hand, transmedia merely defines one specific entertainment format. In this case, the term was never intended to, and should not be used to define a whole new era of convergence content consumption. We therefore propose that convergence, divergence and transmedia be understood and used as separate albeit intertwined situations.

From a transmedia storytelling perspective, we identify convergence as a part of the textual level, in line with Jenkins (2006) notion of "flow of content". Divergence, on the other hand, can be referred to as a fragmentation on the contextual level. Or as Christy Dena outlines in her case description for 'AUTHENTIC IN ALL CAPS', convergence has to do with the intangible and divergence the tangible part of a transmedia solution. In the area of transmedia, we define media disruption and discontinuities as a phenomenon that needs to be integrated into the design of the storyworld in order to create one fictional universe that provides a coherent experience. This could be framed as designing for 'divergent convergence' in the field of transmedia storytelling. Convergence thus allows for transmediation (Siegel, 1995) of the meaning making as well as motivation. The aim is, in other words, to design content that acknowledges, enables and addresses a diverging context. The objective of our transmedia storytelling methodology is to establish a framework for the dynamics of the contextual and content-based elements presented. As we are still establishing the ground work and are subject to an ongoing paradigm change, our proposed methodology readily allows for expansion.

From a psychological perspective, we conclude that we are all innate 'learning beings'. For transmedia designs this comes down to our 'additive comprehension' in relation to content despite diverging context. In a keynote at the AERO 2012 conference, Sir Ken Robinson stated there are two important factors that are characteristic of learners: diversity and creativity. These two characteristics are perhaps the major reason for the evolution of media convergence and divergence. We find this to be particularly relevant for transmedia storytelling, which discards linearity, provides several perspectives to a story, promotes a participatory culture, and often harnesses the added value of collective intelligence.

Janet Murray (2013) argues in a recent publication, 'Transcending Transmedia' that most users no longer separate the TV and web entertainment and she is therefore less convinced of the idea of transmedia productions. She foresees a new form of entertainment or genre that will arise based on an act of imagination rather than a business model that ensures a 360° exploitation of a storyworld. In this chapter, we propose a conceptual transmedia methodology that embraces the paradoxes and complexities of this particular field. We contribute this conceptual methodology to provide foundational work to assist understanding of this emerging participatory genre whatever its label may be.

Acknowledgments We would like to thank Christy Dena, Christian Fonnesbech for insight into their work. We would also like to thank all workshop participants as well as the interviewees for their invaluable insight into their perspectives.

References

AB Europe. (2010). *Multi screeners report* (Media Scope Report 2010).

Abreu, J., & Almeida, P. (2009). From 2BeOn results to new media challenges for social (i)TV. In P. Cesar, D. Geerts, & K. Chorianopoulos (Eds.), *Social interactive television: Immersive shared experiences and perspectives* (pp. 225–246). Hershey, PA: IGI Press.

Benford, S., Giannachi, G., Koleva, B., & Rodden, T. (2009). From interaction to trajectories. In *Proceedings of the SIGCHI Conference on Human Factors in Computing Systems* (pp. 709–718). New York.

Bjork, S., & Holopainen, J. (2005). *Patterns in game design*. Hingham, MA: Charles River Media.

Bolter, D., & Gromala, D. (2005). *Windows and mirrors*. Cambridge, MA: MIT Press.

Castro, E. (2010). *EPUB Straight to the point*. Berkeley, CA: Peachpit Press.

Cheng, B., Stein, L., Jin, H., & Zhang, Z. (2008). Towards cinematic internet video-on-demand. In *Proceedings of the 3rd ACM SIGOPS/EuroSys European Conference on Computer Systems* (pp. 109–122). ACM.

Cooper, A. (2004). *The inmates are running the asylum*. Indianapolis, IN: Sams Publishing.

Czikszentmihalyi, M., & Nakamura, J. (2002). The concept of flow. In C. R. Snyder & S. J. Lopez (Eds.), *The handbook of positive psychology* (pp. 89–92). New York: Oxford University Press.

Dourish, P. (2004). *Where the action is: The foundations of embodied interaction*. Cambridge, MA: MIT Press.

Fauconnier, G., & Turner, M. (2002). *The way we think. Conceptual blending and the minds hidden complexities*. New York: Basic Books.

Fidler, R. (1997). *Mediamorphosis. Understanding new media*. Thousand Oaks, CA: Pine Forge Press.

Fields, S. (1979). *Screenplay* (Rev. ed.). (2005). New York: Delta Publishing.

Freytag, G. (1836). *Die Technik des Dramas*. Leipzig: Verlag von S. Hirzel.

Hassenzahl, M., & Tractinsky, N. (2006). User experience-a research agenda. *Behaviour & information technology, 25*(2), 91–97.

Hassenzahl, M. (2010). *Experience design: Technology for all the right reasons*. La Vergne, TN: Morgan and Claypool.

Humphrey, G. (1924). The psychology of the gestalt. *Journal of Educational Psychology, 15*(7), 401–412.
Jenkins, H. (2006). *Convergence culture: Where old and new media collide*. New York: New York University Press.
Jenkins, H. (2011). Transmedia 202: Further reflections. Accessed January 13, 2013, from http://henryjenkins.org/2011/08/defining_transmedia_further_re.html
Johansson, M., & Porko-Hudd, M. (2013). Smart slöjd med smarta mobiltelefoner?—om didaktiska dimensioner i digitalt lärande (Smart sloyd with smart phones? Didactical dimensions in digital learning). In Marner, A., & Örtegren, H. (Eds.), *KLÄM. Konferenstexter om Lärande, Ämnesdidaktik och Mediebruk (KLÄM. Conference texts on Learning, Didactics and Media use)*. Tilde (skriftserie No. 1). Umeå: Umeå University.
Jones, D. W. (2006). *The tough guide to fantasyland*. New York: Firebird.
Kinder, M. (1991). *Playing with power in movies, television and video games*. Berkeley, CA: University of California Press.
Kolb, D. A. (1984). *Experiential learning: Experience as the source of learning and development*. Upper Saddle River, NJ: Prentice-Hall.
Leonard, D., & Sensiper, S. (1998). The role of tacit knowledge in group innovation. *California Management Review, 40*(3), 112–132.
Lindt, I., Ohlenburg, J., Pankoke-Babatz, U., & Ghellal, S. (2007). A report on the crossmedia game epidemic menace. *Computers in Entertaintainment, 5*(1), 1–8.
Luckin, R., Bligh, B., Manches, A., Ainsworth, S., Crook, C., & Noss, R. (2012). *Decoding learning: The proof, promise and potential of digital education* (Nesta Report). Accessed December 13, 2012, from http://www.nesta.org.uk/press_releases/assets/features/decoding_learning_report
Montola, M., Stenros, J., & Waern, A. (2009). *Pervasive games: Theory and design*. Burlington, MA: Morgan Kaufmann.
Murray, J. H. (2012). Transcending transmedia: Emerging story telling structures for the emerging convergence platforms. In *EuroiTV 12 Proceedings of the 10th European conference on Interactive tv and video* (pp. 1–6).
Murray, J. H. (2013). Transcending transmedia ACM. Accessed October 15, 2013, from cie.acm.org/articles/transcending-transmedia
Negroponte, N. (1995). *Being digital*. New York: Alfred A Knopf.
Netflix Inc. US Patent Nos. 6,584,450; 7,024,381; 7,631,323; 7,403,910; and 7,617,127.
Paas, F., Renkel, A., & Sweller, J. (2004). Cognitive load theory: Instructional implications of the interaction between information structures and cognitive architecture. *Instructional Science, 32*, 1–8.
Prince, G. (1994). Narratology. In M. Groden & M. Kreiswirth (Eds.), *John Hopkins guide to literary theory and criticism* (p. 524). Baltimore: Johns Hopkins University Press.
Pritchard, R., & Ashwood, E. (2008). *Managing motivation*. New York: Routledge.
Säljö, R., & Linderoth, J. (Eds.). (2002). *Utm@ningar och e-frestelser: IT och skolans lärkultur. (Ch@llenges and e-Temptations: ICT and the learning culture of our school)*. Stockholm: Prisma.
Sankey, M., Birch, D., & Gardiner, M. (2010). Engaging students through multimodal learning environments. In C. Steel, M. J. Keppell, & P. Gerbic (Eds.), *Curriculum, technology and transformation for an unknown future: Proceedings ascilite Sydney 2010* (pp. 852–863).
Sanningen om Marika/The truth about Marika. (2007). published by Sverges Television (SVT) and The company P.
Siegel, M. (1995). More than words: The generative power of transmediation for learning. *Canadian Journal of Education, 20*(4), 455–475.
Sir Robinson, K. (2012). Keynote at AERO 2012 conference.
Sternberg, R. J., & Sternberg, K. (2011). *Cognitive psychology* (6th ed.). Belmont, CA: Wadsworth, Cengage Learning.
Tapscott, D. (2008). *Grown up digital*. New York: McGraw-Hill.

Teske, P., & Horstman, T. (2012). Transmedia in the classroom: Breaking the fourth wall. In *Proceeding of the 16th International Academic Mindtrek Conference*. New York: ACM.

The Nielsen Company. Accessed December 13, 2012, from http://www.nielsen.com

Vogler, C. (1992). *The writers journey: Mythic structure for writers*. Studio City, CA: Michael Wiese Productions.

Vyas, D., & van der Veer, G. C. (2006). Experience as meaning: some underlying concepts and implications for design. In *Proceedings of the 13th Eurpoean conference on Cognitive ergonomics: Trust and control in complex socio-technical systems (ECCE 06)* (pp. 81–91). New York: ACM.

Zhang, X., & Urchurtu, E. (2011). A user experience perspective of design for context-aware adaption. In *Proceedings of the 7th International Conference on Intelligent Environments*. IEEE.

Part IV

Convergence and Media Business Outlook

Making Money with and through Information and Knowledge: Reflections on the Inner Context of Competitive Intelligence, Trend Analysis and Revenue Models of Media Companies

Johanna Grüblbauer and Peter Haric

1 Introduction and Problem Discussion: The Shift of Media Companies from Providing Information to Competitive Intelligence

The article dominantly deals with the issues 'information management' and 'competitive intelligence'. It is dedicated to conceptualize how to design and build a convergence application that showcases the big data and information effectively, regardless of which application it came from. Thus, a unified management information system is a result of a successful convergence strategy. Especially in conversational media increases the amount of data—for companies and users. Therefore, it is of even greater importance that media companies understand how they can organize information to save transaction costs in terms of production and usage of the content and to provide value-added-contents from a user's perspective. In the Management of media companies network model thinking is necessary which replaces their former hierarchical organization.

Financial Market Information Services are showing the way to deal with Big Data in the benefit for the consumer in converting data to information and—by contextualizing—to knowledge and action. Instead, in particular in editorial offices it is often only the gut feeling of the editor ruling. Thus, many classic media seem to be one of the last strongholds of an old business operation style which never adapted to modern times. By using Corporate Intelligence as a defined process, a systematic approach to information acquisition, aggregation, processing, analysis,

J. Grüblbauer (✉)
The Austrian Institute of Media Economy, University of Applied Sciences, St. Pölten AUT-3100, Austria
e-mail: jgrueblbauer@fhstp.ac.at

P. Haric
Leading Companies Austrian Institute (Leitbetriebe Austria Institut), Vienna, AUT-1070, Austria
e-mail: Haric@leitbetriebe-institut.at

A. Lugmayr, C. Dal Zotto (eds.), *Media Convergence Handbook - Vol. 2*, Media Business and Innovation, DOI 10.1007/978-3-642-54487-3_17

displaying and distribution could be implemented in the value chain of classic media. The advantages of this CI process and how it could be implemented in Media Companies will be illustrated exemplary in this article.

In terms of the lifecycle of companies, one can observe in various information-oriented media companies a revenue crisis, in terms of the lifecycle of the news industry. With regard to the further development of the company, they have to choose one of two feasible ways: On the one hand, they can actually slip into a revenue crisis, if they maintain their business models. On the other hand, they can focus on their core competencies and relevant innovations to support further growth of the company.

For information-oriented media companies, content production—and thus intelligent information management—is regarded as a core competence. The challenge is that the profitability of the content market is declining sharply and that content on the Internet can be monetized just barely.

But let us explain first, why media companies face a crisis. Therefore we could identify four main reasons:

- Cause of the crisis #1: Media content is no longer tied to carrier media, as used to be the case with print media, so the **expertise and capacity of the production of the carrier media** are no longer competitive advantages. Simultaneously electronically distributed media (such as radio/TV) are no longer used primarily linearly through specific programs (cf. Sjurts, 2010, p. 380).
- Cause of the crisis #2: Even in traditional media content has been largely subsidized by **advertising revenue**—due to the digitization, the opportunities for advertisers to place their messages increased and the best targeting-opportunities are no longer exclusively reserved for mass media.
- Cause of the crisis #3: Traditional media often received **subsidization** by public press funding (e.g. paid newspapers) or license fees (e.g. public service broadcasting); these cross-financing options have not yet been given on the internet.
- Cause of the crisis #4: Digital media can make use of the opportunity of **direct dialogue** and open communication with users (unlike traditional content providers). But simultaneously this option multiplies the number of content providers and thus intensifies competition (cf. Neuberger & Quandt, 2010, p. 67).

The separation of contents from their carrier media, linear distribution, one-way communication and control over the communication to masses shows: despite years of practical application, media companies fail to use and to produce information in a way that recipients are willing to pay for them for their own sake. From this perspective, the following questions arise:

- What is the value of information for recipients and which factors affect this value?
- How to provide information with added value for the recipient?
- What can media companies learn from interdisciplinary approaches about information management to structure their business model? In particular from sophisticated information management approaches such as competitive intelligence or

newer scientific fields such as Information Management and Information Economics?

If this interdisciplinary transfer would work, media companies have not only a new approach to their business model, but also additional benefits for the strategic content management of their company.

Competitive Intelligence is an interdisciplinary approach[1] that is applied especially in contemporary business management and development. CI describes a systematic, continuous and legal[2] collection and analysis of fragmented information (cf. Michaeli, 2005, p. 1) (about competitors, competitors' products, market development, industries, new patents, new technologies and customer expectations).[3] Thus, CI processes are comparable with the tasks of content management, which is focused here as a core competency of the media companies. The objective of CI is to enable and facilitate strategic decision making based on data. The main focus of this study is the relevance of information-oriented media content is in terms of pragmatic benefits to its recipients.

So information must primarily create value for the recipient. To speak in terms of Bourdieu: recipient are striving to incorporate their economic capital (price information) and/or social capital (attention) they used, by means of objectified cultural capital[4] that can be transferred in the form of texts, images, etc., in their own cultural capital (as part of one's personality).[5] As a result, two types of capital can be generated for recipients: on the one hand once again economic capital (economic investment through actions in the economy) and on the other hand, social capital (social investment in relationship management through communicative action based on media content) (cf. Bourdieu, 1992, pp. 55ff.).

2 Literature State of the Art: An Interdisciplinary View on Information

The term "information good" includes "everything that can be digitized [...] displayed and transmitted" (Sjurts, 2010). Thus, information good is characterized by the fact that trading of copies is difficult to control and so tends to lower the value of information good to zero (cf. ibid).

[1] Particularly from military intelligence and market research.

[2] Illegal data collection constitutes espionage, in marked difference from CI.

[3] Even though software can be very useful in collecting, storing and analyzing information and data, it must always be seen in the context of its ultimate purpose, which is supporting human specialists (cf. Michaeli, 2005, p. 473).

[4] This requires editorial transformation of journalists' incorporated cultural capital into objectified cultural capital in order to generate economic capital.

[5] Open communication via social web applications now enables recipients to exchange cultural capital on the same level, i.e. among recipients.

The role of information and media content is viewed differently in communication and media science than in the economic sense.

- In **communication and media science**, content is subsumed as messages and based on its symbolic meaning. Some important questions remain unanswered: How are contents selected? Which display formats are used? And what is their impact?
- **Economic perspectives** consider content as a result or output of a production process of companies or public institutions. In economic terms content has to fulfill two tasks: the product content must be useful to satisfy—directly or indirectly—the needs of users, and the information good has to be scarce, in order to make money (cf. Sjurts, 2010, p. 82). The willingness to pay for information online is generally low, particularly when it is not perceived as a scarce good. This means as long as the recipient perceive contents of other providers as substitute goods and their contents are available free of charge, the willingness to pay tends to zero.

Previously media companies have perceived information good only as a resource within the physical production process. In the Net Economy content receives the value of an independent production and competitive factor. This creates in addition to the level of the physical product, an additional electronic level of digital data or communications networks (cf. Kollmann, 2009, p. 302).

The amount of information that people are facing is, literally exploded in the internet age. When information is cheap, attention becomes expensive. Thus offered information filtering and information search are successful strategies to cope with the flood of information (cf. Gleick, 2011, S. 445). This filter and search functions can be offered for example by search engines like Google, etc. but also by media companies for their recipients—ideally by transforming information on a 'higher level', as they create knowledge for recipients and to generate additional value for them (See Fig. 1).

2.1 Classification of the Types of Information and Their Corresponding Business Models

Information can be described as an intermediate stage from data to knowledge. While an economic benefit based on the value of data is created when knowledge is transformed into action:

Wirtz (2012) categorizes the types of processing of information in different types of transactions, which are following the logic of integrated process steps in the value chain (Fig. 2):

The business model "Content Integrator" is congruent with that value added contribution provided by trend analysis and competitive intelligence to the company: The more extensive the analysis of the business environment and the

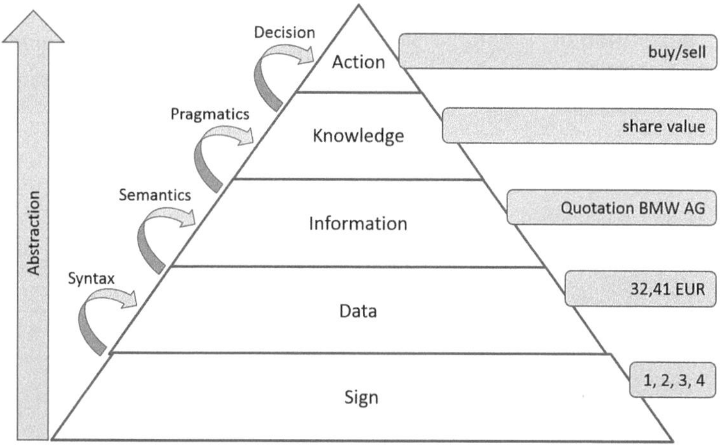

Fig. 1 Information management (Gronau & Lindemann, 2010, p. 12)

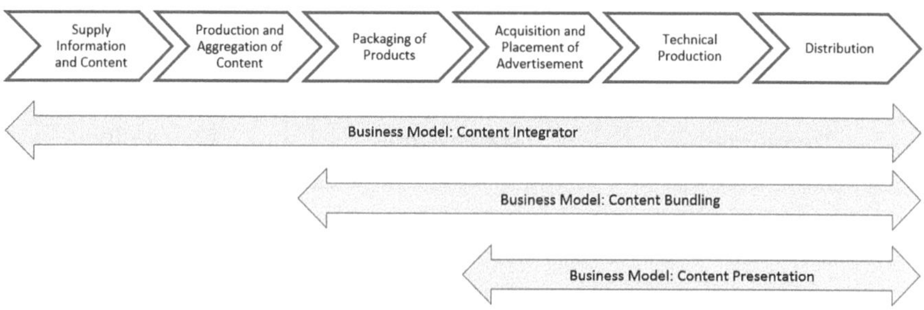

Fig. 2 Content business models (Wirtz, 2012, p. 105)

knowledge of how these are to be interpreted, in the area of technology, society, politics, industry, competitors, customers, etc., the higher the probability that the company is ready to act when the environment of the company changes. This means competitiveness is created by the "refinement" of content and information. The main competitive differentiation potential is created by unique contents.

Competitive Intelligence (CI) and the business model "Content Integrator" follow the same principle: the aggregation of data into information and knowledge. From individual data information can be formed; contextualized information can be transformed into practical knowledge; which will finally enable knowledge-based decision and action. Thus, the value of information is measured based on its support for future action (Fig. 3).

Accordingly, the processes in the CI and the value chain of information-based media companies are identical, as it is their production logic.

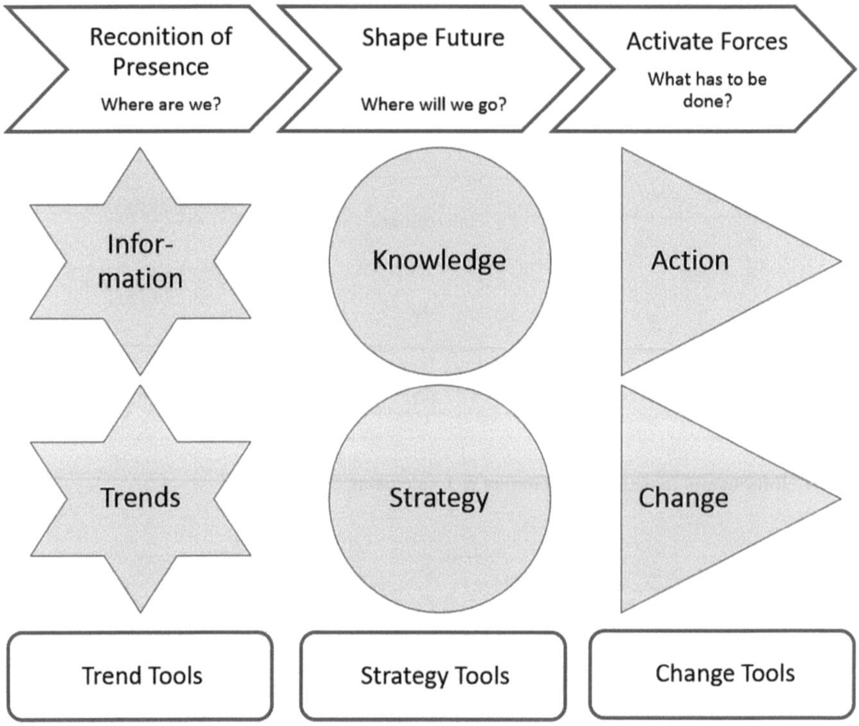

Fig. 3 Change management and actions taken (mod. Scheuss, 2012, S. 141)

2.2 Factors Affecting the Value of Information

Whether revenue can be generated with the produced and offered media product, ultimately depends on the willingness to pay of the market. Do the companies offer goods for which the costs to manufacture and supply them are higher than the potential income? Is there a basically willingness to pay for goods offered on the market? Whether and how much revenue can be achieved depends on the level of exclusiveness and rivalry of consumption of the content. Based on these parameters, content can be classified into common pool resources, (pure) public goods, club goods, and (pure) private goods.

Figure 4 shows the dimensions of supply and distribution of goods. Rival goods (on the left-side) are characterized by the fact that the consumption by a consumer hinders the consumption of other consumers. Non-rival goods (on the right-side) are goods whose use is not limited by the consumption of another consumer. A scarce good is a (pure) public good, which is characterized by the fact that the levels of exclusiveness and rivalry in consumption are high: There is a lack of goods in relation to demand and consumption by one consumer prevents the same benefit for others. On the contrary, a good that is available in abundance is a (pure) public good, such as the radio: There are an infinite number of channels, and the use by one consumer does not limit the use of other consumers.

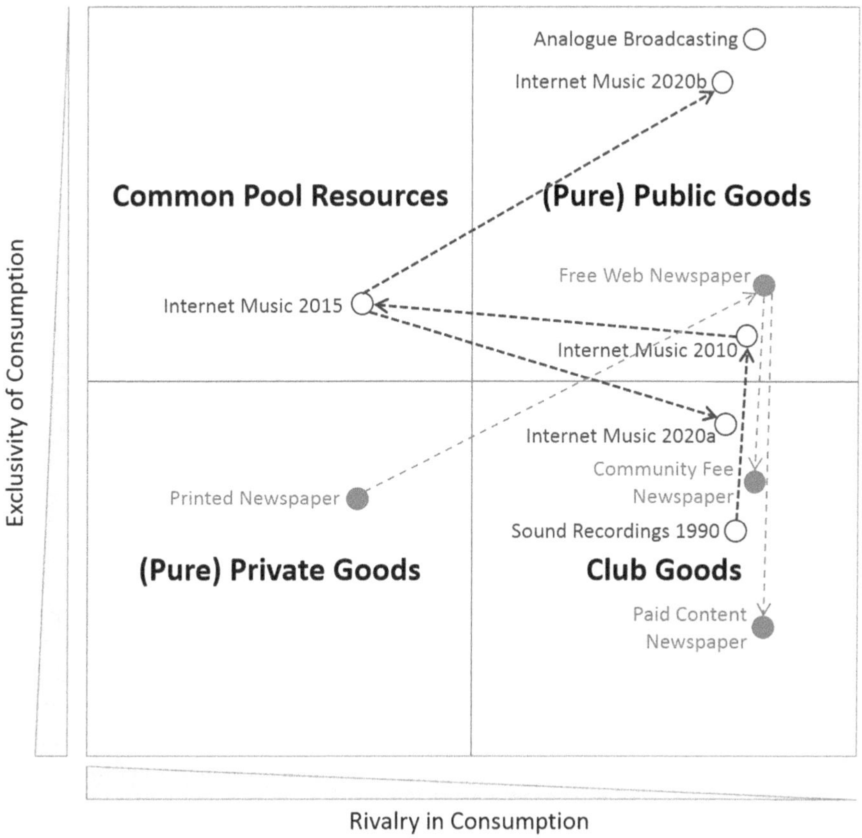

Fig. 4 Types of content by revenue potential (mod. Sjurts, 2010, p. 10)

The classification of "content" as a good by availability and consumption rivalry indicates whether it makes sense to transform a revenue model of contents to a business model: The revenue model "content" on the example of the sale of music can be a business model, if the product can be established as a club good. By, for example, disruptive technologies such as digitization, the situation may change, and the revenue model "content" may be unsuccessful as a business model, because no revenue can be generated, since the characteristics of the good have changed.

Consequently, music industry, newspapers and magazines have a similar problem: their products are converted to public goods as no one is still willing to pay for them. The principle is always the same: added value arises from the contextualization of information For example contents such as business news may achieve a higher price if relevant contents of news agencies are not just displayed (e.g. by content syndication), but if these are packed and put into a context, such as the developments in society, etc.

Consequently, there is little hope that the crisis will pass the printed and online-media, if they do not focus on their core competency: the collection and refining of information to the benefit of their consumers. This is aggravated by the fact that the success of media companies has been determined not solely by the commercial success of content, but significantly on the amount of advertising revenue—based on their target audience (reach).

Pointedly: information that is perceived as exclusive generates the highest revenue. However, with contents which are classified as public goods, because they are not scarce, no revenue will be achieved.

This can be illustrated by the example of the music industry. In the phase of music distribution through e.g. vinyl records as a media carrier, dissemination was necessarily limited. Today the music distribution is almost unlimited possible: It can technically be infinitely reproduced without (or hardly without) changing its quality. Thus, neither the delivery of music is limited, nor is the digital copy is used up by the consumption and so the "benefit" for other consumers is not reduced. As music has evolved from club good ("fee" for LP and CD) to a public good, the music industry now faces the problem that music on the Internet is classified as a pure public good.[6]

2.3 Information and Management

The classical and neo-classical economics consider information and knowledge as a production factor that can be subsumed as work (cf. Lehner, 2009, p. 9). Because people, and thus the production factor work, are the carriers of knowledge and information. This subsumption has been variously criticized, especially in competition-, growth- and game theory (cf. ibid.). In these disciplines, the information and knowledge are seen as the key success factors. Furthermore, the role of information as an economic factor has been moved from the "blind spot" of the classical/neo-classical macroeconomics in the center of the theory by market theory and information economics (cf. ibid.).

In contrast, economists such as Hayek, Schumpeter and Drucker have been recognized early the essential importance of the factor knowledge as a production factor. Especially Schumpeter and Drucker review information and knowledge as the key drivers of innovation and growth.[7] In the so-called "Austrian school of economics", the first approach was established that analyzes the role of information, knowledge and entrepreneurship:

- For Hayek, information is the central function of coordination of anonymous agents. The entrepreneur is according to Hayek, a coordinator of knowledge that is asymmetrically spread in the market process (cf. Hansen, 2009, p. 120).
- For Schumpeter, the entrepreneur is the one who caused the cracks in the economic development in a positive sense. Exploiting gaps in information and the recombination of the factors generated the demand, and is therefore the basis

[6] However, cross-financing in this sector (e.g. radio) is accomplished via broadcasting fees and/or advertising.

[7] Since about the 1990s, information has been accepted as a production factor of its own (cf. Reucher, 2009, p. 46).

of economic development. The driver of change is the entrepreneur, who questions usual facts and ignores conventions: It is not the neo-classical "market", which creates the demand, but the entrepreneur.

It is a basic condition of human action and of institutions such as economy, society, etc., that there is no perfect information, but always incomplete, uncertain, vague and "scattered" information [cf. Hayek 2011(1988), p. 99]. In order to make decisions, entrepreneurs organize the available information on the basis of their subjective models, which are based on experience.

- Kirzner who is the most important disciple of Hayek emphasizes that for each agent, individual i.e. incomplete information is sufficient that they are able to act structured and successfully. Rather, it is even rational—as Hayek pointed out—just not knowing everything: "The general problem is not only to make use of existing knowledge, but to gather exactly as much information as is useful under the given circumstances."[8] (ibid., p. 86) Kirzner and Hayek come to the conclusion: "Coordination of information ensures coordination of action" (Kirzner, 1973, p. 219). Kirzner even goes a step further by almost demanding a shift away from the theory of coordination of markets by prices:

  > "The price system ... is often described as a system of communication—a signal system. When we describe the market process as communication information ... we mean something quite different.... In the market process ... price signals ... are developed ... step by step ... as a process of communication of information." (ibid.)

- The neoclassical approach does not require agents ["In neoclassic theory, the resourceful entrepreneur is not required." (Holl, 2004, p. 48)]. In contrast, for Hayek the entrepreneur is an agent who always collects and processes information and knowledge, to identify opportunities for profit. Consequently, two essential basic concepts can be distinguished: the rational economic man of neoclassical economics and the 'homo agens' in the form of the learning entrepreneur.

According to Hayek, Schumpeter and Drucker the agents are the ones who move the economy. There are not markets that move people. Only the information and knowledge of entrepreneurs lead to an integration of the preferences and actions of market participants (cf. Gray, 1995, p. 82).

Entrepreneurship means "creating new market and new customer" (Drucker 2008(1985), p. 20). For this purpose the "entrepreneur" spends a lot of time collecting and evaluating information (cf. ibid., p. 142): "Robert Bosch spent years studying the new automotive field to position his new company where it could immediately establish itself as the leader. ... in the early stages of a new industry, a new market, or a new major trend, there is the opportunity to search systematically for the special skill opportunity—and then there is usually time to develop a unique skill." (ibid., p. 216).

This "Entrepreneurial Management" is the systematic, continuous scrutinizing of the own products and business model before this happens by competitors or by

[8] Translated from the German original.

the market. Thus create the basis for innovation in the structures of conglomerates that are led by managers (cf. ibid., p. 142).

Due to the growing availability of information, a new situation has arisen: "For many years the information required by a number of professionals ... has grown much faster than the capacity to find it. Professionals have been complaining that they have to spend more and more time hunting for information." (ibid., p. 67) Under these conditions, a paradoxical situation arises: The increasing availability of information, for example through institutional "knowledge professionals" such as market research institutes, information service providers, consultants, etc. does not automatically lead to more knowledge and thus more business success. Rather, the "unexpected outside event" has become the norm. Thus, by a corporate planning, as developed since the 20s at Dupont, Ford and GM and was perfected in the 50s and 60s in the corporations, does not improve planning reliability, but rather weakens it (cf. ibid., p. 48). Thereby information is increasingly becoming a competitive factor. If information is used as the basis of strategic management decisions, the link between information and competitiveness can be represented as follows:

a) Macroeconomic Perspective: The Federal Ministry of Education, Science, Research and Technology, the world's third largest export nation (Germany) (cf. Central Intelligence agency 2011) described the generation and use of relevant data and information already in the late 1990s as the crucial precondition in global competition (cf. BMBF, 1996). Comprehensive and accurate information about markets, products, target groups, competitors and technologies are essential for the success of the company and hence for the business location in the global economy and its increasingly accelerating markets.[9] More than 10 years later, German companies in terms of their corporate intelligence capabilities (CI) are still classified as non-competitive. CI experts come to the conclusion that the United States—as the country of origin of CI approaches—are ahead of the rest of the world by 5–10 years (cf. Baltes, 2011, pp. 9ff.; Frost, Sullivan, 2011, p. 3).
b) Business Administration Perspective: International system vendors (such as IBM, SAS and SAP) offer "big solutions" for the technical implementation of information analysis—so far mainly for Multinationals.[10] Ergo, many companies collect relevant data and information on competitive domestic and foreign markets not systematically but according to situational requirements and scarcely profound.[11] In many medium-sized and larger companies, the collection and analysis of market data and information is used only occasionally—and even large companies

[9] Due to shorter product cycles and a higher intensity of competition.

[10] Conversations with product managers of the companies named above yield the following *reasons*: the solutions are too extensive for the size of even the larger Austrian companies, requiring disproportionate implementation efforts, e.g. additional staff or licenses. In addition, the topic is said to be not yet an issue among Austrian managers.

[11] One of the few studies to empirically prove the lack of systematic data and information gathering in preparation of strategic decisions in Austrian companies: Roitner, 2008

do not use the opportunities extensively, not least for cost reasons.[12] Instead, strategic issues are decided more according to intuition and "gut feeling".[13]

c) Management Perspective: the general management literature defines in the competitive and market analysis various objects (companies, products,...) and information areas (strategy, organization development,...) as important for strategic decision making (cf. Michaeli, 2005, p. 484).
d) IT perspective: Within the last few years, the data management of the strategy departments of companies is extended by new and at the same time for the purpose of CI complementary forms of data management: with the buzzwords cloud computing, social computing and Big Data. These methods complement conventional forms of market and environment monitoring or bring by the availability of data via open interfaces, service-oriented architectures and easy-to-use interfaces entirely new, flexible and cost-effective monitoring and analysis tools. These enable especially SMEs the introduction of a strategy-driven CI (cf. AMD, 2010). So far, though, more and more information is made available—but it depends on their user-oriented curating, i.e. the compilation and integration of information, similar to information-oriented media or information services. Simple data processing capacity, or—in the case of media—unlimited space on websites are not factors of success for themselves.

2.4 CI as a Response to Increasing Complexity and Dynamics

The increasingly turbulent environment of companies can no longer cope with the rigid planning cycles (cf. Burmann, Freiling, & Hülsmann, 2005). So-called "weak signals", i.e. the trends that are opposite to the corporate plan, have to be detected early and have to be incorporated into strategic management (cf. Ansoff, 1975, pp. 21ff.). So an essential competitive advantage is to identify innovation trends earlier than other market participants (cf. also Grüblbauer & Haric, 2013, pp. 671ff.).

Again, the problem is that exploding offers of "information snippets", which are standing side by side unconnected and which are not aggregated to knowledge, provide no added value. But—on the contrary—this creates a scarcity of processing capacity and attention from their consumers.

Porter (1980) created the first systematization of information regarding the competitors, in his study Competitive Strategy: "Techniques for Analyzing Industries and Competitors".[14] This is considered as the beginning of contemporary competitive intelligence.[15]

[12] The prices for reliable, valid market studies exceed €50,000, especially when performed by international business researchers.

[13] Cf. e.g. the extensive studies of strategy culture in Austrian companies by Leitner, 2001; Reisinger, 2007.

[14] Especially in the Anglo-American world these studies have formed the basis of much scientific and praxeological work (e.g. Grüblbauer, 2008).

[15] In Germany, the term "Kompetitive Intelligenz" first appeared (in a title) in a 1998 paper by Salmon & Linarès.

Business models pass through life cycles and have to change. Here individual components can change (product/market combination and transaction relationships, value-added activities, revenue models, customer and competitive benefits) or the overall architecture of the components (Zollenkop, 2006).

The contemporary industrial economy is generally characterized by discontinuities and change. In an "age of discontinuity" (Drucker, 1969), planning in the form of "long-term planning" can be a risk for the development and competitiveness of a company. Instead the entrepreneurial element has to return, which had been lost by the "managerial revolution" and the dominant organizational form of large corporation (cf. Drucker 2008(1985), p. 115).

Moreover, a distinction should be whether if the innovation is endogenous (i.e. from the companies themselves, as it was the case with Robert Bosch) or if innovations are driven or due to exogenous causes, such as technological change. In the case of exogenous innovations companies cannot determine the shaping of the future only themselves, but should react accordingly, so as to be prepared to either adapt the business model or timely countermeasures accordingly (Bhagwati, 2011). This is the situation for the media industry today: Because the need for innovation in media companies is already a reality and traditional media companies were insufficiently prepared, they have to counteract now. But what has changed exactly? Without knowledge measures are difficult to plan. The recent discussions on intellectual property right showed the helplessness of media companies: to focus on getting a share of the Google's profits by searching for contents, instead of seeing the benefits of the Google-given reach of own media and marketing those accordingly.

Media face various strategic challenges: New value chains, new systems of planning aims in strategic management, adaptation of value chains and business models to the rapidly changing technological conditions, global and intense competition for markets, impairment of general information, disruptive technologies, changes in consumer behavior, etc. The main challenge is strategic surprises due to accelerated technological development and increasingly less predictable changes in consumer behavior. But so far, the print and recording industries have only reacted defensively, as they try to defend their oligopolies. The music industry was one of the first industries whose business model has become questionable through the internet at its core—the income from exploitation rights of foreign copyright by sound recording media. The development of the recording industry[16] shows: lack of information leads to strategic surprise and unplanned, helpless reaction.

[16] The term "sound carrier industry" shows the core of the problem: In the real sense, the music industry revenue was generated by the distribution of music. Artists were dependent on the distribution through a "sound carrier" (e.g. CD, LP,...), to distribute their work to the target group. With digitalization, i.e. the faultless, intangible reproduction of music, "sound carrier" have become obsolete and thus the entire business process of the music industry for the distribution of the works of musicians. As artists could sell their works via direct marketing now and consumers could reproduce music almost infinitely without loss of quality (cf. Dolata, 2008, p. 356).

3 Methodology and Approach CI Systems for the Abstraction of Information

"Pioneer premiums" with new products, innovations or established oligopolies can basically be claimed or maintained only limited in the contemporary economy—especially under the conditions of the digital network economy (cf. Stähler, 2002, p. 290).

Companies in dynamic environments are highly dependent on competitive intelligence (CI) in order to minimize the risk of strategic surprise: in terms of new strategic surprises and regarding the maintenance of continuous innovations. "Competitive intelligence is a driver for information." (Rodenberg, 2007, p. 186) According to Peter Drucker the following levers are given: "industry and market changes, unexpected occurrences, incongruities, process needs, demographic changes, changes in perception, new knowledge" (Sharp, 2009, p. 73)—all these factors are part of the search area for Competitive Intelligence, which will be adapted here for media companies.

3.1 Determination of Information Needs

The development of a CI process is identical to the development of the value chain of information-based media companies. In both applications, the central question is: how to transform information into knowledge. The only difference is who will take advantage of the application: In one case, it is useful for the strategic development of the company; in the other case it is useful for the consumers of economic, social or cultural information.

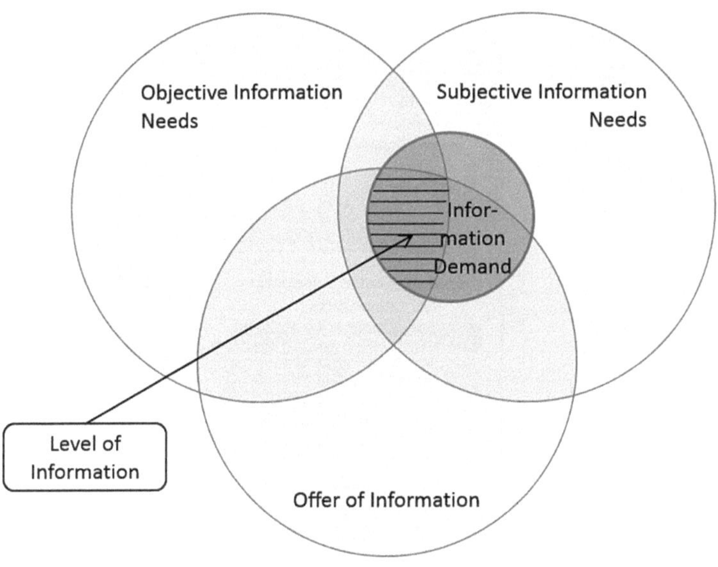

Fig. 5 Determination of information demand (Gronau & Lindemann, 2010, p. 45)

Information management is the creation of a balance between supply and demand for information (cf. Gronau & Lindemann, 2010, p. 45). In order to make information manageable, it is essential to know the needs (demands) of the customer (Fig. 5).

The logic to identify the information needs as an intersection of objective and subjective information needs and the information provided can be applied to the entire information economy. By definition there is no distinction between the "information in the economy" and an economic sector that wants to achieve commercial, profit-oriented goals with information.

3.2 Big Data Management Supported by CI

The basic principle to deal with large amounts of information (Big Data) is to select and to filter. Here the logic of the information value chain has to be adhered to (Fig. 6):

Because of the digitization of data are no longer difficult to collect, rather available in large quantities (Fig. 7).

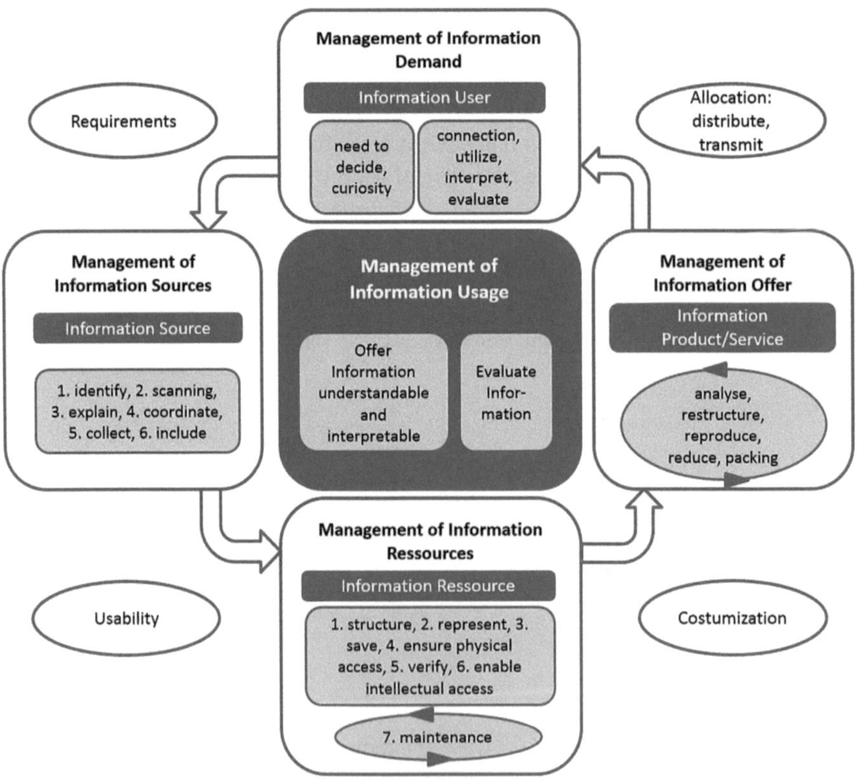

Fig. 6 The information value chain (Gronau & Lindemann, 2010, p. 45)

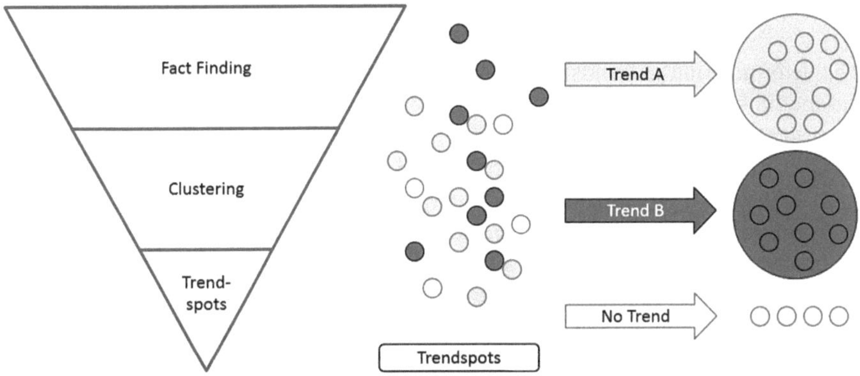

Fig. 7 Trend scouting (modified Scheuss, 2012, p. 55 and 61)

Now the challenge is to select the data, to filter and to structure them.[17] The production process of digital content, during the preparation of editorial content, is comparable with the processes of trend analysis and early warning systems. This basic process of packing of information describes the knowledge and information management, as well as the content production process, particularly in print and online media.

Using statistical and mathematical techniques useful information can be excerpted from data. These can then be transformed into relevant knowledge. Simultaneously, this knowledge can be converted into a set of rules. The aim is to generate knowledge on the basis information, which is interesting but difficult to scout. As a result, unusual data configurations can be identified or future developments can be predicted. Useful tools are graphical methods for pattern recognition, decision trees, cluster analysis and artificial intelligence methods and soft computing (e.g., genetic algorithms,[18] neural networks,[19] fuzzy logic[20]) (cf. Kollmann in: Sjurts, 2010, p. 97).

The generation of data is supported by so-called passive data collection that can be done entirely without direct interaction with the users, such as web log file analysis, cookies, web bugs (small, invisible image files) or additionally installed software (panel spyware/Trojans) (cf. Schoder & Grasmugg in: Sjurts, 2010, p. 628).

[17] Kollmann defines Data Mining as "the filtering of specific data constellations with potential causal explanations on the basis of pattern recognition in the data of an extensive data base." [Translated from the German original] (Kollmann, Tobias in: Sjurts, 2010, p. 97).

[18] Genetic algorithms are stochastic search procedures that attempt to arrive at an optimal solution by starting from one or more acceptable solutions (Cf. Moros, 2012).

[19] Modeling of the structure and information architecture of the brain and nervous system of animals and humans.

[20] Fuzzy Logic refers to a theory developed specifically for modeling the uncertainties and vagueness of colloquial descriptors such as "a little", "rather", "strongly" or "very".

3.3 Process Approach for a CI-System

The determination of trends can be carried out in a variety of ways. Well known is the method of "genius forecasting".[21] More sophisticated methods approach the identification of trends in a multi-stage procedure. The required activities for the identification of issues can be structured in scanning, analysis and monitoring (cf. Heath, 1997, pp. 88ff.).

The figure shows that in the course of the trends and issues management five process steps run circularly: (1) a research problem is defined; (2) to resolve the problem relevant sources are scanned; (3) based on these results spot trends and issues can be identified, analyzed and (4) eventually observed. (5) The aim is to optimize the strategic management and (1) finally to adapt from the given tasks new research questions, etc.

A human being cannot take into account an infinite amount of information for a decision. To create an interactive and user-friendly interface for the analysis is gaining importance, the more extensive the amount of data is that has to be analyzed. Such an interface respectively dashboard does not only facilitate the interpretation of the data, but also allows a customizable use for different problems. Therefore, the requirements planning for a CI system follows the general logic of information management: achieving a balance between information supply and demand for information. The requirements of Fig. 8 can be translated into a technical operation as follows:

The process approach for a CI-System can be divided into three main areas. The data, information and knowledge level are shown in the process approach of implementing a CI system (Fig. 9): scanning and collecting information (1–2), processing and analysis of trends and issues (3–5, 8) and finally monitoring and interpretation of identified trends and issues (6–8). These steps are presented in the following in detail.

- Data level: scanning and collecting information
  - The first step in the process is scanning the data in terms of a non-directional, instrumental observation of the corporate environment. Thereby those events, "weak signals" and—subsequently—trends, topics or (critical) issues are taken into account, which may affect the company. Crucial for the utility of the data for later analysis is the planning of the observation areas, i.e. the identification of all relevant events and issues in the enterprise environment (1).
  - Simultaneously, to achieve an aggregation of information, relevant sources for scanning can be determined. These can include different kinds of digitally published content such as Content of mass media, public authorities, databases or enterprises. The available data from individual sources differ

[21] Experts identify trends based on their competence and experience. Example: the Delphi method.

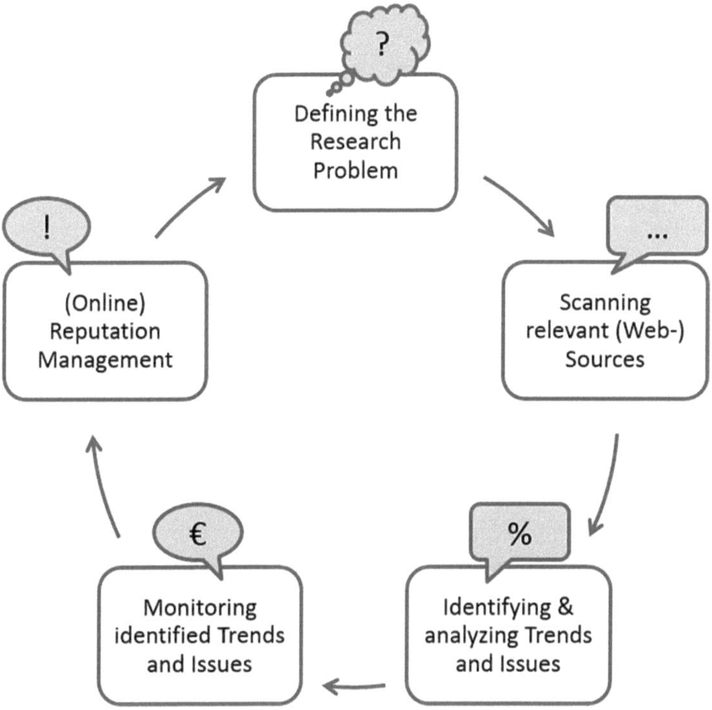

Fig. 8 Competitive intelligence cycle (Grüblbauer & Haric, 2013, pp. 671ff., slightly modified)

in form and quality. The following data types can be distinguished: (a) unstructured data (in forums, blogs, websites, etc.), (b) semi-structured data and (c) structured data (in Open Data networks) (1).
  - The extraction of the data differs according to the source: either via web crawler,[22] through API (Application Programming Interfaces) or web service (queries based on structured data) (2).
- Information level: processing and analysis of trends and issues
  - After various content sources were made available raises the question of the evaluation of the collected data. Hence the so-called trends can be deduced. The actual research work thus lies in the fact to read out the wealth of data target-oriented. This data can be processed, for example by means of text

[22] When scanning, the following two different strategies can be applied. One strategy is to enable the widest possible data base search. The initial search begins with a small number of pages and is then extended to other sites that are in direct connection with it. Another strategy is the type of search that focuses in the first step on one website. Here are links recursively invoked to look more deeply for information on a page. To evaluate the relevance of a page, the following key indicators are determined: backlink count, PageRank, forward link count, location Metric.

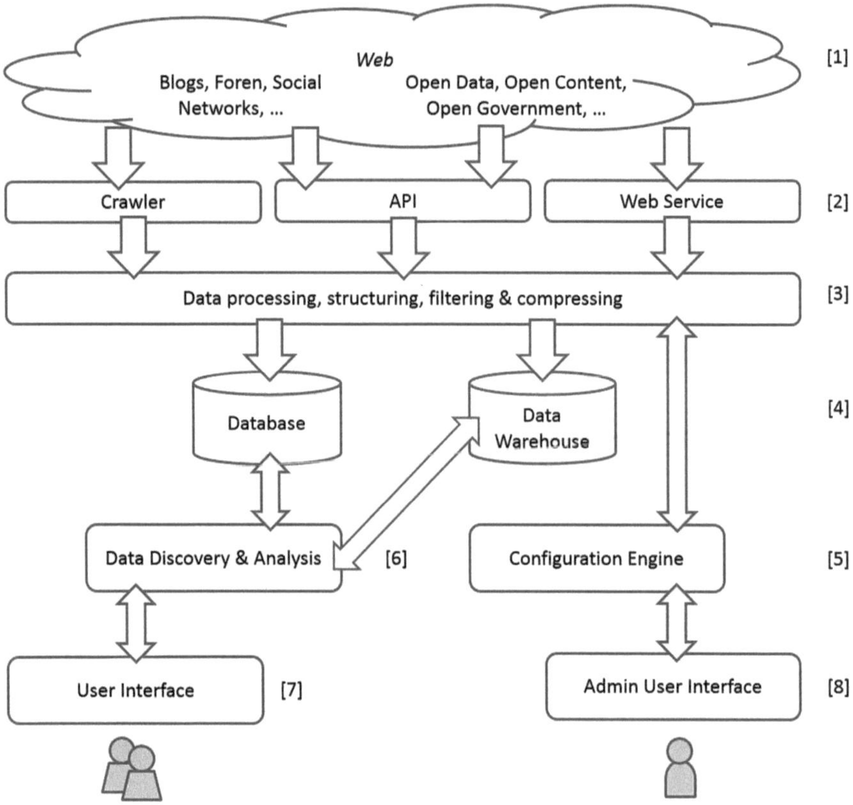

Fig. 9 Process approach of implementing a CI-System. Edited by Gerald Petz, Johanna Grüblbauer, Peter Haric

mining, appropriate categorization and qualitatively in terms of sentiment, keyword clusters, etc.[23] (3)

- After the extraction, the data need to be processed. The processing steps are different depending on the source and extend from structure through text extraction and filtering to aggregation of data (3).

[23] Useful tools for this purpose are a variety of social media monitoring and analytic tools, but also bibliometric analysis and classical sentiment and media resonance analysis. The distinction between the tools begins with the considered data sources and extends to the presentation of the results. Thus, the quality of the tools can not generally be determined, but only with respect to specific research questions. Each tool is particularly well suited for certain aspects. Therefore, good results can be achieved primarily through a combination of several measurement and analysis instruments.

- The extracted and processed data are stored in a data warehouse or a database (4).[24]
- What kind of algorithms and tools are used can be adjusted by the admin via a configuration engine (5 and 8).
- As a result, the extracted and processed data can be searched, aggregated and combined. Up to this stage in the intelligence Circle only information will be generated. These must be compacted and subsequently abstracted to knowledge.

• Knowledge level: monitoring and interpretation of identified trends and issues
  - To what extent (regarding the form and intensity) the monitoring of trends and issues in social media and on news websites can be automated, depending on the specifications of individual tools and differs depending on the provider.
  - In an ideal case the relevant information is already processed automatically, for display in a management cockpit for the user.[25] If such tools are not available, the analysis using human intelligence starts already at this stage of information processing. In this case data from different sources must be manually compressed to information (6 and 7).
  - The bases for the interpretation are digital data which are freely available. But available in the internet these data are only loosely next to each other without nexus and thus have no common meaning. The interpretation aims to generate an informational added value (information and knowledge), for the data that were systematically collected from various sources. To achieve this objective, those data which are relevant to the business, are prepared and related to each other. These can subsequently be used as a strategic decision-making basis for the management or in the case of media content generation for creating professional, user-oriented articles, broadcasts, etc.
  - It is helpful, to compress complex data sets, using appropriate visualizations such as information graphics (e.g. time series, topography, geo-locations or by creating proximity with semantic means,...) to provide a meaningful basis for decisions. The quality of data for valid visualizations and predictions needs to be ensured by a careful selection of sources and the involvement of industry experts.[26]
  - For the monitoring it is essential that the data for longer periods of time (and not just at certain points, such as one-time surveys) are recorded consistently and systematically. Because trends can only be detected early if their changes are observed over certain periods of time.

---

[24] What matters here is which tools are used for analysis. There are numerous monitoring tools, but their suitability can always be determined only with respect to certain requirements (cf. e.g. Grüblbauer & Haric, 2013).

[25] Therefore it is necessary that the selection of data sources, data quality assessment and the definition of "data noise" can be controlled dynamically via a user interface.

[26] "In contrast, a trend that is poorly understood or incorrectly interpreted must be regarded as strategic disinformation—which may be more dangerous than a trend that has been ignored." (Liebl & Schwarz, 2010, p. 325).

3.4 Media Information Management by CI-Based Data Journalism

From the perspective of the value chain, media enterprises can be defined as organizations that combine the intermediates information, entertainment and advertising to a marketable final product (**content packaging process**), the media product (cf. Sjurts, 2005, p. 7). Information services are a specific type of media companies: Information services are a specific type of media companies: they provide information goods—i.e. digitized content—in a combined form using information technologies (cf. Wijnhoven, 2002, p. 50). "The essence of information is the feature of producing value-added changes to its recipients or in the receiving system"[27] (Schädler, 2008, p. 191).

The difference between the **information economy** that serves the internal market and those that act externally is only in its scope, not in the process of sourcing, processing and distribution of information. Further there is a difference in the methodological handling of data and information, i.e. the form of representation in terms of news, reportage, etc. in comparison to data analysis, visualization, etc. But both types share the need for analysis, processing and visualization of connected big data.[28]

So-called **data-journalists** assume here the essential role of the intermediary.[29] Data journalism is a fairly young discipline of journalism and journalistic differs on one essential point from conventional journalism: The target audience is not only the human being but also the machine. Data journalism operates with machine-readable data to analyze, link and provide information. A link to content production is created particularly with computer science and statistics:

- The informatics[30] or computer science itself is related in some way with journalism as it deals with the systematic and automatic processing of information, supported by computer.
- Statistics is a method of dealing with quantitative information.

In the course of information overload, Journalism needs due to its function as an orientation aid—similar to Competitive Intelligence—new instruments that are no longer based only on the direct, written based communication of data and information between transmitter and receiver. The linking of information is increasingly based on metadata that do not offer content for the human recipient, but are

[27] Translated from the German original.

[28] According to a study from the field of knowledge and information management, the global digital data volume doubled every 2 years—a choice between "haystack or treasure chest" loom on the horizon (cf. Schädler, 2008, p. 191).

[29] Other journalistic forms such as the interpretive or investigative journalism intend the procurement of new information or interpretation of existing data.

[30] The term "informatics" is etymologically originated from the synthesis of the terms: Automatic and Information.

machine-readable and their direct target group are computer. Metadata plays an increasingly important role (**HTML**). Only through the links within the database that refer to the text displayed (hyperlink) cross ties and networks of information can occur. Thus, create a contextualization of information with others. Because of hypertext—as a sequential linking of information (text) by cross-reference by a markup language—as one of the key elements of the Internet media, the basis of a long tradition of Western information presentation has changed.[31]

The most radical form of data journalism is now without people—and has been developed in an industry that has been a leader in the media development: the financial industry. **XBRL** (eXtensible Business Reporting Language) is an XML-based language for electronic documents in financial reporting. The aim is an automated evaluation, processing and presentation of data based on a standardized exchange of information and an automated analysis of key figures. This allows that the data are compressed to information; this information is then put into a context in order to generate knowledge that ultimately provides the basis for decisions. XBRL is in a way the idealized information value chain: An automated corporate reporting by corporate IT (essentially so called ERP) systems. For this purpose the system communicates with computers of funds and investment banks, based on such information this enables an automated trading on the exchange. By additional contextualization based on proprietary information (Financial Analysis for discretionary trading), data-driven investment decisions can be made in financial markets.[32]

The news agency Reuters and media company Bloomberg were pioneers in the field of **data journalism**.

- In the early 70s, **Reuters** has implemented a system through which information could not only be shared. Networks and devices that have been made available, allowed on the one hand an exchange of digital data and information, and on the other hand they made it possible to analyze these data and information. Thus, the news agency Reuters successfully entered into the data-journalism business.

[31] It is well known that the Internet has evolved from the Advanced Research Projects Agency Network (ARPANET) in 1969 established by the US Department of Defense. But it is relatively unknown that networking of stock market and financial market participants was developed using electronic media, since the nineteenth Century. This network has been made available to the public market in 1969 in the form of the Alternative Trading System (ATS) to process transactions and to provide information. This "network of scores, telephone, punch card reader, fast printer, telex terminals and disk storage connected the processing of signs, data and information to the decision-making process in financial markets." (Reichert, 2009, p. 116 [translated from the German original]) So in principle, the form of today's Internet has been prepared by the financial network rather than by the military network, because it was basically accessible to public in contrast to the technical network of the U.S. Army (cf. ibid, p. 84). In this respect there is next to the "military" is also an "economic history of the Internet" in the sense of "a mercantile alternative to warfare-oriented media historiography of the Internet." (ibid, p. 88 [translated from the German original]).

[32] Regarding 'revolutionary' importance of XBRL see UK Government Office for Science, 2012, especially pp. 32ff.

- Another outstanding example of a media company that earned money through the networking of information and not with the collection and presentation of information is **Bloomberg** L.P. Bloomberg created in the 80s, computerized network of terminals. This integrated financial reporting, messaging system and transaction system (cf. Grant & Pederson, 1998, pp. 67 ff.; Bloomberg, 1998, p. 60) on one platform. Thus it was in a way the first convergence media. Today, the Bloomberg Professional service is used by around 315,000 "subscribers" in 174 countries (cf. Bloomberg, 2013). This service provides data, information and knowledge, as well as transactions for a large part of the "financial community". Simultaneously, Bloomberg can supply on the same basis approximately 350 financial and economic newspapers and magazines, with news, analysis, background reports and financial data.

Especially in the case of Bloomberg shows that for media companies' two components are relevant for success: human intelligence and technology. Here a close entanglement of data journalism and automated analysis was established already in the 80s.

4 Viewpoint on Convergence: Information Goods as a Business Model and a Subject of the Information Economy

Convergence as a merger of previously separate media offers removes the USPs of individual information-oriented media products. So it creates completely new and different competitive situations.[33] This paper wants to address the convergence problem at the root. For this purpose, the raw data is identified as the common denominator of all information-oriented media products. This raw material has to be refined in order to create meaning. Only by contextualizing, data are transformed into unique content that is valuable, from the perspective of the user.

The differentiation is achieved by the organization and complexity of processing this data, i.e. the transformation of this raw material into information and knowledge.[34] The task of content management is to deal with this complexity.

[33] Traditional media such as newspapers had high barriers to entry: "Printing presses are expensive, and those who owned them enjoyed a collective monopoly over the news and information" (Harkin, 2012, p. 118). Mainly due to the digitization, the market conditions of media companies have changed: Google made information accessible to everyone for free. This service is also used by most traditional media for their research. Since the 80s, changes in the ownership structure of media companies and associated cost-cutting measures—partly at the expense of quality journalism—favored the negative development (cf. ibid, p. 117).

[34] How media companies must act in order not to abolish themselves, because they maintain their traditional business models, which are largely funded from the allocation of the advertising surfaces for years, has been shown e.g. by "The Economist". While the crisis of journalism and the newspaper is ubiquitous, the "Economist" has developed differently: Between 1999 and 2009, the magazine has increased its paid circulation of about 720,000 to 1.4 million. (Harkin, 2012,

Digitalization allows traditional and newly added content providers a large range of behavior, in terms of processing and presentation of data (business models of content presentation to content integration, optimized for various devices and scattered target groups). This increased complexity can only be countered with a comparable complexity by the content providers: "Only variety can destroy variety" (Ashby, 1956). The challenge for media companies and their content management is to plan the use of measures that are necessary in order to control this complexity and to be able to use and to optimize the available resources.

Thus, this increased variety (measure of complexity) is manageable, here is an interdisciplinary deduced human-computer-interaction-solution has been proposed: the CI process approach. In this case both systems focus on their core competence.

- "Keep it Simple": Due to the intelligent use of computer solutions (corporate intelligence) it is possible to cope with the flood of data. This is possible through the automated execution of a maximum number of tasks of data collection and processing.
- "Learn to cope with complexity": There is a need of human intelligence to transform these machine-generated data.[35] Due to the transformation media companies can create unique content offerings, which are contrary to the offers from other suppliers—as added value for the users is generated in the form of knowledge. Only by human intelligence, it is possible to produce context and thereby give meaning to the data.

5 Conclusion

The Process approach of implementing a Corporate-Intelligence-System (Fig. 9) shows an opportunity, how to apply a convergence strategy within a media-company to gain competitive advantages. It is not possible to offer a consolidated

p. 117). Instead of exchangeable messages to the mass, they produce USP-focused content for 'niches'. In the classic manner of the Schumpeterian entrepreneur scrutinize "The Economist" the framing of traditional media: Comparable mass production to lower costs through economies of scale effect, for 80 % of consumers, or quality leadership in narrow niche markets. Hayek and Schumpeter pointed out that it is the entrepreneur as 'game changer', which destabilizes existing equilibria, thereby creating innovation and generates income.

[35] Bloomberg has described the benefits of intelligent data management, in the form of data journalism, compared to a sheer data collection as follows: "The mere input of data remains in the background. Instead, well-educated people can sift through the data for us, who are not only able to store this data, but also to categorize and classify data according to where they belong. [...] The Analysts are those who can provide insight into specific contexts of markets or sectors." (Bloomberg, 1998, p. 63) Thus, the search engine Google does not compete with content-producing media, provided that media companies know their USP. Google News generates lists of search results, but not contextualized information with added value. However, the search engine of Google provides a transparency of the information market, which shows that most of the media offer the same messages more or less identical in wording. Here is the beginning of self-devaluation of journalistic production: human intelligence would be different.

knowledge of markets and the enterprise if the information required by end-users is "trapped in silos" and in different formats. Therefore, data are used as a basis for all further processing of information. But this is only part of the problem; end users need applications which are displaying information in a context, to get a meaning and thus, to gain knowledge. These tasks need human intelligence.[36]

The scope of journalistic work is increasingly analytical. Data journalists are experts who can link the complex data sets by contextualizing these.[37] Without this human intelligence the exploding amounts of data that are available through the digitization remain a "haystack", which can be transformed only by human intelligence into a treasure chest. This paper is thus a plea for intelligent use of the automated processing facilities and to link it with human intelligence. Thus, it is possible on the one hand to organize the exploding data volumes and on the other hand, to create content based USPs of media offers—which cannot be threatened by convergence.

References

AMD. (Ed.). (2010) *The software and compelling economics of big data computing*, White Paper. Available at: sites.amd.com/uk/Documents/Big_Data_Whitepaper.pdf

Ansoff, H. I. (1975). Managing strategic surprise by response to weak signals. *California Management Review, 18*(2), 21–33.

Ashby, W. R. (1956). *An introduction to Cybernetics*. New York: Wiley.

[36] The example of the financial markets illustrates the meaning of competitive intelligence logic. Applied to media companies it could be shown: There is an increasing convergence of human intelligent procurement, processing and analysis of information and its further development through the introduction of new media in combination with statistical processes and big data. It's not just the text "data", but even more the metadata that describes the information. At the same time, there are new ways of contextualizing and analyzing competitive intelligence with which the decisive added value for users offers—especially for information-oriented media company whose core competence is the content.

[37] Data journalists are highly qualified analysts, not robots for copying, moving or storing data (cf. Bloomberg, 1998, p. 63). As long as online editors are even paid less, compared to print editors, the possibilities are limited to operate data journalism. In this respect, the business model of information media applies to the management of 'big data' in competitive intelligence, as well as to data journalism. This is explained in the Editorial Guide from Bloomberg:

- The "Five Fs" of reporting: factual word (adherence to facts), first word (news value), fastest word (subsequent action of an event), final word (complete story), future word (relevance for tomorrow)
- Therefore every story must include additionally the "Five Pieces": markets, economy, government, politics, companies—"These five pieces provide the context and perspective necessary to comprehend the meaning of events." (Winkler, 2011, p. 29)

Data are just raw materials of the media. These must be prepared contextualized to generate information and knowledge as a result. Only through this information- and knowledge-content media companies create non-exchangeable value and benefits for the consumer.

Baltes, V. (2011). *Wettbewerbsforschung zur Entscheidungsunterstützung. Eine Bestandsaufnahme zur Competitive Intelligence in der deutschsprachigen Unternehmenswelt.*
Bhagwati, M. (2011). *Geschäftsmodelle—das Wirtschaftslexikon.* Accessed February 14, 2013, from http://www.daswirtschaftslexikon.com/d/gesch%C3%A4ftsmodelle/gesch%C3%A4ftsmodelle.htm
Bloomberg, M. (1998). *Bloomberger über Bloomberg.* München: Börsenbuchverlag.
Bloomberg LP. (2013). *Bloomberg Facts and History* | Press Room | Bloomberg L.P. Bloomberg. Accessed February 16, 2013 http://www.bloomberg.com/pressroom/facts/
BMBF. (Ed.). (1996). Information als Rohstoff für Innovation. Programm der Bundesregierung 1996–2000. *Bibliotheksdienst, 30*, Jg.(H 10), 1739–1753.
Bourdieu, P. (1992). *Die verborgenen Mechanismen der Macht.* In Schriften zu Politik & Kultur (Hrsg.), *Margareta Steinrücke.* Hamburg: VSA.
Burmann, C., Freiling, J., & Hülsmann, M. (Eds.). (2005). *Management von Ad-hoc-Krisen. Grundlagen—Strategien—Erfolgsfaktoren.* Wiesbaden: Gabler.
Central Intelligence Agency. (Ed.). (2011). *The World Factbook.* Accessed November 12, 2012, from https://www.cia.gov/library/publications/the-world-factbook/rankorder/2078rank.html
Dolata, U. (2008). Das Internet und die Transformation der Musikindustrie—Rekonstruktion und Erklärung eines unkontrollierten Wandels. *Berliner Journal für Soziologie, 18*(3), 344–369.
Drucker, P. (1969). *The age of discontinuity.* New York: Harper and Row.
Drucker, P. (2008/1985). *Innovation and entrepreneurship.* Oxford: Harper and Row.
Frost, & Sullivan. (Eds.). (2011). Market research and competitive intelligence 2011.
Gleick, J. (2011). *Die information: Geschichte, theorie, flut.* München: Redline Wirtschaft.
Grant, T., & Pederson, J. (Eds.). (1998). *International directory of company histories* (Vol. 21). Detroit, MI: St. James Press.
Gray, J. (1995). *Freiheit im Denken Hayeks.* Tübingen: Mohr.
Gronau, N., & Lindemann, M. (2010). *Einführung in das informationsmanagement.* Berlin: GITO mbH.
Grüblbauer, J. (2008). Wie man die Konzentration von Mediaagenturen berechnet. Das Beispiel Österreich. In W. J. Koschnick (Ed.), *Focus-Jahrbuch 2008* (pp. 507–538). München: Focus Magazin.
Grüblbauer, J., & Haric, P. (2013). Social media monitoring tools as instruments of strategic issues management. In M. Friedrichsen & W. Mühl-Benninghaus (Eds.), *Handbook of social media management* (pp. 671–688). Berlin: Springer.
Hansen, H. (2009). *Gründungserfolg wissensintensiver Dienstleister—Theoretische und empirische Überlegung aus Sicht der Competence-based Theory of the Firm.* Wiesbaden: Gabler.
Harkin, J. (2012). *Niche: The missing middle and why business needs to specialize to survive.* New York: Little, Brown Book Group.
Hayek, F. (2011/1988). *Die verhängnisvolle Anmaßung.* Tübingen: J.C.B. Mohr.
Heath, R. (1997). *Strategic issues management. Organizations and public policy challenges.* Thousand Oaks, CA: Sage.
Holl, C. (2004). *Wahrnehmung, menschliches Handeln und Institutionen.* Tübingen: Mohr Siebeck.
Kirzner, I. (1973). *Competition and entrepreneurship.* Chicago: University of Chicago Press.
Kollmann, T. (2009). *Gabler Kompakt-Lexikon Unternehmensgründung: 1.500 Begriffe nachschlagen, verstehen, anwenden* (2nd revised and extended edition). Wiesbaden: Gabler.
Lehner, F. (2009). *Wissensmanagement—Grundlagen, Methoden und technische Umsetzung.* München/Wien.
Leitner, K. H. (2001). *Strategisches Verhalten von kleinen und mittleren Unternehmen. Eine empirische Untersuchung an österreichischen Industrieunternehmen vor einem industrieökonomischen und organisationstheoretischen Hintergrund*, Wien. Accessed September 13, 2012, from systemforschung.arcs.ac.at/Publikationen/5.pdf

Liebl, F., & Schwarz, J. (2010). Normality of the future: Trend diagnosis for strategic foresight. *Futures, 42*, 313–327.

Michaeli, R. (2005). *Competitive Intelligence: Strategische Wettbewerbsvorteile erzielen durch systematische Konkurrenz-, Markt- und Technologieanalysen*. Berlin: Springer.

Moros, R. (2012). *Genetische Algorithmen—Kurzeinfuehrung in Genetische Algorithmen/ Grundlagen*. Accessed November 12, 2012, from http://techni.tachemie.uni-leipzig.de/jsga/gakurz.html

Neuberger, C., & Quandt, T. (2010). Internet-Journalismus: Vom traditionellen Gatekeeping zum partizipativen Journalismus? In W. Schweiger & K. Beck (Eds.), *Handbuch Online-Kommunikation* (pp. 59–79). Wiesbaden: VS Verlag für Sozialwissenschaften.

Porter, M. E. (1980). *Competitive strategy: Techniques for analyzing industries and competitors*. New York: Free Press.

Reichert, R. (2009). *Das Wissen der Börse—Medien und Praktiken des Finanzmarktes*. Bielefeld: Transcript.

Reisinger, S. (2007). *Strategisches Management in Klein- und Mittelunternehmen des produzierenden Sektors*. Linz. Accessed September 13, 2012, from www.ism.jku.at/lehre/diplomarbeiten/DA_Schoeppl_Karin.pdf

Reucher, E. (2009). *Information und Wissen im Spannungsfeld—Von der Informationstheorie und Betriebswirtschaftslehre*. Frankfurt: Peter Lang GmbH.

Rodenberg, J. (2007). *Competitive intelligence and senior management*. Delft: Eburon.

Roitner, A. (2008). *Competitive intelligence in Austria*. Accessed September 13, 2012, from http://othes.univie.ac.at/756/

Salmon, R., & de Linarès, Y. (1998). *Global denken, global gewinnen: die kompetitive Intelligenz—eine Kombination aus Verstand und Menschlichkeit*. Wiesbaden: Gabler.

Schädler, N. (2008). Datenfriedhof oder Schatzgrube—Jeder hat es selbst in der Hand. In F. Keußer & F. Neumann (Eds.), *Wissens- und Informationsmanagement—Strategien, Organisation und Prozesse* (pp. 189–196). Wiesbaden: Gabler.

Scheuss, R. (2012). *Trend tools*. Regensburg: Walhalla.

Sharp, S. (2009). *Competitive intelligence advantage—How to minimize risk, avoid surprise, and grow your business in a changing world*. Hoboken, NJ: Wiley.

Sjurts, I. (2005). *Strategien in der Medienbranche: Grundlagen und Fallbeispiele* (3rd revised and extended edition). Wiesbaden: Gabler.

Sjurts, I. (2010). *Gabler Lexikon Medienwirtschaft* (2nd revised and extended edition). Wiesbaden: Gabler.

Stähler, P. (2002). *Geschäftsmodelle in der digitalen Ökonomie*. Hamburg: Josef Eul.

UK Government Office for Science. (Ed.). (2012). *Standards in computer-based trading—a review*. London.

Wijnhoven, F. (2002). Design components for information market services. A framework for research and practice. In J. L. Monteiro, P. M. C. Swatman, & L. V. Tavares (Eds.), *Towards the knowledge society: ECommerce, Ebusiness, and EGovernment* (pp. 49–66). New York: Springer.

Winkler, M. (2011). *The Bloomberg way: A guide for reporters and editors*. Hoboken, NJ: Wiley.

Wirtz, B. W. (2012). *Medien- und Internetmanagement* (8th ed.). Wiesbaden: Gabler.

Zollenkop, M. (2006). *Geschäftsmodellinnovation: Initiierung eines systematischen Innovationsmanagements für Geschäftsmodelle auf Basis lebenszyklusorientierter Frühaufklärung*. Wiesbaden: Springer.

A Model for Evaluating Converging Media for Advertising Purposes

Mattias Svahn, Richard Wahlund, Marie Denward, Claudia Rademaker, and Patrik Nilsson

1 Introduction and Problem Discussion

Converging media lends marketers greater media choice opportunities and also places new demands on the marketer's strategic analysis. There are established media planning models that aid the marketer, but how shall the marketer reason when it comes to converging media?

This chapter is about aiding buyers of media for marketing purposes who want to understand and include converging media in their media planning. We analyse existing theory on converging media and media planning, and propose aspects of importance when evaluating converging media. This results in a new model for categorizing and defining different types of converging advertising media.

Electronic supplementary material The online version of this chapter (doi:10.1007/978-3-642-54487-3_18) contains supplementary material, which is available to authorized users.

M. Svahn (✉) • R. Wahlund
Stockholm School of Economics , Stockholm, Sweden
e-mail: Mattias.svahn@hhs.se; Richard.Wahlund@hhs.se

M. Denward
The Interactive Institute, Stockholm, Sweden
e-mail: marie.denward@tii.se

C. Rademaker
University of Stockholm, Stockholm, Sweden
e-mail: claudia@stics.se

P. Nilsson
Stics Research, Stockholm, Sweden
e-mail: patrik@stics.se

A. Lugmayr, C. Dal Zotto (eds.), *Media Convergence Handbook - Vol. 2*, Media Business and Innovation, DOI 10.1007/978-3-642-54487-3_18

1.1 Problem Discussion

Media planning decisions are often a great concern for advertisers since the largest portion of marketing communication budgets is spent on buying media space. Figures from the IRM Institute for Advertising & Media Statistics[1] show that about 80 % of the aggregated marketing communication budget is spent on buying media space. Wahlund, Rademaker, Nilsson, and Svahn (2013) found that this accounts for almost 50 % of all the media industry's incomes in Sweden. That indicates the utmost importance of advertising media decisions to both brand holders and to the media industry. Media planning, including media selection is also becoming increasingly complex and challenging as media convergence necessitates a greater range of decisions (Percy & Elliot, 2005; Rademaker, 2013). Nevertheless, advertisers have to make sure that the media budget is spent effectively (Iyer, Soberman, & Villas-Boas, 2005).

2 Literature Review on Converging Media

Media convergence has been defined and explained as a technological and as an industrial phenomenon (cf. Dwyer, 2010; Pool, 1983). The smartphone currently used as a phone, music player, camera, mini surf pad, social media station, micro-blog publisher, e-mail client, game console, GPS navigation device, or anything else that an app or web browser may allow the smartphone to be used for, is an example of technological convergence.

In Pool's, 1983 view convergence is a process where modes converge, blurring the distinctions between point-to-point communications (e.g. post, telephone and telegraph) and mass communications (e.g. press, radio and television). This works both ways. A single physical device could carry services previously provided separately, for example cable TV, telephony and IP-TV. Or a medium specific service like broadcasting could be supplied in different physical ways. Even today the concept is defined similarly, although more looked at as an (on-going) process where new technologies are accommodated by existing media and communication industries and cultures (Dwyer, 2010: 2). The development towards media convergence puts an onus on planning models to take in changed business practices.

Baldwin, McVoy et al. (1996) found that media convergence may be researched in three ways: (a) as consolidation through industry alliances and mergers, (b) as the combination of technology and network platforms, and (c) as the integration between markets and services. Jin (2009) found that the close relationship between media structure and content, and the integration of production of old and new media is most usually researched.

The consumer driven convergence process is equally important, and less researched. It is a social and cultural process; social in that the tactics of grassroots incorporate some of the circulation of content, and cultural since it relies on the

[1] http://www.irm-media.se/about_irm [accessed 2016 01 15].

active participation of consumers as consumers are stimulated and encouraged to search for new and more information, and make a variety of connections among media content, dispersed on different media platforms (Jenkins, 2006: 3).

Converging media shapes how the world is perceived and so sets new consumer habits (Du Gay, 1997), which is why media planning models (cf. 2 this chapter) need to be developed.

Broadcast media markets have become unpredictable and fast moving (Deuze, 2007) and media producers have to continuously come up with new creative products (Hesmondhalgh, 2002; Küng-Shankelman, 2000; Lorentzen & Frederiksen, 2005; McQuail, 2005). What used to be broadcast television is nowadays an online interactive, co-producing media environment with infinite information-providing capacities (Küng, Picard, & Towse, 2008). This puts advertising media planning models (e.g. Fig. 1) in the danger of becoming irrelevant.

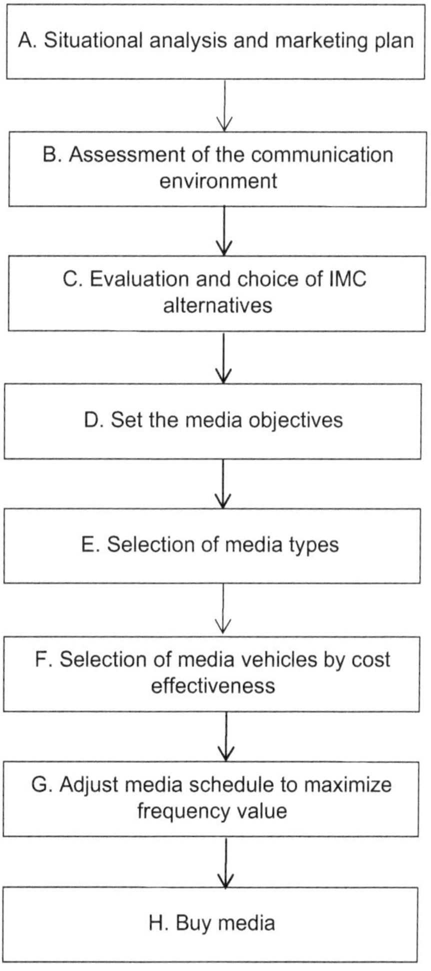

Fig. 1 A basic model for media planning [Wahlund et al. (2013), developed from Cannon, Leckenby, & Abernethy (2002), and De Pelsmacker, Geuens, & Van den Bergh (2007)]

Some researchers have focused on scrutinizing the core principles of converging and pervasive advertising media, for example Müller, Alt, and Michelis (2011). They did it primarily from a pervasive computing perspective, relating to advertising and human-computer interaction, which does not lessen the strain on the theory of media planning models. Another group, The Pervasive Media Studio in Bristol UK, has presented "The Pervasive Media Cookbook", a showcase of several interesting cases expanding the boundaries of media design[2] from a design and art perspective. However, advertising media planning perspectives are not covered.

Svahn (2005) developed a media planning categorization of digital games made for marketing purposes, which was further developed by Wahlund et al. (2013). It contains a media planning perspective. Even if the analysis is limited to games it can be and is used as a point of departure for the converging advertising media categorization presented in this chapter.

2.1 Literature Review of Advertising Media Planning

The media planner must consider the size of the target audience at a given time (*reach*), how often the target audience is expected to be exposed to the advertiser's message during a given time period (*frequency*), the qualitative value of an advertising message in a specific medium compared to the same advertising message in another medium (*impact*), how the selected media complement and support each other within the media mix (Coulter & Sarkis, 2005), the way the messages compete with each other for attention (Rosengren, 2008), and last but not least the budget.

2.1.1 Media Planning and Selection

Media planning is a series of decisions about finding the most optimal way(s) to deliver a message to its' target audience (Barban, Cristol, & Kopec, 1993; De Pelsmacker et al., 2007; Sissors & Baron, 2002; Sissors & Bumba, 1996). Media convergence and integrated marketing communications (IMC) now force a reassessment of the decision process.

IMC is a "a cross-functional process for creating and nourishing profitable relationships with customers and other stakeholders by strategically controlling or influencing all messages sent to these groups and encouraging data-driven, purposeful dialogue with them" (Duncan, 2002: 7; Kitchen & Schultz, 1999). Duncan and Mulhern (2004) refer to IMC as *one voice, one look*. IMC is a process that converges media strategies over different media, and a force to be taken into account when producing new media selection theories for converging media.

Hoffman and Novak (2000) proclaim that conventional approaches to media planning cannot be applied for digital media. Cannon et al. (2002), on the other hand, take a contrary view by arguing that all media can and should be held to a universal

[2] http://pervasivemediacookbook.com/ [accessed 2016 01 15].

standard. Pasadeos, Barban, Yi, and Kim (1997) speculate that the development of new technologies and new media could create new demands for media modelling.

2.1.2 Fundamentals of Media Planning Models

Figure 1 presents our point of departure, a basic media planning model. The model starts off with a general market situational analysis (step A in Fig. 1). Colley (1961) formulated it the in the classic DAGMAR acronym—Defining Advertising Goals for Measured Advertising Results. It refers to a model where objectives of an advertising campaign are set so that the campaign can be quantitatively measured.

Step B is an assessment of the communication environment, for example legal aspects, local media consumption habits and competitors' communication efforts, the brands' share of voice and assessing possible trends in media mix compositions of the competitors.

Step C is an evaluation and selection of IMC alternatives. It determines the type of marketing communication activities[3] that are required in order to reach the communication objectives (Cannon et al., 2002; Wahlund et al., 2013).

Step D is to set the media objectives in terms of *reach*, *frequency* and *impact*, *cost*, *weight*[4] and *continuity*, i.e. a continuous,[5] pulsing[6] or flighting schedule[7] (De Pelsmacker et al., 2007).

Steps E to G are the media selection, resulting in a media schedule. First (step E), the communication objective(s) are allocated to different types of media. Next (step F), the specific media vehicles to be used are identified (for example particular programs and/or magazines within each type of media). These should be the most cost-effective in performing each communication objective (De Pelsmacker et al., 2007). The next step (G) consists of adjusting the media schedule to make sure each task is supported and the frequency value is maximized.

The media planning process ends with buying media space (H).

3 Aspects of Converging Media for Advertisement

In this section we propose three qualities to consider when evaluating converging media for the advertising media mix. The qualities are (1) dominance of media or brand message, (2) degree of media efficiency, and (3) degree of pervasiveness.

[3] For example advertising, sales, sales activities, PR, events and sponsoring.

[4] Usually expressed by GRPs (Gross rating points, i.e. a percentage/amount of the target group).

[5] The total amount is continuously spent throughout the entire ad campaign.

[6] A set level of advertising is executed during the whole campaign period, with exception during specific periods when higher levels of advertising are executed.

[7] No advertising takes place during specific periods while higher levels of advertising takes place during peak demand periods.

3.1 Dominance of Media or Brand Message

The first quality is media experience versus brand message carried by the medium. At one end of this spectrum, the medium is the dominant party and the advertising is incidental. At the other end, the medium is fully designed to serve the brand rhetoric. We have split this aspect into five distinct steps, which will also serve as the basis for our categorization of converging media. These five steps are of specific relevance when evaluating and choosing IMC alternatives. They are important also when selecting media types and media vehicles by cost-effectiveness, which must be taken into account when tuning the media schedule to maximize frequency efficiency of advertising. The five steps are:

1. *Attractors:* Something that draws attention to the advertising medium but is in no way related to the ad messages. This quality does not offer the media buyer any precise targeting and attractors are often of a rather simple (though not unsophisticated) media design.
2. *One-shots:* The defining quality of the medium is that it is part of an integrated marketing campaign and designed to expose one single brand in a typically short, simple and generic way. The consumer does not interact in a learning way with the brand exposed, nor are connotations or qualities of the media itself employed in ways that communicate the brand. The rhetorical message has thus moved well inside the media space but remains outside the interaction that the one-shot offers. The main point of difference from step one is that the media experiences are made with the single explicit purpose of distributing a brand message; that is their whole *raison d'être*. This differs from an *attractor,* as the media experience of such would exist whether or not it was employed for a campaign.
3. *Dynamic ad-insertion*: This concerns ad placements in a "top shelf" media experience. There were no brands inside the content space in step one. In step two there was *one* brand inside the content space. In step three there are many brands inside the content space. The media buyer shares the attention space of the content with several other advertisers.

   The brands presented in the media space in step three are interchangeable. It is a superficial way of interaction as the brands do not interact with the media content to any particular degree. If they did they would not be so easily interchangeable. The frequency of exposure to the message is dependent on the users' movement through the media content. An augmented reality browser where local bars, restaurants and points of interest are layered onto a view through the smartphone's camera is an example of this. The view and the frequency of exposure are dependent on position and direction of the user, therefore the brands must be exchangeable. This offers wide opportunities for geographic segmentation and the timing but limits the interaction design possibilities of the ad-insertion (cf. Sect. 2 in this chapter).
4. *Haute couture*: This step also concerns product placements, though without the dynamic insertion of step three. Step four is defined by the brand message being built in at the design stage of the media product. So there are next to no limits on

the interaction between the media user and the ad, but little room for geographic segmentation and timing (cf. Sect. 2 in this chapter).

5. *Full experience*: Step five concerns experiences that are entirely customized. Everything from the customized media experience to the type of media is chosen to communicate the intended rhetorical message. The brand strategy lies at the basis of the media design already at the outset. The consumers' interaction with the commercial message can be intense and the media psychological impact can be large. The rhetorical message might be inside the media space but there is no longer any need for any brand to be obviously graphically visible since the media product itself is the rhetoric.

The first three steps are characterized by the lack of or limited chance for the media user to interact with the brand in a meaningful or learning way. The brand message would simply have to adapt to the media. In steps four and five the media-content space adapts and is shaped according to the needs of persuasive interaction with the brand message.

3.2 Degree of Media Efficiency

The second quality concerns the degree of efficiency of the medium in forwarding the message. De Pelsmacker et al. (2007) postulate certain efficiency criteria that should be evaluated before deciding which media are to be included in a media plan (cf. Fig. 1). The criteria are:

- *Qualitative criteria*, i.e. the medium's image-building capability and emotional impact, the audience's involvement with the medium, the medium's suitability for demonstrating the advertised product and the amount of information that can be conveyed.
- *Quantitative criteria*, i.e. the medium's intrinsic reach, frequency, selectivity, geographic flexibility, speed of reach, message life and seasonal influence.
- *Technical criteria,* i.e. production costs, media buying characteristics and media availability (De Pelsmacker et al., 2007). While qualitative and quantitative criteria are criteria that are directly related to a medium's characteristics, technical criteria are instead calculated costs related to the production of communication messages placed in the medium and the characteristics of buying space in the media. Thus, production costs are considered technical criteria. For example, when TV is selected as the advertising media type because of its image-building capacity (qualitative criteria) and its intrinsic reach (quantitative criteria), then the production costs of the TV ad, i.e. technical costs, should be taken into consideration.

The reader may realize that the dominance quality discussed earlier has an overlap with the qualitative criteria suggested by De Pelsmacker et al. (2007). And so has the next quality to be discussed, the degree of pervasiveness. The solution is that we subtract overlapping elements from the criteria suggested by

De Pelsmacker et al. (ibid). The reason for this is that while the overlapping aspects have not been much recognized when applying qualitative criteria for media selection, they have become more salient for converging media. The 'media efficiency' discussed in this quality is thus less than, and only part of, the total media efficiency strived for by De Pelsmacker et al. model.

We have chosen to fuse all of De Pelsmacker et al. criteria—and one then only partly—in one aspect or dimension of 'general media efficiency'. This we do despite the fact that it is possible for a medium to have for example strong technical criteria, while being weak on qualitative and quantitative criteria. However, since the communicative efficiency of a medium is dependent on all three of De Pelsmacker et al. criteria evaluated simultaneously they need to be considered together as one dimension. When a medium scores high on this measure it has a degree of efficiency on all three of De Pelsmacker et al. criteria, indicating a medium that may be attractive for buying advertising space.

3.3 Degree of Pervasiveness

The degree of pervasiveness is a design factor that must not necessarily be present for a medium to be defined as a converging media. However, when it is present it is a strongly salient element of convergence. This quality concerns the extent to which the medium is present in the variations of time, place and social context that comes from the media consumer moving around in daily life (Montola, 2011). We have taken these three components—originally suggested as components of pervasiveness of games (Montola, Stenros, & Waern, 2009)—as point of departure for the third quality aspect to be used for evaluating converging media for advertising purposes.

For the purpose of this paper we have developed the three components of Montola et al. (2009) into something different than the original. Still we have kept the original headings in order to mark the conceptual connectedness to the original constructs.

- *Spatial Expansion* is when an experience, originally a game, metaphorically leaves its platform and takes place in the vicinity of the player, employing characteristics of the surroundings to shape a media experience that changes with the surroundings. The most obvious non-game converging media example is augmented reality. Spatial Expansion in this chapter concerns the degree to which the media can allow the advertising to change along with the daily movements of the media user.
- *Temporal Expansion* is when a converging media experience has no on or off button. The user experiences no set start of the media session, neither a set end to it. It concerns to what degree the converging media experience allows for timing of advertising with the users' rhythms of life, and the withdrawal and reinsertion of advertising elements at given moments chosen by the media. For example social media can be said to have a degree of temporal expansion as there is no designed start or end to a social media session, it goes on independently of the

user. Facebook, for example, alerts the user even when the individual is not in a media use session.

- *Social Expansion* is when the design of a converging media experience takes the player's own social resources into account for the design to be fully realized and experienced by the user. This relates to Jenkins (2006) notion that converging media are not necessarily an industrial or technically design phenomenon, but what can result from media users relating to each other and also to others (non-media users), in ways not originally intended. Social expansion can also involve the circulation of content on grassroots level.

Although different, we propose that the three components are merged into one general quality degree of pervasiveness. The reason is that the pervasiveness in converging media experiences is usually larger than the sum of its parts, and thus should be considered a unity. When a medium scores high on this measure, it has a high degree of pervasiveness indicating a medium that fully employs all three design qualities.

4 A Model for Categorizing and Evaluating Converging Media for Advertising Purposes

Based on the three qualities described above, we propose the model in Fig. 2 to be used to categorize and define converging media for evaluation of their efficiency in the media mix, as well as the media mix as a whole, for an advertising campaign.

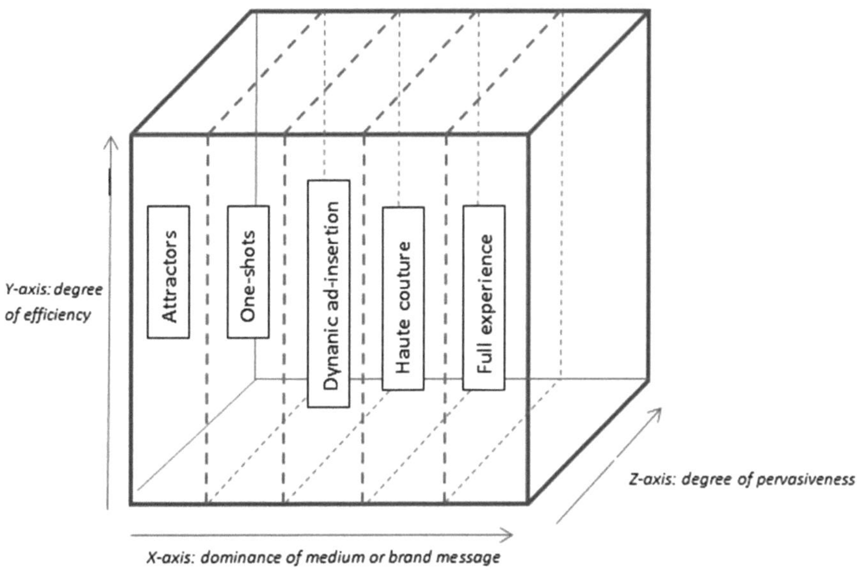

Fig. 2 A model to be used to categorize and define converging media and media mix for evaluation of their efficiency for an advertising campaign

The model is to be integrated with the classic model in Fig. 1 to bring media planning practices up to par with converging media. The proposed model is presented as a cube, with the three qualities described above as axes. The three axes are the x-axis, dominance of medium or brand message, the z-axis, degree of pervasiveness and thirdly the y-axis, degree of efficiency.

The cube is then divided into five blocks, or categories, based on the aspect concerning the weighing up of the medium and the brand message. The five blocks, or categories, are (1) Attractors, (2) One-shots, (3) Dynamic ad-insertion, (4) Haute Couture, and (5) Full experience. Within each of these categories, converging media can be defined and evaluated based on degree of pervasiveness and degree of efficiency. This is then done in the steps C, E, F and G in the model for media planning in Fig. 1. In the final section of this chapter, we apply the proposed model to some examples.

To evaluate the whole media mix, the media planner should make a judgement of the combination of the different media used with respect to the three aspects or dimensions of the model. Although the media planner should strive for a high degree of efficiency for all media in the media mix, they should complement each other on the other two dimensions, rather than be of the same kind.

5 Application of the Proposed Model on Examples of Converging Media

In the previous section we proposed a model to be used to categorize, define and evaluate converging media considered for use in an advertising campaign. The cubic structure of the model, based on the three aspects or dimensions discussed above and shown in Fig. 2, makes it salient how the axes are connected vessels. The basis for the categorization is the quality concerning the weighing up of the medium and the brand message, shown on the x-axis. To be evaluated, a medium or converging media is measured with a design analysis and positioned in the cube with respect to the three dimensions, and then evaluated from its position.

In the following final section of the chapter we apply the proposed model to a number of examples in order to show whether it can be used as intended. Roman numerals have been used to show the positions of the exemplified media plotted onto the cube in Fig. 3.

5.1 Category One: Attractors

The first category of media on the x-axis was called "Attractors". Attractors do not offer the media buyer any precise targeting and are often of a rather simple (though not unsophisticated) design. Attractors are high or low in efficiency (the y-axis), and high or low in pervasiveness (the z-axis). We here define and discuss four positions: I, II, III and IV.

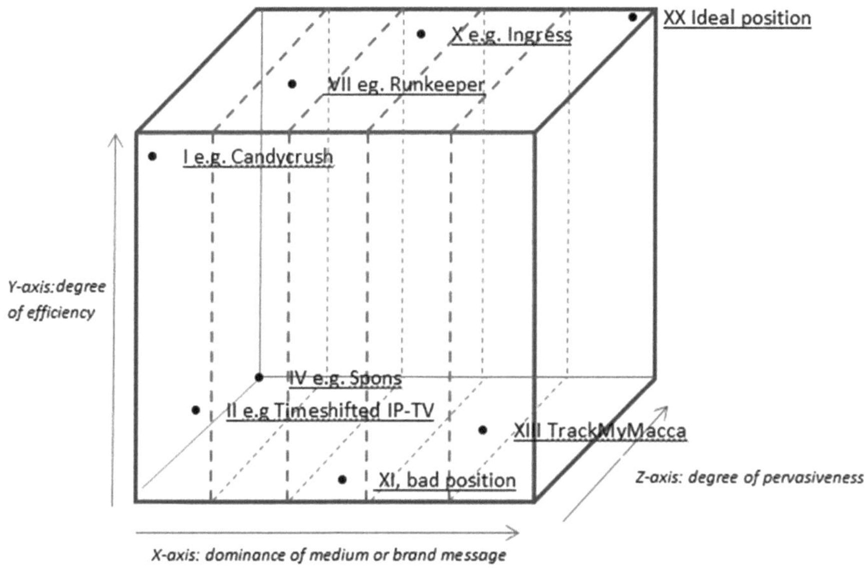

Fig. 3 A model for categorizing, defining and evaluation media for advertising purposes, applied to some examples of converging media

Position I is defined by being far to the left on the x-axis (the medium dominates), high on the y-axis (efficient) and low on the z-axis (non-pervasive). In such a position the medium dominates over the brand, is highly efficient quantitatively, qualitatively and economically, and not very pervasive. Such an attractor is also rather simple in the aesthetical aspects of advertising design with a degree of convergence that is less obvious, but has a good reach.

Converging media with these qualities are for example games that are not really advertising products per se, but exist to draw eyes to the advertising surrounding the games. Some examples from non-digital converging media are paper magazines devoted to crossword puzzles and games (print, games, text, pictures), or sports games, like for example football, that draws viewers to the rink-side banners surrounding the playing field. Digital media examples are Facebook games like Candycrush Saga.[8]

Position II is defined similar to position I still being far to the left on the x-axis and still high on the y-axis but has moved some distance on the z-axis. Such a position would still let the medium dominate over the brand (x-axis), still be highly efficient quantitatively, qualitatively and economically (y-axis), and have some degree of pervasiveness (z-axis). A medium with such a position would give a rather simple media experience to the end user but yield a degree of strategic

[8] King.com (2013) Candycrush Saga https://apps.facebook.com/candycrush/ [accessed 2016 01 15].

sophistication to the strategist as it offers the media planner greater options as regards choice of time and place when the advertising is to be received by the end receiver.

Converging media with position II qualities are for example TV ads in time shifted IP-TV such as provided by for example BBC in the UK, YLE in Finland, and SVT and TV4 in Sweden. These allow the viewer to see a show that was once broadcast, at a time and on a device of the viewer's choice.

That may seem like state-of-the-art converging media, and from a technological standpoint it is. However, the ads are the same that were shown in the broadcast show, and inserted in the same ways. But the time shifting implies a temporal expansion and thus some degree of pervasiveness on the z-axis.

A further example of a medium with position II qualities is rink side advertising that is not really 'there' at the place of the play. Instead, the advertising is added in the TV transmission and adapted to the place of the viewer. This implies a spatial expansion and thus, again, some degree of pervasiveness on the z-axis.

Position III is defined similar to positions I and II, by still being far to the left on the x-axis, but now low on the y-axis and low on the z-axis. Such a position would let the medium dominate over the brand, be inefficient qualitatively, quantitatively and economically, and have little pervasiveness. A medium with such a position would give the marketers little strategic options and the viewer only little experience. A medium that falls in this position is to be shunned.

Position IV is also similar to positions I and II, defined by still being far to the left on the x-axis and still low on the y-axis, but now high on the z-axis. Such a position would still let the medium dominate over the brand, be overall inefficient but have some degree of pervasiveness, for example sponsored Geocaching.[9]

5.2 Category Two: One-Shots

The second media category on the x-axis was called One-shots, intended to expose one single brand. The player does not interact with that brand in a learning way, nor are intrinsic connotations or qualities of the medium itself employed in ways that communicate the brand. We define and discuss four positions for this category: V, VI, VII and VIII.

Position V is defined by being second to the left on the x-axis (the medium still dominates over the brand, but not so heavily), high on the y-axis (efficient) and low on the z-axis (non-pervasive). Such a position would let the medium to some limited extent dominate over the brand, be highly efficient quantitatively, qualitatively and economically and not very pervasive. An example of a one-shot

[9] National Park Service (2013) http://www.nps.gov/miss/parknews/cacheintrashout.htm [accessed 2016 01 15].

non-converging medium is the classic 30-second TV-spot. Converging one-shot media are for example advergames.[10]

Both these examples have in common that they are made for one brand of mostly rather generic game or film design templates, the rhetoric is inside the media space, and they offer a short play or viewing time (x-axis). Both can have a large range, and none of them must necessarily be expensive to produce. Both are somewhat deficient in the precision of the targeting (y-axis). If the brand advertised is for a fast moving consumer good, the small degree of precision may not be such a large problem as long as the range is large.

Position VI is defined by being second to the left on the x-axis (the medium still dominates over the brand, but not so heavily), low on the y-axis (inefficient) and low on the z-axis (non-pervasive). Such a position would let the medium to some limited extent dominate over the brand, be inefficient quantitatively, qualitatively and economically, and not very pervasive. Like position III it is an unattractive position due to the lack of efficiency, and no pervasiveness to make up for the lack of efficiency with the strength of the impression.

Position VII is defined by being second to the left on the x-axis (the medium still dominates over the brand, but not so heavily), high on the y-axis (efficient) and high on the z-axis (pervasive). Such a position would let the medium to some limited extent dominate over the brand, be highly efficient quantitatively, qualitatively and economically and very pervasive. An example of this position is one of the location based gamification services, like for example Runkeeper[11] or Nike.[12]

Position VIII is defined by being second to the left on the x-axis (the medium still dominates over the brand, but not so heavily), low on the y-axis (inefficient) and high on the z-axis (pervasive). Such a position would let the medium to some limited extent dominate over the brand, be highly efficient quantitatively, qualitatively and economically and have pervasive qualities. This is for example the branded corporate events produced by the company Green Hat People[13] as these are spatially, temporally and socially expanded corporate events produced for the clients' branding purposes, on a generic template. Their range may be limited (quantitative criteria on the y-axis) but the events are not that costly (economic criteria on the y-axis).

5.3 Category Three: Dynamic Ad-insertions

Category three concerns ad placements in a premium converging media experience. We define and discuss four positions for this category: IX, X, XI and XII. Position IX is defined by being third to the left on the x-axis, i.e. when the medium and the

---

[10] See https://en.yeeply.com/blog/advergaming-in-mobile-applications/ for some current examples on mobile platforms. [accessed 2016 01 15].

[11] http://runkeeper.com/ [accessed 2016 01 15].

[12] http://nikeplus.nike.com/plus/ [accessed 2016 01 15].

[13] http://www.greenhatpeople.com/ [accessed 2016 01 15].

brand focus are balanced. The medium has been adapted to accommodate brands in general but not one particular brand. The position is high on the y-axis (efficient) and low on the z-axis (non-pervasive), for example a traditional product placement in a popular TV series (cf. Russell, 2002). In the case of in-game advertising such placements were originally found in premium console games and took the shape of posters or similar on flat virtual surfaces, usually in-game virtual billboards, vending machines or similar. With the advent of smartphones it can also be found in smaller games for smart phones as a part of the *freemium* business model.

Position X is defined by being third to the left on the x-axis. The medium has been adapted to accommodate brands in general but not one particular brand. The position is high on the y-axis (efficient) and high on the z-axis (pervasive). One example is the branding of a live sports event, like football matches, when the branding on the players shirts is exchangeable and added in the live mix,[14] giving a degree of spatial expansion. The difference from position I rink side advertising is that in position X, the rhetoric has moved inside a pervasive media-space.

A further example is an augmented reality browser like Wikitude[15] or Layar,[16] where local bars, restaurants, apartments for sale, and other points of interest are layered onto a view through the smartphones' camera. As the view is dependent on position and direction of the user they are intrinsically exchangeable. The frequency of exposure to the message is dependent on the users' movement through the geographic area.

Position XI is defined by being third to the left on the x-axis. The medium has been adapted to accommodate brands in general but not one particular brand. The position is low on the y-axis (inefficient) and low on the z-axis (non-pervasive). This is basically a useless position, and if a medium is found to be in this place the medium should be shunned, whether it is converging or not.

Position XII is defined by being third to the left on the x-axis. The media has been adapted to accommodate brands in general but not one particular brand. The position is low on the y-axis (inefficient) and high on the z-axis (pervasive). This position is similar to position X. The examples given are the same ones; the difference is the efficiency measure (y-axis). One example is the branding of the spatially expanded pervasive game Ingress, where the player moves in the real world with a game world layered upon it.[17] The difference is the degree of business success the medium has. Ingress is, however, still experimental and available to US players only. Today Ingress is in Position XII. Should Ingress become a global success with technical criteria that has the ease of use of for example Google Adsense then it will be in Position X.

[14] http://fayteq.com/products.html [accessed 2013 09 05].

[15] www.wikitude.com [accessed 2016 01 15].

[16] www.layar.com [accessed 2016 01 15].

[17] https://fevgames.net/ingress-and-axa-redefine-mobile-game-advertising/ and http://www.ingress.com/ [accessed 2016 01 15].

5.4 Category Four: Haute Couture

Category four also concerns product placements, though without the dynamic insertion of category three. We define and discuss four positions for this category: XIII, XIV, XV and XVI. All four positions are defined by the medium being heavily adapted to accommodate one particular brand in a way that may seem unique though on second thought need not be so. Other brands in the same larger product category could also find use for the advert in the medium. Category four positions could drive category, as well as a particular brand.

Position XIII is low on the y-axis (inefficient) and high on the z-axis (pervasive). An example is Conqwest (Lantz, 2007), a team-based outdoor treasure hunt type of game featuring large amounts of spatial and social expansion. In Conqwest a mobile phone was used as a bar-code reader to upload game-status information to the coordinating game centre. The game was designed to show the technical competence of the telecom company Qwest. It could of course be used as a medium for any telecom company.

A further example is TrackMyMacca.[18] In early 2013 in Australia, McDonald's launched an app that brings together GPS, supply chain info, and Augmented Reality visualization technology to allow app users to get an image of the origin of the ingredients in the fast food they have just purchased. This may seem unique for McDonald's, but is only unique for the category "*prepared edible goods eaten out-of-home*". The app design could just as well work for Burger King, Subway or a prime brand gourmet restaurant. The efficiency of Conqwest and TrackMyMacca may not have been so high, while the qualitative criteria of both were likely excellent, and the quantitative and the technical criteria of both were likely rather poor.

Position XIV is low on the y-axis (inefficient and low on the z-axis (non-pervasive). Once again this is a position to be shunned.

Position XV is high on the y-axis (efficient) and high on the z-axis (pervasive). This is an attractive position close to the ideal. As converging media are still in an experimental phase it is hard to give an example of a case that is both highly efficient and highly pervasive. The only flaw of a case in this position is that its technical efficiency would be marred by the way a medium in the category can drive both category and brand. That is like having a fire for the crows.

Position XVI is high on the y-axis (efficient) and low on the z-axis (non-pervasive). A somewhat old example but still very telling of the strengths and weaknesses of converging media is the way Sony Ericsson had a smartphone placed in the game Tom Clancy Splinter Cell (Calvert, 2004). The image and interface of the phone was the interface of the game. The teams and tasks were maneuvered through an image of the phones' interface. This offered opportunities for Sony Ericsson to both present learning about a complex consumer product and

[18] Mcdonalds (2013) www.youtube.com/watch?feature=player_embedded&v=zBnqg_KTSIc [accessed 2016 01 15].

to give subcultural connotations, and to use the medium in a way that employed its unique characteristics. The downside was that the popularity of the advertising carrier—the game—lasted much longer than the product, in this case longer even than the brand Sony Ericsson. The advertising, i.e. the integrated product placement, could not be withdrawn at a point when that would have been necessary for the brand strategy.

This is a position that offers some degree of freedom for the brand (x-axis) as the game was adapted to the product advertised, a low degree of pervasiveness (z-axis) as we are talking about a conventional game, and some degree of efficiency (y-axis) as the qualitative characteristics were very strong, and the quantitative acceptable. The technical characteristics are unknown to us.

5.5 Category Five: The Full Experience

Category five is experiences that are entirely customized. Everything from the converging media experience to the type of converging media is chosen to communicate the intended rhetorical message. We define and discuss five positions for this category: XVII, XVIII, XIX and XX. All five positions are defined by being fifth from the left on and the aphelion of the x-axis.

Position XVII is low on the y-axis (inefficient) and high on the z-axis (pervasive). One example is Art of the H3ist[19] It was an alternate reality game made to promote the Audi A3. An alternate reality game—*ARG*—is an interactive narrative that uses the real world as its media platform and uses storytelling "bread-crumbed" across several media to deliver a story that is altered and impacted in real time by the participants' ideas and actions (Dena, 2008; Denward & Waern, 2008; Werner, 2008). Instead of presenting the car at an officially scheduled car show, all that was shown was an announcement that it had been stolen and a request that the public report clues to an Audi website.

In fact, the car had not been stolen. Audi wanted the public to participate in an alternate reality game where clues were planted in public places to hint at where the thieves (actually actors) had taken the car. On websites they posted evidence, and the public was encouraged to watch the website as actors followed the car over a period of time. This is a case that is highly pervasive (z-axis), and wholly designed for the brand (x-axis), so from media and communication design perspectives it is a prominent case, however given that it was probably expensive to produce with a rather limited range, resulting in low efficiency (y-axis).

A further example is the series of Power Games, produced by the Energy Design Studio of the Interactive Institute[20] (Bång, Svahn, & Gustafsson, 2009; Gustafsson, Bang, & Svahn, 2009, Svahn 2014). These were a series of pervasive games

[19] Mckinney.com (2008) Art of the H3ist http://mckinney.com/work/clients/audi/art-of-the-h3ist [accessed 2016 01 15].

[20] Nowadays called The Interactive Institute Swedish ICT, www.tii.se

designed to stimulate families' interest of conserving household electricity. A series of games were produced that in different ways made the real life household electricity consumption into a parameter of the game (spatial expansion). Family households have one electricity contract as a shared resource, so a player could not be successful without engaging the other family members (social expansion), driving a viral discussion about the campaign topic.

Position XVIII is low on the y-axis (inefficient) and low on the z-axis (non-pervasive). An example is a coffee-table give away book telling the story of a brand.

Position XIX is high on the y-axis (efficient) and low on the z-axis (non-pervasive). An example is the *Landlords Game* (Parlett, 1999) which illustrates that converging media can be a matter of design and usage and not only technology. This almost 105-year-old board game is the precursor of Monopoly and looks and is played very much like Monopoly. However, it is a piece of political advertising proposing the socialist notion that private ownership of land leads to a highly unequal distribution of wealth. Technically it consists of a bit of cardboard, with graphic design and some tokens. That is the obvious media channels, but the small seemingly inconspicuous parts come together to produce a converging whole that is a larger political message.

This example has its core central design round a message (X-axis), was highly efficient qualitatively, quantitatively and technically (the production and distribution costs are likely to have been minuscule), and while non-pervasive (z-axis) is still a case of some very simple media (cardboard, graphic design and tokens) converging to a larger whole.

Position XX is high on the y-axis (efficient) and high on the z-axis (pervasive). This is the ideal position, like position XV it is hard to present a real life example that fully reaches this position, even more so for position XX than position XV.

6 Conclusions

This is a theoretical model, but with a practical potential. It is an early exploration as to what extent established theory with a converging interpretation has the potential to improve the media planners' work as described in Sect. 2 in this chapter.

We find that we agree with Cannon et al. (2002) that while all media can and should be held to a universal standard, we also adhere to Pasadeos et al. (1997) who claimed that new analysis tools are necessary if all media are to be held to a universal standard. The proposed model is to be integrated with the established media selection process illustrated in Fig. 1. The integration of the models in Figs. 1 and 2 needs to be validated by empirical work. For now, the proposed model can illuminate for example how a flighting schedule shall be designed in converging media, how the communication environment is to be assessed (B in Fig. 1), and how the evaluation and choice of IMC-alternatives are to be done (C in Fig. 1) or the media selected (E in Fig. 1).

This is a theoretical high-order model. It needs to be fleshed out with further details. The theory needs to be iterated against the empirical business reality to mature. That may also lead to finding conflicting goals, or if the different subcategories carry unequal weights. It is also necessary to find the design language that can measure the design of cases brought onto the model. That is to be done in further research.

References

Baldwin, T. F., McVoy, D. S., et al. (1996). *Convergence: Integrating media, information and communication*. Thousand Oaks, CA: Sage.

Bång, M., Svahn, M., & Gustafsson, A. (2009, September 1–4). Persuasive design of a mobile energy conservation game with direct feedback and social cues. In *Proceedings of DiGRA 2009*.

Barban, A. M., Cristol, S. M., & Kopec, F. J. (1993). *Essentials of media planning: A marketing viewpoint* (3rd ed.). Lincolnwood, IL: NTC Business Books.

Calvert, J. (2004). Pandora Tomorrow Starring Sony Ericson. Accessed February 4, 2013, from http://www.gamespot.com/news/pandora-tomorrow-starring-sony-ericsson-6091084

Cannon, H. M., Leckenby, J. D., & Abernethy, A. M. (2002). Beyond effective frequency: Evaluating media schedules using frequency value planning. *Journal of Advertising Research, 42*(6), 33–47.

Colley, R. H. (1961). *Defining advertising goals for measured advertising results*. New York: Association of National Advertisers.

Coulter, K. S., & Sarkis, J. (2005). Development of a media selection model using the analytical network process. *International Journal of Advertising, 24*(2), 193–216.

De Pelsmacker, P., Geuens, M., & Van den Bergh, J. (2007). *Marketing communications: A European perspective* (3rd ed.). Essex, England: Prentice Hall, Pearson Education Limited.

Dena, C. (2008). Emerging participatory culture practices: Player-created tiers in alternate reality games. *Convergence: The International Journal of Research into New Media Technologies, 14*, 41.

Denward, M., & Waern, A. (2008). Broadcast culture meets role-playing culture. In M. Montola & J. Stenros (Eds.), *The book of Solmukohta 2008: Playground worlds: Creating and evaluating experiences of role-playing games* (pp. 248–261). Helsinki: Ropecon ry.

Deuze, M. (2007). *Media work*. Cambridge: Polity.

Du Gay, P. (1997). *Production of culture, cultures of production*. London: Sage.

Duncan, T. R. (2002). *IMC: Using Advertising and Promotion to Build Brands*. New York: McGraw-Hill (International Edition).

Duncan, T. R., & Mulhern, F. (2004). *A White Paper on the Status, Scope and Future of IMC*. New York: McGraw-Hill.

Dwyer, T. (2010). *Media convergence*. Maidenhead: Open University Press.

Gustafsson, A., Bang, M., & Svahn, M., (2009). Power Explorer—A casual game style for encouraging long term behavior change among teenagers. In *Proceedings of the ACM SIGCHI International Conference on Advances in Computer Entertainment Technology (ACE 2009), October 09, 2009, ACM International Conference Proceeding Series.*

Hesmondhalgh, D. (2002). *The cultural industries*. London: Sage.

Hoffman, D. L., & Novak, T. P. (2000). How to acquire customers on the Web. *Harvard Business Review, May–June*, 179–188.

Iyer, G., Soberman, D., & Villas-Boas, J. M. (2005). The Targeting of Advertising. *Marketing Science, 24*(3), 461–476.

Jenkins, H. (2006). *Convergence culture: where old and new media collide*. New York: New York University Press.

Jin, D. Y. (2009). *Deconvergence: A shifting business trend in the digital media industries.* The International Communication Association ICA 2009, Chicago, Illinois.

Kitchen, P. J., & Schultz, D. E. (1999). A multi-country comparison of the drive for IMC. *Journal of Advertising Research, 39*(1), 21–38.

Küng, L., Picard, R. G., & Towse, R. (2008). *The Internet and the mass media.* Los Angeles: Sage.

Küng-Shankelman, L. (2000). *Inside the BBC and CNN. Managing media organisations.* New York: Routledge.

Lantz, F. (2007). Conqwest an urban treasure hunt. In F. V. Borries, S. P. Walz, & M. Böttger (Eds.) (ÅRTAL), *Space time play. On the synergy between computer games, architecture, and urbanism* (1st ed.). Basel: Birkhäuser.

Lorentzen, M., & Frederiksen, L. (2005). The management of projects and product experimentation: Examples from the music industry. *European Management Review, 2005*(2), 198–211.

McQuail, D. (2005). *McQuail's mass communication theory.* London: Sage.

Montola, M. (2011). A ludological view on the pervasive mixed-reality game research paradigm. *Personal and Ubiquitous Computing, 15*, 1.

Montola, M., Stenros, J., & Waern, A. (Eds.). (2009). *Pervasive games: Theory and design.* Germany: Morgan Kaufman.

Müller, J., Alt, F., & Michelis, D. (2011). *Pervasive advertising.* London: Springer.

Parlett, D. (1999). *The Oxford history of board games.* Oxford: Oxford University Press.

Pasadeos, Y., Barban, A., Yi, H., & Kim, B.-H. (1997). A 30-year assessment of the media planning literature. *Journal of Current Issues and Research in Advertising, 19*, 1.

Percy, L., & Elliot, R. (2005). *Strategic advertising management* (2nd ed.). Oxford: Oxford University Press.

Pool, I. D. S. (1983). *Technologies of freedom: [on free speech in an electronic age].* Cambridge, MA: Belknap.

Rademaker, C. A. (2013). *Green media: Exploring green media selection and its impact on communication effectiveness.* Dissertation, Stockholm School of Economics, Stockholm.

Rosengren, S. (2008). *Facing clutter: On message competition in marketing communication.* Stockholm: EFI The Economic Research Institute Stockholm School of Economics.

Russell, C. A. (2002). Investigating the effectiveness of product placements in television shows. *Journal of Consumer Research, 29*, 3.

Sissors, J. Z., & Baron, R. B. (2002). *Advertising media planning* (6th ed.). Crawfordsville, IN: McGraw-Hill.

Sissors, J. Z., & Bumba, L. (1996). *Advertising media planning* (5th ed.). Lincolnwood, IL: NTC Business Books.

Svahn, M. (2014). *Persuasive pervasive games: The case of impacting energy consumption.* Doctoral Dissertation, Stockholm School of Economics. Retrieved from: http://urn.kb.se/resolve?urn=urn:nbn:se:hhs:diva-2299

Svahn, M. (2005). Future-proofing advergaming: A systematisation for the media buyer. In *Proceedings of the Second Australasian Conference on interactive Entertainment, November 23–25, 2005, ACM International Conference Proceeding Series, 187.*

Wahlund, R., Rademaker, C. A., Nilsson, C. P., & Svahn, M. (2013). Mediernas roll i marknadskommunikation. In G. Nygren & I. Wadbring (Eds.), *På väg mot medievärlden 2020* (5th ed., pp. 165–188). Lund: Studentlitteratur.

Werner, M. (2008). Fringe's ARG Upsets Nature, Messes with Your Mind. Accessed February 4, 2013, from http://io9.com/5046329/fringes-arg-upsets-nature-messes-with-your-mind/gallery/1

The Live Event Ecosystem: Local Media Against Global Ticket Sellers?

Raphael Giesecke

1 Introduction

With the rise of Internet based music distribution and sharing, sales of music CDs have been stalling for years and thus artists' revenues decline. As a counter measure, live music events have increased in both number and size. Putting it simple, such events offer unique live experiences to their audience, who in return pay artists and organizers through buying a ticket. From a more holistic viewpoint however, the live event ecosystem is much larger and, for instance in Finland, just 30 % of the money spent for an event is for the ticket (Valli, Jansson, Koivistoinen, Peltola, & Mynttinen, 2012). But what actors actually form a live event ecosystem, which of them are media companies, and what roles do they play today and in the future? More specifically, when live events are typically discovered via friends' recommendations (Valli et al., 2012) and tickets bought from globally active agencies, what roles remain for other media companies? It is public knowledge that many, if not most news publishers see revenues drop whereas a few event related media businesses flourish in their respective niche. The question is why, and how is this related to media convergence or divergence?

1.1 Continued Disruptions in Media Industry

The main empirical context of this research is the media industry, and the topic of interest is the media industry's organisational behaviour (Buchanan & Huczynski, 2010) in the live event ecosystem. It is obvious that media industry business models are in a state of change, if not disruption. Drivers of this global business change have been studied and described by, e.g. Giesecke and Immonen (2010). They

R. Giesecke (✉)
School of Science, Aalto University, P.O. Box 15500, 00076 Espoo, Finland
e-mail: Raphael.Giesecke@aalto.fi

A. Lugmayr, C. Dal Zotto (eds.), *Media Convergence Handbook - Vol. 2*, Media Business and Innovation, DOI 10.1007/978-3-642-54487-3_19

propose to develop new business models by "establishing and nurturing network relations with key stakeholders, users and audience, and (advertisement) customers" in order to "combine new technologies and new meanings into a truly innovative business epiphany" (ibid. 6). For both practical and research purposes, it makes sense to study and describe such network of stakeholders and customers of the live event ecosystem. Furthermore, effects of media convergence and divergence should be revealed by examining the roles of ecosystem actors and their interactions.

1.2 Events and Their Ecosystem

In this context I use the term ecosystem as *business* ecosystem (Peltoniemi & Vuori, 2004). However, I do not necessarily suggest that actors behave following the mechanics of population ecology of organisations (Hannan & Freeman, 1977). I prefer to use the term ecosystem more freely for a set of actors who are connected in some way, but who do not necessarily know all other actors, nor their roles or intentions in the ecosystem. The particular business ecosystem around live events and their management is part of what Castells (2011: 785) describes as "cultural production network"—in my terms cultural production *ecosystem*. The ecosystem includes at least the event providers, event calendar providers, but also ticket providers, media companies and a range of further stakeholders as well as the audience, the end-users. Having located live events within culture, the focus is on events organised by professionals, associations or laypersons with the aim that interested parts of the general public spend their leisure time by experiencing the event. Examples include music concerts, festivals, art fairs and, on the fringes, sports events.

2 Prior Research

The management of live events was described by, e.g. Getz (1997), Watt (1998), Allen, O'Toole, Harris, and McDonnell (2002) and Shone and Parry (2004). Besides these handbooks for practitioners, for instance Getz (2002) studied reasons why festivals fail and linked his findings to, e.g. population ecology (Hannan & Freeman, 1977). The limitation of this literature is that live events were studied and described from the event organiser's point of view, and often set in a context of tourism instead of media. Research on media convergence abounds (Dwyer, 2010 provides a comprehensive overview), however perspectives were often cultural/ artistic, consumer driven (Jenkins, 2006) or technology based (Zheng, Fu, & Lugmayr, 2012). When business perspectives were taken, adaption of business models to economic theory predictions was favoured.

2.1 Media and Live Events

Research linking organisational behaviour, media convergence and live events, is rare, published results are even rarer—I am not aware of any. Thus in 2010 I suggested a joint academic and industrial research project which would supply a few key elements necessary to better understand the event ecosystem. I started with sketching an initial network of actors and their roles (Mensonen et al., 2011: 7–10). Based on these, project members Andelin et al. (2012) investigated municipalities and professional (news) media in local event ecosystems, focussing on public-private cooperation. A key finding was that municipalities engage in event management in various roles, if necessary, but with only a few core motivations: culture production, attracting tourists and keeping their local event ecosystem alive during the long low seasons. In parallel, Valli et al. (2012) studied the event experience from the audience point of view. Their most interesting findings relate to why people actually attend events (the reason is primarily the artist), how they find out about the event (primarily through friends and *Facebook*), and how much they spend on average per event in Finland (EUR 110.20, of which EUR 33.35 for the ticket). Finally, Cheung (2012) evaluated the business models in the event ecosystem (ibid. 36–49), and refined the description of actors' roles and their relations (ibid. 50). Cheung also addressed the sources of value creation in the ecosystem, following a business model oriented approach. In a final step I analysed the event *management* ecosystem. Sections 4 and 5 are based on the resulting project report (Giesecke, 2012).

2.2 Ecosystem Stakeholders

In business ecosystems in disruption, a significant number of stakeholders are unknown to at least some actors. Often the problem is that some of these yet unknown stakeholders are needed to achieve radical innovation breakthroughs (Verganti, 2009). Business ecosystem stakeholders can be, e.g. investors, entrepreneurs, firms and the public sector. A typical stakeholder's aim is to cooperate—and later collaborate—as ecosystem actor. However, ecosystems in disruption are fuzzy, difficult to describe and even more difficult to be understood. To make things worse, it is in many cases unclear, which roles in an ecosystem, stakeholders are willing to take on. Exemplary roles are inventor, developer, provider, enabler, governor or innovator (Chesbrough, 2003). In case of a converging ecosystem, its typical shared goal is to improve the overall performance of the ecosystem actors. Ecosystem divergence is typically caused by competing actors in similar roles.

2.3 The Choice of Theory

The question remains, which is the best way to describe a business ecosystem, especially when investigating phenomena of media convergence and divergence. I considered three initial options: (1) actors' business models, both on individual and collaborative levels; (2) value chains; and (3) organisational networks. For all options, the status quo needs to be examined, as well as the future plans, i.e. the actors' intentions for change. Option (1) is usually related to economic theories, such as transaction costs (Ghoshal & Moran, 1996). Its core problem is that very few firms are willing to share their collaboration models. Option (2) works for ecosystems with a slow change rate, typically in a state of continuous improvement. In such ecosystems, the value chains in use are commonly known. However, in the case of the event ecosystem, there may be too many disruptive effects, and existing value chains might be outdated and replaced soon. Option (3) is based on relations between actors in the ecosystem and thus may provide a good overview of actors' interactions; however the explanatory and predictive power could be limited. Still, using organisational network theory allows a flexible approach for examining the actors in the event ecosystem, their current and planned roles, and their interactions. Thus I chose this option as my main theoretical base.

2.4 Network Theory and Organisations

In times of globalisation more and more organisations become part of networks, in most cases as members of supply chains or value chains. In an ideal world network members would collectively anticipate, describe and develop the strategy of how they want to collaborate in the future. This could be done by using a few sets of rules on how to describe the networks and the roles, in a highly iterative and fast process. In reality, competition and the absence of modelling rules mean that network members (i.e. organisations) rarely find the time to describe their (intended) roles, and hardly ever jointly model the network of which they are a part.

Still, researchers have applied network theory to business contexts, as presented in the comprehensive overview of Nohria and Eccles (1992). One particular discourse relates to the function of *weak* (Granovetter, 1973, 1983, 1985) and *strong* network ties, which Gilsing and Nooteboom (2005) applied to innovation networks. Concrete proposals exist for how the structure of interfirm networks should look like, under various conditions (Uzzi, 1997; Smedlund, 2009). Moreover, strategic aspects of network forming were studied (Jarillo, 1988, 1993). These studies were based on network structure, tie properties (strength and relational qualities), but not on node properties. This is consistent with general network theory considering a node as just a "point" in a network (Freeman, 1979).

For this research, I may need to distinguish some node properties, as a node represents a network actor, and actors are different. These differences can be related to the power of an actor over others in the network. Whereas network theory can explain this power through node centrality, based on network structure (Freeman, 1979), there are further forms of power in ecosystems, which cannot be related to structure (Castells, 2011). Examples are dependencies between actors and the potential effects of in- and exclusion from networks (Nahon, 2011) or the roles of "switches" (Castells, 2011). Switches are actors bridging different parts of a network. In that respect there is discord between structural network scholars (Granovetter,

1973; Freeman, 1979; Uzzi, 1997) on the one hand and more relationship quality and power oriented scholars (Barzilai-Nahon, 2008; Nahon, 2011; Castells, 2011) on the other. Finally, for the purpose of knowledge exploration and exploitation, different dimensions of proximity between actors (Boschma, 2005) could be investigated. The proximity framework, originating from geography, may add too much complexity to network theory, however it may help in understanding the behaviour of clusters of actors in the event ecosystem. My intention in this study is to identify which of the two frameworks, qualitative power and dimensions of proximity, complements structural network theory best. In conclusion, all three theories/frameworks have clear benefits and deficiencies—but is there a way to combine the strengths? Can a combination of the theories tease out new insights that support our understanding about concrete media convergence and divergence within the live event ecosystem? I aim to find out.

2.5 Research Questions

The research questions consequently start with examining the specific ecosystem—live events—, the relation to the phenomenon of media convergence, and finally address the intended theoretical contribution of this study. The questions are:

1. What roles do actors in the event ecosystem take on, and which are the relations between roles? How and why do roles and relations change?
2. How is organisational behaviour in the event ecosystem related to media convergence or divergence—is it cause or effect?
3. Which combination of theories supports the understanding of these phenomena best?

3 Methodology

Being interested in understanding rather than explaining and in processes rather than isolated events, I decided to use qualitative data. As part of a scientific business benchmark, my project[1] partners and I performed 16 semi-structured interviews in 16 event ecosystem organisations from four countries. The interviews were conducted on-site and took an average of 1.5 h. Responses were recorded on an A3 size business model canvas[2] (Osterwalder, Pigneur, & Clark, 2010), which was later validated by the interviewee. As an exception, one person filled the canvas by himself and sent it by e-mail. The interviewees were typically in higher management or CEO positions. A total of 14 canvasses were also used for a Master's thesis (Cheung, 2012) I instructed, focussed on business models. When analysing the

[1] Located within the Tivit Next Media programme—http://www.nextmedia.fi

[2] http://www.businessmodelgeneration.com/downloads/business_model_canvas_poster.pdf

interviews it turned out that eight interviewees mentioned *Facebook* explicitly, e.g. when referring to their relations to customers. Being unable to interview a *Facebook* representative, I included a public business profile of *Facebook* by Reuters[3] in my data. Table 1 provides an overview of the data sources.

Table 1 Interviewed organisations (i.e. actors) and *Facebook*

Actor	Main Role	Description
Urban music and arts festival	Event Provider	Large urban music and arts festival in Helsinki
Dance association		Helsinki based non-profit association with the aim to promote Flamenco through events, information and courses
Archipelago municipality		A small municipality in the Finnish archipelago
Mid-size municipality		A mid-size municipality on the southern coast of Finland
Large municipality		A large municipality close to Helsinki
Small size ticket seller	Ticket Seller & Distributor	A small Finnish ticketing solution provider focusing on small and medium size events
Medium size ticket seller		A medium size Finnish ticket solution provider
Event feed provider	Event Feed Provider	Swiss event feed provider offering easy access to various Swiss event calendars
Gig booking service	Event Related Service Provider	Gig booking service connecting artists and venues in Finland
Regional web portal	Event Calendar Provider	Finnish regional (archipelago) information web portal, covering events, restaurants, tourism etc.
Urban event publisher		Magazine and web service for event discovery targeted at young urban Finnish adults
Regional event calendar		Publicly funded event calendar in Flanders (Belgium)
Bilingual event calendar		Event calendar covering selected bilingual regions in Finland
Urban culture publisher	Ad Space Seller, Marketer	Publishing house based in a large North-German city with the aim to facilitate discovery of local city culture
Low cost car rental service		Low cost car rental service, using advertisements on rental cars as a source of revenue. Operates in various countries
ICT[a] provider	ICT[a] Provider	Finland based international provider of content management systems based on semantic web
Facebook	Social Media	"*Facebook* [. . .] is engaged in building products to create utility for users, developers, and advertisers."

[a]*ICT* = Information and communications technology

[3] http://in.reuters.com/finance/stocks/companyProfile?symbol = FB.O. Accessed 21 May 2011.

My overall research setting followed the inductive case approach based on contextualisation (Ketokivi & Mantere, 2010); using thematic analysis (Boyatzis, 1998) per role, network analysis (Freeman, 1979) for relations between roles and grounded theorizing (Strauss & Corbin, 1997) for establishing new roles. The unit of analysis is the ecosystem actor, its role(s) and its inter-organisational relationships in one, focussed context (Yin, 2009).

The analysis as such was performed in four steps:

1. In the baseline analysis I allocated generic roles per actor and identified their key partners, customers and enablers. I also checked relationship consistency.
2. In a first synthesis I propose a generic event *network*.
3. Further analyses involved network power constellations, gatekeepers and the proximity between roles, leading to the generic event *ecosystem*.
4. Concluding, I discuss media conversion and diversion in the event ecosystem.

4 Actors, Roles and Relations: Who, What and with Whom?

As my data consists of one interview-based sheet per specific actor—and not per generic role—I needed to synthesise the data from specific to generic level, first. On the generic (role based) level, I then performed three relationship consistency checks: I investigated (1) whether customers or enablers are key partners; (2) possible circular relations—i.e. customers being enablers; and (3) cooperation within the own role.

4.1 Actors' Roles

In a first step I allocated generic roles per actor, based on the section "Value Propositions" of the business model canvasses. The roles were initially based on Mensonen et al. (2011: 8) and Cheung (2012), and then developed further during my iterations with the data. Table 2 shows the results. Most of the interviewed actors take on several roles. For clarity I chose one main role per actor, which reflects its core value proposition. Additional roles, marked "X" are those in which the actor is active and which provide a significant value to the actor. Roles marked "S" are secondary roles in which the actor has been acting, but with either low or unclear value contribution.

A few roles in Table 2 are not taken by any of the interviewed actors. However, they have been mentioned as key partners, customers or enablers. Furthermore, "Public Sector" is not a role in the same class as the other roles. It is an *additional* role that distinguishes three municipalities and a public sector project from industry and associations. However, "Public Sector" illustrates clearly that the interviewed municipalities take on rather many roles—one could suspect a lack of focus, which Andelin et al. (2012) confirm. Regarding construct validity it would have been beneficial to have three or more actors per role. This did not work out for "ICT

Table 2 Actors' and *Facebook*'s roles in the event ecosystem.

Actor	Role																
	Event provider	Ticket seller & distributor	Event feed provider	Event infrastructure provider	Event related service provider	Event calendar provider	Media content provider	Ad space seller, marketer	ICT provider	Social media (Facebook)	Consumer, end user	Artist	Funding provider, investor	Sponsor	Advertiser	Governor/ Authority	Public sector
Urban music and arts festival	M																
Dance association	M			S													
Archipelago municipality	M	S		S									X	S			X
Mid-size municipality	M			S	S	X							X	S F			X
Large municipality	M			X		X							X				X
Small size ticket seller		M						F									
Medium size ticket seller		M				X		X									
Event feed provider			M			X											
Gig booking service		F			M			F									
Regional web portal		F	X			M		X									
Urban event publisher			F			M	X										
Regional event calendar			X			M											X
Bilingual event calendar			X			M		F									

Urban culture publisher	S F				(F)	X	X	M									
Low cost car rental service					S			M									
ICT provider									M								
Facebook					X	X		X	S	M	X						

M = main role; X = additional role; S = secondary role; F = intended future role

Provider". However, the role of ICT providers have been investigated and described rather broadly and deeply by others, already. Another role "Media Content Provider" was validated by journalists and editors in my project team.

The interviewees were also asked about a future outlook, i.e. how their answers would change in five years. The intended future roles are thus marked "F". Three actors wish to extend their business towards the "Ad Space Seller, Marketer" role. This role formerly required broad covering of audiences, but as of today, social media companies have changed it successfully to very targeted, metadata based user engagement. It is the role in which social media takes the most revenue away from traditional media. Also ticket selling and distributing seems to be attractive, as two actors wish to engage in this role in the future. However, this is a role in which strong competition has led to a concentration of most ticket sellers in three globally active ticket selling companies. Finally it is noteworthy that none of the actors plan to drop any role—focus seems not to be the core issue in this ecosystem.

4.2 Actors' Key Partners, Customers and Enablers

Another field in the Business Model Canvas is allocated to "Key Partners". Here, interviewees were to name their most relevant partners—for instance those who would be hard to replace. I established a matrix in the same format as Table 2 for these key partner relationships. That matrix is the first one supplying information for a network analysis.

Besides asking interviewees directly about their connections to other roles it is also possible to analyse their customers and enablers. For customer relations I analysed the fields "Customer Relationships", "Customer Segments" and also "Channels" of the individual business model canvases. Similarly, I analysed enabler relations. Enablers usually provide tangible and intangible assets, human resources and financial support. I analysed the fields "Key Resources" and "Cost Structure" and cross-checked with "Key Activities" of the individual business model canvases. The results were recorded in a Customers and an Enablers matrix, respectively.

Based on the three matrices I analysed how actors consider their customers and enablers and how much they cooperate with other actors performing the same role. Through this approach I checked the consistency of the interviewees' answers and my interpretation.

Customers as Partners The question here is whether actors regard their customers purely as customers or also as key partners. This question addresses the actors' mind-set and the depths of their customer orientation. Interestingly, 10 out of 14 actors do not regard the role Consumer, End User as their key partner, even if all 14 regard that role as customer.

Enablers as Partners Similarly, here the question is whether enablers are regarded as key partners, or whether they are rather easily replaceable. Note that

7 out of 8 actors do not regard the role Social Media as their key partner, even if all 8 regard it as enabler. On the opposite the role Media Content Provider is considered by six actors as key partner, but only by one actor as enabler.

Enablers as Customers This analysis investigates whether there are circular relationships. The outcome is that the urban event publisher has 4 of these relations and the regional web portal has 3, all others have none, 1 or 2.

Own Role Cooperation Finally, this last analysis evaluates whether actors cooperate with other actors in the same role. This cooperation would indicate that competition has been reduced due to, e.g. strong customer segmentation. The result is quite clear in that respect that only four actors cooperate in their main role. These roles are Event Feed Provider and Public Sector.

5 From Event Network to Event Ecosystem

For the first synthesis it is necessary to aggregate from specific actors to general roles. Thus whatever data are available per actor need to be aggregated into the various roles that the respective actor may take on, as shown in Table 2. Based on these aggregations, I built new matrices, this time based on role-to-role relations.

5.1 A Network of Key Partners

The aggregations can be performed in three ways: (1) based on the main role choice, (2) based on main and additional roles and (3) based on future roles. Figure 1 shows the resulting network based solely on main roles and thus provides the least dense network. As the role Public Sector is in fact an aggregation of roles it has been omitted from Fig. 1 and all further figures for reasons of clarity. Ties with an aggregation level > 1 are labelled 'mutual' in the network.

Taking into account additional roles taken by the interviewed actors it is possible to refine Fig. 1. In a second step, I also augmented the network with ties anticipated for the future, leading to Fig. 2.

At this point it is worthwhile considering the 'de-cluttering' of the network illustration, as too many roles connected by too many ties can become confusing. However, no information relevant to network analysis shall be lost. I suggest two 'filters' for data reduction: (1) splitting the network into mainly one-time earning and multi-time earning networks; and (2) removing, re-defining or merging roles:

Focus on Multi-Time Earnings basically means the same network in use for every event. Mainly one-time earnings (e.g. the sales of ICT tools) are excluded. This leads to the exclusion of the ICT Provider role, as this role is focussed on selling, supporting and maintaining its ICT, which is typically subject to one

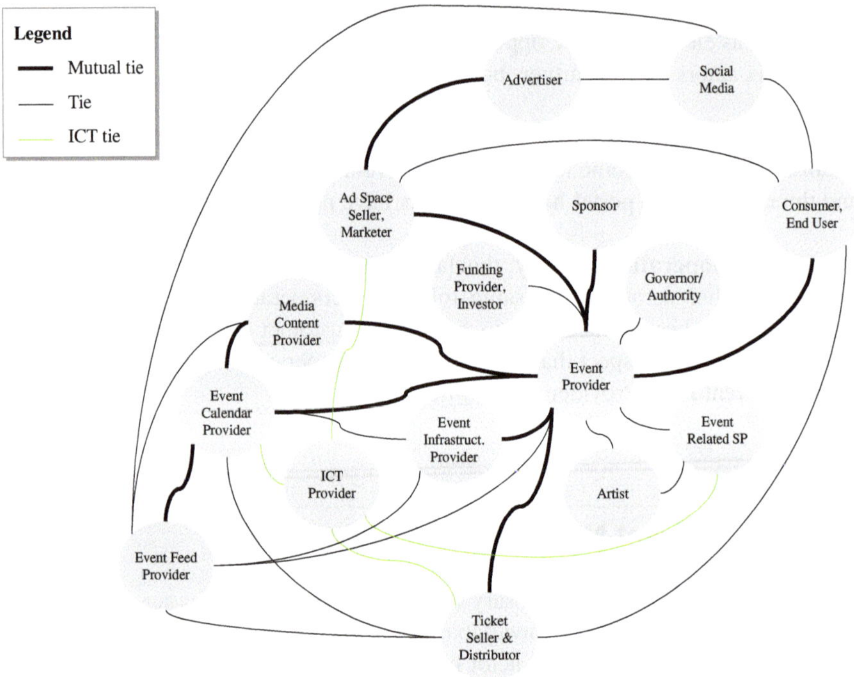

Fig. 1 Event network based on main roles' key partners

contract per software, not one per event, even if the earnings can be scaled to the number of events processed.

Decomposing the Sponsor Role A role that can be considered as aggregation of roles—like Public Sector—is the Sponsor role. Basically, it can be regarded as a combination of the roles Advertiser and Funding Provider, Investor. Thus in the following network figures, all ties to Sponsor will be attached both to Advertiser and Funding Provider, Investor.

The double 'filtering' leads to a slightly less cluttered network illustrated by Fig. 3.

Whereas Fig. 3 illustrates already quite clearly the network, its topology and the centrality of the roles (Freeman, 1979); the ties are without 'directions', yet. It is unclear whether partnerships are really mutual between roles and it is also unclear, what the 'content' of a tie might be. Is it for instance supportive, enabling, governing or providing? Thus, I investigate the enabler and customer relationships per role in the next sections.

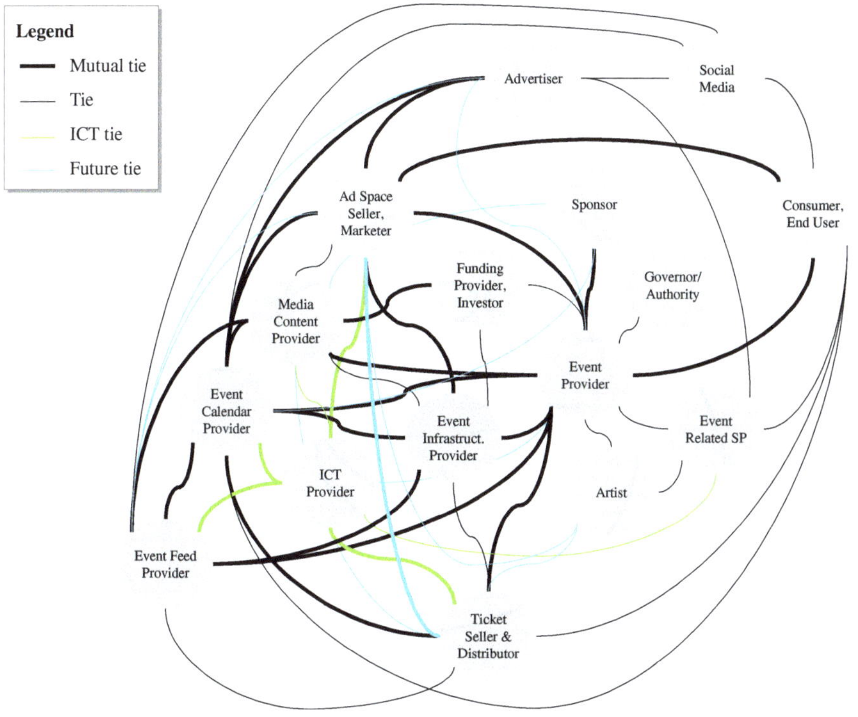

Fig. 2 Event network with main, additional and future key partners

5.2 The Overall Network Including Customers and Enablers

Also the roles' customer and enabler relations can be aggregated in three ways, based on the main role choice, additional roles and future roles. I aggregated all and Fig. 4 shows the resulting network. One observation based on Fig. 4 is that three roles wish to enable the Event Feed Provider. This enabling would be such that the Event Feed Provider discovered events by scanning ticket sales offers, getting early alerts from the Event Related Service Provider about upcoming events being prepared, and finally end users tip off about events they plan to join.

A second observation reveals that enabling is a source of value creation. For instance the Event Calendar Provider needs to pay for the enabling services (i.e. the actual event feeds) of the Event Feed Provider.

Whereas Fig. 4 provides the most 'complete' network view it is also overloaded with information. One way to reduce the amount of information is to eliminate the roles Consumer, End User and Social Media, as their relations (consumption and feedback) are quite clear. Similarly, advertisers and funds provide money, and authorities govern, in general. This leads to Fig. 5.

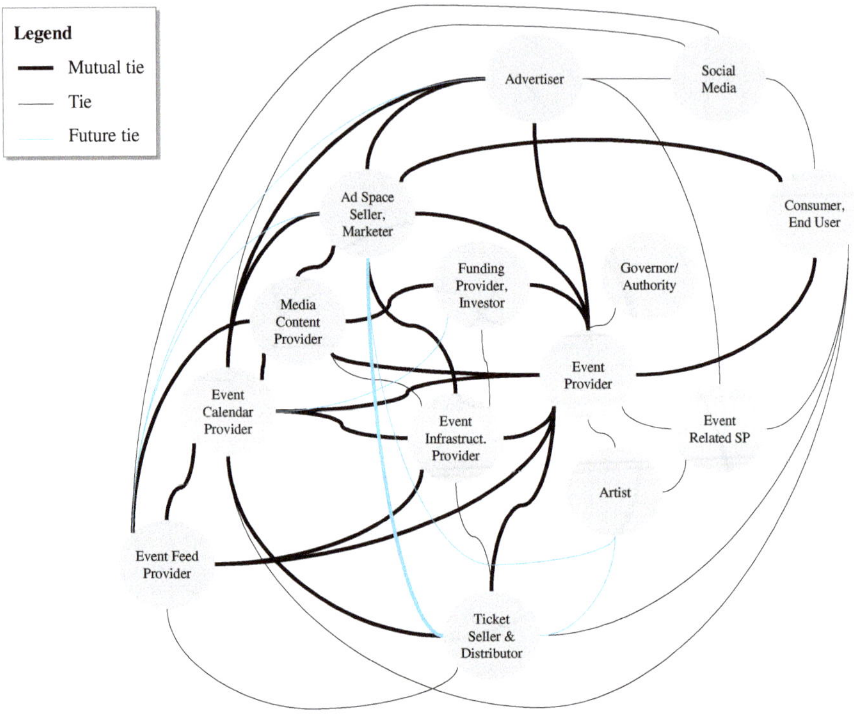

Fig. 3 Revised event network illustrating key partner ties

With Fig. 5 I present a network illustration that contains only event-based roles plus the news media industry in its roles of Media Content Provider and Ad Space Seller. In the absence of payments from customers, advertisers and funding providers it is interesting to investigate how many enabling services need to be paid in this network. Or in other words, which roles are 'money drains'?

The most obvious result in Fig. 5 is the confirmation that artists expect payments from event providers. Secondly, event calendar providers will need to pay event feed providers. The role Event Calendar Provider shows a clear dependency on other roles, with six roles as enablers. No other role is so dependent. What becomes obvious next is that ticket sellers can exist without all other roles but the Event Provider. The news media industry however, wishing to sell both content and advertisement space, is dependent on many more roles.

As a consequence of this observation I decided to divide the professional event ecosystem into two main clusters: (1) the *event provision* cluster, including Event Provider, Artist, Event Infrastructure Provider, Event Related Service Provider and Ticket Seller & Distributor; and (2) the *event information* cluster, including Event Feed Provider, Event Calendar Provider, Media Content Provider and Ad Space Seller, Marketer.

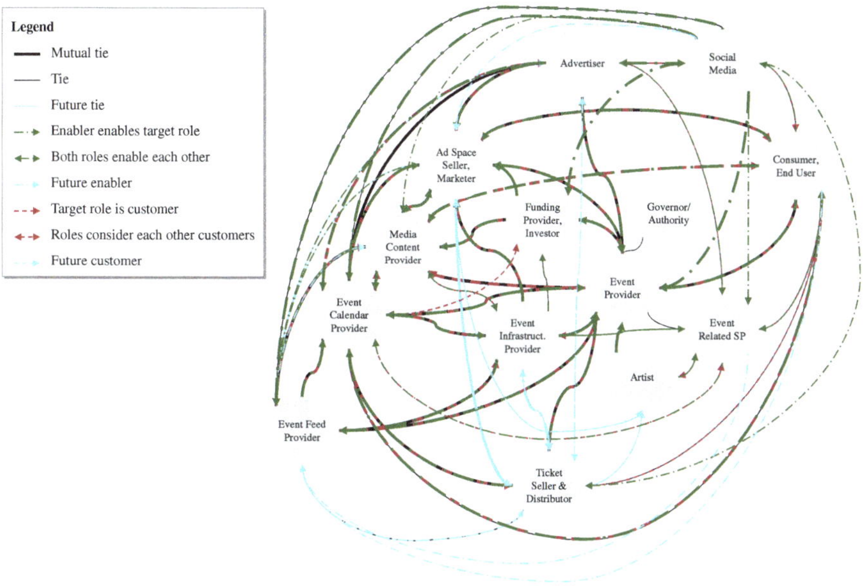

Fig. 4 Overall event network including key partner-, customer- and enabler ties

5.3 Network Power Constellations

In this section I investigate the event network as illustrated in Figs. 4 and 5 regarding different forms of power.

Networking Power is the joint power of core roles *within* in a network over roles *outside* (Castells, 2011). In a basic sense, this power is about in- and exclusion of a network. It is based on Network Gatekeeping Theory (Barzilai-Nahon, 2008). This theory may not yet help understanding the *overall* event ecosystem as the core roles should not be interested in keeping further roles out of the network. In the past, one such role to be kept out could have been Social Media; however it is now part of the network simply through massive consumer use and thus acceptance.

However, when looking into the two ecosystem clusters (event provision and information), the situation differs: the Ticket Seller & Distributor role could have an interest to keep other roles out of the event provision ecosystem and thus prevent the event information ecosystem from earnings from Sponsors and Advertisers. Although such interests would only be valid for events which are under full ticket seller control, it is important to investigate how ticket sellers build and maintain their networks and with whom (which roles) they engage and who they try to keep out.

It is noteworthy that neither my team nor I were able to interview any of the three global ticket sellers or their local subsidiaries. At least three subsidiaries were not allowed to give interviews and even appointments earlier agreed were cancelled.

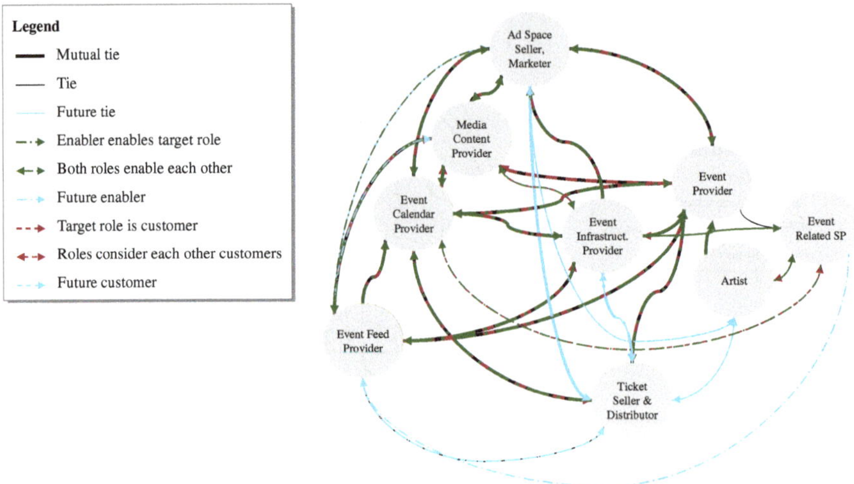

Fig. 5 The professional event network

As a conclusion, only the role Ticket Seller & Distributor could be interested in exercising the substantial networking power of the event provision ecosystem. However, more research is needed to substantiate this tentative finding.

Returning to the Social Media role, Nahon (2011) states

> These collective patterns of usage constitute *Facebook* as a major gatekeeper in the Internet. *Facebook* would not be a gatekeeper without the consent, support, and active participation of the gated.

The problem here is that both Facebook Inc. (the company) and *Facebook* as a social media tool can be regarded as gatekeepers. Facebook Inc. attracts advertisers but keeps advertisements away from *Facebook* users whose profiles do not match the advertisement. Thus advertisers pay only for matching, targeted ads. In contrast, the conventional mass media industry, even today, mostly offers 'crowds' to advertisers, with only little metadata for customisation.

On the other hand the *Facebook* users themselves act as gatekeepers for their own pages and communities. Their combined 'I-like' power can very effectively bring an event to massive attention—or keep it gated by simply not mentioning it in the community.

Network Power is the power of standards/rules in a network (Castells, 2011). Within the event ecosystem, standards and rules apply, both set by roles between each others as well as rules set by authorities (e.g. general event permits, environmental permits, maximum electricity supply, facilities needed per attendees etc.). Whereas the rules set by law (and enforced by authorities) are explicit, the network standards and rules might be implicit.

In the long term, potential customers are more powerful in changing rules than potential enablers. Standards as well need at least the acceptance of consumers, even if they are of utmost importance to enablers. *Facebook* has effectively managed to standardise a platform for consumer interaction—both in terms of informing and reaching out (with micro-targeted advertisements) as well as feedback collation. This network power by platform standardisation redirects already a considerable amount of advertisers' spending away from the conventional media industry towards the Social Media role. Until now, the conventional media industry has not taken any step in preventing this platform standardisation—they rather prefer to play by the new rules, so to speak, and build their presence and 'I like' buttons directly in *Facebook*, thus even more establishing the standard that hurts their advertisement income most.

More standards have been built around electronic ticket sales, preventing outside roles from taking on the Ticket Seller & Distributor role. Online ticket purchase is becoming the rule, with 42% of all purchases (Valli et al., 2012: 11), and thus provides network power to the ticket seller.

A role that could develop its network power much more is the ICT Provider. A prerequisite would be the agreement on certain standards for event feeds, event description, event-to-ticket links etc. and especially for event related metadata, which can be shared with and exploited by advertisers. This metadata should ideally also include consumer data (e.g. from event related user tweets, pictures, notes and links).

Networked Power is the power of roles over other roles in a specific network (Castells, 2011). Measured by the centrality (Freeman, 1979) of roles in Figs. 4 and 5 the networked power per role is typically quite high. Besides the roles Governor/Authority, Artist and Sponsor all roles have four or more key partner ties, and most of them are strong ties, as defined by Granovetter (1973). Thus the networked power theory does not single out any specific role as being in a state of power over others through networked power. However, authorities have institutional power, artists have nearly a monopoly on experience provision and sponsors (through conditional funding) have undisputed acceptance within the network.

Network-Making Power is the power to build or orchestrate (Castells, 2011 uses "program") specific networks or to bridge/connect ("switch") networks according to the interests of one or more roles. This should be the most interesting approach for the media industry, as Castells repeatedly links network-making power to the media communication networks, most of them owned by the media industry. He also proposes the "hypothesis of the centrality of communication networks to implement the power-making process of any network" (ibid. 786). And he suggests "switching [i.e. bridging] different networks is a fundamental source of power" (ibid. 786). The downside is that he also attributes this to social media, such as *Facebook*, and mentions explicitly Facebook Inc.'s Mark Zuckerberg as one of the owners and controllers of media corporations (ibid. 781).

In the context of the event ecosystem some power to build or orchestrate the network might be hidden in the conventional media industry's roles; however my data does not provide a clear picture. The Social Media role is exercising its power already in that it extends the concept of 'event' also to people's private lives and their own private, even intimate events. Also, social media have established themselves as gates to consumers and end-users for other roles. Third, the Ticket Seller

& Distributor role has linked event discovery quite successful with the event as such, the artists and finally the actual ticket purchase.

When it comes to the power of bridging networks, the picture looks somewhat different. Basically the role Event Calendar Provider constitutes a bridge already, bridging media industry with the event provision ecosystem. Similarly, the Event Provider role bridges culture, arts, even sports with commerce (ticket sales, advertising, sponsorships), authorities (laws) and infrastructure (venues, tourism, gastronomy etc.). Based on the interviews, even if many other roles are keen on relating and connecting with the Event Infrastructure Provider role, until now the Event Provider as such has the most experience and bridging power.

Another role in favour of the conventional media industry is the Ad Space Seller, Marketer. This role maintains the bridge to all the creative and marketing agencies that have great influence on advertisers' and sponsors' budget allocations. According to my data, between 25 and 50 % of the event information ecosystem turnover passes through this role.

5.4 Roles' Proximities

In this section I investigate the proximity between roles in the event network. For the analysis I map sections of the interview notes to five dimensions of proximity, as illustrated in Table 3.

This analysis obviously involves a large degree of interpretation as the interviewees have not been asked explicitly about their proximities, nor has the data been recorded in a way that proximity was explicated. Thus the results in Fig. 6 should be regarded on overall level, as trend providing, not in their details.

Neither too close nor too far proximity is regarded as beneficial (Boschma, 2005). Too close proximity in the social, organisational and institutional dimensions may lead to opportunism, whereas too little cognitive proximity leads to a lack of understanding. The effects of too much proximity are (ibid. 71):

Table 3 Dimensions of proximity with explanation, and data mapping

Proximity dimension	Explanation (Boschma 2005)	Interview notes section[a]
Cognitive	Width of the knowledge gap	Key activities and key resources
Organisational	"The extent to which relations are shared [...] within or between organisations"	Between roles which are typically taken by the same actor
Social	"Trust based on friendship, kinship and experience"	Key activities. Additionally between end users and social media
Institutional [or cultural]	Shared "institutional rules [...] as well as a set of cultural habits and values"	Shared rules and cultural values based on Value Proposition
Geographical	Spatial distance	Customer Segments

[a]Mapping to section of the business model canvasses

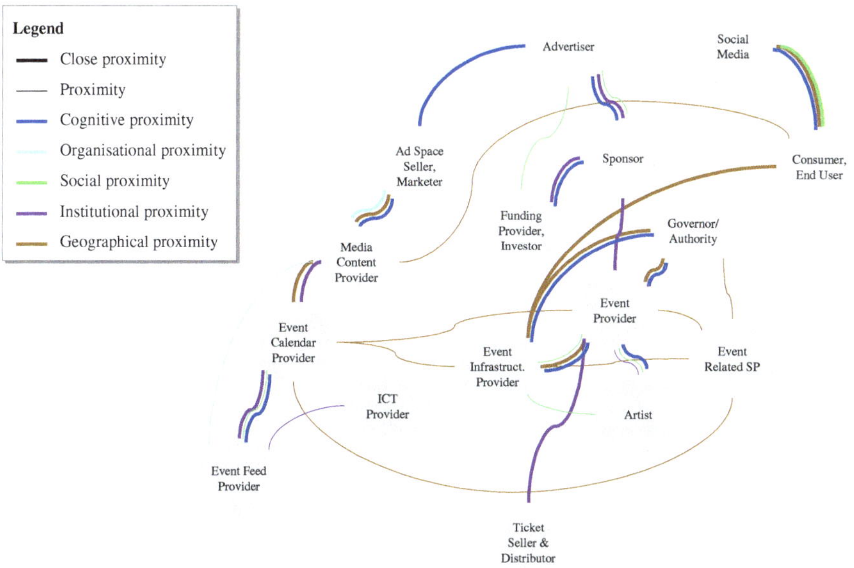

Fig. 6 Close proximities in the event ecosystem

- Cognitive: lack of sources of novelty
- Organisational: bureaucracy
- Social: no economic rationale
- Institutional [or cultural]: lock-in and inertia
- Geographical: lack of geographical openness

Keeping the risks of too little and too much proximity in mind, in Fig. 6 I propose a tentative overview of proximities between roles.

5.5 The Live Event Ecosystem

Based on the live event network (Figs. 1–5) and the roles' proximities in the network, it is possible to synthesise the live event ecosystem as such. The synthesis involves a graphical combination of Figs. 5 and 6 towards a higher degree of abstraction. For clarity I bundled proximity ties and added only key partner ties. I also changed the spatial location of two roles in order to avoid crossing ties. The topology remains the same. The result, the live event ecosystem, is shown in Fig. 7.

The Clusters in the Event Ecosystem, indicated by ellipsoids in Fig. 7, split the event ecosystem into four separate sub-ecosystems:

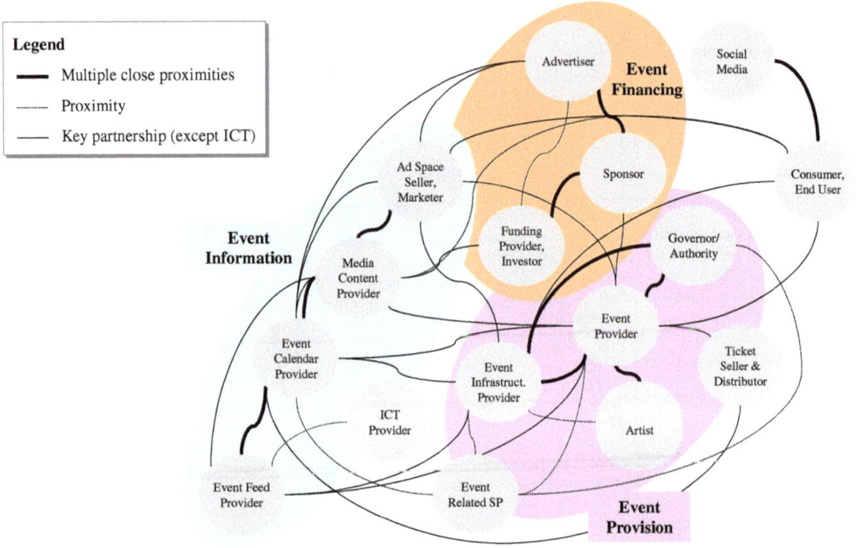

Fig. 7 The live event ecosystem, based on proximity and key partnerships

- The *event provision* cluster, including Governor/Authority, Event Provider, Artist, Event Infrastructure Provider and Ticket Seller & Distributor; and to a lesser degree the Event Related Service Provider.
- The *event information* cluster, including Event Feed Provider, Event Calendar Provider, Media Content Provider and Ad Space Seller, Marketer. This cluster contains event discovery, as well as event news and analysis (i.e. pre- and post event information).
- The *financing* cluster, incl. Advertiser, Sponsor and Funding Provider, Investor.
- The *event sharing and consumption* cluster, including Consumer, End User and Social Media.

How to Read the Ecosystem Fig. 7 should be interpreted as follows:

- The financing cluster may provide monetary input to all roles, directly
- The ICT Provider may enable all roles with ICT and also with technical support
- The Social Media role aggregates user opinions and 'I-like' behaviour etc. in a way that influences user decisions considerably. Users seem to be happy in engaging between each other through the social media platform, which allows them to keep all their friends literally in their breast pocket (i.e. in their mobile phone).
- The Governor/Authority role sets rules and establishes laws which are—in one way or the other—applicable to each of the roles. However, the closest control is exercised over the Event (Infrastructure) Provider.

6 Conversion in Diverging Event Ecosystem Clusters

In this section I synthesise my findings from a viewpoint of media convergence and divergence in the event ecosystem. I present practical implications on how to manage convergent divergence in general and in particular, per cluster and per role. The section is mainly, but not exclusively, targeted towards practitioners working in organisations such as companies, public sector (especially municipalities), associations and non-profit organisations as well as individuals (e.g. volunteers, artists and enthusiasts). I recommend that readers reflect their own organisation's value proposition, core competence and customers before continuing reading.

6.1 General Causes and Effects of Conversion

Role Choice Determines Convergence—and Vice Versa Actors in the event ecosystem may take on one or more roles (Figs. 4–7) or parts of roles (e.g. only Ad Space Seller out of Ad Space Seller, Marketer). The main driver of the role choice is the intended value offering. This may be focussed or packaged. A typical focussed value offering is for instance an event feed for cultural events in *one* city. The point of convergence in that case is a geographical space (within the city limits), certain types of urban culture, and a certain kind of cognitive (i.e. intellectual) proximity. This convergence is being exploited by the interviewed urban culture publisher. A large package offering would for instance include event discovery, ticket acquisition, transport to and accommodation at the event location, the event itself and a post-event, social media based, analysis. Such package would require an actor to take on rather many roles, which are located in divergent ecosystem clusters. Consequently, the actor would need to make inherently divergent roles converge, which puts considerable stress on its organisation.

Key Partners Selection based on Convergence A value proposition needs certain competences, which need to be resourced. Key resources are kept inside the organisation and selected enabling competences outside. As a general rule, in a situation of convergence, embeddedness (Uzzi, 1997) of enablers makes sense (and thus key partnerships), whereas in a situation of divergence, enablers can be kept at arm's length distance (Uzzi, 1997). The close proximity ties in Figs. 6 and 7 suggest embeddedness and thus confirm my proposition that inside converging clusters embeddedness is typical, whereas between diverging clusters many relations are on arm's length distance (i.e. driven by formal contracts, not developing into key partnerships). Enablers as key partners are able to provide strongly customised tools, e.g. ICT, and thus need much more details about an actor's goals and routines and feedback for improvement. This calls for mutual trust, another feature of embeddedness.

Divergent Customers Support Innovation Customers may become key enablers and/or key partners, depending on value proposition and convergence. However, even in a case of divergence, customers are a source of ideas, feedback and improvement proposals and thus support innovation.

Convergence Through Value Chains Value chains connect actors, their enablers and customers. Usually they continue to the end customer and also contain feedback loops. Meaningful convergence means also to investigate the enabler's enablers in order to build the most complete value chain. A convergent value chain may lead to the discovery of possible extensions of an actor's network and thus new value offering opportunities, markets or innovative re-use or re-combination of existing service ideas. Also, value chains may contain actors, which should be developed into key-partners (i.e. through embedded relationships).

Close Proximity may Lead to Suffocating Convergence Close proximity in any of the five dimensions (Boschma, 2005) leads to convergence. However, too much proximity means that an actor may lack sources of novelty or geographical openness; suffer bureaucracy, lock-in or inertia; or become naïve regarding economics. This is a form of convergence which is typical for many mid-sized media houses. In such case it is useful to occasionally break system boundaries and explore other points of view, other (business) ecosystems and research which is only indirectly related to an actor's own activities, and thus impose some degree of divergence. However, too much exploration, without exploitation, leads to a lack of focus with lack of implementation and consequently lack of results and income.

6.2 Struggles in the Event Information Cluster

The term media industry is rather fuzzy, as it may include news media, entertainment media, the games industry or even social media. In this subsection the main focus is on news media. Thus, the focus is on event discovery as well as event news (i.e. reports) and analysis (other forms of journalism). From such point of view an event calendar in itself could be considered a logical value proposition: the consumer can discover information about upcoming events. This proposition follows the news-feed logic of news agencies such as Reuters or Finland's STT. The feed arrives in a certain format or structure, is being edited and/or translated if necessary, and channelled to the audience. Thus event calendar work processes have similarities to news production work processes. Consequently, the media industry in the event information cluster has the following potential problems of close proximity: lock-in (lack of business ideas, new concepts, and openness) and inertia (continuously replicating similar work processes, ideas, concepts etc.). In hindsight it is no surprise that successful social media platforms, such as *Facebook*, *Twitter*, *Google* or *YouTube* have not been invented and realised by the traditional media industry. The following subsections provide some recommendations on how to address the problems the (news) media industry is facing.

News Media to Learn from Entertainment Media Obviously the media *entertainment* industry is a very active part of the event provision ecosystem. It acts as Event Provider (e.g. TV shows, certain sport events, cultural programmes focussed on events), Event Infrastructure Provider (e.g. TV and movie studios), Event Related Service Provider (e.g. live TV coverage from all sorts of events) and it has countless artists under contract. The competences, the knowledge and the related ideas of theses roles need to be explored by the *news* media industry. For the success of media companies convergence is urgently needed between the divergent event information and event provision clusters. News media should not infringe or take over entertainment industry roles; however they should inform themselves within their own organisations or within their own media industry comfort zone about how to package an event.

Actively Build and Bridge Networks The media industry has network-making power, still (Castells, 2011). Now it is time to establish and *use* this power. Mid-sized news media actors should investigate how the large, global news and entertainment networks perform this task. The starting point is that the Event Calendar Provider role constitutes a bridge already, between the news media and the event provision ecosystem. A more powerful role in favour of the conventional media industry is the Ad Space Seller, Marketer. This role maintains the bridge to all the creative and marketing agencies that have great influence on advertisers' and sponsors' budget allocations. It is important to (re-) establish a sufficient flow of money coming from the financing cluster of the ecosystem. This needs a tightening of relations with that cluster, and the Ad Space Seller, Marketer role has the best competences to do so. As a precondition, the use of event metadata needs to be enforced, thus offering as much customisation opportunity as existing in social media. This would need a strong conversion of content and advertisement tagging.

The media industry also needs to offer its network building and bridging ability actively to other actors, i.e. offer them access to networks otherwise inaccessible. In the words of the CEO of a German urban culture publishing house in my interview (translated into English):

> We consider the network as such as a value proposition. And we build and maintain the network.

Focus Your Business Focus is the necessary complement of the two first recommendations. The more networking and collaboration is involved, the more the final service needs focus. This is even more important for a packaged service. For instance the German urban culture publisher offers many value propositions around events and even provides certain types of events as such. It thus needs a considerable network of enabler-customers, many of them considered key partners. However, the event focus is strictly on culture (i.e. no sports) and strictly local: only one city, even *excluding* the immediate city *surroundings*. Within this focus however, 'anything goes': events, accommodation, dining, arts etc. Due to this focus the collaboration with local authorities and the municipality works very well, as all partners in the networks are very much aware of each other's core competences and also their limits.

City Events are More Profitable The local, geographical focus has been mentioned in 14 out of 16 interviews for a good reason: covering a fuzzy geographical area does not pay off. Covering a city does, as cities are both centres of cultural events and densely populated, thus matching event offers with a potentially large audience. This logic indicates that municipalities lacking large centres, especially if they are sparsely populated, cannot be as attractive to the event ecosystem as a large city. And, although sad from a societal point of view, the media industry may need to draw the consequences and focus on the profitable areas. According to the empirical evidence, 8 out of 16 interviewees focus on cities, only. All eight work profitable. In summary, focus means focus on the type of event (as narrow as possible) and *city*.

Affinity between Content and Ads: Strong Need for Convergence Events are mostly entertaining and belong to the cultural production ecosystem. Thus, the strict ethics of news content production do not apply. Consequently, affinity between event and news content on the one hand and advertisements on the other is a 'must have'. All social media build on exactly this affinity: content-advertisement-user. Thus metadata about the event, its content, the ads, and the audience need to be gathered and matched as closely as possible. There is a need for process-, technology- and mindset convergence. The interviewed urban culture publisher extends the affinity even further by building actively on the cognitive proximity of its advertisers, event providers and customers (audience). All of them are interested in certain cultural events, want to discover them, discuss them, experience them and review them, sometimes even together. This is as close to social media as possible, without acting as social media. Note that Facebook, Inc. cannot organise a local event and organise dedicated face-to-face network sessions—but a local media company can, and should.

Event Calendar Providers in Troubles This role, with six enabling ties, is the most dependent role in the ecosystem, making it rather vulnerable. The main problem of this role is that it needs to ideally discover an event before most others do—otherwise it produces 'old news'. Thus it needs a tight network of informants or a good event feed provider. The feed provider, however, will need to be reimbursed. Based on the empirical data, other roles regard the Ad Space Seller, Marketer and Media Content Provider as more central as the Event Calendar Provider, making this role even less attractive.

Event Feed Providers are in a similar position as Event Calendar Providers, with a key difference: usually they have a clear, payment oriented, customer relationship with Event Calendar Providers. Thus this role is more attractive, as indicated by one interviewee, wishing to provide event feeds in the future.

Media Content Providers need to make sure content has affinity to events, ads and audiences. In the final consequence, with full convergence, this means that event related content metadata need to be available to interested advertisers *long before* the event.

Ad Space Seller, Marketer is a role that three interviewed actors wish to take on in the future—obviously as another source of income. This means that ad space revenue will not only be taken away from the conventional media industry by social media, but also by other non-media roles in the event ecosystem. One way for the media industry to protect itself against ad income losses is to ensure high content-ad affinity.

6.3 The Prospering Event Provision Ecosystem

The event provision cluster of the event ecosystem, as illustrated in Fig. 7 may exist on its own, without the *news* media industry, however in that case it may lack event discovery opportunities for the consumers. Note that several roles in the event provision ecosystem can be taken on by the *entertainment* media industry. The Event Provider role as such is a kind of melting pot of sub-roles. These include event organiser, coordinator, manager, but also, on the fringes, artists' managers. The common denominator of these sub-roles is that they are supportive, enabling roles for the *event as such*, not for event *related services*. This bandwidth also explains why so many municipalities act in the Event Provider role: even if they do not provide the actual experience, which is the job of the artist, they arrange it. In the following, I present two key recommendations.

Exercise Bridging Power There are several powerful bridge roles in this ecosystem. The most powerful is the Event Provider role that bridges culture, arts, even sports with commerce (ticket sales, advertising, sponsorships), authorities (laws) and infrastructure (venues, tourism, gastronomy etc.). As a case in point, 8 out of 16 interviewed actors were dependent on the Event Provider role. Moreover, in many cases it is the Event Provider that possesses a gatekeeper function towards the Event Infrastructure Provider role. There is also empirical evidence that the Event Provider role has the best customer-partner and enabler-partner embeddedness (i.e. regards its customers and enablers as key partners) and has the most circular relationships, i.e. considers other roles as both customers and enablers. This indicates a high awareness of the need for collaboration, and also the applied capability to bridge different networks and ecosystems.

Second, the Artist role represents artists, performers, entertainers and with them the whole cultural ecosystem. These actors are full of ideas, creativity and innovation potential. It makes sense to be linked to them already from a point of view of innovation alone—with or without events. The commercially successful individuals know this fact and exercise their bridging power between arts/culture/creativity and commerce.

Third, the Event Infrastructure Provider is quite often linked to tourism (Getz, 1997; Watt, 1998) and/or trade fairs and in most cases partners with catering suppliers. Tourism and trade are two different ecosystems, both larger (in terms of turnover) than the event ecosystem. Here, the Event Infrastructure Provider role ideally offers a bridge between ecosystems.

Fourth, the Event Related Service Provider role hosts a large mixture of sub-roles. Services can include pre-event services, such as provided by the interviewed gig booking service, simply connecting artists with venues; live-event services, such as broadcasting, digital media platform provision (for, e.g. picture or comment sharing); and also post-event services, such as archived discussions, pictures, videos etc. There is no limit for ideas on event related services and a successful services provider will offer new, valuable and exciting services.

Fifth, the Ticket Seller & Distributor builds the most standardised and commercially successful bridge between the event providing roles and the actual, paying audience. As each and every ticket gets invoiced, the selling business model is ridiculously easy, often taking a fee out of both event provider and spectator. This may explain why two interviewed actors wish to take on this role in the future. The advice here is simple: grow the amount of sold tickets while concurrently standardising all business processes. In the final consequence this means going global and standardise all user interaction, if possible online. Three ticket sellers are globally present already: *Ticketmaster* (a Live Nation Entertainment brand), *Eventbrite* and *Tickets.com* (a subsidiary of MLB Advanced Media, L.P.). The next candidate is Europe's *CTS Eventim AG*. According to my data, three roles wish to enable the Ticket Seller & Distributor role in the future and thus envision taking part in one way or the other in the ticket selling profits.

Finally a note to all other roles: the Ticket Seller & Distributor role may have an interest to keep other roles out of the event provision ecosystem and thus prevent the event information ecosystem from earnings, e.g. from Sponsors. This is a case of *intended* divergence.

Offer Focussed, Packaged Events A consumer rarely spends money only on the event as such. According to Valli et al. (2012: 15f) on average in Finland only EUR 33.35 are spent on tickets, with EUR 10.11 for transportation, EUR 15.30 for food, EUR 26.15 for beverages, EUR 14.12 for merchandise, EUR 1.27 for additional services and EUR 9.90 for accommodation, totalling EUR 110.20.

In conclusion, all this potential spending should be as much bundled as possible in order to get as much possible fees, ads, rewards, bonus points etc. out of the EUR 110.20 as possible. Thus event discovery already needs to offer links to transportation, caterers and tourism (i.e. accommodation) plus all thinkable additional services. Finally, sponsors need to be able to position their brands, especially beverage brands, close to the event information. And pay for doing so.

Packaging events, taken serious, needs various forms of convergence—at least processes, content based metadata, technologies and interfaces. The Event Provider, in spite of its centrality and connectedness, will rarely have the network in place to offer such a comprehensive package. However, a small network of actors may locally agree on such packaging, with, e.g. local media industry providing the package with its link to the potential audience. With a focus on only specific events, such package could be commercially viable.

6.4 Public Sector and Municipalities: Size Matters

All my empirical data suggest that size matters regarding a municipality's needed efforts in the event ecosystem. Whereas big cities hardly ever are forced to take matters into their own hands, and on top receive considerable tax income from several actors in the event ecosystem, small, sparsely populated municipalities need to invest considerable resources and take on quite many roles by themselves. Thus especially for the small- and midsize municipalities, less for the larger ones, the following recommendations apply:

- Municipalities should change from being funding provider to being sponsor—attracting tourists needs a strong brand.
- Concentrate on core public functions: culture production, attracting tourists and keeping the local event ecosystem alive. Outsource other roles to professional actors.
- Build on your institutional, cultural and local proximity. Build on the mutual trust you have, and that global companies do not. This works well with local volunteers, artists, associations, SMEs etc.
- Use your power in the Governor/Authority role to attract favourable events and bar off unfavourable events.
- Feed your own local event information into regional, national and surrounding cities' calendars. It is more lucrative to take on the Event Feed Provider role than the local Event Calendar Provider role.

6.5 Convergent Divergence for Other Roles?

ICT Providers enjoy a strong potential of network power. By supplying integrated ICT solutions, providers cause technology convergence in their customers' networks. From a business point of view, ICT Providers should develop their bridging (and thus convergence) power between actors and ecosystems: they should identify collaborators by investigating Figs. 2–7 and supply software that enables, or at least supports, their collaboration. From a technology convergence point of view this would need the agreement on standards for event feeds, event description, event-to-ticket links etc., and especially on event related metadata, which can be shared with and exploited by advertisers. This metadata should ideally include consumer data, e.g. from event related user tweets, pictures, notes or links.

As a concrete convergence example, an ICT Provider could connect Event Feed Providers with Event Calendar Providers; then continue with Ticket Sellers & Distributors and eventually connect Ad Space Sellers and Marketers.

Financing It is a peculiar finding that my interviewees selected Funding Providers, Investors and Sponsors only twice as key partners. Even Advertisers were selected only four times. To keep the potential main income generator at arm's length may be rooted in considerations of journalist ethics, however from a business point of view it makes more sense to converge to an embedded relation.

Innovators seem to be closest to the financing cluster and to the individual actor's research organisation. Future research should try to locate innovation explorers, innovation merchants, innovation architects and innovation missionaries (Chesbrough, 2003: 39) in the event ecosystem.

Social Media is the strongest enabler and cause of user engagement, and social media led to the strongest affinity between content (often user generated), ads and users. In the live event ecosystem, Ticket Sellers & Distributors exploit social media far more effective than the event information ecosystem cluster. News media need to catch up in understanding the core mechanisms of social media, and how to exploit them.

7 Conclusions

In this study I aimed to describe a generic live event ecosystem with a focus on media industry and its convergent divergence. The main finding is that the *event provision* cluster and the *event information* cluster diverge. This divergence is based on various effects—proximities of actors in the clusters, the degree centrality of the event provider in its cluster, a lack of closeness of media companies towards the *event provision* cluster actors and (in some cases) the application of news journalism ethics in a context of entertainment. Moreover, too close proximity in the *event information* cluster may lead to suffocating convergence, which translates into antiquated business models, lack of innovation and lock-in to obsolete working processes in media companies. At the same time global ticket sellers attract advertisers and sponsors away from traditional news media, use their own Internet portals alongside social media for event discovery and ticket sales, and exploit standardised work processes on global scale.

Whereas many media companies stand by and watch, some small and medium size enterprises flourish in their niches. My findings indicate that these agile actors typically learn from the entertainment media industry; package events; build and bridge networks; focus stronger; and ensure affinity between content and advertisements. Meanwhile, successful municipalities change from being funding provider to being sponsor and concentrate on their core public functions: culture production, attracting tourists and keeping the local event ecosystem alive. They actively build on their institutional, cultural and local proximity. They also feed event information into regional, national and city event calendars.

Supporting convergence, media businesses typically connect various ecosystems or networks and thus should investigate in and develop their potential network bridging power. This is especially valid for media technology providers whose software platforms are crucial for convergent operations of otherwise diverse media actors. In summary, active exploring and exploiting of the ecosystem enables media businesses to keep up with global actors, confirming Giesecke and Immonen's (2010) network building propositions.

Contribution to Theory Whereas the main purpose of this Chapter is to illustrate the live event ecosystem and its divergence in practice, I propose some implications on theory. By applying network theory I was able to visualise the event ecosystem as a network of generic roles and I present the centrality (Freeman, 1979) per actor. The network power dimensions (Castells, 2011) allowed interpretations of power over others inside or outside of the network (cluster), including the divergence between the two main clusters of the event ecosystem. However, only the consideration of five dimensions of proximity between actors (Boschma, 2005) confirmed this divergence between the *event provision* and *event information* clusters. Thus the main contribution here is that network theory, augmented by *proximity as a tie quality* supports our understanding of ecosystems better than either of the two theories (network theory and proximity framework) on their own. In that regard my findings support Nahapiet and Ghoshal's (1998) call to distinguish between structural (i.e. network), cognitive (proximity) and relational (a mix between tie quality and proximity) capital. Another implication is that network nodes do not necessarily need to represent the actors as such: data can be aggregated in such way that a network represents a *generic* ecosystem of *roles*.

Limitations I admit that Castells (2011) power dimensions are somewhat speculative. However, three of the dimensions are rooted in network theory already, and they help interpreting findings in complex qualitative data.

The data of this study was collected in four European countries, from 16 organisations and—in case of Facebook Inc.—from Reuters. Thus it might not represent the global live event ecosystem in its entirety. However, interviewees provided large amounts of partners and relationships of their organisations, and in total a high degree of saturation of possible roles in the ecosystem was reached. For instance the final two interviews did not add any new actors or roles to the ecosystem. Furthermore, interviewees reviewed their data sheets and some key informants reviewed a prior report of this study. In hindsight, the structuring of the data sheets (as business model canvas) was suboptimal, as various types of relationship needed to be interpreted. The interview questions could have been more concrete towards relationships and their qualities. Still, much of the evidence confirms prior studies from various fields into a more sense making synthesis.

Acknowledgments Cordial thanks to (in alphabetical order) Jan-Erik Andelin, Zeerim Cheung, Stina Giesecke, Janne Saarela, Camilla Stjernvall-Malmberg and Jesse Valli. This work was partially supported by the Finnish Funding Agency for Innovation TEKES as part of the Tivit Next Media programme.

References

Allen, J., O'Toole, W., Harris, R., & McDonnell, I. (2002). *Festival and special event management*. Queensland, Australia: Wiley.
Andelin, J.-A., Stjernvall-Malmberg, C., Berg, K., Lehtimäki, H., & Strömdahl, K. (2012). *Public-private cooperation: Municipalities and media in the local event ecosystem*. Helsinki: Tivit Next Media Programme.
Barzilai-Nahon, K. (2008). Toward a theory of network gatekeeping: A framework for exploring information control. *Journal of the American Society for Information Science and Technology, 59*(9), 1493–1512.
Boschma, R. (2005). Proximity and innovation: A critical assessment. *Regional Studies, 39*(1), 61–74.
Boyatzis, R. E. (1998). *Transforming qualitative information: Thematic analysis and code development*. Thousand Oaks, CA: Sage Publications, Inc.
Buchanan, D. A., & Huczynski, A. (2010). *Organizational behaviour* (7th ed.). Harlow, England: Pearson Education Limited.
Castells, M. (2011). A network theory of power. *International Journal of Communication, 5*, 773–787.
Chesbrough, H. (2003). *Open innovation: The new imperative for creating and profiting from technology*. Boston: Harvard Business School Press.
Cheung, Z. (2012). *Evaluating the fit of business models in the event management ecosystem*. Master's thesis: Aalto University.
Dwyer, T. (2010). *Media convergence*. Berkshire, England: Open University Press, McGraw-Hill.
Freeman, L.C. (1978–1979). Centrality in social networks: Conceptual clarification. *Social Networks, 1*(3), 215–239.
Getz, D. (1997). *Event management & event tourism*. New York: Cognizant Communication Corp.
Getz, D. (2002). Why festivals fail. *Event Management, 7*(4), 209–219.
Ghoshal, S., & Moran, P. (1996). Bad for practice: A critique of the transaction cost theory. *The Academy of Management Review, 21*(1), 13–47.
Giesecke, R. (2012). *Event management ecosystem*. Helsinki: Tivit Next Media Programme.
Giesecke, R., & Immonen, S. (2010). Media sector business drivers in 2020. Paper presented at the *Third NEM Summit*.
Gilsing, V., & Nooteboom, B. (2005). Density and strength of ties in innovation networks: An analysis of multimedia and biotechnology. *European Management Review, 2*(3), 179–197.
Granovetter, M. S. (1973). The strength of weak ties. *American Journal of Sociology, 78*(6), 1360–1380.
Granovetter, M. (1983). The strength of weak ties: A network theory revisited. *Sociological Theory, 1*(1), 201–233.
Granovetter, M. (1985). Economic action and social structure: The problem of embeddedness. *American Journal of Sociology, 91*(3), 481–510.
Hannan, M. T., & Freeman, J. (1977). The population ecology of organizations. *American Journal of Sociology, 82*(5), 929–964.
Jarillo, J. C. (1988). On strategic networks. *Strategic Management Journal, 9*(1), 31–41.
Jarillo, J. C. (1993). *Strategic networks: Creating the borderless organization*. Oxford: Butterworth Heinemann.
Jenkins, H. (2006). *Convergence culture: Where old and new media collide*. New York: New York University Press.

Ketokivi, M., & Mantere, S. (2010). Two strategies for inductive reasoning in organizational research. *Academy of Management Review, 35*(2), 315–333.

Mensonen, A., Giesecke, R., Bäck, A., Saarela, J., Stjernvall-Malmberg, C., Andelin, J.-E., et al. (2011). *Event management: Functional requirements of the ecosystem and roles in the ecosystem.* Helsinki: Tivit Next Media Programme.

Nahapiet, J., & Ghoshal, S. (1998). Social capital, intellectual capital, and the organizational advantage. *Academy of Management Review, 23*(2), 242–266.

Nahon, K. (2011). Fuzziness of Inclusion/Exclusion in networks. *International Journal of Communication, 5*, 756–772.

Nohria, N., & Eccles, R. (1992). *Networks and organizations: Structure, form and action.* Boston (MA): Harvard Business School Press.

Osterwalder, A., Pigneur, Y., & Clark, T. (2010). *Business model generation : A handbook for visionaries, game changers, and challengers.* Hoboken, New Jersey: John Wiley & Sons.

Peltoniemi, M., & Vuori, E. (2004). Business ecosystem as the new approach to complex adaptive business environments. *Frontiers of e-Business Research 2004, FeBR 2004, Conference Proceedings,* pp. 267-281.

Shone, A., & Parry, B. (2004). *Successful event management: A practical handbook* (2nd ed.). London, UK: Cengage Learning Business Press.

Smedlund, A. (2009). Social network structures for explicit, tacit and potential knowledge. *International Journal of Knowledge Management, 5*(1), 78–87. doi:10.4018/jkm.2009010106.

Strauss, A., & Corbin, J. (Eds.). (1997). *Grounded theory in practice.* Thousand Oaks, CA: Sage Publications.

Uzzi, B. (1997). Social structure and competition in interfirm networks: The paradox of embeddedness. *Administrative Science Quarterly, 42*(1), 35–67.

Valli, J., Jansson, P., Koivistoinen, L., Peltola, E., & Mynttinen, T. (2012). *What was your last event experience?* Helsinki: Tivit Next Media Programme.

Verganti, R. (2009). *Design-driven innovation: Changing the rules of competition by radically innovating what things mean.* Boston, MA: Harvard Business Press.

Watt, D. C. (1998). *Event management in leisure and tourism.* UK: Addison Wesley Longman Ltd.

Yin, R. K. (2009). *Case study research: Design and methods* (4th ed.). Thousand Oaks, CA: Sage Publications.

Zheng, H., Fu, Y., & Lugmayr, A. (2012). The development of ambient TV in 2016. *Proceeding of the 16th International Academic MindTrek Conference,* pp. 233–236.

Media Business Drivers and Concepts 2020: The Case of David and Goliath Revisited

Stina Giesecke and Raphael Giesecke

1 Introduction

How many times have you read the same headlines over and over again, and how often do you feel that the same content, just differently packaged, keeps popping up from medium to medium? The media sector is compelled to look for new business opportunities as its current business models are more and more infringed by expanding market participants, such as *Facebook*, *Twitter* or *Spotify*. But the outcomes we see are mostly manifestations of convergence—the same content through different channels. And the audience seems not to like content convergence as subscriptions are massively quit. Thus news media profits keep falling, media flagships are purchased on a shoe-string budget and social media flourishes. So, what's wrong?

In this Chapter we argue that the media sector, especially mid- to large size companies, suffers from a conservative approach to convergence.[1] More daring approaches to convergence and especially divergence may unlock vast amounts of underutilised business potential in media businesses. But what are these approaches, and how could they be implemented? To find an answer, we combined media scenarios for the year 2020 (our data) with theories of system innovation and business modelling. Addressing media company executives, we also propose a framework for consequential organisational development.

[1] Tameling and Broersma (2013) provide an illustrative example about "de-convergence" (ibid. 20–21) in a Dutch newsroom. After vertical convergence in 2006 (integrated news production for multiple platforms within one brand) the company changed in 2011 to horizontal convergence (news production in separate newsrooms per platform, all serving multiple brands).

S. Giesecke (✉) • R. Giesecke
School of Science, Aalto University, P.O. Box 15500, 00076 Aalto, Finland
e-mail: Stina.Giesecke@aalto.fi

A. Lugmayr, C. Dal Zotto (eds.), *Media Convergence Handbook - Vol. 2*, Media Business and Innovation, DOI 10.1007/978-3-642-54487-3_20

We know from practical evidence that radical innovations supported by radically new technology, offering new meanings to the media audience (Verganti, 2009), have strongly supported small, agile, risk taking businesses in their niche markets. Such actors address divergent, fragmented media audiences. On the other hand global actors, such as Internet providers, telecom operators and independent content brokers or distributors have entered the broadcasting and media industry market (Zheng, Fu, & Lugmayr, 2012). Mass markets are more commonly covered by big media conglomerates, spanning national, cultural and language borders. As a consequence, prospects for mid-sized media companies are bleak.

Regarding theory, we understand a business system as an evolutionary guidance system (Banathy, 1998: 168). We know that business modelling as such is open to both the most conservative and most innovative business drivers and value propositions. However, in order to make a real difference in the converging media industry, a media company or company network needs open-minded business system re-thinking and intuition to find new ways to innovations. This requires the courage for allowing divergent, nonconformist business strategy thinking; at least the capability as a company to swim upstream; plus inspiring leadership and proactive organisational development. If these criteria are met the probability to find promising business models will increase.

What we do not know, is the future. However, based on scenario building, the most important business drivers for divergent convergence in the media sector can be identified. During a Finnish collaboration initiative between industry and universities, media scenarios were crowd sourced (Giesecke et al., 2010; Giesecke & Immonen, 2010). This data, from contributors of a wide range of expertise and nationalities, allows us to synthesise a spectrum of business drivers we expect to be essential for both media convergence and divergence.

In summary, we investigate divergent convergence by applying a system approach to media business. We consider media business as a dynamic and complex whole interacting as a structured community situated in a particular environment. We intend to contribute to the understanding of business systems from the viewpoints of knowledge intensiveness and organisational development, with knowledge considered as one of the most important professional assets and social values of an individual. Consequently, the research questions that guide our work are:

- Which are the strongest media industry's business drivers by 2020?
- Which drivers can be used for technological breakthroughs combined with radically new meanings?
- Which business concepts do we envision for 2020?
- How should media companies develop themselves for the future?

2 Theoretical Considerations for Media Business Development

We define the media sector as a knowledge intensive industry (Hipp, 1999) that uses knowledge work (Blackler, 1995) in its production (converting information into knowledge) in order to create user experiences and novel knowledge for users. In this context we especially emphasise our understanding that *knowledge* is exclusively created, acquired, processed and shared by *humans*. Information may become *experience* after a human being cognitively and emotionally processed it. The hardware and software of digital gadgets process only *data* that can be converted to *information* for the user. Such information should be considered as just a 'raw material' for the industry but not as high value adding end-product or service.

The essentiality of creating new user experiences and thus flourishing new business possibilities cannot be reached by applying some minor adjustments in the existing product or service offering. Likewise, incremental technological improvements in producing, publishing or receiving media products will not provide sufficient foundation for new business and earning opportunities for the media sector. Instead, we propose using a holistic, integrated business ecosystem approach (see Chapter *The Live Event Ecosystem*). Hence, the combined (and optimised) value of the business ecosystem of users, industry, ICT providers and further stakeholders needs to be taken more seriously into account. Firstly, we suggest that the media sector as a whole (and not only in some selected parts, e.g. gaming) needs to progress more aggressively towards new technologies and especially towards new meanings for the audiences. By doing so, the media sector will increase its possibilities to discover new breakthrough business models. Second, we suggest that an essential means towards business ecosystem thinking is to broaden the focus from profit models to the entire business model. Third, we suggest that realisation of the holistic, vision and strategy based business model is not possible without influential organisational development.

2.1 Divergent Radical Changes Through System Innovation

We are intrigued by Verganti's (2009) suggestion that merging technological breakthroughs with radical innovation of meanings is a particularly effective type of innovation strategy. According to Verganti, incremental changes, whether they happen in technology or in providing new meanings for audiences, will lead only to market pull based business. Generally, related to product or service development, users' imaginations rarely significantly extend beyond their daily experiences. Thus putting the user in the centre, results sometimes in important but mostly in limited ideas on how to incrementally improve product functionality or service quality.

Similarly, the 'professionals' (as opposite to customers) rarely step out of their own comfort zones voluntarily and thus rarely invent something radically different that would drastically change the content and context of their daily work. When building on the current knowledge of the markets and immediate customer needs only incremental improvements and some new market pull will be created.

Also, just pushing technological advancements or improving the familiar technology without designing its meaning, will lead to technology push markets that usually are in the high risk of fading away after the fore runner technology enthusiasts discovered the next gadget. In extreme cases the offered technology could also be so much ahead of its time that simply nobody is able to understand its meaning in the current society.

Based on the reasoning that neither market pull or technology push driven design alone is capable to produce radical innovations, one should strive for both a radical change of the technology in use and in making sense of things, i.e. offer new meanings for the audiences.

2.2 Divergent Business Models and Business Systems

A business model usually reflects a company's understanding about its value propositions, its current and potential new customers, and how they acquire and consume the product or service in question. The typical challenge for a company is to overcome the conservative preconceptions about what business models should look like and not to constrain itself from developing radically different ones (Casadesus-Masanell & Ricart, 2011). Fritscher and Pigneur (2010) developed a business model ontology, refined by Osterwalder, Pigneur, and Clark (2010) into a business model canvas, consisting nine elements describing essential components of the company's business. The central component is the value proposition. Relevant value proposition is possible only if your customer segments, customer relationships and distribution channels match your value proposition. The revenue flow mirrors the customers' willingness to pay for the products or services. The company needs key resources, key activities and a partner network to be able to deliver the value proposition. These form the basis for the cost structure.

In order to be competitive, companies strive for business models that differentiate them in the market and that are not easy to copy. Applicable also for media sector, e.g. Anderson (2009) has described the future, online economy to be based on processing power, hard drive storage, and bandwidth. The costs of these elements are heading towards zero and thus constitute an engine behind the new Free Economy. Consequently, successful business models are based on abundance, not on controlling scarcity. Anderson distinguishes three types of business models: (1) direct cross subsidies—offering a product or service for free as an inducement to pay for something else, (2) three party markets—a third party pays for participating in a market created by free exchange, and (3) 'freemium'—offering a basic product/product for free but more value added components for a fee.

Unlike Anderson (2009), we do not see the above mentioned three earning models as ready made business models to be implemented. Instead, we will present ideas for doing business, which are better labelled as business concepts, i.e. one step short of a real business model. Conceptually, Itami and Nishino's (2010) definition

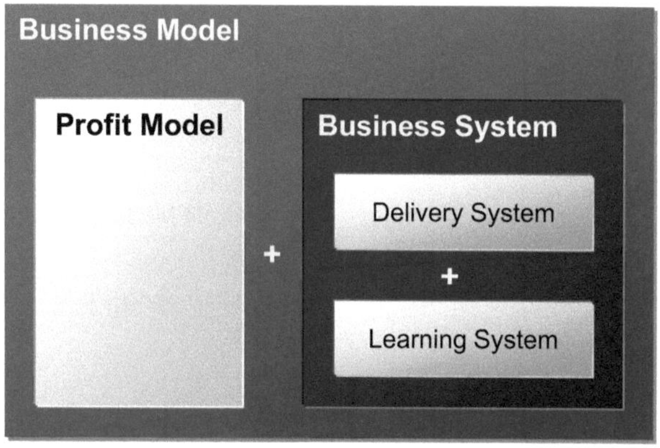

Fig. 1 The elements of a business model (Itami & Nishino, 2010)

of a business model (Fig. 1) fits well to our holistic view and system approach. As commonly accepted, a business model has two elements: a business system and a profit model. The profit model is the more visible and often the more glamorous part due to its direct link to the firm's bottom line. However, the authors note, "...the business system is more often the real 'meat', because it performs twin roles—first as the system that does the 'real work' in terms of realizing a firm's strategic differentiation intent, and secondly in accelerating its learning for the future"(ibid. 364).

A business system is the production and delivery system that a firm designs—within and beyond its boundaries—to deliver its products and services. This also supports Giesecke's and Immonen's (2010) view on system approach based business concepts that are developed with key stakeholders, users and audiences, and (advertisement) customers.

2.3 Business Model Based Organisational Development

A successful realisation of a business model is not possible without thorough understanding of the business model's implications on the company's entire business system and the profit model. Implementing a business model is company dependent and provides the possibility to form a competitive edge in the market. A firm needs a vision and mission for positioning itself in its business ecosystem and for communicating its purpose to itself as well as to the stakeholders. This vision leads to a strategy on what to do in-house, what to outsource and with whom to collaborate on what. Consequently, describing the elements of the business model will also provide insight for the organisational development needs (Fig. 2). Moreover, we support Casadesus-Masanell and Ricart's (2011) suggestion to carefully consider the possible managerial choices and their consequences when the executives make decisions about how the organisation should operate.

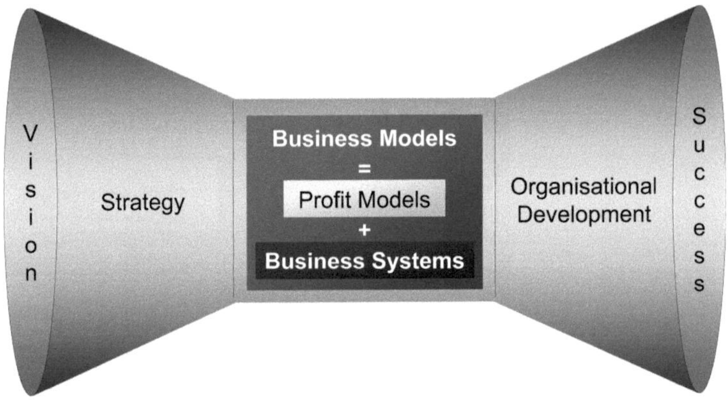

Fig. 2 Organisational development based on vision, strategy and business model

When designing its business system (i.e. delivery and learning system), a firm usually has to address at least the following three issues (Itami & Nishino, 2010: 365–366):

- the division of labour between the firm and its trading partners (typically a decision between outsourcing as well as crowd sourcing and internal procurement),
- internally, how the firm organises its in-house working system and
- externally, how the firm controls and integrates the activities of its trading partners.

Organisational development is commonly understood as the systematic use of managerial, leadership and organisational behavioural principles and practices to increase the organisational effectiveness. Therefore, with a clear view about the connections to the vision, strategy and business models, organisational development will effectively and positively influence on the business success of a firm.

3 Methodology

Our research design of defining media business drivers and concepts 2020 consists of four steps. The first step is about analysing and defining the most influential business drivers shaping our society and media business. Along nine business driver dimensions we discuss convergence and divergence. In the second step we map these business driver dimensions against Verganti's (2009) concept of technology epiphany, i.e. the conveyed new meanings for people, enabled by radically new technology. In the third step we use the most promising dimensions and the business drivers attached as a basis for synthesising possible business concepts. In a final step we draw our practical conclusions for companies. We apply and translate the business model ontology into a general business processes description of a knowledge intensive media company. The outcome presents a framework that facilitates media companies to map their own business processes and organisational development needs.

3.1 Data and Methods for Compiling Media Business 2020 Drivers

Our analysing methodology for identifying the most significant media business drivers, and their links to media conversion and diversion, is based on thematic analysis of our qualitative data (Matthyssens & Vandenbempt, 2003; Remenyi, Williams, Money, & Swartz, 1998). First, our Vision 2020 team[2] of 10 media researchers from different expertise areas identified dimensions, into which possible business drivers could be classified (Giesecke et al., 2010). The next section is clustered into these nine dimensions.

Next, the drivers as such were identified. The researcher team created trigger questions per each dimension to stimulate most versatile views about future business drivers. For gathering these views, a crowd sourcing approach was chosen. About 300 possible contributors from both academia and industry worldwide were invited via e-mail to submit their input.

Within 6 weeks 38 volunteers, 31 from within the media sector and seven from adjacent areas, such as ICT, futures research, design and humanities, developed 42 possible views of media 2020. The submissions from the contributors typically consisted of one to two A4 pages of a narrative or mini scenario of how, e.g. media user behaviour or media business will look like in 2020. We used thematic analysis to integrate these contributions, defining converging and diverging views per dimension. A saturation point per dimension was reached after a certain amount of inputs. Saturation means that new contributions do not deviate from the envelope of the existing contributions. The focus on the analysis was on the long term future. Drivers were not meant to be short/mid term extrapolations (trends) of what we already know for certain.

3.2 Methods for Deriving Media Business 2020 Concepts

In order to bridge the most promising business drivers to business concepts 2020 we used the submissions from the previously described crowd sourcing (i.e. the data) and views of future earning logics from the literature. We analysed this qualitative material thematically by categorising the potential business concepts as conventional or radical in terms of the potential to help the companies to diversify themselves in their markets. In this way, we were able to identify three mutually

[2] As part of the Tivit Next Media programme—http://www.nextmedia.fi

complementary approaches for new business concepts. We consider these as key enablers for companies to develop their own, company specific, business models.

3.3 Methods for Describing the Organisational Development Framework

As we adopted the system approach for developing the future media business, we embedded Osterwalder et al.'s (2010) business model canvas elements into Itami and Nishino's (2010) definition of a business model. The core of all organisational activities, whether they will be connected to the profit model, learning or delivery system of a company, is the value propositions. Thus, with value propositions as the starting point, we 'translate' the business model canvas elements to fit media industry organisations. We describe the relations between the elements with activities necessary to fulfil the value propositions. The resulting chart serves as a framework for organisational development.

4 The Nine Dimensions of Media Business Drivers in 2020

In this section we present an outlook on possible media business drivers in 2020. We specifically analysed our data towards trends of media conversion, diversion, and combinations of both. The nine dimension types, which cluster our analysis, were defined by our project team before data gathering.

4.1 Social Values as Key Guidance for New Meanings

By social values we mean a broad area of qualities or beliefs that are commonly shared and that guide our behaviour as community members. We use Whitman's (2009) definition: a social value is "an enduring, normative belief that describes a preferred mode of social conduct or end-state in society and justifies action to attain or sustain a preferred social order" (ibid. 309). This dimension provides the general context for the eight more specific dimensions. We identified knowledge, networking, sharing and role of humanised technology to be powerful change agents of our social values.

Knowledge Strongest Key to Power and Success By 2020 an expected four Billion humans and a similar amount of sensors of things will produce digital data (some as information) at least once a day. Already today there are plenty of initiatives to aggregate information to knowledge, mainly by putting information into relevant context. There are claims that the knowledge society has been achieved already in Europe, soon accompanied by an Internet of knowledge (EIFFEL, 2009).

Our data indicates that in 2020 knowledge as such is regarded as one of the most important professional assets and social values of an individual. Thus competition for the 'true' sources of knowledge will be intense. Due to the sheer volume of new information people rely more on their personal networks (i.e. personal socio-graphs) and on recommendations within their shared socio-graphs (friends of friends) to find relevant information. Thus, similar to intellectual property, there will be a change in intellectual capital which includes not only human capital (human intellect), but also structural capital (networks or other organisational structures that link people to each other) and relational capital (the types and characteristics of the relationships) (Bontis, 1998; Nahapiet & Ghoshal, 1998).

Still, much information retrieval will be based on search, and consequently the public will want free access to news sites. The EU has already put forward the notion of 'free movement of knowledge' (available to all, no matter where it is situated) as a fifth freedom (Missikoff et al., 2010). Digital literacy and broadband access have been declared essential to successful living in Europe and the US.

The business perspective on knowledge is that in 2020, competition will be about who is most effective at sharing content that matters most with the audience.

Networks and Networking as Intrinsic Values Both today and in 2020 people connect based on shared mindsets, lifestyles and business interests. However, the dynamics in the networks will grow exponentially. What once has been referred to as 'status update', will become in 2020 a more and more dynamic, fluid (ultimately real-time) stream of information, including information per person in the network, about the location (real or virtual), others present (real or virtual), subject of activity, expected duration, and finally, the reasoning: why are you there, for what purpose? Additionally, the available recommendation systems will be interlinked to this, as to provide dynamic recommendations.

The reason behind this is simply that people trust and learn more from their socio-graphs. We also assume that people cultivate personal networks as a hobby, and that the importance of something is defined by trusted peers. Moreover people actively store their memories (micro-histories) for their network.

In a culminated scenario, beyond 2020, people indeed identify themselves less by nationality, profession or class status and more by their position in networks. Social success is measured by the size and the quality of personal networks, as well as the 'value' (recommendations, updates, 'competences' etc.) that individuals add to the respective network. These networks will cover activities in professional, private, real, virtual and public life, respectively.

If this scenario (most likely for younger generations) will realise, it will certainly create divergent counter movements (at least 'anti real-time', 'anti big-brother' initiatives), based on traditional values of human relations.

More Sharing, Participation and Collaboration As networks as such are a highly regarded social value by 2020, people will be even more motivated to *use* them. Sharing, participation and collaboration has started already. Community building and, as a consequence of the preference of 'sharing' versus 'owning', communal ownership will increase.

Regarding sharing, privacy will remain the biggest issue. Privacy is still a truly important (social) value in 2020; however views on publicity vs. privacy may differ between generations. People want to control how they interact and remain in control of publicity vs. privacy of the content they generate. This continues to pose challenges to those who own personal content. People will remain strict about private information not being mixed up. Still, more personalisation and control of media content are welcome, leading to a faster convergence between content and advertisements, based on (user) metadata.

Regarding collaboration, some people most likely become citizen editors and publishers. In a business context, we see increasingly 'flat' social interactions and structures. This implies new forms of ICT-enabled collaboration outside hierarchical structures and controlled media, of which open innovation, collective intelligence and agile organisations are examples. The social culture in 2020 is more participative as well as more collaborative (Missikoff et al., 2010).

Convergence of Physical and Digital By 2020 portable cocktail identities allow and facilitate the expression of multi-faceted (virtual) personalities. Multi tasking as practiced by the younger generations already today converges concurrently the real (e.g. home, work, phone call) and virtual (web, TV). People who are used to multi tasking will not likely get rid of this habit. Boundless surfing blurs physical and digital borders between sites, (virtual) places, groups and communities.

Better immersion closes further the gap between physical and digital: information is presented in real-time modus in a way in which real persons, real places and real news play important roles. The users will interact with this information in the same way as they interact today in a multi-player online role play game. Knowledge sharing and other serious activities will be facilitated by game-like interaction.

Care robots (developed as solutions for the diminishing population and problems of an ageing society) are the first evidence of things behaving like humans. This will change our perceptions of what is 'human' and what is not.

Due to lack of better wording and as hyper-real is taken; we label this convergence phenomenon *Über-Reality*.

4.2 Social Media Provide the Non-intrusive Ad

The most important driver, due to spreading social media, is semantic knowledge in semantic profiles. Semantics increase the convergence of (i.e. affinity between) content, ads and user interests. Other main drivers are mentioned in the social values dimension in Sect. 4.1.

4.3 User and Audience Behaviour Does Not Converge

While the baby boomers are still significant in numbers, generation Y and millenniums dominate the cutting edge media field and will not likely return to any earlier, narrower, meanings of media. Quality media will be demanded as much as today, but users co-create more content than today. Users engage automatic search bots and knowledge curators, but also allow packaged, 'walled garden' media. Thus we anticipate a significant degree of behavioural divergence.

4.4 User and Audience Interfaces Become Intuitive and Ambient

Navigation by 2020 is more image based, easier, more intuitive and less dependent on language and keywords. Augmented reality is commonly used in mobile devices; interfaces expand to advanced (bio) sensors; voice and gesture recognition allows searching of videos. Media interfaces are ubiquitous—media surrounds and accompanies people everywhere: as digital, location aware billboards, decoration media, including interactive wallpapers; even 'passive media' has an interactive channel available; mobile gadgets are aware of anything the user wants to stay up-to-date with. Libraries become information integration and sharing centres with personal, virtual librarians. In conclusion we anticipate 'ambient convergence': media interfaces integrated into their environment, in big and small, virtual and real scales.

4.5 Diverging Media Content

Segmentation (i.e. content divergence) plays a key role, and consequently the role of general, mass media is less significant by 2020. In parallel, there will be a clear division between freely accessible standardized news information and paid for, high quality information services. More services based on publicly available open data, including maps and statistics of all kinds, appear.

4.6 Professional Journalism

The key issue is that professional journalists need to manage the information flow, analyze it fast, tutor or guide the citizen journalist, develop the work in different professional-amateur constellations, and initiate user content co-creation and ultimately crowd sourcing campaigns.

4.7 Gaming and Working Converge

The Real World Takes on Games Logic By 2020 political discourse, group working, simulations, decision-making and open innovation systems will all be influenced by lessons learnt from game design and game interaction. Educational content, too, will be augmented with playful interactivity. This simply means that what we conceive as reality will become gradually more game-like.

Games as such will be conceptualized more widely than today through a wide variety of playful and interactive contents. The largest changes will happen in video and book-like content, which will contain interactive components that are part of the original design and story.

The diversity in games is reflected in game interfaces, which vary from advanced sensors to voice and gesture recognition. The sensor systems will read various bio-signals more easily and comfortably and enable new types of game and interaction adaptations and biofeedback for increased self-awareness.

Games Discover the Real World We reckon that virtual worlds and mirror worlds, as well as augmented reality and lifelogging as outlined in the Metaverse Roadmap (Smart, Cascio, & Paffendorf, 2007) will strongly converge. The reasons are as follows:

1. New online games and virtual worlds will increasingly take the physical reality as such as their 'playground', e.g. a possible successor of *Second Life* or *World of Warcraft* will be located in real world places. Real authors were quoted in *Alan Wake* and real brands were used already by 2010.
2. The integration of real-time news into such games and worlds will lead to an even stronger convergence with daily reality.
3. Game figures (avatars) will be rewarded if their owners prove that they accomplished a meaningful task in the real world.
4. This setting will be supported by devices allowing people to follow the game situation wherever they are (or move) through convincing augmented reality.

In consequence we assume that by 2030 (not yet 2020) *Google Earth* will be re-developed to assist such environments. Billions of sensors and millions of enthusiastic users will digitally mirror on distributed servers their local, physical environment until, what we call 'e-Earth', a federated, digital, online and partially real-time model of the globe, will be ready to serve as the ultimate immersive, virtual, mirror *and* game platform.

Note the business opportunities for real world physical places such as cafés and shops displaying game situations online, in real-time. Customers will certainly visit places with much 'action' and many avatars around. We know already today that communities used for movie sets successfully attract tourists. This will expand to game sets.

4.8 Media Companies and Their Features

Divergent Service Proposition and Delivery Users and audience will demand professional services whatever interactive media is in use or however sovereign the users will be in accessing information. It is unrealistic to delegate to search engines other than routine services.

For instance, the users will evaluate the quality of journalism by the experienced reliability, credibility, trustworthiness and transparency of the content production and editing processes. The common expectation is that there will always be a real demand for 'official' (trusted) news both for an elite audience willing to pay and for the masses expecting the content for free.

Human editors, professional journalists, and other professionals will be needed for 'fine-tuning' the contents for different customer segments. Professionalism includes new types of roles that indicate also new types of interaction with the users and audience. The roles of information gate keeper, filter and contextualiser are devoted to structuring information. Additionally needed roles are curator, who will decide what content is made available; activator, who will initiate user generated content and crowd sourcing campaigns; and customer intelligence officer, who investigates the needs of the user and audience groups as well as customer groups. New skills and competences such as the ability of making creative leaps in the analysis of information, ability to produce different types of journalism for different types of users and audience, and ability of creating stories and narratives around the subject in focus, will be needed.

However, the number of permanently employed journalists and professionals in 2020 will be dramatically less than some 10 years earlier. Journalists will work mainly as self-employed freelancers with project type assignments. Smaller organisational units set new requirements for permanent personnel. Since not everything can be done alone with a small amount of staff, specialisation and scoping will take place alongside with networking and collaborating with other content providers, also with non-professionals. At the same time, the demand for multi-skilled journalists working mobile (and thus remote) without support of the 'home-base' work community will increase.

Divergent Meaning of *Brand* As sharing and consuming of contents takes place within user networks, company and product brands gradually lose significance for the users. This trend will be strengthened if the users will be technically and legislatively (less IPR regulations) more empowered to have access and control over the contents by their own. An additional view on brand is whether it is established around a company or individual content providers such as journalists, bloggers etc. This will have clear implications on the possible business models and thus service production processes of the media company.

Divergent Meaning of *Content* Another feature deals with the nature of the content demanded in the future. Whereas one view emphasises that entertainment is taking over so called serious content, the other expects good quality, serious journalism to hold its position, even if only for ‘elite’ customers. In between is the third view supporting the idea that several different types of journalisms will co-exist. In an extreme view, new contents will not be professionally produced any more while the citizen generated content will be effectively re-used in the production.

4.9 Converging Technology

This dimension covers technology related drivers which are not yet mentioned in the dimensions above.

Devices Rather Converge They are typically mobile, as only large screens remain stationary. A typical mobile device knows its location, directions and its user’s biometric information. First notebooks with flexible 3D displays appear. All premium devices allow mobile access to ‘rich’ media content available nearly everywhere while the border between applications and communications is blurred. A case of convergence.

Tools Converge Tools for (citizen) journalism include mobile editing and audio/video recording while being capable of real time publishing via online services. From an industrial point of view, tools for automated planning of articles are broader and smarter. Automated editorial tools check the relevance and importance of news and validate the user generated news, with auto-summarization, headline generation, filtering and auto-generation of links.

Finally, printing presses allow printing of newspapers and magazines in various qualities—e.g. a typical tabloid format can easily be converted into an A4 sized glossy magazine. Local print on-demand supported with global digital publishing will serve widely dispersed, fragmented niche markets

5 Conclusions

Having an idea about the possible business drivers we investigate how to make use of this knowledge. We map the drivers vs. technology and meanings, translate the learnings into business concepts and, finally, suggest a tentative framework for organisational development. We emphasise the framework as a tangible tool for media companies to put the learnings of this chapter into practice.

5.1 Innovation Potential of the Business Drivers

Whereas the business drivers illustrate the future, we still need to synthesise their possible impacts on media businesses. Our chosen viewpoint is innovation. Thus our task is to map the nine business driver dimensions into areas of incremental and radical change, corresponding to low and high innovation opportunities. As a 'map' candidate we choose Verganti's (2009) design-driven innovation "technology epiphany" (ibid. 5). Whereas we replace 'design-driven' by 'system-driven', the map axes remain the same (see Fig. 3). Verganti distinguishes technology- and meaning-driven innovation and provides evidence that innovation scoring high on both (shown in the upper right hand quadrant) will lead to maximum success. Our candidates for radical change, both in technology and conveying new meanings to the users/audience, are *gaming* (see Sect. 4.7); *media companies* (see Sect. 4.8) and *social values* (see Sect. 4.1). We see gaming as *the* innovation driver for the media sector as games penetrate the real world and the real world starts to be 'gamified'. Both the meanings for humans as well as the involved technology by 2020 can be used for disruptive innovation, especially by the more conservative news media industry. With less emphasis on technology, this is also valid for *social values*. By 2020 they will have changed the meaning of life as such probably more than existing social media have done within the last 10 years. We admit we want to provoke with our choice of *media companies*. However, with the tiny incremental development steps we have witnessed in the past years, we anticipate plenty of opportunity for radical innovation—if company executives succeed in changing their staff's conservative mindset—a Goliath's task.

We have reserved the least innovation prone (lower left) quadrant for *user & audience behaviour*. With this choice we follow Verganti in his view that user studies—by definition—cannot lead to radical innovation. How many 'users' have requested the automobile, Facebook or Google maps *before* their invention?

We mapped the remaining dimensions according to their innovativeness—either towards more radical change in technology or meaning. See Fig. 3.

The following two subsections are dedicated especially to the dimension *media companies & their features*. The individual company's executives need to decide

Fig. 3 Map of business driver dimensions

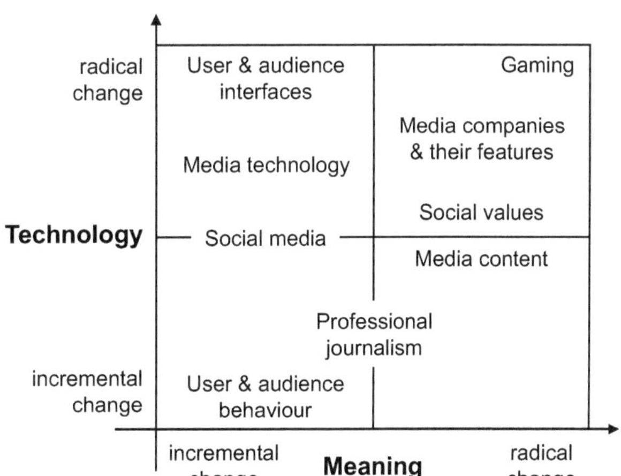

whether they want to position their firm as agile player in the niche markets as 'David' or whether they want to step into 'Goliath's' shoes.

5.2 Possible and Plausible Business Concepts by 2020

Based on the theoretical considerations of the need for divergent business models and the narrative material from the contributors we envision preliminary scenarios for the media companies that could enable them to diversify themselves in the local or global market. Here, our focus is especially on media companies' own visions and strategic choices, i.e. issues where the companies should be very proactive instead of being satisfied to be reactive.

Utilising Customer Intelligence Remarkably, even if the challenges of the media sector are as disruptive as presented in Anderson's view, media companies are still rather conservative in suggesting future, novel business concepts, i.e. how the sector will create genuine and competitive value for their shareholders and customers. Nevertheless, in our analysis we were able to recognise six types of core business concepts, representing a continuum from conventional, incremental business development to more radical, creative solutions, as following: new business through (1) providing targeted professional and user generated content to targeted audiences, (2) versioning content especially for mobile devices, (3) selling user information to a third party, (4) facilitating user's identity creation and expression—enabling cocktail identities, (5) providing services both for end users (individuals) and a third party (another business or non-profit organisation)—towards new meanings for the customer, and (6) providing content dependent devices and virtual environments for creative media use.

The first logical step when developing content business is transforming the traditional media from mass media (same content for many) to user centred media (content tailored to individual needs). User and audience groups must be much more effectively and accurately segmented. Accordingly, also the content could be offered with augmented reality or specific user specified attributes such as relaxation, dramatic details etc. even in the genre of news and narrative reporting.

Moreover, users and audience are expected to become more mobile, and maybe less consciously *consuming* media but rather *living* with media. This leads towards a more precise versioning of the content driven by the end user devices. These contain intelligent technologies that recognise the situation of the user.

Selling user information to a third party acknowledges the content as a means for gathering user and audience groups and valuable data about them, which can be merchandised further for various purposes, mainly for advertisers.

Facilitating users' identity creation and expression, for example by using their affiliation to brands, exploits the users' new social value—*you are what you share*. Supporting identity building provides possibilities for new revenues. The natural extension of this is to offer also additional identity consulting products and services.

Taking service business opportunities seriously shakes the industry (and its staff) considerably. A service is intangible, and basically an experience for the user. When the quality or performance of a media service is expressed with attributes such as immersion, the service providers need to know about their customers much more than the traditional reader statistics.

The most unconventional business concepts are those that are not content-centred, that concede the control to the user or to companies providing user interfaces and concentrate on devices, platforms and virtual environments, whether they be new hardware and software solutions or extended experience services for the customers. The core competence of media companies lies still firmly on their knowledge about their users and customers, and on the ability to gather this information. This knowledge will also attract collaboration partners outside of the media industry, which in turn facilitates the offering of extended services.

Earnings from Advertisers Advertising has traditionally been the most significant source of income for media companies. Advertising impact and users' experiences are evaluated with demand for more reliable and accurate methods. Advertising revenue losses are an evident consequence of being unable to provide value-adding services for the advertisers. Advertisers look for means to reach clearly defined target groups using, e.g. the possibilities of social media.

Advertising is related also to brands—and consequently to the capability to establish, maintain and capitalise on a brand. A brand is a platform to attract audience and thus advertisers. Advertisers are nevertheless confronted with the new social or cultural value of 'word of mouth' or viral marketing. This is connected to user or audience behaviour relying on trustworthy informants. Here, the media companies need to take on an active role of educating advertisers about the media users and their attributes.

Earnings from Users and Audience At least six different types of user payments were identified. Ordered from most conventional payments to more creative they are: (1) traditional income from subscription fees, advertisements, copy sales, access fees, and even some type of media tax; (2) micropayment based on actual, real-time consumption of media; (3) payment for on demand hyper-local contents; (4) payment for desired attributes of the content such as 'lean-back and relax', 'dying to know more' etc.; (5) payment for services such as filtering, organising, investigating information; and (6) payment only for the device or delivery platform for content receiving and sharing.

The user by 2020 is expecting to have more content for free. However, this trend can be complemented with micropayments. Examples could be payment by number of clicks, uploaded files or searches in archives. The first views are free but the more interesting, contextually tailored, content would need payment.

Adding local content to the information stream is, at least to a certain degree, also possible to being automated by using metadata. However, the more (hyper-) local we go, the more probable it is that also human, local knowledge is required. Locality or even hyper-locality offers chargeable value for the users. Combined with the micro-payments it would be possible to have targeted information just at the right time (occasion) and in the right place.

A more refined earning model is based on payments connected to the desired attributes of the content. This is already a familiar concept for customers when selecting a product for purchase, but still these payments could be developed more in line with the media experience that the mobile user is looking for in a particular moment and content.

Payment for attributes of the content is already very close to the idea of payment for the services. The first logical step seems to be to offer personal media assistance and content provision services. Media consumers will need guidance but also filtering of information and this is a clear service. We already discussed user identity creation as a service. Collecting and sharing revenues between different services and platforms will make the payment models for companies more complex, and as such this also represents a major opportunity for new technologies to emerge.

5.3 From Business Concepts to Organisational Development

A traditional way to develop new business models is to diversify or extrapolate from the current business status to a predicted or even unknown future with a set of questions such as *who will be our future customers*, *what will they value*, and *how will we deliver*? All these questions cover explicitly and implicitly not only organisational practices but relations to customers as well as partners and other stakeholders. We expect that companies able to answer these questions in a context of a holistic business ecosystem will eventually have a stronger market position, whether in a niche or a global market. As a first step of outlining a company's position in a business ecosystem, we present in Table 1 a 'translation' of the well known business model canvas elements into elements of a company's delivery and learning system.

Following Itami and Nishino (2010) argument that the people who are working in the companies to deliver, learn and realise the strategic intentions of the executives are the essence of a company's business model (see Fig. 1), it is not possible to neglect the significance of organisational development. As a practical conclusion, we visualise in Fig. 4 a general, business model based framework for organisational development linking the different elements of a business model as well as the description of the influence between the connected elements. Moreover, we associate a company's learning system with the 'activity perspective' of the framework, and the 'customer perspective' with delivery system. In the centre of all activities are the value propositions.

The activity perspective describes how the value proposition is produced. It consists of a company's key activities (tasks, work flows), company job descriptions and roles of partner network members, and key resources (competences, tools and processes). The value proposition production requires

Table 1 Mapping of the business model canvas (Osterwalder et al., 2010) to organisational development aspects in a company's business system

Business model canvas field	From business model to organisational development	Business system: delivery and learning system aspects
Value propositions	The value we provide	Value propositions
Key partners	Who we work with	Roles, job descriptions
Key activities	What we need to do	Tasks, work flows
Key resources	What kind of resources we need	Competences, tools, processes
Customer relationships	How we relate to target groups	Service experiences
Channels	How we reach the target groups	Distribution channels
Customer segments	What are our target groups and stakeholders and their needs	Users, co-creators, advertisers
Cost structure	Activity based costs	Cost structure
Revenue streams	Delivery based revenues	Revenue streams

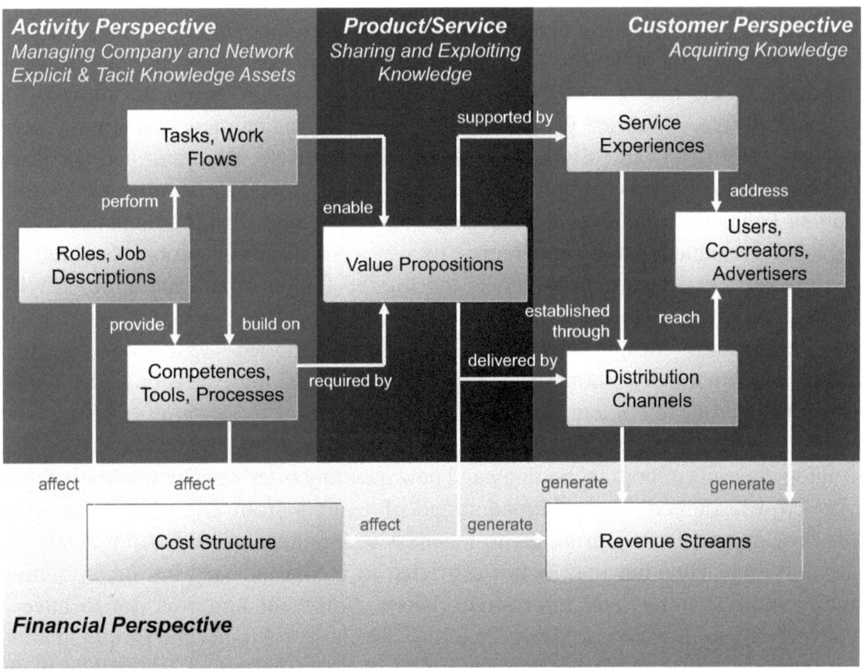

Fig. 4 A framework for organisational development based on the business model canvas. http://www.businessmodelgeneration.com/downloads/business_model_canvas_poster.pdf

from a company or a business ecosystem effective and efficient leadership and management of the knowledge assets.

The customer perspective describes how the value is consumed. This consists of service experience as the basis of customer relationships, users, co-creators and advertisers as customer segments, and distribution channels as means of delivering the service. Meaningful value proposition requires from a company or a business ecosystem effective and efficient acquisition and utilisation of knowledge about service users.

The financial perspective considers what inherent costs are the most significant for the business system; what competences, tools and processes are the least cost efficient; what tasks and workflows are the least productive; which roles are the least/most important in the business system and what will be the consequences of the choices that are made.

6 Discussion

Our Chapter addressed the concepts of system driven innovation, business system and organisational development in the purpose of provoking a more holistic approach to the media sector's value propositions. Future media business is, in essence, divergent service provision that—based on radical new converging technology—will offer radically new meanings for the users. We identified the three most powerful business driver dimensions—social values as framing the overall human behaviour, gaming as extending the reality, and media companies themselves. We concluded that convergence may be a benefit for the global 'Goliaths' in the market. Divergence, however, is the lifeline for the small and agile 'David' companies whether operating locally or globally. Moreover, we concluded that the business models—despite how well they may be designed—will never be effectively implemented without respective organisational repercussions and actual organisational development actions.

We selected our approach to be holistic, embracing divergence and convergence from viewpoints of both technology and new meaning offering. Our qualitative data was crowd sourced from a broad range of experts from engineering to work psychology and from journalism to business economics, active both in academia and industry. Thus our data is not restricted to media industry but incorporating future societal trends and innovative visions. Since our aim was not to have a converged, top-down, view about the future, we chose not to use the Delphi method for our data gathering. Our interest was first in gathering the most possible divergent descriptions and mini scenarios and then analysing them from a viewpoint of media business drivers.

Our theoretical suggestion in the context of media sector is to focus on business concepts based on a system approach. If business concepts are developed by establishing and nurturing network relations with key stakeholders, users and audience, and (advertisement) customers the possibility to combine new technologies and new meanings into an innovative business epiphany will increase.

As a practical implication our main claim is that the business concepts in use have a strong impact on core business processes and the entire organisational system with its technology, work tasks and processes, as well as personnel and their roles and competences. There is a threat that by betting on the wrong business concept, a company will fail to develop. Generally, media companies will need to consider their value propositions much more carefully. When growing into the size of 'Goliath' the mental distance from a company's daily activities to customers is in danger to grow too much. Once the focus on meanings for customers is lost, the picture of the value propositions becomes very vague if not destructive for the company.

We recognise also limitations due to our approach. We collated our data from 27 Finnish and eleven international contributors from all continents. Still, the relatively small number of contributors is counterbalanced with the variety of stakeholder groups their represent and the rich qualitative data they provided. When analysing the data and developing future business concepts, we needed to maintain a delicate balance between speculation and conservatism. Speculation can be counterbalanced by, e.g. good reasoning and convincing narratives. Conservative views are mostly based on vague fears of the future. Academic research as such encourages a certain degree of conservatism, with its inevitable focus on the state of the art.

Moreover, we acknowledge that other researchers may arrive at different interpretations than we did. Still, we are confident that concentrating on learning from gaming, trends in social values and organisational development are well reasoned themes for future research for media business development.

Acknowledgments We thank the two anonymous reviewers for their constructive comments. We cordially thank all the future scenarios contributors and the Vision 2020 team, especially Man-Sze Li, Esa Sirkkunen and Kai Kuikkaniemi. This work was partially supported by the Finnish Funding Agency for Technology and Innovation TEKES as part of the Tivit Next Media programme. A preliminary version was presented at the Third NEM Summit, 2010.

References

Anderson, C. (2009). *Free: The future of a radical price*. New York: Hyperion.

Banathy, B. H. (1998). Evolution guided by design: A systems perspective. *Systems Research and Behavioral Science, 15*(3), 161–172.

Blackler, F. (1995). Knowledge, knowledge work and organizations: An overview and interpretation. *Organization Studies, 16*(6), 1021–1046.

Bontis, N. (1998). Intellectual capital: An exploratory study that develops measures and models. *Management Decision, 36*(2), 63–76.

Casadesus-Masanell, R., & Ricart, J. E. (2011). How to design a winning business model. *Harvard Business Review, 89*(1/2), 100–107.

EIFFEL. (2009). *Think tank on future internet 2009 report*. www.fp7-eiffel.eu

Fritscher, B., & Pigneur, Y. (2010). Supporting business model modelling: A compromise between creativity and constraints. In P. Palanque, J. Vanderdonckt, & P. Wild (Eds.), *England D* (pp. 28–43). Springer Berlin Heidelberg: Task models and diagrams for user interface design.

Giesecke, R., & Immonen, S. (2010). *Media sector business drivers in 2020*. Paper presented at the Third NEM Summit.
Giesecke, R., Immonen, S., Sirkkunen, E., Grenman, K., Bäck, A., Helle, M., et al. (2010). *Media Scenarios 2020 Version 2*. Helsinki: Tivit Next Media Programme.
Hipp, C. (1999). Knowledge-intensive business services in the new mode of knowledge production. *AI & Society, 13*(1), 88–106.
Itami, H., & Nishino, K. (2010). Killing two birds with one stone. Profit for now and learning for the future. *Long Range Planning, 43*, 364–369.
Matthyssens, P., & Vandenbempt, K. (2003). Cognition-in-context: Reorienting research in business market strategy. *The Journal of Business and Industrial Marketing, 18*, 595–606.
Missikoff, M., Drissi, S., Giesecke, R., Grilo, A., Li, M.-S., Mehandjiev, N., et al. (Eds.). (2010). *Future internet enterprise systems (FinES) cluster. Research roadmap. Final report*. Brussels: European Commission, European Communities.
Nahapiet, J., & Ghoshal, S. (1998). Social capital, intellectual capital, and the organizational advantage. *Academy of Management Review, 23*(2), 242–266.
Osterwalder, A., Pigneur, Y., & Clark, T. (2010). *Business model generation : A handbook for visionaries, game changers, and challengers*. Hoboken, NJ: Wiley.
Remenyi, D., Williams, B., Money, A., & Swartz, E. (1998). *Doing research in business and management: An introduction to process and method*. London: Sage Publications Ltd.
Smart, J. M., Cascio, J., & Pappendorf, J. (2007). *Metaverse roadmap overview*. Retrieved January, 2013, from http://www.metaverseroadmap.org/overview/
Tameling, K., & Broersma, M. (2013). De-converging the newsroom: Strategies for newsroom change and their influence on journalism practice. *The International Communication Gazette, 75*(1), 19–34.
Verganti, R. (2009). *Design-driven innovation: Changing the rules of competition by radically innovating what things mean*. Boston, MA: Harvard Business Press.
Whitman, J. R. (2009). Measuring social values in philanthropic foundations. *Non-profit Man-agement and Leadership, 19*(3), 305–325.
Zheng, H., Fu, Y., & Lugmayr, A. (2012). The development of ambient TV in 2016. *Proceeding of the 16th International Academic MindTrek Conference*, pp. 233–236.

Media Convergence is *NOT* King: The Triadic Phenomenon of Media "Convergence-Divergence-Coexistence" *IS* King

Artur Lugmayr and Cinzia Dal Zotto

1 Introduction

Convergence is a heavily discussed term in the media field today. Where many marketing departments are tagging almost any new digital activity to 'convergence', the underlying meaning of this term in a larger context is rather poorly understood. The multiplicity of uses of the word "convergence" already creates diverging views, conceptualizations and understandings (Arango-Forero, Roncallo-Dow, & Uribe-Jongbloed, 2015). With the introduction of digital TV in Europe, one single standard called DVB was ought to emerge and convergence was to bundle all Internet based services on one single set-top-box. Even radio was ought to be digitalized under a single standard called DAB to provide the consumer with a single platform in enjoying digital services. Reality is far away. Consumers diverge and use multiple platforms such as mobile TV, IP-TV, or Internet TV services to enjoy their interactive experiences. Marketing departments brand this service eco-system as convergence; however, this never really happened as consumers diverge in their use, habits and perceptions.

Discussing convergence or divergence within a media industry context requires a more general and abstract discussion about the evolution of media and its principles. For this purpose we extend our reflection to a larger context, and consider evolution from a biological point of view. In biology evolution is understood as a "process of continuous change from a lower, simpler or worse to a higher, more complex or better

A. Lugmayr (✉)
Visualisation and Interactive Media, School of Media, Culture, and Creative Arts (MCCA), Curtin University, Bentley/Perth, Western Australia, Australia
e-mail: lartur@acm.org; artur.lugmayr@lugymedia.com

C. Dal Zotto
Academy of Journalism and Media, Faculty of Economics and Business, University of Neuchâtel, Neuchâtel, Switzerland
e-mail: cinzia.dalzotto@unine.ch

A. Lugmayr, C. Dal Zotto (eds.), *Media Convergence Handbook - Vol. 2*, Media Business and Innovation, DOI 10.1007/978-3-642-54487-3_21

state" (Merriam-Webster, http://www.merriam-webster.com/dictionary/). Evolution theory tries to explain all phenomena related to the development of species, and how these species reached their current state. Two main theories attempt to explain the evolution process (Wikipedia, http://en.wikipedia.org/): The first theory (*convergence theory*) assumes, that the present state of a specie was determined, and was bound to develop as it is today. Additional features of individual species developed independently and had to emerge due to natural preconditions (e.g., wings of species had to develop due to the existence of air). The second theory (*contingency theory*) assumes that the present state was determined by randomness and particular contingent events. Thus, new features of species would emerge accidentally, and the present state would not be replicable, even if the same pre-conditions would exist. No matter which theory better describes the phenomenon, sound investigations confirm that the biological evolution led to (Wikipedia, http://en.wikipedia.org/) processes of adaptation, co-evolution, cooperation, specialization, and extinction.

Comparing media evolution with the process of biological evolution seems now to be far-fetched. Nevertheless, we believe it is an excellent starting point for gaining more insights and understanding media industry developments. As following step, we thus link the idea of biological evolution to the world of media, and in particular to the phenomena of convergence, divergence, and coexistence. Indeed, each new form of media has a new set of features that either develops independently or in dependence to other media forms. According to McLuhan's law of retrieval (McLuhan, 1994; McLuhan & Lapham, 1994), each new media retrieves features from an old media. For instance, the World Wide Web inherited features from textual and visual media in form of web pages. Following this argumentation, our main consideration is—can we consider this media evolution as a matter of convergence, divergence, or coexistence? Are we dealing with a parallel development of media, where each media form develops similar additional features independently of other media? Are we dealing with a convergence evolution process, where each new media form is inheriting certain features from other existing media forms? Or is media evolution a matter of contingency, according to which new media forms emerge accidentally? Or is convergence and divergence a phenomenon on markets how new media environments are perceived by customers, following the argumentation in Montpetit (2015), where convergence "is disrupting [and] engendered dislocation and divergence in the way next generation media dissemination and consumption are perceived and marketed"?

As mentioned above, no matter if the root of evolution is convergence or contingency, from a biological perspective the process of evolution leads to adaptation, co-evolution, cooperation, specialization, and extinction (Wikipedia, http://en.wikipedia.org/). We can find numerous examples for these phenomena within the media industry: considering the history of mass-communication and media the biological phenomenon of *specialization* well resembles that of fragmentation, which characterized the development of mass-media towards niche media (see e.g., Kueng, 2008). TV content evolved from fixed program schedules to video on demand options; TV content, if previously delivered through a few channels only to masses, now reaches infinite niche audiences by delivering highly specialized content. The notion of *adaptation* can be compared to the needs of media to adapt to technology developments. Newspapers are finally adapting and moving from print to online publishing, allowing micropayments and online subscription models. *Cooperation* is clearly identifiable with the merger and acquisition activities taking place to allow firms to compete within a changing media landscape or to gain knowledge about how to successfully operate in the digital domain (Lugmayr & Dal Zotto, 2013; Dal Zotto, Galichet, & Lugmayr, 2011). Traditional media houses that are too slow in adapting to the new digital environment might face *extinction*, as Manroland and Schlott in Germany or Border Books in the USA well show. And we could enlist many more examples at this place. Media are currently experiencing a rather complex evolutionary step in their development, a step which was initialized by the advent of digital technologies and the Internet and of which convergence is just one single aspect. Adaptation, specialization, co-evolution and extinction of media are phenomena resulting not only from media convergence but also from divergence and coexistence processes, as depicted in Fig. 1.

The aim of this book chapter is to analyze all contributions to the two volumes of this edited handbook, and extract the trendsetting ideas within the media convergence research field. On the basis of the chapters, we will thus discuss examples of convergence, coexistence and divergence processes at different levels. For doing so we will proceed as follows. First of all, we will describe the method that we used to detect the most frequently discussed topics within the book chapters, and thus to create topic clusters. Then, we will continue with an in depth analysis of the sub-themes that within those clusters mostly reflect phenomena of media evolution and of the interplay between media convergence, divergence and coexistence processes. In particular the tag cloud produced from all book chapters (presented in Fig. 2) demonstrates the one sided approach in research in dealing with the interplay between these three processes. This emphasizes the fact, that scholars need to address more research in all three directions rather than focusing only on investigating one. Finally, an overall discussion will conclude the chapter, raise new questions and open the horizon to further research in the field.

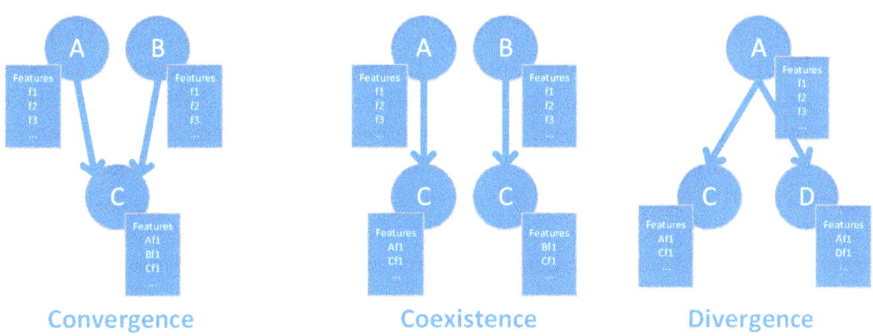

Fig. 1 Convergence, coexistence, divergence: the three processes characterizing the evolution of media

Fig. 2 The triad of media progress: convergence–coexistence–divergence

2 Methodology and Approach

All 38 book chapters of the two edited volumes have been examined within the scope of this study (please see Ala-Fossi, 2015; Arango-Forero et al., 2015; Damasio, Henriques, Teixeira-Botelho, & Dias, 2015; Deliyannis, 2015;Denward, 2015; Dey, 2015; Duh, Meznaric, & Korošak, 2015; Ebner, Schon, & Alimucaj, 2015; Fleury, 2015; Foster & Brostoff, 2015;Gershon, 2015; Georgiades, 2015; Giesecke, 2015; Grüblbauer & Haric, 2015; Hamacher, 2015; Ibrus, 2015; Innocenti, Pescatore, & Rosati, 2015; Immonen & Giesecke, 2015; Karmasin, Diehl, & Koinig, 2015; Leminen, Huhtala, Rajahonka, & Westerlund, 2015; Matteucci, 2015; McKay, 2015; Montpetit, 2015; Mustic & Plenkovic, 2015; NS, Varghese, Pal, & NK, 2015; Reichl, 2015; Sabiha, Morrison, Annika, & Obal, 2015; Sacco, 2015; Sousa, de Lemos Martins, & Zagalo, 2015; Spyridou & Veglis, 2015; Svahn, Wahlund, Denward, & Rademaker, 2015; Tuomi, 2015; Veglis, Dimoulas, & Kalliris, 2015; Villi, Matikainen, & Khaldarova, 2015; Vukanovic, 2015; Wellbrock, 2015; Tavares & Schofield, 2015; Zlatanov & Koleva, 2015). Together with the literature review that we have incorporated in the introduction to the edited volume and to this chapter, the corpus of the 38 handbook chapters provided us with sufficient material to identify key discussion clusters within the media convergence research field and to gain deeper understanding of the thematic area to conduct:

- investigation of several contributed book chapters to understand the underlying phenomena, driving forces, and their impact on the transformation of media industry;
- contribution with a new theory to the field of media studies, in particular how convergence-divergence-coexistence implicates in the field of media development;
- understanding traditional perception of media business, and the impact on business emerging from new media ecosystems;
- raising questions about current discussions in research, and pinpointing to areas that are rather scarcely emphasized.

To conduct the study, we aimed at an exploratory process, where we emphasized quantitative methods based on text mining techniques. We adopted the methods presented in Jockers (2014) and Williams, G. Available: http://handsondatascience.com/, resulting into the process illustrated in Fig. 3. Our main tool for conducting the study was the statistical software package R which includes various text mining packages such as tm, NLP, LDA, and worldcloud (R-Project, http://www.r-project.org). We utilized R including its text mining packages in combination with the text mining software package KHCoder (Coder, K. Available: http://khc.sourceforge.net/en/) to simplify the execution of the study. We also experimented with the RQDA package of R, but conducted the qualitative analysis thereafter based on *Keyword in Context Concordance* (*KWIC*) (see Jockers, 2014; Williams, G. Available: http://handsondatascience.com/) searches to speed up the process. Our basic data material was represented by the 38 book chapters contained in the two edited volumes. The analysis of the book chapters and the review of existing

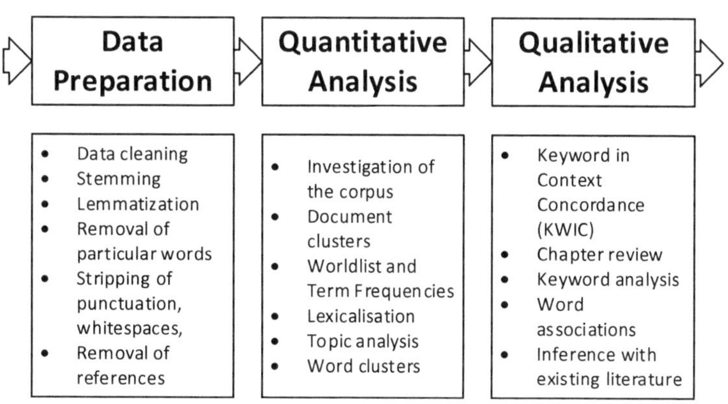

Fig. 3 Methodology and approach

literature in the field of convergence allowed us to synthetize at the end of this final chapter a new viewpoint on this thematic area.

Data Collection and Preparation The corpus of the study is based on the 38 book chapters, which provided us with sufficient material to perform the study and gain a deeper understanding of the thematic area. In the data preparation phase, we performed low level steps such as conversion to machine readable formats, removing special characters and references from each paper. Particular words, that should not be part of the analysis process, have also been removed from the corpus. For the experiments conducted in R, we performed essential corpus pre-preparation tasks such as stripping whitespaces, punctuations, and non-alphanumeric and numeric characters. After some experimentation in R, we decided to avoid stemming, as it led to some misleading results at a later stage of the study. While using KHCoder we relied on the preparation functions of this tool, and the lemmatization function via the Stanford POS tagger in particular became a very handy solution.

Quantitative Analysis This analysis focused on the investigation of the corpus through the generation of word clusters, word frequency lists, the term frequency distribution, the document frequency distribution, TF-DF plot, word association calculations, cluster analysis, and an LDA analysis. The methods applied for the quantitative analysis are based on descriptions in Jockers (2014) and Williams, G. Available: http://handsondatascience.com/. The results of the quantitative analysis are presented in the appendix of this book chapter. These results were analyzed, and led was the basis of the qualitative analysis.

Qualitative Analysis The main tool for the qualitative analysis was the keyword in context (KWIC) technique, according to which we analyzed the context of the keywords *convergence, coexistence,* and *divergence*. In this part of the analysis, we tried to understand the different key-ideas brought forward by the different chapters which we summarized in the following sections of this book chapter.

3 The Dynamics of Media Evolution, or the "Convergence–Divergence–Coexistence" Triad

As a result of our analysis and with reference to the content which the book chapters have been based on, we defined six main topics. Within the scope of the identified topics, we further focused on a few sub-themes more clearly reflecting the media evolution phenomena discussed within the book chapters, as illustrated in Table 1.

In the following subsections the identified sub-themes are extensively discussed through direct references to the related book chapters.

Table 1 Topics and themes across the contributed book chapters

Themes	Discussion topics	Chapter numbers and references
Business, market, and quality	Divergence of traditional media firms' perception of business, and the emergence of convergent media ecosystems	Wellbrock (2015), Ebner et al. (2015), Matteucci (2015), Karmasin et al. (2015), Vukanovic (2015), Leminen et al. (2015), Reichl (2015), Immonen and Giesecke (2015)
Social, interaction and networking	New modes of interaction and media consumption	Arango-Forero et al. (2015), Gershon (2015), Foster and Brostoff (2015), Tavares and Schofield (2015), NS et al., (2015), McKay (2015), Sabiha et al. (2015)
Content and technology	The "convergence divide"—convergence through divergence, divergence through convergence	Montpetit (2015), Veglis et al. (2015), Deliyannis (2015), Tuomi (2015), Damasio et al. (2015), Fleury (2015)
Media form and audience	Emergence of new media forms: transmedia, social media, ambient media, and the challenges for advertising	Arango-Forero et al. (2015), Tuomi, (2015), Villi et al. (2015), Denward (2015), Sousa et al. (2015), Innocenti et al. (2015), Ibrus, (2015), Sabiha et al. (2015), Svahn et al. (2015), Sacco, (2015)
Information, intelligence and networks	Effects on communication and information processing: data, information, knowledge, and networks of communication	Hamacher (2015), Mustic and Plenkovic (2015), Duh et al. (2015), Grüblbauer and Haric (2015), Giesecke (2015), Immonen and Giesecke (2015)
Media organizations and profession	Driving forces, power, and coexistence of evolution of traditional policies and media	Dey (2015), Spyridou and Veglis (2015), Ala-Fossi (2015), Zlatanov and Koleva (2015), Georgiades (2015)

3.1 Divergence of Traditional Media Firms' Perception of Business, and the Emergence of Convergent Media Ecosystems

Today's media industry requires a rethinking of existing managerial and organizational models. Organizational practices within media organizations seem to diverge and thus not to be appropriate for the reality in which media firms are currently operating. As Karmasin et al. (2015) states, there is a "need [for convergence mastering organizations and collaboration oriented management competences. This requires] that companies [...] not only have to understand [...] changes in society, consumer behavior and business models, but [...] also [...] have to be able to react appropriately and proactively". The divergence between managerial strategies and business reality appears to be a major dilemma for the media industry, engaged since more than a decade in a structural transformation process. This dilemma has been well discussed in Vukanovic (2015). The paper underlines the need for media companies to embrace emerging technologies and recognize that media divergence can "successfully perform as vendor lock-in, top-down corporate process and a bottom-up consumer driven process". On the same path Leminen et al. (2015) points out that the divergence between media strategies and business reality increases if existing business models are not abandoned and new models introduced into the market. According to Leminen et al. (2015) the rigid belief system of traditional media firms prevents them to innovate and develop their business models according to industry and technological changes. New entrants coming from other industries foster cross-industrial convergence as well as industry specific convergence of different business models elements such as technology platforms, revenue logic, content production and value creation as the publishing industry case shows.

A market driven business model convergence can though have a reverse side and lead to journalistic quality divergence of content in traditional and online media (Wellbrock, 2015). The authors of this chapter argue that, because advertising represents their main source of income, online news media are acting as "audience production plants" so that serious biases with regard to high quality content and minority issues might arise. An extension of the activities of public service broadcasters to the online market is suggested as a possible solution to regulate that market, and stimulate news media to engage in fulfilling their public service mission. This is a clear argument driving all news media to converge in ensuring their watchdog and democracy enhancing function.

The case of traditional Television operators seems to minimize the divergence between their perception and the reality of the business situation. TV operators have acknowledged rather early the incoming Internet threat and have been working hard, in several domains to prepare themselves and accommodate convergence while exploiting their TV market incumbency. As Matteucci (2015) concludes, despite a high technological turbulence established TV and telecom operators strive to keep market shares through walled garden solutions and by lobbying for stronger IPR protection. However, audience trends signal already a new storm coming with online TV viewing progressing at high rates and at the expenses of traditional TV consumption. Standardization and regulation have thus been enabling technological and business convergence for a while, but they might not be sufficient to prevent a diverging business trend with the emergence of hybrid media forms in the future.

A media ecosystem where convergence and divergence coexist seems thus an emerging reality within the media industry. Nevertheless, how the industry will cope with the dynamics of such an evolution from a management point of view still remains an open question. A cross-disciplinary approach appears to be the starting point to tackle this issue, and for this purpose the concept of communication ecosystems has been introduced by Reichl (2015). According to the authors that concept represents a multidisciplinary framework to simultaneously address convergence issues from an economic, technical and user perspective. By opening up communication with users and taking into consideration their input, the quality of user experience can be improved and problems of divergence at consumption level might be overcome.

The importance of establishing and nurturing network relations with key stakeholders and in particular with users and customers is underlined also in Immonen and Giesecke (2015). Such relations would increase the possibility to combine new technologies and new meanings into innovative business product ideas. Following a holistic approach the paper claims that applied business concepts have a direct impact on the entire organizational system including the technologies in use, tasks, processes as well as staff roles and competences. Clearly in line with classic organizational development theories, Immonen and Giesecke (2015) points out that losing the customer focus—thus diverging from the customer view—might blur the value proposition picture and bring a company to develop in a wrong if not destructive direction. The impact of customer oriented applications of new technology on business models has been addressed and discussed also in Ebner et al. (2015): the possibility of printing a personalized scientific book is seen as a solution to bridge the gap between open access and printed journals, and thus as a further step towards media convergence. This again requires organizational forms to adapt accordingly and develop into network like converging structures, as stated in Gershon (2015). A cross-disciplinary discussion seems to be appropriate to better understand audiovisual media production, and in particular serial narratives, where planning and practice diverge (Innocenti et al., 2015).

3.2 New Modes of Interaction and Media Consumption

The way how we interact with media and information technologies has been greatly developing and will continue to evolve. If technological convergence is allowing this development, along this path we can already see a divergence process emerging as always more specific and individualized solutions appear with regard to spaces, devices, and modalities in which interaction takes place. One example of divergence in this field has been discussed in Foster and Brostoff (2015), where the authors analyze the application of media interaction in game design and urban planning, two distinct areas which though share the goal of building up a cohesive and user-friendly visual narrative about alternative realities. The dynamics of media evolution extends the impact of interaction modes' development towards other domains, such as urban planning, and considers for instance how people interact with objects in urban spaces. Divergence effects are clearly visible in interaction design and user experience both in urban planning and game design. More and more multisensory devices stimulate human senses, it is thus essential to develop strategies which allow humans to interact with technology in a multimodal way and to "design systems that are sensitive to what the user wants without having been given explicit commands" (NS et al., 2015). As a consequence, video game simulations might be applied to urban planning and urban planning data might enrich game design. As authors state in Foster and Brostoff (2015), it may only be a matter of time for "real world urban planning and game simulation to converge to the point where both are subsets of some larger field of study". As embedded media are diverging, the development of user interfaces will help to cope with the personal and divergent media stimuli. A solution might be multisensory interaction design, as addressed in Tavares and Schofield (2015). This chapter enhances the need for user interfaces to be designed taking "plurality, adaptability, and cognitive ability aspects" into consideration (Tavares & Schofield, 2015) so to help the consumer cope with a diverging media landscape.

A quite far more complex issue is information system design—in particular the information system design within media industries. The question is to which extent information systems at corporate level can be designed to fulfil the needs of user experience and new media consumption modes, while capturing the tacit intellectual capital within knowledge intensive firms. This issue has been addressed in McKay (2015). If we further consider, as we have already discussed, that for the media industry the key product is still content and convergence leads to the emergence of new forms of media such as transmedia storytelling, then corporate information systems should be designed to support those new media forms as well as the interaction and user-friendly modalities they require as e.g., described in Sabiha et al. (2015).

The change of media consumption behavior is critical issue in the evolution of media, to which chapter Hamacher (2015) devotes particular attention by questioning if traditional media houses are targeting a specific selected audience through their content offerings. Besides presenting an excellent methodology to analyze textual online content, Hamacher (2015) presents a solution for media houses to optimize their portfolios to target particular audience niches.

3.3 "The Convergence Divide": Divergence through Convergence, Convergence Through Divergence

The evolution of media has brought about convergence processes. However, as we have already mentioned above, simultaneous divergence processes can be detected and the issue of a potential "convergence divide" has emerged. Looking at evolution of technology, McLuhan already emphasized (McLuhan, 1994) that technological development has clear social implications. Thus, when discussing media convergence, discussing if technology convergence may or may not lead to social convergence seems to be inevitable. This illustrates the fact that convergence on one level does not essentially lead to convergence on another level. Within the scope of this section, a few examples shall illustrate the "convergence divide", which has been underlined by many chapters in this book. In particular, we recall here the tension arising between "technological convergence" and "social divergence" in mobile Internet discussed in Damasio et al. (2015).

The issue of a convergence divide is especially visible when considering convergence at technology level leading to divergence at cultural and economic level. Such a divergence process is evident in the field of networking, which has been excellently described in Montpetit (2015): the cornerstone of the Internet is network neutrality—each user on the Internet is treated equally, as well as all her/his traffic. Thus, the Internet is a place where traffic is "not managed" and coordinated by a non-central authority. With network providers pushing towards new digital payment models, and the idea of prioritizing consumer traffic related to online paying services, network neutrality seems to be an issue of the past, while the time of managed internet services has come. As stated in Montpetit (2015), Netflix, Instagram, Amazon, Hulu, and Vines are applications that are built on a principle of network neutrality. However, for other services such as IPTV, TV interactive service offerings, and mobile phone TV services the "friction between managed and unmanaged services will continue" (Montpetit, 2015). Even if we know today that on February 26 the US Federal Communication Commission has approved a law enforcing net neutrality, we also know that in the rest of the world the net neutrality issue is still open. Thus, this book chapter nicely demonstrates a good example of convergence divide showing that, while technologies converge, culture and business models are diverging. Thus, convergence on one layer (technology) is leading to divergence on another (social).

An interesting aspect of convergence has been discussed in Deliyannis (2015) where the author states that "interactive TV broadcasting provides [...] backward compatibility that avoids divergence". Following this argumentation, media consumption does not essentially diverge based on interactive content; in fact, content with novel interaction requirements may act as a driver for convergence through the development of interactive television systems featuring advanced interactive and presentation methods. In this example, divergence may thus lead to convergence. A similar line of argumentation has been followed in Tuomi (2015), where this relation has been examined through the example of television and mobile devices acting as second screens and facilitating television watching experiences.

A counter example, convergence as a result of divergence, has been presented in Mustic and Plenkovic (2015). Here the authors discuss the diverging requirements of consumers in terms of visual presentation to increase information credibility, and show how the use of convergent graphic media represents a convergent answer to consumers' diverging perspectives. A discussion around the thematic area of digital TV, which has been conducted in Matteucci (2015), highlights again the interrelation between technological convergence and divergence. The chapter argues that technological standardization led to technical convergence on one hand; however, it might have contributed to the development of different platforms and thus to divergence on the other hand. Technological standardization could thus be considered also as enabler of diverging consumers. This thesis is sustained also in Ala-Fossi (2015) where the convergence divide has been exhaustively discussed with reference to television broadcasting. According to the authors, the development of standardized television platforms did not essentially lead to convergence on other levels. On the contrary, with the digitalization of terrestrial television an increased divergence at content production, distribution and consumption level—rather than convergence—could be observed. Even the number of terrestrial broadcast standards increased from three to five, and with it—following Ala-Fossi (2015)—the consumer divergence.

This phenomenon has also been illustrated in Karmasin et al. (2015), where the authors state that "while content is used on multiple platforms [. . .] and convergent, it takes more devices to display it", meaning that technology is diverging. To cope with these developments, and allow media organizations to apply the latest technology in distributing their content across platforms, a coherent organizational model for cross-media publishing is required. A framework solution for creating such a model has been presented in Veglis et al. (2015).

To conclude the discussion about the interrelationship and mutual influence of convergence and divergence processes, maybe a completely different approach is required to cope with the convergence divide. A user centered approach is proposed in Fleury (2015), in which authors ask themselves if it would not make more sense to find a solution to understand and meet end users' needs to benefit from the evolution of media rather than to statically analyze phenomena of convergence divide. Thus, "how [is it possible] to understand and support end users [. . .] in making sense and benefiting from the evolution [of] media" (Fleury, 2015). Taking the example of convergence between television and mobile technology, the authors show how media consumption is becoming increasingly tight to individual preferences and activities, while media consumers are still bound to content providers as communication channel. Challenges that need to be tackled here are thus mostly of usability nature such as content synchronization across devices which is relevant both for end users and broadcasters. Eventually, a coherent model for cross-media publishing is required to allow media organizations to utilize latest technology in distributing their content across platforms, and a solution framework has been presented in Veglis et al. (2015).

3.4 Emergence of New Media Forms: Transmedia, Social Media, Ambient Media, and the Challenges for Advertising

The history of information processing (see e.g., Available: http://www.historyofinformation.com/index.php) for a complete overview] goes in parallel with the emergence of new media forms. Indeed, the introduction of new technologies brought to the emergence of new forms of media. McLuhan (1994) understood this phenomenon long ago, as also others did when discussing the effects of new media forms. The evolution of media implies the continuous re-invention, testing of new ideas, adapting these to the current state, and the development of new theories in the field of media studies to describe the phenomenon. New media features are added, other features mutate and some others get completely removed within processes of both convergence and divergence. A lengthy discussion illustrating the multidimensional nature of convergence, as well as the coexistence of convergence and divergence processes within the dynamic evolution of media is featured in Ibrus (2015). Convergence as part of an historical media evolution process has been underlined also in the introduction section of this book chapter and in Arango-Forero et al. (2015), where the multifaceted use of the word convergence has been addressed.

A consequence of technological convergence and media convergence processes in terms of new media forms is *transmedia storytelling*, i.e., the possibility to tell stories though and across different media. As stated in Sousa et al. (2015), transmedia storytelling requires the addition of new information pieces via each of the different communication channels engaged in telling the story. This is what makes the difference from traditional storytelling where one complete content piece is then published across different media forms. Alice in Wonderland published as film, novel, videogame, website, and illustrated book is an example of traditional storytelling (Sousa et al., 2015). The emergence of new cross-media technology enabling new ways to develop content and tell stories across media is discussed also in Ibrus (2015) and Innocenti et al. (2015), as well as in Denward (2015) where a practical example of transmedia storytelling has been presented: *Sanningen om Marika* (*The Truth about Marika*), a program which has been produced by the Swedish public broadcaster SVT in collaboration with a pervasive games startup company.

Such examples emphasize the fact that media evolve also at a narrative and semiotic level and media convergence affects not only the industry as such but also culture and society (Sabiha et al., 2015). This is why a cross-disciplinary approach is necessary to understand nature of media convergence, the impact of which is evident also in the development of notions such as narrative and/or story as discussed in Innocenti et al. (2015). The need for clearer understanding and definition of narrative as we know it today, and in particular of transmedia storytelling as a result of media convergence, is highlighted also in Sabiha et al. (2015) by focusing on the importance of considering user experience design early in the story writing process. According to the authors of this chapter transmedia storytelling "discards linearity, provides several perspectives to a story, promotes a participatory culture, and often harnesses the added value of collective intelligence". Diversity and creativity, the two main characteristics of "all innate learning beings" (Sabiha et al., 2015), are thus at the very base of media convergence and divergence processes: learning beings participate, share, thus converge; however, by adding new perspectives to a story they stimulate divergence as well. A contemporary example of converging different perspectives into one channel is represented by today's social networks such as Twitter. The convergence of news distribution in such networks has been discussed in Villi et al. (2015), and authors conclude that delegating news diffusion through social networks enables the news corporations to connect to a diverging audience on diverging devices.

Social media have a further effect that is to be included in the dynamics of media evolution. By illustrating the complex relationship existing between professional journalists and social media, Sacco (2015) points out how journalists can act as moderators within social networks so that multiple functions such as those of producer, mediator and curator of news converge within the role of a professional journalist.

Another example well representing another side of the media evolution process is *ambient media* (also known as ubiquitous media or pervasive media). This term describes media embedded throughout our environment, integrating sensor data such as location based services, and adapting to the context of the consumer (Lugmayr, Serral, Scherp, Pogorelc, & Mustaquim, 2013; Lugmayr, Zou, Stockleben, Lindfors, & Melakoski, 2013). One very basic example of ambient media is Foursquare, or Facebook displaying location based information. User interfaces enabling interaction with content are becoming part of our daily living environment and constitute a base for further developments of interaction forms and towards "ambient media convergence" (Immonen & Giesecke, 2015). Understanding the dynamics of media convergence is or utmost importance in order for media evolution processes to be sustainable. Thus, the discussion about the evaluation and development of advertising within a converging media world, as well as the inclusion of converging media solutions in media and advertising planning, as problematized in Svahn et al. (2015), becomes fundamental.

3.5 Effects on Communication and Information Processing: Data, Information, Knowledge, and Networks of Communication

To further elaborate the argumentation of Zlatanov and Koleva (2015), technology convergence leads to wider and easier accessibility to information, and thus to information overload. In order to cope with information overload and to interpret the information stream of individuals new social structures emerge. Technology savvy brokers of information and influencers alter the structure of existing social networks within the process of information diffusion (Zlatanov & Koleva, 2015). As a consequence, we can say that technology convergence leads to social divergence. Nevertheless, the divergence of information overload into different social network information streams requires a minimum of visual convergence in representing information and knowledge. As suggested in Mustic and Plenkovic (2015), visual communication allows for acting as carrier, and for easy interpretability of media messages. Thus, if information overload is one aspect of convergence, the way how information spreads through networks and peers is both an aspect of visual convergence and of consumption divergence. To make it more predictable new models are required and one possible solution has been presented in Duh et al. (2015). This paper defines a predictive model describing how information can spread between nodes within a communication network. From a communication theory point of view, measurements and in particular quantitative metrics for building theoretical models of information are a necessity. An excellent overview and introduction into the basic metrics that can be used in gaining quantitative results from empirical data have been presented in Hamacher (2015).

So far theoretical aspects related to information processing and communication have been discussed. However, to a larger extent, information processing and communication have a direct impact on organizational practice. For instance, convergence through company mergers most probably leads to divergence in content handling—due to different ways of processing information and different company cultures—and thus to a higher content management complexity (Grüblbauer & Haric, 2015). This is a rather tricky issue, as content and information typically represent the main product of any media firms. In Grüblbauer and Haric (2015) authors discuss this aspect of convergence by problematizing the complexity of business intelligence and corporate information systems. In Gershon (2015) this discussion is extended to transnational corporations. The author highlights the importance of having an intelligent network system in place to allow international business communication, as well as the fact that intelligent network systems directly affect the spatial design and operations of transnational organizations. In other words, convergent communication networks allow organizational structure divergence. This diverging organizational process, which can be defined as an ecosystem of which media firms are part, is discussed in Giesecke (2015) where diversification processes through the example of live event ticket selling are presented. The emergence of ecosystems where convergence and divergence processes coexist have been also discussed in Immonen and Giesecke (2015).

3.6 Driving Forces, Power, and Evolution of Traditional Policies and Media

Where does convergence come from, and who is the driving force behind convergence? Is convergence a phenomenon that media and IT firms are pushing forward in order to expand their business and reach a dominant position in the digital world? According to Dey (2015) "The inherently disaggregated, divergent structure of the Internet has spawned a democratic culture [where] it is unlikely that even media conglomerates will find it easy to dominate the media". This perspective is shared also in Immonen and Giesecke (2015), where the authors argue that convergence is the driving force allowing large media companies to stay in power, and divergence empowers small and innovative newcomers to flourish in niche markets. This might explain the current dilemma of mid-sized media companies attempting to compete with emerging digital companies such as music services as e.g., Spotify, which are pushing audiences to diverge (https://www.spotify.com). This last viewpoint leads to another argumentation, which has been presented in Ala-Fossi (2015) citing the case of the integrated digital broadcasting system in Japan: "convergence [must] be pursued and initiated by people in power [as] it [does not happen by itself] as an automatic process".

The case of information spread discussed in Zlatanov and Koleva (2015) represents a good example of how technological convergence may lead to divergence. The chapter points out that while the quantity of available information rises and converges into certain social media networks, technology savvy actors' ability to process information leads to their superiority in influencing information spread and stream direction. Information spread, which is fundamental to form public opinion, depends on how information is altered and interpreted by individuals within communication processes. When people are confronted with an information overload, information brokers interpreting the flood of information become necessary. Furthermore, media channels' convergence together with the dominance of social media as information diffusion tools leads to divergence at social level, as the structure of existing information networks is altered in the information diffusion process and technology illiterate actors are unconsciously segregated. As a practical case, we would like to refer to Zlatanov & Koleva (2015), where this effect has been illustrated on the example of the introduction of Blue-Ray and HD DVD.

The superiority of technology literate actors is well described also in Veglis et al. (2015) where the transformative power and effects of convergence processes within news media firms are discussed. If on one hand technological convergence seems to open up new opportunities for established media to consolidate and new media to emerge, on the other hand its effects on journalistic content quality, newsroom structures and working conditions are diverging from traditional standards and raise serious doubts. Organizational issues have also been discussed in Georgiades (2015), where the author problematizes the media managers' understanding of information provision/communication. He argues that media managers provide information in order to achieve employees' trust and focus on ensuring employees' independence in their job instead of trying to build up a convergence enhancing dialogue. This leads not only to diverging understandings but also to diverging organizational behaviors. Especially the pressure to transform towards digital media houses and work more efficiently causes divergence on organizational level. As pointed out in Spyridou and Veglis (2015) on the example for journalism—the pressure to create a super-journalist, thus a journalist literate in all new technologies, causes a divergence between journalistic quality and efficiency. A similar issue is raised in Georgiades (2015), which emphasizes employee involvement, and communication to achieve convergence rather than pure efficiency.

4 Discussion

Browsing and analyzing the content of all chapters within the two volumes of this edited handbook on media convergence we cannot conclude that convergence is the only process actually characterizing and shaping the evolution of media. Media evolution appears to be more the result of convergence, divergence and coexistence processes at different levels: technological, organizational, and social, to name the ones that have been mostly approached in this handbook. There is a strong interaction between the processes taking place at those levels, so that convergence processes are not independent from divergence processes and vice versa. More appropriately we can speak of an evolution where media convergence and divergence processes happen simultaneously, coexist and often influence each other building up on and developing existing features, forms and structures. Furthermore, sometimes media convergence hinders divergence in terms of business diversification as for instance discussed in Deliyannis (2015). Here the authors point out that interactive television should offer new features of television experience but always allow the consumer to jump back to an old passive non-interactive content mode at the same time. Such an offer is based on modular software components enabling the support of new interactive features by simplifying the "tuning" process. In this case convergence is basically hindering business diversification—such as the creation of a new device—through the requirement of e.g., backward compatibility within an existing device, i.e., the television. This would lead to the conclusion, that convergence might be a hurdle in the development of innovations in the world of media.

Most contributions to this edited handbook focus on the thematic area of convergence, only a few touch upon divergence, and almost none on the notion of co-existence. This clearly demonstrates a general lack of research in wider and more diverse directions for gaining understanding on the phenomenon of media evolution. Media divergence processes characterized by the development of additional features, and thus of new media forms emerging from similar forms, in each evolutionary step are still poorly investigated. Coexistence processes, in which new media rather develop in parallel to similar media forms, are poorly investigated, too. Chapters' contributions further highlight the notion of media ecosystem opposed to the one of "single" media entities; they underline the complexity of a dynamic media context, the evolution of which is characterized by the interplay of media convergence, divergence and coexistence processes.

We believe that focusing research on the sole phenomenon of convergence, thus neglecting other perspectives explaining media evolution, will hinder the building of theory in media studies. On the contrary, considering media convergence, coexistence, as well as divergence processes as part of a larger context such as the wide and dynamic evolution of media—where new features emerge, disappear, or are combined with one another as a matter of convergence or contingency—is the real key to understand them as well as their future development. Biology evolution docent and so do the chapters included in this edited handbook. We can thus conclude that future research within media divergence and coexistence processes as well as combining media convergence, divergence and coexistence phenomena is therefore very much needed.

Appendix A: Quantitative Analysis

Topic Analysis (LDA)

The topic analysis has been performed based on R, (R-Project, http://www.r-project.org), and consisted on the building of a lexicon, whose vocabulary was used as input for the *Latent Dirichlet Allocation* (*LDA*) analysis. We used a Gibbs sampler and built six topic clusters, which have been examined for topic clusters to identify higher level phenomenon based on the techniques described in Jockers (2014) (Table 2, Fig. 4).

Table 2 Topic analysis

	Business, market, and quality	Social, interaction, and networking	Content and technology	Media form and audience	Information, intelligence, and networks	Media organizations, and profession
1	Business	Interaction	Mobile	Media	Information	Journalism
2	Market	Design	Content	Transmedia	Data	Broadcasting
3	Model	User	Social	Audience	Event	Information
4	Media	Multimodal	Interactive	Story	Media	Organizational
5	Quality	Human	Television	News	Network	European
6	Services	Information	Users	Game	News	Radio
7	Models	Tools	Devices	Games	Roles	Employees
8	Open	Speech	Internet	Storytelling	Content	Policy
9	Markets	Knowledge	Services	Production	Ecosystem	Digital
10	Industries	System	Access	Medium	Social	Employee
11	Company	Experience	Networks	Jenkins	Role	Network
12	Product	Intelligent	Video	Narrative	Model	Regulation
13	Companies	Virtual	Media	Social	Business	Standard
14	Industry	Interface	Phone	Series	Article	Managers
15	Service	Systems	Screen	Audiences	Thus	Ownership
16	Publishing	Gesture	Convergence	Twitter	Actors	Regulatory
17	Billion	Learning	Network	Brand	Results	Management
18	Book	Game	Interaction	Facebook	Provider	Competition
19	Content	Computer	Publishing	Stories	Table	Speech
20	Economics	Video	Activities	Participation	Sources	Involvement

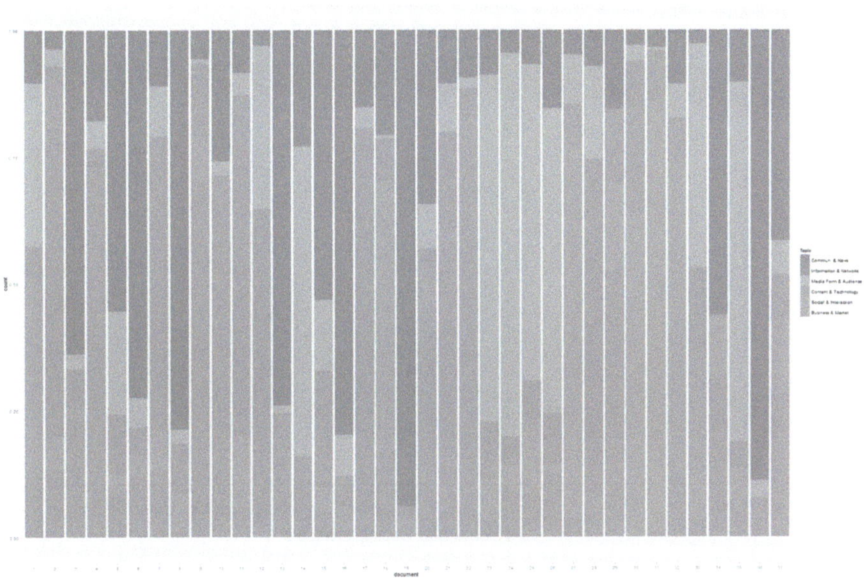

Fig. 4 Dominant topics within the 37 contributed book chapters

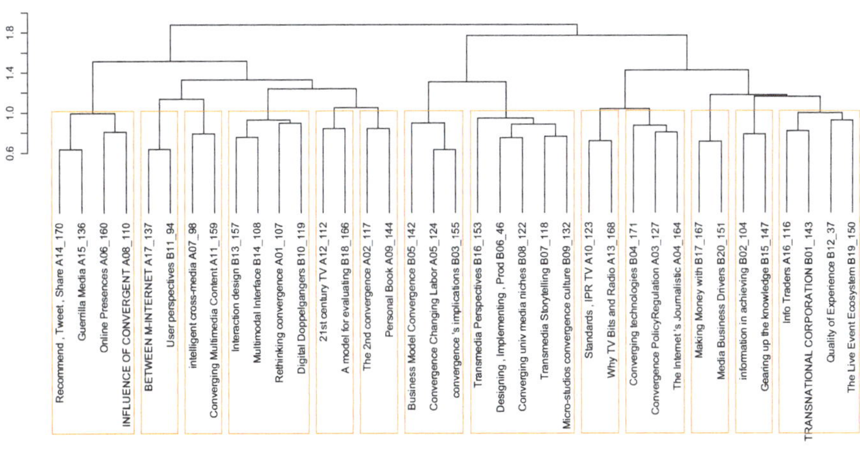

Fig. 5 Document cluster analysis

Document Cluster Analysis (Incl. Agglomeration Stages)

In the following picture the cluster analysis of all book chapters is presented. The parameters were: method (ward), distance (Cosine), Standardized by words, TF-IDF, min. term frequency of 150 (resulting into 127 selected words), and 13 optimal clusters (Figs. 5 and 6).

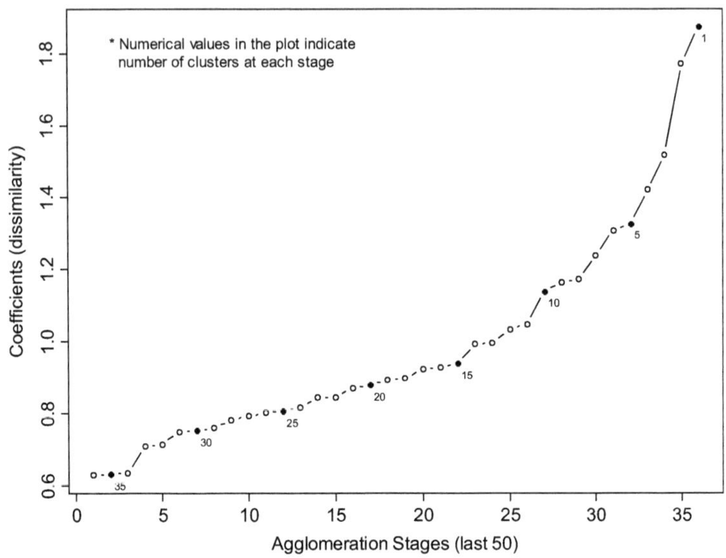

Fig. 6 Agglomeration stages of the document cluster analysis

Term Frequencies (TF) (Fig. 7)

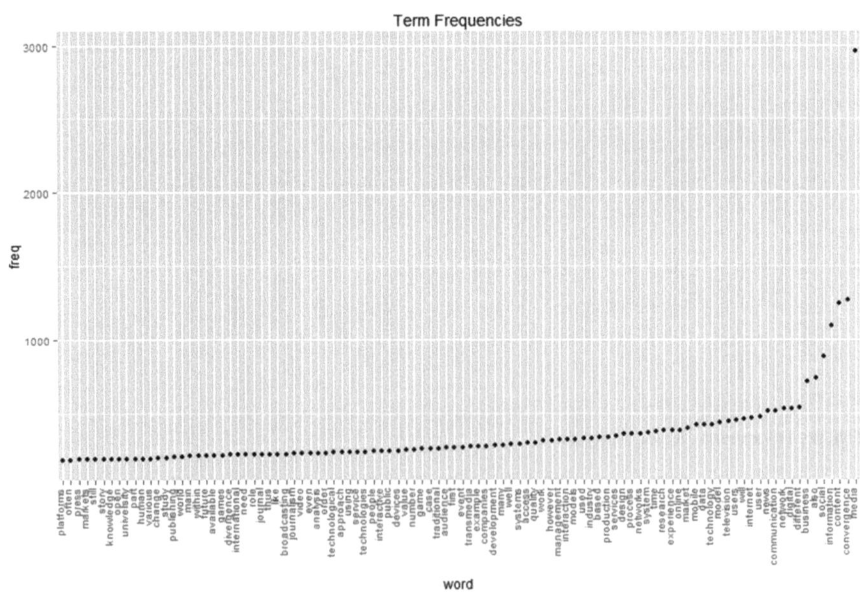

Fig. 7 Term frequencies of the most 40 used terms

Word Clusters

Word Clusters: Convergence

Media convergence/41459.593
Convergence of media/5276.675
Technological convergence/4830.857
Organizational convergence/3734.403
Technology convergence/2502.125
Digital convergence/2411.864
Market convergence/2034.277
Convergence process/1986.008
Network convergence/1985.531
Convergence era/1833.161
Process of convergence/1390.205
Device convergence/1250.46
Convergence of content/1247.204
Industry convergence/1115.899
Convergence culture/1091.031
Media convergence/960.024
Digital media convergence/898.199
Business model convergence/886.534
Content convergence/831.469
Tv convergence/817.11
Cultural convergence/807.233
Global convergence/796.413
Media convergence/753.811
Different levels of media convergence/753.34
Convergence of media content/645.631
Convergence media convergence/591.892
Convergence policy/552.556
Model of convergence/507.953
Convergence of news media/489.921
First convergence/475.537
Model of media convergence/464.839
Convergence of business models/462.427
Power of convergence/440.672
Convergence research/417.651
Phenomenon of media convergence/410.04
Introduction media convergence/403.226
Process of media convergence/394.543
Effective media convergence/380.561
Era of convergence/366.632
Management of media convergence/356.183
Use of media convergence/354.155
Case of convergence of media/352.602
Models of convergence/351.545
Divergent convergence/346.535
Technological convergence of media/343.065
Effects of convergence/263.659
Technical convergence/257.737
Digital broadcast media convergence/255.824
Future of convergence/255.563
Form of convergence/244.492
Media convergence results/244.399
Industrial convergence/243.142
Concept of convergence/242.388
Convergence trends/242.388
Economic convergence/240.419
Impact of media convergence/240.212
Specific convergence/238.91
Issue of convergence/235.675
Abstract media convergence/232.006
2nd convergence/231.612
Age of media convergence/229.792
Effects of media convergence/229.116
Convergence of media industries/226.303
Data convergence/224.521
Network convergence success/223.393
Interactive tv convergence/222.39
Convergence problem/214.309
Convergence experience/211.342
Experience of convergence/211.342
Media convergence issues/209.786
Regulatory convergence/209.321
Media convergence predictor/208.712
Effect of media convergence/205.61
Media convergence phenomenon/205.02
Service convergence/200.684
Result of media convergence/196.813
Digital convergence design/196.399
Impact of convergence/188.695
Convergence of users/185.315
Convergence phenomena/184.748
Driver of media convergence/183.952
Kind of media convergence/183.952
Phenomena of media convergence/180.75
Word convergence/180.665
Media convergence of today/177.237
Convergence age/176.552
Convergence effects/175.773
Debate of media convergence/174.685
Different types of convergence/173.845
Convergence of industries/172.546
Forces of media convergence/171.932
Convergence management/170.353
Other media convergence dynamics/170.02

Media convergence analysis/331.954	Convergence of networks/169.961
Successful media convergence/307.611	Use convergence/168.9
Global convergence of media/301.6	Use of convergence/168.9
Digital tv convergence/298.539	Case of convergence/167.79
Aspects of media convergence/298.344	Media research the concept of convergence/164.647
Term convergence/296.863	Literature review media convergence/163.93
Digital convergence issues/294.413	Modern media convergence/161.03
Area of media convergence/289.837	Technological convergence/161.029
High convergence media content/288.358	Structural convergence/160.957
Concept of media convergence/283.853	Problem discussion media convergence/158
Evolution of media convergence/283.766	Viewpoint of media convergence/157.9
Social convergence/280.315	Cross-industrial convergence/155.362
Idea of convergence/280.267	
First convergence media/280.236	
Cultural media convergence/274.605	
Age of convergence/264.828	

World Clusters: Divergence

Media divergence/1347.552	Issue of divergence/28.087
Social divergence/901.992	Divergence issues/27.531
Divergence of media business models/890.854	Effect divergence/26.713
Convergent divergence/331.053	Various divergence issues/26.137
Media convergent divergence/303.159	Divergence of broadcast media/25.901
User divergence/186.746	Further divergence processes/24.827
Divergence of media forms/167.753	Possibilities of divergence/24.367
Convergent divergence media/151.579	Jensen-shannon divergence/21.379
Content divergence/148.638	Specific convergence-divergence aspects of qoe/20.385
Divergence of content/148.638	Audience convergence/divergence/19.554
Divergence of media business drivers/148.027	Divergence operating/19.242
Divergence models/125.688	Situation of divergence/18.6
6.5 convergent divergence/94.587	Terminal divergence/17.387
Technological advantage of media divergence/92.315	Content-based divergence/12.712
Market divergence/90.915	Intended divergence/11.83
Complementary process of media divergence/88.307	Divergence cross-technology issues/11.487
Divergence trends/86.661	Behavioural divergence/11.486
Divergence processes/75.066	So-called Kullback–Leibler divergence/9.491
Processes of divergence/75.066	Convergence/ divergence issue/9.196
Technological impact of media convergence/divergence/60.609	Strong divergence ion/9.071
Case of divergence/59.99	Outlined divergence/8.989
Ecosystem divergence/53.987	Convergence/ divergence effect/8.746
Kullback–leibler divergence/50.848	Phenomenon of convergence/ divergence/8.708
Divergence projects media disruption/49.475	Conditioned divergence/8.122
Post divergence/44.437	Divergence transpire/8.122
Divergence phenomena/44.035	Supports divergence/8.122
Degree of divergence/42.978	Main convergence/ divergence trade-offs/7.985
Cultural divergence/41.23	
One source of divergence/39.845	

Divergence paradigm/39.721	Convergence-divergence dilemma of qoe/6.685
Need of user divergence/39.236	Convergence/ divergence trough/6.223
Various divergence processes/32.136	Jensen-shannon divergence of eq./5.447
Convergent divergence framework/31.626	Convergence-divergence mechanism/ 3.798
Additional divergence/29.328	Pro-divergence incentives/3.13
Divergence dynamics/28.612	Space/ time divergences/1.414

Full World Clusters

Social media/69484.384	Value of information/1260.623
Media convergence/41459.593	Several media/1255.266
Media content/35209.445	Device convergence/1250.46
News media/24409.342	Convergence of content/1247.204
Media companies/18968.177	Digital broadcasting/1223.402
Media industry/17979.539	Journalistic quality/1220.407
New media/16171.599	New business model/1208.572
Digital media/13949.863	Online content/1159.214
Different media/13385.665	4.2 social media/1158.073
Business model/11168.862	Video games/1149.24
Traditional media/10582.376	Social networking/1145.295
Business models/8281.9	Media owners/1142.52
Social networks/7963.842	Media business development/1138.263
Media business/7103.481	Social media features/1137.947
TV content/6516.154	Social interaction/1126.77
Social tv/6276.593	Cross-media content/1122.023
Converged media content/5565.936	Industry convergence/1115.899
Convergence of media/5276.675	New ways/1097.223
Digital tv/4860.404	Media landscape/1093.938
Technological convergence/4830.857	Convergence culture/1091.031
Interactive tv/4340.214	Media studies/1090.474
Convergent media/4098.611	Information stream/1087.57
Mass media/3941.827	Media producers/1070.026
Media experience/3929.058	Media market/1050.532
Social network/3921.883	User interaction/1049.775
Organizational convergence/3734.403	TV broadcasting/1036.183
Online media/3363.022	Belief systems/1008.26
TV series/3360.4	Content types/1000.741
Same time/3314.651	Business model change/997.96
Other media/3250.895	Digital information/985.2
Other hand/3192.07	Convergent media content/984.679
2.1 media/3114.236	Information architecture/975.07
New business models/3025.946	Current business model/973.303
Mobile tv/2971.688	Content management/970.359
Media space/2833.878	Content consumption/966.376
User experience/2722.473	Media convergence/960.024
Transmedia storytelling/2631.679	Media forms/954.362
Mobile phone/2629.286	Cross-media publishing/951.904
Media types/2540.364	Business ecosystem/948.63
News content/2504.56	Media content provider/944.804
Technology convergence/2502.125	User-generated content/943.604

Media managers/2497.945	New model/939.412
Digital convergence/2411.864	New content/922.635
Experience design/2398.807	TV set/920.299
Media markets/2391.198	Digital radio/919.156
Media industries/2316.741	New media corporations/917.241
Content production/2278.271	Traditional tv/912.466
Event ecosystem/2272.082	Multiple media/910.394
New media business models/2242.509	Publishing industries/905.305
New information/2205.451	Social divergence/901.992
New technologies/2195.902	Media design/900.632
Media models/2178.52	Digital media convergence/898.199
Media product/2157.365	Social media networks/894.864
Mobile devices/2123.649	Divergence of media business models/890.854
Media consumption/2044.274	Business model convergence/886.534
Market convergence/2034.277	Revenue model/874.517
New technology/2005.228	Content distribution/873.892
Convergence process/1986.008	Media tools/871.117
Network convergence/1985.531	Information management/869.823
TV experience/1892.884	Publishing industry/862.818
Media sector/1892.244	Provision of information/861.371
Business system/1890.668	Video content/860.841
Media production/1835.985	Media platforms/859.422
Convergence era/1833.161	TV market/854.06
TV screen/1804.645	Business concepts/852.939
News media industry/1764.942	Information technology/851.725
Mobile phones/1750.489	New business/847.976
Media technology/1722.848	Different types of media/845.909
Media planning/1719.023	Internet media/839.119
Content of media/1717.534	Social media role/836.478
Media company/1712.974	Content convergence/831.469
Public media/1708.607	Other roles/818.495
Information content/1698.196	Tv convergence/817.11
Digital content/1648.606	Selected media/809.551
Social media content/1624.557	Advertising market/808.469
Media user/1618.407	Cultural convergence/807.233
2.0 media/1557.118	New services/806.275
Media users/1531.194	Open access/800.774
Media consumers/1511.248	Global convergence/796.413
Interaction design/1502.207	Social value/790.198
Social media platforms/1482.328	Intelligent network/783.903
Distribution of media content/1449.303	Multimedia experience/783.543
Online market/1418.07	Mobile media/783.28
Process of convergence/1390.205	New models/780.18
Media divergence/1347.552	Employee involvement/774.3
Content providers/1339.509	Same content/767.689
New service/1336.13	(the rest of the list is suppressed)
Media organizations/1329.929	
Mobile access/1322.896	
Interactive media/1281.277	
Quality of experience/1279.314	
Business model evolution/1275.07	

References

Ala-Fossi, M. (2015). Why TV bits and radio bits did not fit together? digitalization and divergence of broadcast media. In A. Lugmayr & C. Dal Zotto (Eds.), *Convergent divergence—Crossdisciplinary viewpoint on media convergence*. Germany: Springer.

Arango-Forero, G., Roncallo-Dow, S., & Uribe-Jongbloed, E. (2015). Rethinking convergence: A new word to describe an old idea. In A. Lugmayr & C. D. Zotto (Eds.), *Convergent divergence—Crossdisciplinary viewpoint on media convergence*. Germany: Springer.

Damasio, M., Henriques, S., Teixeira-Botelho, I., & Dias, P. (2015). Between m-internet users and stakeholders: Convergence or divergence? In A. Lugmayr & C. D. Zotto (Eds.), *Convergent divergence—Crossdisciplinary viewpoint on media convergence*. Germany: Springer.

Deliyannis, I. (2015). Converging multimedia content presentation requirements for interactive television. In A. Lugmayr & C. D. Zotto (Eds.), *Convergent divergence—Crossdisciplinary viewpoint on media convergence*. Germany: Springer.

Denward, M. (2015). Designing, implementing and producing for participation: Media convergence practice. In A. Lugmayr & C. D. Zotto (Eds.), *Convergent divergence—Crossdisciplinary viewpoint on media convergence*. Germany: Springer.

Dey, S. (2015). Convergence policy and regulation—A free speech perspective. In A. Lugmayr & C. D. Zotto (Eds.), *Convergent divergence—Crossdisciplinary viewpoint on media convergence*. Germany: Springer.

Duh, A., Meznaric, S., & Korošak, D. (2015). Guerrilla media: Interactive social media. In A. Lugmayr & C. D. Zotto (Eds.), *Convergent divergence—Crossdisciplinary viewpoint on media convergence*. Germany: Springer.

Ebner, M., Schon, S., & Alimucaj, A. (2015). Print your personal book. In A. Lugmayr & C. D. Zotto (Eds.), *Convergent divergence—Crossdisciplinary viewpoint on media convergence*. Germany: Springer.

Fleury, A. (2015). User perspectives on media convergence: Results from quantitative and qualitative studies. In A. Lugmayr & C. D. Zotto (Eds.), *Convergent divergence—Crossdisciplinary viewpoint on media convergence*. Germany: Springer.

Foster, S., & Brostoff, J. (2015). Digital doppelgangers: Converging technologies and techniques 3D world modeling, video game design and urban design. In A. Lugmayr & C. D. Zotto (Eds.), *Convergent divergence—Crossdisciplinary viewpoint on media convergence*. Germany: Springer.

Georgiades, S. (2015). The effects of information in achieving employee involvement and thus organizational convergence in a media organization. In A. Lugmayr & C. D. Zotto (Eds.), *Convergent divergence—Crossdisciplinary viewpoint on media convergence*. Germany: Springer.

Gershon, R. (2015). Redefining business space: Intelligent networking and the transnational organization. In A. Lugmayr & C. D. Zotto (Eds.), *Convergent divergence—Crossdisciplinary viewpoint on media convergence*. Germany: Springer.

Giesecke, R. (2015). The live event ecosystem? local media against global ticket sellers? In A. Lugmayr & C. D. Zotto (Eds.), *Convergent divergence—Crossdisciplinary viewpoint on media convergence*. Germany: Springer.

Grüblbauer, J., & Haric, P. (2015). Making money with & through information and knowledge? competitive intelligence, trend analysis and revenue models of classic media. In A. Lugmayr & C. D. Zotto (Eds.), *Convergent divergence—Crossdisciplinary viewpoint on media convergence*. Germany: Springer.

Hamacher, K. (2015). Online presences of traditional media vs. blogs—Redundancy and unique news coverage. In A. Lugmayr & C. D. Zotto (Eds.), *Convergent divergence—Crossdisciplinary viewpoint on media convergence*. Germany: Springer.

Ibrus, I. (2015). Micro-studios meet convergence culture: Clustering, dialogues, innovation. In A. Lugmayr & C. D. Zotto (Eds.), *Convergent divergence—Crossdisciplinary viewpoint on media convergence*. Germany: Springer.

Immonen, S., & Giesecke, R. (2015). Media business drivers 2020—The case of David and Goliath revisited. In A. Lugmayr & C. D. Zotto (Eds.), *Convergent divergence—Crossdisciplinary viewpoint on media convergence*. Germany: Springer.

Innocenti, V., Pescatore, G., & Rosati, L. (2015). Converging universes and media niches in serial narratives. An approach through information architecture. In A. Lugmayr & C. D. Zotto (Eds.), *Convergent divergence—Crossdisciplinary viewpoint on media convergence*. Germany: Springer.

Jockers, M. (2014). *Text analysis with R for students of literature*. New York: Springer.

Karmasin, M., Diehl, S., & Koinig, I. (2015). Media convergence? implications for businesses: Is there a demand for new business models, organizational structures and management/employee competencies? In A. Lugmayr & C. D. Zotto (Eds.), *Convergent divergence—Crossdisciplinary viewpoint on media convergence*. Germany: Springer.

Kueng, L. (2008). *Strategic management in the media—From theory to practice*. London: Sage.

Leminen, S., Huhtala, J.-P., Rajahonka, M., & Westerlund, M. (2015). Business model convergence and divergence publishing industries. In A. Lugmayr & C. D. Zotto (Eds.), *Convergent divergence—Crossdisciplinary viewpoint on media convergence*. Germany: Springer.

Lugmayr, A., Serral, E., Scherp, A., Pogorelc, B., & Mustaquim, M. (2013). Ambient media today and tomorrow. *Multimedia Tools and Applications*, 1–31 Available: 10.1007/s11042-012-1346-z

Lugmayr, A., & Dal Zotto, C. D. (2013).Convergent divergence—Cross-disciplinary viewpoint on media convergence. In *European Media Management Education Association Conference 2013—Digital Transformations and Transactions in Media Industries*. Bournemouth.

Lugmayr, A., Zou, Y., Stockleben, B., Lindfors, K., & Melakoski, C. (2013). Categorization of ambient media projects on their business models, innovativeness, and characteristics evaluation of Nokia Ubimedia MindTrek Award Projects of 2010. *Multimedia Tools and Applications, 66*, 33–57. doi:10.1007/s11042-012-1143-8.

Matteucci, N. (2015). Standards, IPR strategies and digital TV convergence: The EU experience. Part I: Theories. Part: Empirical evidence. In A. Lugmayr & C. D. Zotto (Eds.), *Convergent divergence—Crossdisciplinary viewpoint on media convergence*. Germany: Springer.

McKay, E. (2015). Gearing up the knowledge engineers: Effective experience design through effective human-computer interaction (HCI). In A. Lugmayr & C. D. Zotto (Eds.), *Convergent divergence—Crossdisciplinary viewpoint on media convergence*. Germany: Springer.

McLuhan, M. (1994). *Understanding media: The extensions of man*. Cambridge, MA: MIT Press.

McLuhan, M., & Lapham, L. H. (1994). *Understanding media—The extentions of man*. Cambridge, MA: MIT Press.

Merriam-Webster. *Merriam-Webster online dictionary*. Available: http://www.merriam-webster.com/dictionary/

Montpetit, M.-J. (2015). The 2nd convergence: A technology viewpoint. In A. Lugmayr & C. D. Zotto (Eds.), *Convergent divergence—Crossdisciplinary viewpoint on media convergence*. Germany: Springer.

Mustic, D., & Plenkovic, M. (2015). Influence of convergent media on the perception of information credibility. In A. Lugmayr & C. D. Zotto (Eds.), *Convergent divergence—Crossdisciplinary viewpoint on media convergence*. Germany: Springer.

Reichl, P. (2015). Quality of experience convergent communication ecosystems. In A. Lugmayr & C. D. Zotto (Eds.), *Convergent divergence—Crossdisciplinary viewpoint on media convergence*. Germany: Springer.

Sabiha, G., Morrison, P. A., Annika, W.-E., & Obal, D. (2015). Transmedia user experience design. In A. Lugmayr & C. D. Zotto (Eds.), *Convergent divergence—Crossdisciplinary viewpoint on media convergence*. Germany: Springer.

Sacco, V. (2015). How does social media shape media convergence? The case of journalists covering war and conflict. In A. Lugmayr & C. D. Zotto (Eds.), *Convergent divergence—Crossdisciplinary viewpoint on media convergence*. Germany: Springer.

Sreekanth, N. S., Varghese, N., Pradeepkumar, C., Pal, V., Prasad, G., Pal, S., et al. (2015). Multimodal interface for effective man machine interaction. In A. Lugmayr & C. D. Zotto (Eds.), *Convergent divergence—Cross disciplinary viewpoint on media convergence*. Germany: Springer.

Sousa, M. N., de Lemos Martins, M., & Zagalo, N. (2015). Transmedia storytelling: The roles and stakes of the different participants in the process of a convergent story, in divergent media and artifacts. In A. Lugmayr & C. D. Zotto (Eds.), *Convergent divergence—Crossdisciplinary viewpoint on media convergence*. Germany: Springer.

Spyridou, L.-P., & Veglis, A. (2015). Convergence and the changing labor of journalism: Towards the 'super Journalist' Utopia. In A. Lugmayr & C. D. Zotto (Eds.), *Convergent divergence—Crossdisciplinary viewpoint on media convergence*. Germany: Springer.

Svahn, M., Wahlund, R., Denward, M., & Rademaker, C. (2015). A model for evaluating converged media for advertising purposes. In A. Lugmayr & C. D. Zotto (Eds.), *Convergent divergence—Crossdisciplinary viewpoint on media convergence*. Germany: Springer.

Tavares, T., & Schofield, D. (2015). Interaction design for convergence medias and devices: A multisensory challenge. In A. Lugmayr & C. D. Zotto (Eds.), *Convergent divergence—Crossdisciplinary viewpoint on media convergence*. Germany: Springer.

Tuomi, P. (2015). 21st century television—Gathering around second screens. In A. Lugmayr & C. D. Zotto (Eds.), *Convergent divergence—Crossdisciplinary viewpoint on media convergence*. Germany: Springer.

Veglis, A., Dimoulas, C., & Kalliris, G. (2015). Towards intelligent cross-media publishing: Media practices and technology convergence perspectives. In A. Lugmayr & C. D. Zotto (Eds.), *Convergent divergence—Crossdisciplinary viewpoint on media convergence*. Germany: Springer.

Villi, M., Matikainen, J., & Khaldarova, I. (2015). Recommend, tweet, share: User-distributed content (UDC) and the convergence of news media and social networks. In A. Lugmayr & C. D. Zotto (Eds.), *Convergent divergence—Crossdisciplinary viewpoint on media convergence*. Germany: Springer.

Vukanovic, Z. (2015). Converging technologies and diverging market trends of internet/web and Traditional media. In A. Lugmayr & C. D. Zotto (Eds.), *Convergent divergence—Crossdisciplinary viewpoint on media convergence*. Germany: Springer.

Wellbrock, C. M. (2015). The internet's impact on journalistic quality. In A. Lugmayr & C. D. Zotto (Eds.), *Convergent divergence—Crossdisciplinary viewpoint on media convergence*. Germany: Springer.

Wikipedia. *Wikipedia, the free encyclopedia*. Available: http://en.wikipedia.org/

Zlatanov, B., & Koleva, M. (2015). Info traders, innovation and information spread. In A. Lugmayr & C. D. Zotto (Eds.), *Convergent divergence—Crossdisciplinary viewpoint on media convergence*. Germany: Springer.

Dal Zotto, C., Galichet, G., & Lugmayr, A. (2011). Managing media convergence: An economic and editorial project, In *European Media Management Conference 2011*, Moscow: European Media Management Association (EMMA).

About the Authors

Jacob M. Brostoff holds a B.A. in Linguistics from McGill University and a Master of Urban and Regional Planning from Portland State University. He was the urban planner on the award-winning másTransit team (mastransit.net). Jacob lives in Portland, Oregon where he works for the City of Portland.

Prof. Dr. Cinzia Dal Zotto is Professor of Media Management at the Faculty of Economics and Business at the University of Neuchâtel, Switzerland. Before that and until December 2008 she was Research Director at the Media Management and Transformation Centre and Associate Professor at the Jönköping International Business School in Sweden. Dal Zotto received her Ph.D. in Organizational and Human Resource Development at the University of Regensburg in Germany. Between 2001 and 2003 she had a post doc scholarship funded by the German Ministry of Education and her research dealt with new ventures' growth in the media sector. In that program she supervised doctoral students and was a consultant for start-up firms. Dal Zotto has published three books, various book chapters, and papers in the fields of organization, human resources, entrepreneurship and strategy. She has taught at the universities of Regensburg and Passau in Germany, at the University of Westminster in London, at Trento and Bolzano University, as well as at the ESC Toulouse in France. She further held seminars for the Joint Research Centre in Ispra, Italy. Dal Zotto was a visiting scholar at UC Berkeley in 2002. She also worked as Research Analyst at International Data Corporation in London during 1995 and at Reuters Venture Capital in Munich in 2001.

A. Lugmayr, C. Dal Zotto (eds.), *Media Convergence Handbook - Vol. 2*, Media Business and Innovation, DOI 10.1007/978-3-642-54487-3

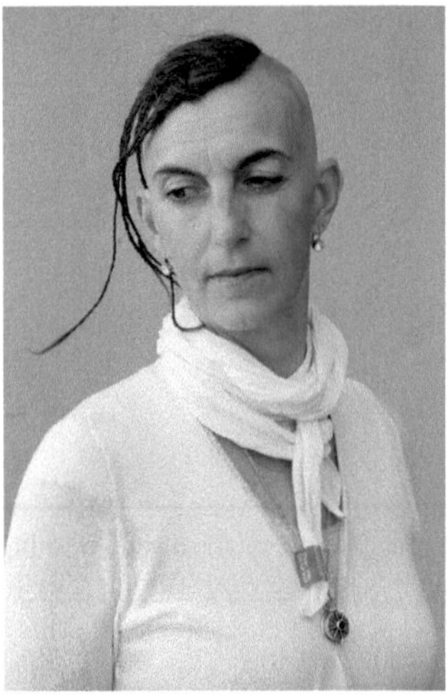

Marie Denward is employed as a senior researcher (Ph.D.) at The Interactive Institute Swedish ICT, an experimental IT and design research institute that conducts applied research and innovation. Her research interest comprises media use, different forms of game and play activities, transmedia storytelling, participatory culture and how new media imposes upon people's everyday life in the rapidly changing media landscape. She has been in leadership and management positions at the Interactive Institute since 2001. Marie has been involved in interdisciplinary research projects focusing on youth culture, media use, media production, media convergence, participatory culture and design research.

Sandra Diehl is Associate Professor at the Department for Media and Communications at the Alpen-Adria University of Klagenfurt, Austria. She received her Ph.D. and her habilitation in Business Administration from Saarland University in Germany. Her research interests include media and convergence management, international and intercultural advertising, CSR and health communication. Sandra Diehl has published in numerous journals, such as the International Journal of Advertising, Advances in International Marketing, Advances in Consumer Research, the International Marketing Review, and the European Advances in Consumer Research. She has authored and edited several books, among them the Advances in Advertising Research. She is also board member of the European Advertising Academy.

Alexandre Fleury first obtained a Master's degree in ICT from the french engineering school Ecole Centrale d'Electronique. He then moved to Aalborg University in Denmark where he obtained firstly a Master of Science in Engineering majoring in Intelligent Multimedia, and secondly a Ph.D. under the Wireless Communication programme during which he established near future trends in interactive media convergence while revisiting user experience evaluation methods. Throughout this education Alexandre published a number of peer-reviewed journal and conference articles.

His educational and professional experiences place Alexandre at the crossing of information technology, human-computer interaction, user research, and studies. Well versed in user-centred design, he prefers acting early in the product development cycle and enjoys exploring bourgeoning concepts, developing prototypes and testing them out with potential users. In parallel he finds user research in itself fascinating, and believes that people's experience with media products should be investigated using the largest set of methods, from standard usability to art-inspired ones.

Selected Publications

Fleury, A., et al. (2013). Evaluating user preferences for video transfer methods from a mobile device to a tv screen. *Pervasive and Mobile Computing*, *9-2*, 228–241 (Elsevier).

Fleury, A. (2013). *Near future trends in interactive media convergence using quantitative and qualitative approaches*, PhD thesis, Aalborg University.

Fleury, A. (2012). Drawing and acting as user experience research tools. In *Proceedings of APCHI 2012* (pp. 269–278). ACM.

Fleury, A., et al. (2011). Evaluating ubiquitous media usability challenges: Content transfer and channel switching delays. In: *LNCS 6770* (pp. 404–413). Springer.

Fleury, A., et al. (2011). Acceptable channel switching delays for mobile tv. In *Proceedings of EuroITV 2011* (pp. 195–198). ACM.

Shaun Foster is an assistant professor of 3D Digital Design at the Rochester Institute of Technology in Rochester, New York. He received his Masters in Fine Arts at RIT in 2002. He then went to four Directions Productions as manager of Animation and was also Lead Effects Artist there. He worked on multiple award winning, national and regional TV, advertising and the highly successful animated short "Raccoon and Crawfish". Shaun has developed two educational multimedia exhibits for museums prior to joining RIT. Foster is also the webmaster for the Professional SIGGRAPH chapter. He was recently awarded several grants (internal and external) dealing with next generation technologies (including Kinect and Augmented Reality tourism).

Mr. Ganga Prasad R. received the B.Tech. degree in Computer Science Engineering from Jawaharlal Nehru Technological University, Hyderabad, in 2008. His research interests include natural language processing, automatic speech recognition, augmented reality, machine learning and pattern recognition.

Dr. Stavros Georgiades is a Lecturer of Management at the Department of Economics, Accounting and Finance, Frederick University. He holds a Ph.D. in Management and a B.A. in Accounting and Economics from the Henley Business School, University of Reading, UK. Prior to his current appointment he was a Business Consultant in London, UK, for several years, his clients including the BBC.

Stavros has authored a monograph on "Employee Engagement in Media Management—Creativeness and Organizational Development" published by Springer in 2015. His work cuts across the fields of Organization Studies and Media Management and has been published in the Business and Management Research Journal and the Journal of Business and Economics.

He serves as an ad hoc book reviewer for Routledge Publisher, as a reviewer for the journals Organizational Change Management and Business and Economics, and for conferences like the Academy of Management. Stavros is currently a member of the Editorial Board of the Journal of Business and Economics, the Organizational Behaviour and the Human Resource divisions of the Academy of Management and the Organizational Behaviour division of the Brazilian Academy of Management (ANPAD).

He has carried out research projects in association with European, US and South American Media Organizations, including Channel 4 UK and Globo Brazil, the second-largest commercial TV network in annual revenue worldwide, and is an associate member of EU funded projects being responsible for their business plans. His most recent research deals with sensemaking in organizations, placing emphasis on such topics as the practicality of organizational change theory and the causal explanations of creativity assessment. In January 2014 he was appointed to the Board of Directors of the Cyprus Broadcasting Corporation (CyBC) by the President of the Republic of Cyprus.

Richard A. Gershon, Ph.D., (Ohio University) is Professor and Co-Director of the Telecommunications and Information Management program at Western Michigan University where he teaches courses in Media Management and Telecommunications. He is the author of multiple books, including *Media, Telecommunications and Business Strategy* 2nd ed. (2013). Dr. Gershon is a Fulbright scholar having held visiting appointments at the University of Navarra, Spain and Nihon University, Japan. Dr. Gershon has twice been selected for national teaching honors, including the Steven H. Coltrin Professor of the Year Award (2000) by the International Radio and Television Society (IRTS) and the Barry Sherman Award for Teaching Excellence (2001) by the Management and Economics division of the Association for Education in Journalism and Mass Communication (AEJMC). He is a founding member of the Information and Telecommunications Education and Research Association (ITERA) and currently serves as the Chairman of the Board of Directors.

Raphael Giesecke after enjoying a 17 years long career in industry, he returned to academic work in 2007. He has directed various, often large-scale, multi-national and multi-disciplinary research projects at Airbus, the leading aircraft manufacturer, and at the department of Industrial Engineering and Management at Aalto University's School of Science in Finland. His research interest is in innovation networks on two levels—individual persons and organisations. Giesecke has studied a variety of innovation ecosystems in change and disruption, including electric mobility, the news media industry and live events ecosystems and, most recently, digital services.

Sabiha Ghellal is an assistant Professor for Experience and Interaction Design at the Stuttgart Media University. Her research explores the creation of new interactive designs to support future cultural and creative experiences, with particular emphasis on transmedia storytelling. As a former manager for Research and Innovation at Sony Europe, she has worked on several large productions in the area of mobile entertainment, game design and interactive television, in addition to researching mixed reality and pervasive gaming in the context of EU funded projects such as IPerG and IPCity on behalf of Sony Europe.

Johanna Grüblbauer, Deputy Head of The Austrian Institute of Media Economy and Senior Lecturer at Department of Media Economy, University of Applied Sciences St. Pölten. Degree in Media-Management, PhD in Media- and Communication Studies, University of Vienna, has worked as Manager for Market and Media Research. Current research projects: Digital Business & engaged Users as a Challenge for Content-Management; Strategic Content-Communities-Monitoring for Mainstream- and Social Media; Strategic Management of Traditional and Digital Media Companies.

Dr. Peter Haric, Degree and Ph.D. TU Berlin (Economic and Social Sciences, Economic History and Military History); Certified Rating Analyst (CRA); Certified Financial Manager (CFM). Since 1992, management consultant (i.e. for Evonik, Roche, Volksbank International, Commerzbank, Pimco, KfW, various middle-sized companies) with specialization on Strategic Management, market-oriented corporate management, Corporate Communications, restructuring/turnaround management (SMEs). Establishment and Management of the Leitbetriebe Austria Institut (Leading Companies Austria Institute); Lecturer Media Management FH St. Pölten (Austria); Editor Gabler Wirtschaftslexikon (Gabler Encyclopedia of Business and Economics). Work areas/researchinterests: Strategic Management (focus: growth companies—Hidden Champions); Company analysis and valuation; Industry and sector analysis; Media Business and Managerial Economics.

Indrek Ibrus, Ph.D., is currently a Researcher at the Estonian Institute of Humanities at Tallinn University. He is also an advisor on audio-visual affairs at the Estonian Ministry of Culture. His main strand of research investigates the co-evolutionary effects of textual, social and institutional dynamics in shaping the new textual forms of multiplatform media and the associated institutional forms of production. He recently co-edited (with Carlos A. Scolari) "Crossmedia Innovations: Texts, Markets, Institutions" (2012). Dr. Ibrus received his Ph.D. in 2010 from London School of Economics and Political Science.

Stina Immonen, D.Sc.(Tech) works as senior lecturer at Aalto University's School of Science in Finland. Her area is work psychology and leadership at the department of industrial engineering and management. Her teaching and research involves organisational and human capital development in ICT-enabled business.

Veronica Innocenti is Assistant Professor at the University of Bologna, where she teaches History of Broadcasting and Film Marketing. She holds a Ph.D. in Film studies from University of Bologna. She has been a visiting scholar at UCLA and a speaker to several national and international conferences (among others, University of Lugano, University of Glasgow, University of Huddersfield, University of Nottingham). She has been the organizer of the international conferences Media Mutations 3 and 4, dedicated to narrative ecosystems. She is author of several publications, including books, books chapters and articles. She co-authored, with Guglielmo Pescatore, a book on television serial narratives: *Le nuove forme della serialità televisiva. Storia, linguaggio e temi* (2008). She is the co-editor, with Alice Autelitano, of a special issue of the journal *Cinéma and Cie* dedicated to TV serial narratives in Europe.

Matthias Karmasin is full professor and chair for media and communications sciences at the Alpen-Adria University of Klagenfurt, Austria. He is corresponding member of the philosophic-historic class of the Austrian Academy of Sciences and Director of Institute for Comparative Media and Communication Studies. He is an expert for media accountability and he has published extensively on media ethics and media management, political communication, communication theory, and media practice. Furthermore, Matthias Karmasin has developed the media stakeholders' theory. Recently, he has published an empirical study about the Austrian media and their managers. His professional experiences include business consultancy as cofounder of Medienhaus Wien (http://www.mhw.at) as well as several stations abroad.

Isabell Koinig is a Postdoctoral Researcher at the Department of Media and Communications at the Alpen-Adria University of Klagenfurt, Austria. She just finished her doctoral dissertation that investigated how different advertising appeals in non-prescription (OTC) drug promotions were received in a cross-cultural context. Her research interests predominantly concern the areas of health communication, intercultural advertising, organizational developments and communication practices, CSR reporting as well as media and convergence management.

Artur Lugmayr is A/Prof. at Curtin University, Australia, where he teaches and supervises students in visualization technologies, interactive media, media management, and digital humanities. Artur was Professor for digital media management at the Tampere Univ. of Technology (TUT), Finland 2009–2014 establishing the Entertainment and Media Management Lab. and the New Ambient Multimedia Lab. 2004–2009. Artur holds a Dr.-Techn. degree (Information Technology, TUT), and is pursuing his Dr.-Arts studies at Aalto Univ., Helsinki, Finland in motion pictures. He was visiting scientist in Singapore, Brisbane, Austria, Ghana; since 2000 raised/involved in 1.7+ MEUR funding (excl. 2014/2015 applications); 170+ publications; 24+ invited keynotes; and 27+ invited guest lectures. He founded and chairs the Association for Information Systems (AIS) Special Interest Group "AIS SIG eMedia" and the International Ambient Media Association (iAMEA); and is active member of the ACM TVX steering board, IFIP TC 14 for Entertainment Computing, MindTrek Association, Academic MindTrek steering board, and IEEE IG MENIG. Previously he was international advisor for the NeoTV Lab/Buenos Aires/Argentina, founder/chair of the ISO/IEC SC29 WG 11 MPEG "MPEG-21 in Broadcasting". He won the Nokia Award of 2003 with the text-book "Digital Interactive TV and Metadata—Future Broadcast Multimedia" published by Springer-Verlag in 2004; ELISA award for the text-book "Ambient Media and Beyond"; a special mention for his NAMU research group of the ELISA Ubimedia Award at MindTrek in 2007 for the open source Portable Personality (P2) software. Artur has founded the Academic MindTrek Conference series, the Semantic Ambient Media (SAME) workshop series, the Nokia Ubimedia MindTrek Award series, the EuroITV Grand Challenge series, and the Interactive Digital TV in Emerging Economies workshop series. He organized the Academic MindTrek Conference for 7 years, as well as EuroITV conference in Tampere in 2010. He also is publisher and founder of the open access journal and series entitled International Journal on Information Systems and Management in Media Industry (ISMMI). He acted as review/committee member and editorial board member of several publishers and journals, as e.g. SERC Press, Wiley&Sons, IGI-Global, Springer-Verlag, IEEE, Inderscience, Int. Journal on Media Management; and as part of the review/committees of several conferences, as e.g. ACM-MM, AmI, ICEC, MobileCHI, ICME, EuroITV, NordiCHI, UIC, MMEDIA, Interact, uxTV, ACM-CHI, EBRF, IEEE-IE, VC, PDP. However, in the future he will only accept these "positions" when knowing the organizers. He organized 8+ int. conferences, 20+ int. workshops at conferences, held 13+ conference tutorials, organized 6+ media awards, edited 6+ books; and edited 12+ special issues.

Moisés de Lemos Martins is Full Professor at the Department of Communication Sciences of the University of Minho (UM). He directs the Communication and Society Research Centre (CECS), which he founded in 2001; the journal Comunicação e Sociedade, also founded by him in 1999; and the Revista Lusófona de Estudos Culturais, which he created in 2013 with Maria Manuel Baptista. Currently, he is the Director of two doctoral programs, in *Cultural Studies* and *Communication Studies: Technology, Culture and Society*, promoted jointly by the University of Minho and other Universities and Higher Education Institutions. He is President of the SOPCOM—Associação Portuguesa de Ciências da Comunicação, since 2005; the Lusocom—Federação Lusófona de Ciências da Comunicação, since 2011; and Confibercom—Confederação Ibero-Americana das Associações Científicas e Académicas, since 2012. Among other books, he published *Crise no Castelo da Cultura—Das Estrelas para os Ecrãs* (2011), *A Linguagem, a Verdade e o Poder. Ensaio de Semiótica Social* (2002), *Para uma Inversa Navegação. O Discurso da Identidade* (1996); *O Olho de Deus no Discurso Salazarista* (1990). He edited, with other authors, *Comunicação e Lusofonia. Para uma análise crítica da cultura e dos media* (2006) and *Comunicação e Cidadania* (Proceedings of the V Portuguese Congress on Comunication Sciences, Braga, CECS, UM, 2008).

Elspeth McKay is an Associate Professor of Information Systems (IS) at the RMIT University, School of Business IT and Logistics, Melbourne, Australia. She is passionate about designing effective eLearning resources for the education sector and industry training/reskilling programmes, including: investigations of how individuals interpret text and graphics within Web-mediated learning environments. She has designed e-Learning tools implemented through rich internet applications; including: ARPS—an advanced repurposing pilot system, COGNIWARE—a multi-modal e-Learning framework, GEMS—a global eMuseum System, eWRAP—Electronic work readiness awareness programme, EASY—Educational/academic (skills) screening for the young, offering enhanced accessibility through touch screen technologies. Over the last decade Dr. McKay has published extensively in the research fields of HCI and educational technology. In recognition of her contribution to the professional practice of IS research, she was elected as a Fellow of the Australian Computer Society (FACS).

Ann Morrison is Associate Professor at the Department of Architecture, Design and Media Technology, Aalborg University, Denmark. Her prior work with tangible computing, Mobile AR and natural user interfaces addresses situated use and the urban environment. Morrison's current research investigates wearable computing solutions while advancing open-ended styles of explorative interaction. Morrison is currently working with wearable (tactile and audio) computing in the EU funded CultAR project.

N. K. Narayanan received his Ph.D. degree under the Faculty of Technology from Department of Electronics, Cochin University of Science And Technology in 1990, in the area of Speech Signal Processing. From 1981 December to 2003 he worked as Lecturer in various Government Colleges in Kerala, India. He has published more than hundred research papers in reputed journals and conference proceedings in the area of Speech processing, Image processing, Neural network and Bioinformatics. He also guided many Ph.D. dissertations in the above said area. He was the founder Professor and Head of the School of Information Science and Technology, Kannur University, Kerala, India. Currently he is working as Senior Professor in the Department of Information Technology, Kannur University, Kerala, India.

Carl Patrik Nilsson holds a Ph.D. from the Umeå School of Business and is the CEO of the Stockholm Institute of Communication Science and researcher and lecturer at the Stockholm School of Economics Centre for Media and Economic Psychology. His research focus is on consumer behaviour and marketing management. He has published widely on consumer psychology and marketing management. He currently researches customer complain management.

Marta Noronha e Sousa is a doctoral student in Communication Sciences, in the Center for Communication and Society Studies (CECS), University of Minho (UM). Her field of inquiry is located at the threshold of adaptation, transmedia storytelling and other intertextual exercises, which extend through a diversity of media. Her main focus lies on narrative dynamics, formats and aesthetics in the new media environment, particularly on audiovisual and digital media, using a multimedia approach. Among other participations in conferences and journals, Marta has published *Narrative at a Crossroad: The Question of Fidelity in the Adaptation of Literary Works to the Cinema* (2010, CECS, UM, Braga), based on her Master's thesis, and participated in the research project *Processing Social Signals in Human-Computer Interaction* (engageLab, UM), working specifically on non-verbal communication.

Damjan Obal is a recent Ph.D. at the University of Maribor where his research focus was in the intersection of human-computer interaction, Interaction Design, Participatory Design, Human-centered design and ubiquitous computing. He is an interdisciplinary researcher and a UX designer who opened his own studio. He worked with numerous international clients such as the Mayo Clinic, Challenge: Future, CNN and others. In 2013 Damjan co-founded a startup called Edgar that helps people tell and share stories online.

Ms. Supriya N. Pal holds a Master's degree in Computer Science from the University of Mumbai. She is a Technical Lead in applied research projects of Software Engineering division in C-DAC, Electronics City, Bangalore. Her research interests include Software Engineering, Web technologies, Cloud and Mobile Computing and its applications.

Guglielmo Pescatore is full professor at the University of Bologna. His work is characterized by the attention to the phenomena of mutation, both for what concerns the textual forms, as well as the social valences of audiovisual media. His research on TV series gave rise to a line of study dedicated to narrative ecosystems and "vast" narratives, that is to say extended in time and space. Recently he devoted several essays, articles and speeches to this topic. The study of narrative ecosystems led him to adopt an interdisciplinary approach that ranges from studies in media economics and marketing to those related to information technologies, up to an emerging discipline such as information architecture. This has fostered interdisciplinary reconsideration of film studies and audiovisual media in a new key, which aims to promote dialogue between different disciplines, in order to account for the complexity and degree of articulation achieved by contemporary media. The study of narrative ecosystems has found its expression in the international conference Media Mutations (of which he is promoter and member of the Scientific Committee) devoted to this topic.

Mr. Pradeep Kumar Ch. received the B.Tech. degree in Computer Science Engineering from Jawaharlal Nehru Technological University, Vizianagaram, in 2011. Since 2011, he has been working with center for development of advanced computing. His research interests include Image Processing, Computer Vision, Artificial Intelligence, Machine Learning and Internet of Things.

Claudia Rademaker is Assistant Professor at Stockholm Business School, Stockholm University. Claudia holds a Ph. D. degree in business administration from Stockholm School of Economics. Her doctoral thesis entitled "Green Media: exploring green media selection and its impact on communication effectiveness" was conducted at the Center for Media and Economic Psychology. She has published her academic work in peer-reviewed journals such as Journal of Cleaner Production. Before conducting academic research, Claudia has worked internationally for over 15 years with marketing and international business development.

Peter Reichl has been studying mathematics, physics and philosophy in Munich and Cambridge (UK). After finishing his Ph.D. studies in Computer Science at RWTH Aachen and ETH Zurich, he became Key Researcher at FTW Telecommunications Research Center Vienna, Austria, where he was responsible for the research area "User-centered Interaction and Communication Economics". Dr. Reichl has published more than 100 journal and conference papers in the areas of telecommunication economics, user perception of telecommunication services and Quality-of-Experience, next generation networks, and future mobile networks and services. Following his habilitation at TU Graz, he became holder of the SISCOM and RBUCE WEST International Research Chairs on "Network-based Information and Communication Ecosystems" (NICE) at Université Européenne de Bretagne/INRIA/Télécom Bretagne in Rennes, France. In 2011 he has been appointed Professor for Networking Technology at Aalto University Helsinki, Finland, and from 2013/14 on he is affiliated with the University of Vienna as Full Professor for Computer Science and head of the Research Group "Cooperative Systems (COSY)".

Luca Rosati is a practitioner information architect and assistant professor of Information Architecture at the University for Foreigners of Perugia, in Italy. He has a long practice in designing complex information systems for large companies and organizations in order to ensure a graceful user experience. Luca spoke at several international conferences, is member of the editorial staff of the Journal of Information Architecture and of the board of the first Italian Master of Information Architecture (University for Foreigners of Perugia in partnership with Information Architecture Institute). His last book, with Andrea Resmini, is *Pervasive Information Architecture: Designing Cross-channel User Experiences* (Morgan Kaufmann 2011). He supports a holistic approach to information architecture, and its application to media studies. His website is lucarosati.it.

Professor Damian Schofield is the Director of Human Computer Interaction at the State University of New York (SUNY) at Oswego, USA. He also currently an Adjunct Associate Professor at Edith Cowan University, Perth, Western Australia. Professor Schofield also remains a director of Aims Solutions Ltd., a UK based simulation company.

Mr. N. S. Sreekanth, is a Graduated in Physics from University of Calicut, Kerala, India in 1999. He secured his Masters Degree in Computer Applications from Visveswariah Technological University, Belgam, Karnataka, India, in 2003. His research interest includes Speech recognition, Image processing, Gesture recognition, Multimodal Computing and Human Computer Interaction. Currently He is working as Senior Technical Officer in Center for Development of Advanced Computing—Bangalore, India.

Mattias Svahn is a research fellow at the Stockholm School of Economics Institute of Research, Visiting Professor at the Stockholm School of Economics in Riga, and senior consultant to We Are Insight: http://weareinsight.se/. He researches the media psychology of pervasive games and gamification. His work bridges consumer behaviour sciences, game design sciences and play theory, see e.g. http://nordicgame.com/games-can-persuade-families-to-save-energy/. He has published and presented at many international conferences on the consumer psychology of ambient media and pervasive advertising, and is currently working on a book on gamification that blends science with short stories from fiction authors to illustrate the future of gamification and transmedia https://www.facebook.com/gamificationframtiden to be published 2016.

Professor Tatiana Aires Tavares is a researcher of Digital Video Applications Lab at the Federal University of Paraiba (UFPB) at Joao Pessoa, Brazil. She is also a Visiting Professor at State University of New York (SUNY) at Oswego, USA. Her research interests include design interaction, accessibility and multimedia.

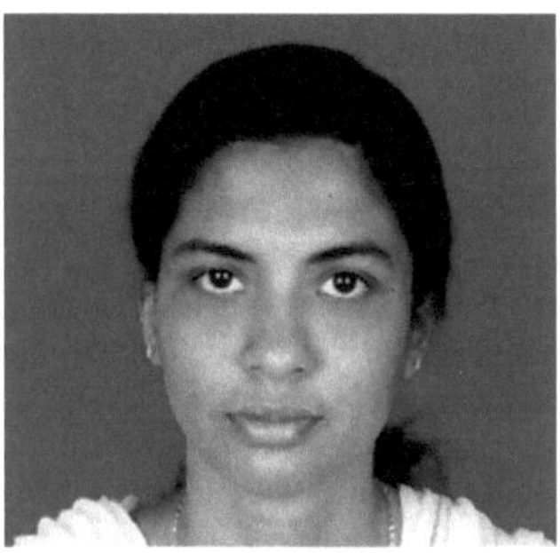

Ms. Nobby Varghese received the B.Tech. degree in Computer Science and Engineering from University Of Kerala, India in 1999. Her research interests include Natural Language processing, Human Computer Interfaces and Accessibility. Currently she is working as Senior Technical Officer in Center for Development of Advanced Computing—Bangalore, India.

Vaishali Pal has received the B.Tech. degree in Information Technology from West Bengal University of Technology and has been working in C-DAC, Bangalore since 2011. Her area of interest include Machine learning, data analysis, Image processing and speech processing. She also worked in e-learning projects.

Zvezdan Vukanovic is an Associate Professor and the Director of Master's Program in Media Economy at the University of Donja Gorica (UDG), Podgorica, Montenegro where he teaches courses in New Media Management and Economy and Network Economy in Media Industry and Business at the Faculty of International Economics, Finance and Business, Faculty of Humanities and Faculty of Information Systems and Technology. He obtained a postdoctoral fellowship at the McLuhan Program in Culture and Technology, Coach House Institute, Faculty of Information, University of Toronto in 2008. Vukanovic received a Ph.D. in economics at the Faculty of Business—The Megatrend University of Applied Sciences, Belgrade in 2006; M.A. in Arts and Media, The University of Arts, Belgrade 2004 and B.A. in Arts and Media, University of Victoria, British Columbia, Canada in 1999. During 2008–2009 he was a visiting lecturer at the School of Business and Economics, Thompson Rivers University, Canada. Vukanovic is the author/editor of seven books and numerous articles and book chapters related to topics in media management.

Richard Wahlund is The Bonnier Family Professor in Business Administration, especially Media, at the Stockholm School of Economics (SSE). He is also Head of the Department of Marketing and Strategy and of the Research Center for Media and Economic Psychology at SSE. In addition to media, his research and publications concern economic behaviors (e.g. financial and tax behavior, and consumer behavior), economic expectations, the psychology of decision making, branding and corporate image, other aspects of marketing communications, gender issues, business ethics and market oriented risk management, where economic-psychological theories and methods is the uniting link.

Annika Wiklund-Engblom, M.A. in developmental psychology (Åbo Akademi University, Finland), and IT-pedagogics (Kalmar University, Sweden) has worked as an educational consultant for e-learning development projects in the industry and is now the research coordinator at MediaCity at Åbo Akademi University, where the focus lies on user experience design and research using cross media and transmedia design approaches. Her current Ph. D. thesis investigates learning experiences and self-regulation of corporate e-learners.

Nelson Zagalo is Assistant Professor at the University of Minho (UM). His main research is centered at the edge meeting points between film studies and entertainment virtual reality, developing knowledge on the adaptation of cinematic emotions to videogames, and employing a multidisciplinary approach bringing together cognitive psychology, art theory, new media, and computer science. He created and coordinates the master program on Interactive Media at UM; founded and co-heads the laboratory EngageLab; founded the Portuguese Society for Videogames Sciences and is an integrated researcher at the Communication and Society Research Centre (CECS). Nelson has published the book *Interactive Emotions, from Film to Videogames* (2009), edited the book *Virtual Worlds and Metaverse Platforms: New Communication and Identity Paradigms* (2011), and in 2012 has created the ebooks imprint engagePress.

MIX
Papier aus verantwortungsvollen Quellen
Paper from responsible sources
FSC® C105338

If you have any concerns about our products,
you can contact us on
ProductSafety@springernature.com

In case Publisher is established outside the EU,
the EU authorized representative is:
Springer Nature Customer Service Center GmbH
Europaplatz 3, 69115 Heidelberg, Germany

Printed by Libri Plureos GmbH
in Hamburg, Germany